Biology for a modern society

Biology for a modern society

LOUIS LEVINE, Ph.D.

Professor of Biology,
The City College,
City University of New York,
New York, New York

with 353 *illustrations*

The C. V. Mosby Company

SAINT LOUIS 1977

Color scintiscans of body organs on cover courtesy

Dr. Nicholas Verde, Deaconess Hospital, St. Louis, Missouri

Printed in the United States of America

Distributed in Great Britain by Henry Kimpton, London

The C. V. Mosby Company
11830 Westline Industrial Drive, St. Louis, Missouri 63141

Library of Congress Cataloging in Publication Data

Levine, Louis, Date.
Biology for a modern society.

Includes bibliographies and index.
1. Human biology. I. Title. [DNLM: 1. Biology. QH308.2 L665b]
QP34.5.L48 612 76-46318
ISBN 0-8016-2990-X

GW/VH/VH 9 8 7 6 5 4 3 2 1

To the memory of

Abraham J. Goldforb

who always challenged his students to explain
how organisms survive and reproduce in a world of limited resources
and fierce competition

Preface

With the ever changing pattern of life in our modern society, the content of many of our traditional courses is being modified. In our society today, we are witnessing a stress on improving the quality of life of all the world's people and a broad and real concern for our environment. This has led instructors and students in beginning biology courses to look for different kinds of information and to place less emphasis on some of the more traditional topics. There has developed an increased need to present material that is meaningful to students who come from varying backgrounds and who are faced with complex problems peculiar to our twentieth century way of life. It is toward meeting these needs that this book has been written.

This book is designed for a biology course in which the central theme is man, both as an individual and as a species. We shall examine man's well-being as an individual in all its aspects: physical, mental, and emotional. We shall be concerned with man's relationship as a species to the physical environment and to the other organisms on earth. In addition, we shall discuss man's past evolutionary history in order to better understand how we became as we are today. We shall also examine the role of technology in shaping our modern society and its possible influence on the future of mankind.

We begin with an examination of the success of our society in extending the life expectancy of the individual and in increasing the number of individuals of our species. There follows a discussion of the need of an ever increasing world population for an adequate food supply and the means available to achieve this. We then examine the functions of food, its various nutrient components, and its availability throughout the world.

At this point, our attention turns to the processes by which food is made into more living material. This includes a study of the digestive process, the absorption of its end products into the circulatory system, and their transportation to the various types of cells that make up the body. We shall look at the cell's genetic material, which specifies the code for the arrangement of amino acids into either structural or enzymatic proteins. This is followed by a study of cell division and the various patterns of inheritance.

We then consider the organ systems of the body that give us protection, support, and body movement. After this comes a discussion of the body system that obtains oxygen and the one that rids the body of metabolic wastes. This is followed by an examination of the coordinating systems of the body, which leads us to a study of reproduction and the stages of individual development. Next, we turn our attention to the changes that have occurred in our species over many generations and close our study of biology for a modern society by examining the role played by our technology in changing the earth and the organisms that live on it.

Each chapter is preceded by a list of learning objectives presented as questions to the student. These learning objectives indicate the major themes to be covered and serve to alert the student to the broad general questions to which the chapter will address itself. At the end of each chapter is a list of suggested readings for those interested in learning more about a particular topic. These readings include both general articles, covering broad aspects of the particular subject, and more detailed references, providing an in-depth study of the topic. At the end of the book is a glossary defining the technical terms used in this book as well as referring to the chapters in which these terms are discussed. Finally, there is a series of appendices detailing the structure of atoms, the formation of various types of chemical bonds, the chemistry of acids and bases, and tables of conversion of measurements from the English to the metric system and vice versa.

I would like to avoid any misunderstanding about the use of certain terms in this book. Along with the changing attitude of people toward the previously accepted roles of males and females in our society has come a resentment on the part of some toward the use of such terms as "man" and "mankind" when referring to groups that include both sexes. I would like to point out that I use these terms with no implication of gender, but strictly as generic terms that describe a group, in the same manner as we use the term "lion" or "tiger" to describe a particular group within the cat family, although in both cases the term can also refer to the male of the species.

This book has benefited from the efforts of many people. I am especially indebted to Dr. Norman M. Saks, who read portions of the manuscript, and offered valuable suggestions for its improvement. Any shortcomings of the book, however, are my responsibility alone. Finally, I wish to express my appreciation to Mr. Joseph T. Fevoli for preparing all the illustrations and to Mrs. June April Grossfield and Mrs. Mary Lou Thompson for performing the arduous task of typing the manuscript. Credits for tables and figures from other publications are given in the legends according to the wishes of the author or publisher.

Louis Levine

Contents

A

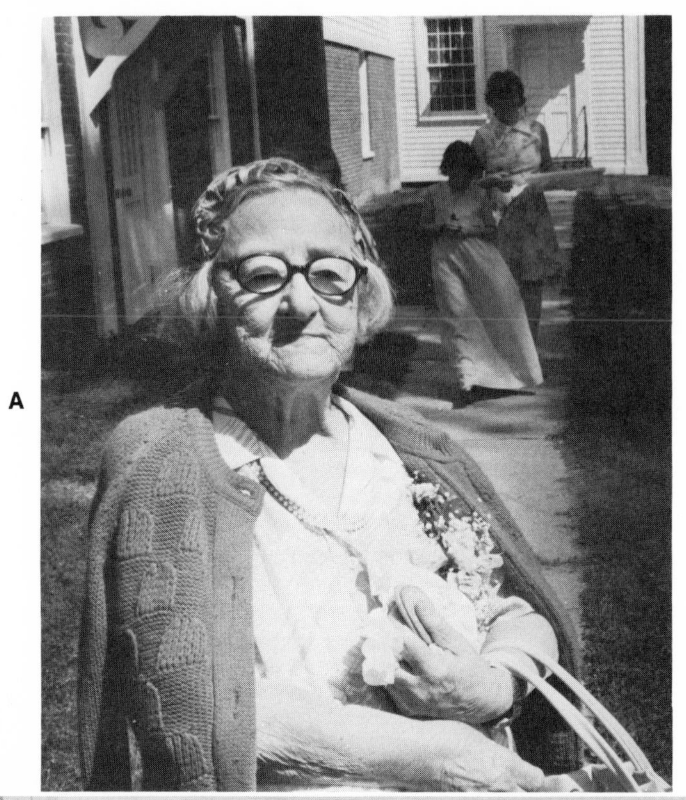

B

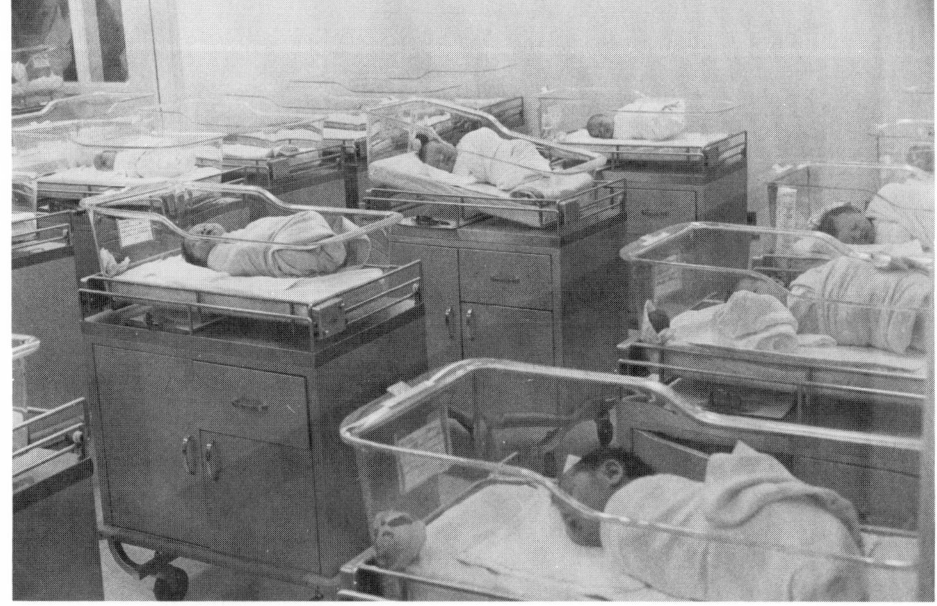

Two aspects of biological success. **A,** Extension of the individual's length of life. **B,** Increase in the number of members of the species. (Copyright © 1976 by Theodore R. Lane.)

CHAPTER 1 Biological success and mankind

LEARNING OBJECTIVES

- What constitutes biological success for an individual?
- What constitutes biological success for a species?
- How has the life expectancy of the world's human population changed over the past 2000 years?
- What is the distribution of current life expectancy figures over the various continental populations?
- What has happened to the size of the world's human population over the past 2000 years?
- What has happened to the doubling time of the world's human population over the past 2000 years?
- How is the world's human population distributed over the various continents?
- What is the rate of growth of the human population on each continent?
- What is the relative density of the human population on each continent?
- How is the world's human population distributed among the differently sized urban communities?
- How are the differently sized urban communities distributed over the various continents?

In the study of biology as it relates to our modern society, the central theme must be man, both as an individual and as a species. As an individual, man's well-being, in all physical, mental, and emotional aspects, is of paramount importance. As a species, his relationships both to his physical environment and to other species, plant and animal, become matters of real concern. We are also interested in knowing as much as possible about our species' past evolutionary history in order to better understand how we became as we are today. Such knowledge should aid us in understanding the possible future evolutionary pathways our species can follow. Above all, we are intensely concerned with the problems that man faces today, both from an individual and a species point of view, and with the available solutions to these problems. Where solutions to our problems are not available, we need to know the most promising avenues of research that will yield the desired information.

On both personal and group levels, human beings are involved in achieving **success,** and it is important to understand the meaning of the term "success" in the biological context. For the individual, biological success means extending the length of his life for the longest possible period of time. For the species, biological success consists of two factors: (1) the production of as many offspring as possible by the members of each generation and (2) the spread of the members of each generation over as much of the earth as possible.

It is apparent at once that biological success for the group may not coincide with biological success for the individual, and vice versa. The bearing of large numbers of offspring may be detrimental to the health and well-being of both the parents and their children; yet such a situation would represent biological success for the species. On the other hand, the continuation of life far beyond the age of reproduction and child rearing may be most desirable from the individual point of view, but it can be detrimental to the species, because it limits the amount of food and other natural resources available for the production of more offspring by other members of the group. Therefore in formulating social, economic, and even political policies, members of our modern society must consider the biological consequences of such policies on the success of both the individual and the species.

BIOLOGICAL SUCCESS OF THE INDIVIDUAL

If we examine our modern society with respect to how well it has achieved biological success for the individual, we find that it has done a truly magnificent job in extending the average length of life. This is clearly illustrated by the data in Table 1-1, showing the estimated average life expectancy throughout the world at various times during the past 2000 years. The average life expectancy throughout the world has tripled during the past 2000 years and, more dramatically, has doubled during the last 325 years. However, to obtain a more meaningful picture of this high degree of biological success for the individual, we must ask whether all populations of the

Table 1-1
Average worldwide life expectancies during last 2000 years

Date (AD)	Life expectancy (years)
1	22
1650	33
1850	41
1930	60
1973	67

Table 1-2
Average life expectancies in various regions (1973)

Region	Life expectancy (years)
Africa	45
Asia*	56
Europe†	69
Latin America	62
North America	71
World‡	67

*Includes Australia and New Zealand.
†Includes Soviet Union.
‡Based on variances in regional populations.

earth share equally in this great achievement. The answer to this question is given by the data in Table 1-2. It is quite clear that those born in North America and Europe can expect to live longer than the world average, whereas those born in Africa, Asia, and Latin America cannot. It is important to ask why this is so, because in the answer may be found the means to increase the life expectancy of everyone on earth, even including those who now enjoy the longest average life span.

A major cause of the increase in life expectancy throughout the world in recent time has been the drop in infant mortality. As an example of this phenomenon, we shall first examine the situation in North America and Europe. In 1800 20% of the children born did not survive the first year of life. This high rate of infant mortality diminished gradually throughout the century, and by 1900 the infant mortality rate was about 10%. It then declined more rapidly, and by 1950 only 2% of the children born failed to survive their first year of life. This rate of infant mortality has remained relatively stable up to the present time. In contrast, the countries of Africa, Asia, and Latin America were still experiencing an infant death rate of about 20% as recently as 1900. Even today as many as 16% of the infants in Africa, 12% of the infants in Asia, and 7% of the infants in Latin America die during the first year of life.

Although a reduction of the infant death rate has been the main reason for our increased life expectancy, there have also been reductions in the death rate during all the early years of life. We can see this in the following statistics for North America and Europe: In 1840 about 65% of those born lived to age 15 years, whereas in 1950 about 75% lived to age 60 years (Fig. 1-1).

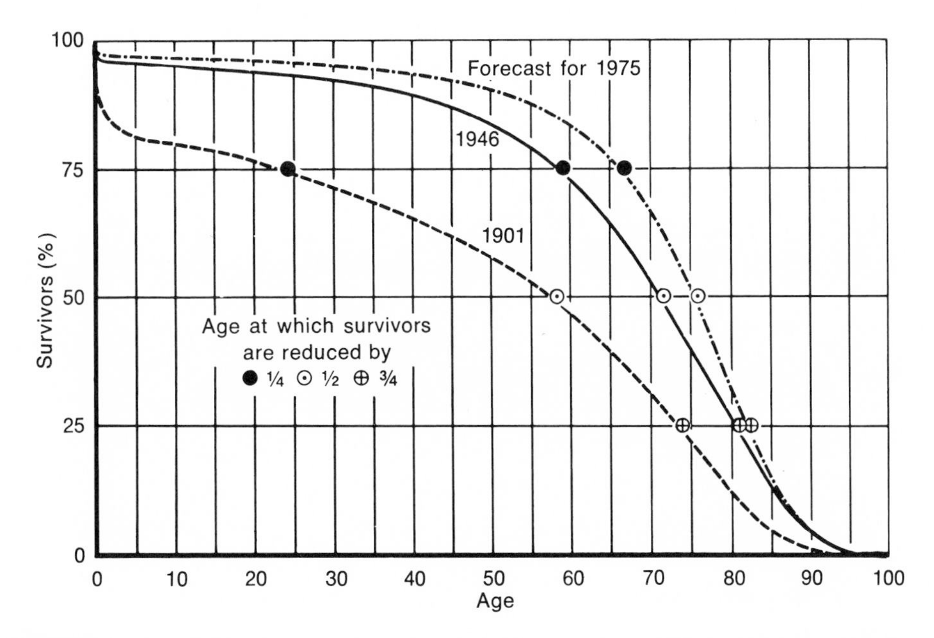

Fig. 1-1
Survivors from birth to successive ages according to life tables for United States 1901, 1946, and forecast for 1975. Forecast made on basis of low mortality predicted by Bureau of the Census, 1970. (Modified from Turner, C. E. 1971. Personal and community health, ed. 14. The C. V. Mosby Co., St. Louis; based on data from Metropolitan Life Insurance Co.)

What conditions led to this increase in life expectancy throughout the world? Interestingly enough, the dramatic event that led to our present high level of biological success was the Industrial Revolution. It began in England in 1750 and gradually extended to western Europe and North America over the next hundred years, but only relatively recently has it reached Africa, Asia, and Latin America. The term "Industrial Revolution" refers to the changes that marked the transition from a mainly agricultural society with its emphasis on relatively simple tools to an industrial society with its emphasis on complex machinery (Figs. 1-2 and 1-3). The biologically important features of this great transformation include the following:

1. An expansion of agriculture, both in terms of the amount of land cultivated, especially in North America, and the productivity of agricultural workers. This expansion was achieved as a result of the mechanization of farming procedures, the abundant use of fertilizers, and the genetic improvement of crops.
2. A growing trend toward urbanization, as agricultural changes made it possible to support more people with a smaller proportion of the population involved in farm labor.
3. A switch from the use of biologically renewable sources of power (human labor, domestic animals, and wood) to the use of nonrenewable sources of power (coal and oil) for manufacturing and transportation, and later the use of coal and oil to produce electricity for many other purposes in our modern society.

Fig. 1-2
Plowing of field, using domestic animal as source of power. (USDA photograph; from Arnett, R. H., Jr., and Bazinet, G. F., Jr. 1977. An introduction to plant biology, ed. 4. The C. V. Mosby Co., St. Louis.)

Fig. 1-3
Grain harvest on large midwestern farm, using modern mechanized equipment. (USDA photograph; from Williams, S. R. 1973. Nutrition and diet therapy, ed. 2. The C. V. Mosby Co., St. Louis.)

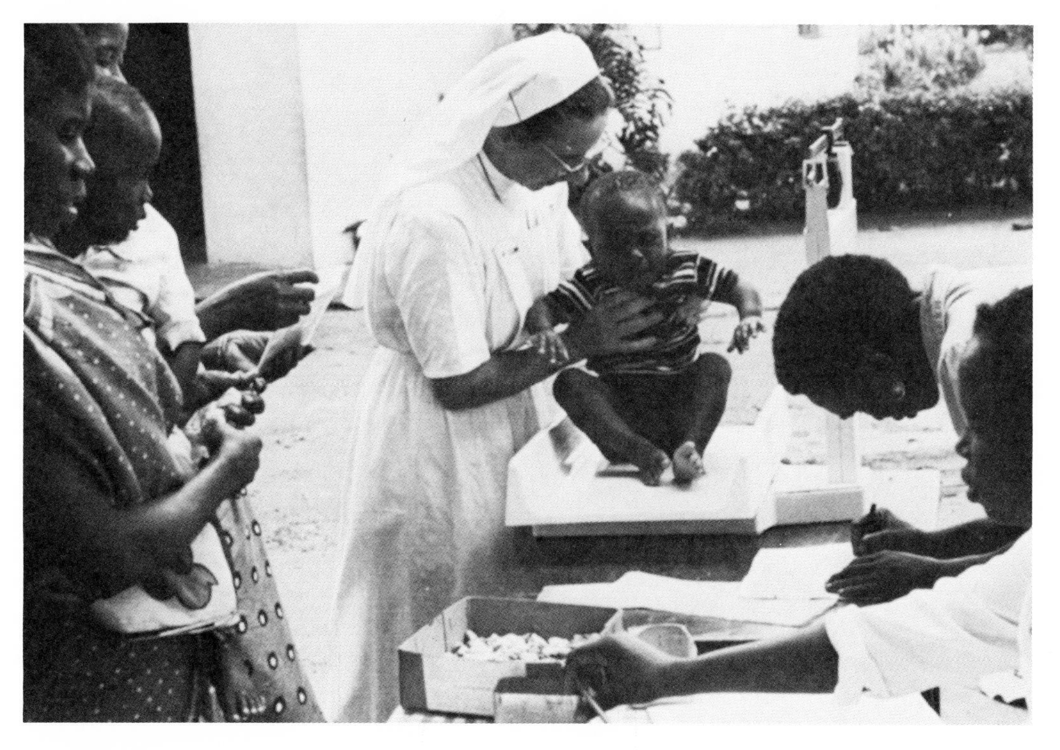

Fig. 1-4
Tanzanian infant is given weight check at the nutrition clinic at Pugo, near Dar es Salaam. The clinic was established to fight malnutrition and raise health standards. Maternal and child health and family planning services are part of the country's health program. (From U.S. Agency for International Development. 1976. World population growth and response. Washington, D.C.)

Accompanying the processes of industrialization and urbanization was a great expansion of scientific knowledge and technology, including dramatic changes in the understanding, treatment, and prevention of disease. There developed a new approach to personal cleanliness, the purifying of food and water, and the conquest of specific infectious diseases that had long plagued man. Surprisingly, the need to separate water supplies for drinking from those for sewage was not firmly established until about 1850. It was even more recently that the causative agents of yellow fever, cholera, typhoid fever, typhus, amebic dysentery, tuberculosis, and malaria were identified and the means of controlling them developed. Effective control of malaria, involving the widespread use of DDT against the mosquitoes that spread the disease, did not begin until 1945. The success of the battle against malaria has been most significant in increasing the life expectancy in parts of Africa, Asia, and Latin America, where just prior to 1945, malaria was the single most potent cause of illness and death. With the near elimination of most of these serious infectious diseases, the causes of death have shifted to the chronic and stress diseases, such as cancer and heart disease, which more frequently strike older individuals.

As we consider the truly remarkable achievement that our modern society has made in promoting the biological success of the individual, we should ask whether an even greater measure of well-being for the individual can be obtained and how this can be accomplished. One area for improvement obviously lies in bringing the benefits of our medical knowledge and technology to the people of Africa, Asia, and Latin America, so that their life expectancies can be raised to the level now enjoyed by the people of North America and Europe (Fig. 1-4). Also, having eliminated most of our serious infectious diseases, we should now turn our research efforts toward understanding and conquering the chronic and stress diseases that still plague us. In this effort, we must not only consider the physical disabilities of advanced age, but also its mental and emotional problems. A further aim must be to improve the quality of life at all ages. In our study of the different systems of the body and how they contribute to the individual's well-being, we shall consider the problems that arise in the functioning of these systems and the means available for correcting the various types of malfunctions.

■ BIOLOGICAL SUCCESS OF THE SPECIES

How well has our modern society achieved biological success for man as a species? We find that our way of life has made man the most successful species on earth today. As noted earlier, biological success for the species consists of two factors: (1) the production of as many offspring as possible by the members of each generation and (2) the spread of the members of each generation over as much of the earth as possible. We shall now examine each of these critieria to see just how successful the species man has become and what are some of the problems that his success has brought him.

□ Population size

When considering biological success in terms of numbers of individuals produced per generation, the general rule is that the more individuals, the greater is the biological success of the species. The number of individuals is a reflection of the species' reproductive efficiency. Reproductive efficiency takes into account the abilities of the members of a species to survive the rigors of the physical environment, to obtain sufficient nourishment for growth and energy, to escape from enemies, to find a mate, and to produce offspring.

In order to understand just how successful mankind is today with respect to number of individuals, we must examine Fig. 1-5, which illustrates the growth of the world's human population during the past 2000 years. It is very clear that the number of human beings in the world is increasing at a tremendous rate. The greatest increase in the rate of growth has occurred during the last 150 years. This rate of growth can be better appreciated by evaluating the data in Table 1-3, which gives the estimated size of the world's human population at various times throughout history. The table also indicates the number of years required at each point for mankind to double in number, given the birth and death rates of the world population at each time period. According to United Nations' estimates, the world population is presently increasing by 8000 people every hour, or by approximately 70,000,000

Table 1-3
World's human population size and doubling time at various dates

Date (AD)	Estimated size	Doubling time (years)
1	300 million	2000
1650	500 million	200
1850	1 billion	80
1930	2 billion	45
1975	4 billion	35
2000	6-7 billion	?

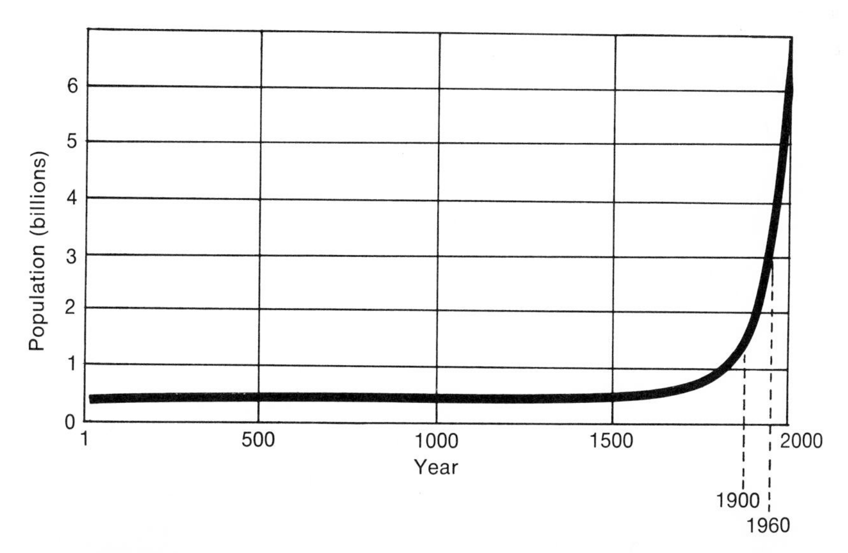

Fig. 1-5
Growth of the world's human population over the past 2000 years.

people each year. In the United States, the present net increase in population is 1,700,000 people each year.

The striking fact that is apparent in Table 1-3 is that the "doubling time" for mankind has been decreasing at a very rapid rate. This has been the result of a dramatic decrease in the death rate over the last 200 years because of advances in medicine coupled with the relatively high birth rate during the same period of time. In 1973 the overall world birth rate was estimated to be about 33/1000 people, whereas the death rate was about 13/1000 people. Thus the annual rate of increase (the difference between birth rate and death rate given in percent) in the world population was 2.0%. Should that growth rate continue, there will be about 14 billion human beings on earth by the year 2040 and 28 billion by the year 2070.

Whether it will actually be possible to sustain human populations of such large sizes is, of course, one of the important problems of our modern society. If, however, this can be achieved, it will demonstrate the biological success of the human species in exploiting its environment and producing large numbers of individuals. In later discussions, we shall analyze the effects of such a population explosion on the quality of life. However, even at present population levels, as we can see from Table 1-3, the world's human population of 1975 is eight times larger than that of 1650. This increase has taken place in slightly more than 10 generations (30 years being considered the average time between human generations) and represents a very high level of biological success.

Up to this point in our discussion, we have considered the world's human population as a single unit and have described its overall growth. However, we realize that the human species is composed of a number of different geographical populations and that they are not all of equal size. Fig. 1-6 indicates the main geographical areas of the world and the percentage of the world's population contained in each area. The

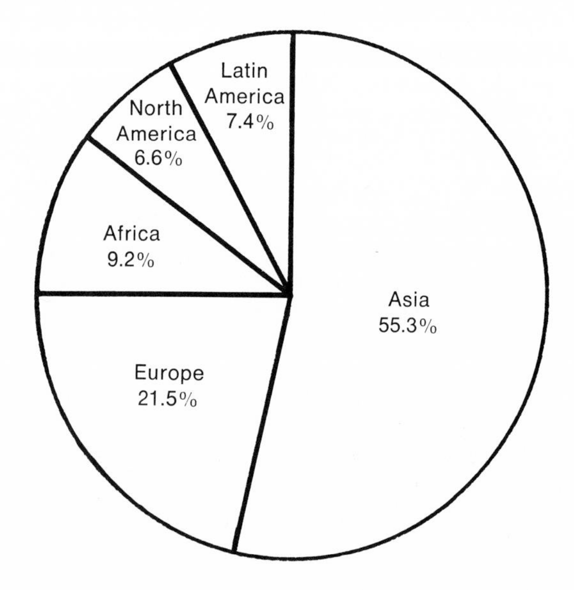

Fig. 1-6
Percentage of the world's human population in each of the major geographical areas. Asia includes Australia and New Zealand; Europe includes Soviet Union.

continent of Asia contains more than half the people of the world, whereas any one of the other continents contains significantly fewer people. When we analyze the growth pattern of each region of the earth from 1650 to 1970, as seen in Fig. 1-7, we find that the Asian population has consistently been the largest segment of the world's population.

Will the relative proportions of the various regional segments of the world's population remain as they are today? Table 1-4 lists the population size and the percentage of annual rate of population growth in the different regions of the world. A zero percent growth rate would mean that the population in question is completely stationary in size: its birth rate equals its death rate. A 1% annual rate of increase means that the population will double in size in 70 years, whereas the comparable doubling time for 2% is 35 years, and for 3% it is 24 years. Populations with higher annual percentage rates of increase will, given sufficient time, be larger than those with lower rates of increase. Therefore Table 1-4 illustrates that at the present rates

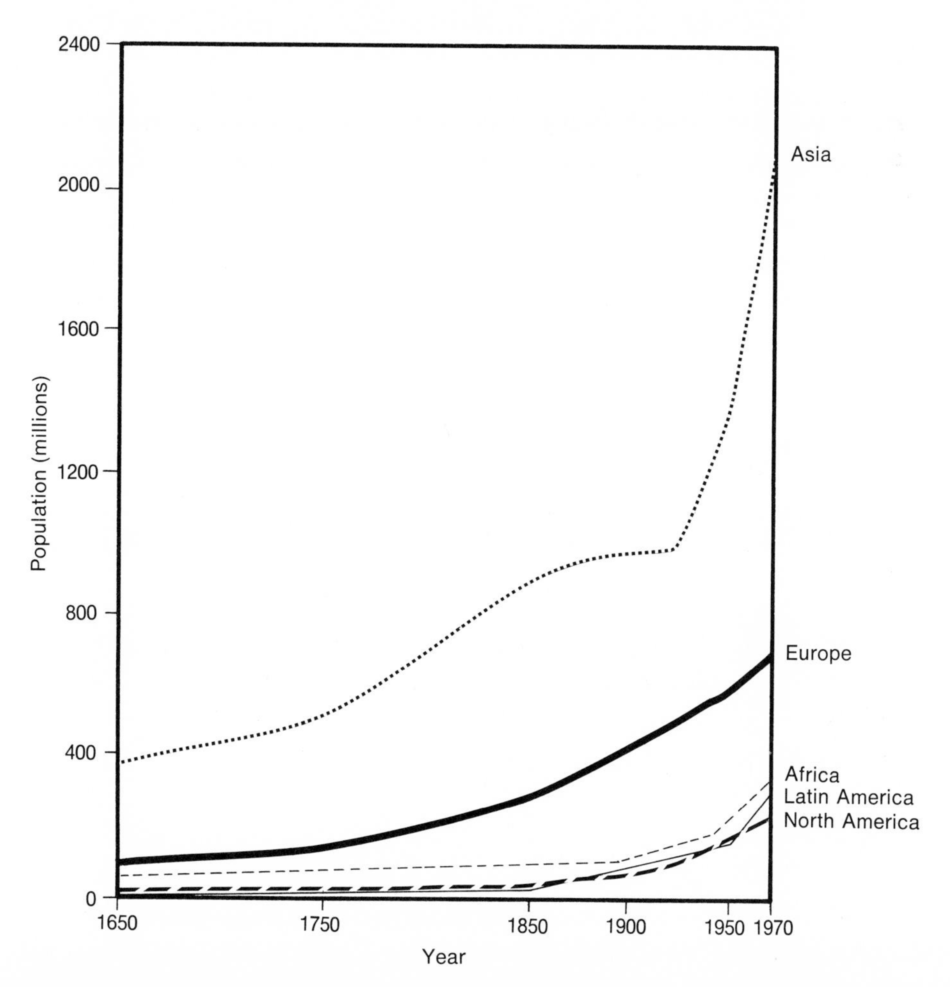

Fig. 1-7
Growth of human population in each major geographical area of the world. Asia includes Australia and New Zealand; Europe includes Soviet Union.

Table 1-4
Population size and annual rates of population growth (1973)

Region	Population size (millions)	Rate of growth (%)
Africa	374	2.5
Asia*	2225	2.3
Europe†	722	0.7
Latin America	308	2.8
North America	233	0.8
World‡	3862	2.0

*Includes Australia and New Zealand.
†Includes Soviet Union.
‡Based on variances in regional populations.

of increase, the populations of Africa and Latin America will eventually become larger than those of Europe and even Asia. Within 50 years the populations of Africa and Latin America will each be larger than the population of Europe, and the Latin American population will have just about equaled that of Africa.

Based on the information contained in Table 1-4, we conclude that the African and Latin American populations are exhibiting the greatest degree of biological success as measured by their reproductive efficiency. However, as mentioned earlier in this chapter, before a society can establish a sensible population policy, it must examine the effects of a population explosion on the quality of life of its members.

Population distribution

When considering the geographical distribution of man throughout the earth, we are again impressed with the high degree of success that our species has achieved. There are very few places on the earth that have not been populated by human beings. No other independently migrating species has a distribution approaching that of mankind. In fact, the only organisms that are equally widespread over our planet are those that live in or on the bodies of human beings, such as the bacteria in our intestinal tracts and certain body parasites.

Another important point emerges when we consider the distribution of other types of animals. For example, the group of animals known as the bears are also spread rather widely over the earth. However, each region is occupied by a different species of bear. We find the polar bear in the Arctic regions of the world, the grizzly bear in western North America, the black bear in most of the rest of North America, the brown bear in Asia and Europe, the spectacled bear in Latin America, and the atlas bear in Africa. Each species of bear represents an evolutionary adaptation to its particular area, and the members of the different species cannot normally mate with one another. Man is quite different in that his ability to live in the diverse regions of the world is not based on the development of different species, but rather on his ability to create technologies that meet the challenges of the various areas. Individuals from the different human populations are all members of the same species; they can mate with one another and produce fertile offspring.

Table 1-5
World's land areas, populations, and relative population densities (1973)

Region	Land area (%)	Population (%)	Relative density (population/land area)
Africa	22.1	9.2	0.4
Asia*	26.6	55.3	2.0
Europe†	20.2	21.5	1.0
Latin America	15.2	7.4	0.5
North America	15.9	6.6	0.4
World	100.0	100.0	

*Includes Australia and New Zealand.
†Includes Soviet Union.

Thus far we have only considered the overall distribution of man on the surface of this planet. But if we are to understand the basis of many of the environmental problems of our modern society, we must also examine the pattern of man's distribution. We are all generally aware of the fact that mankind is spread unevenly over the earth. A greater appreciation of the situation can be obtained by examining Table 1-5, which shows the relationship between land area and human population for each region of the world. The two extreme cases are Asia and North America. Asia has more than half of the world's population, but only a little more than a fourth of the land area. In contrast, North America has one sixteenth of the world's population living on one sixth of the earth's surface. One result of this unequal distribution of people and land is that there are now five times as many people per unit of land in Asia as in North America. As a point of reference, we should remember that in the United States, there is an average of 58 people per square mile of land area.

The pattern of uneven distribution of human beings over the earth has been further complicated by one of the outstanding features of the Industrial Revolution, namely, the growth of cities. As mentioned earlier in this chapter, the use of machinery on farms resulted in an increase in food production and a decrease in the need for farm labor. This situation caused a general migration of people to the towns and cities where jobs were available in the adjacent factories and mines. The extent of this shift in population distribution is evidenced by the fact that in 1800 only 2% of the world's population lived in communities containing 20,000 or more inhabitants, whereas by 1960 this had grown to more than 25%. The extent of urbanization has not been the same in all the major regions of the earth, as can be seen in Fig. 1-8, which shows the urbanization levels of the various parts of the world in 1960. In North America, northern Europe, Australia, and New Zealand, most of the people live in large towns and cities. In contrast, most of the people of tropical Africa reside in rural communities and small towns. Each of the other regions of the world has experienced some intermediate level of urbanization.

Why is a community of 20,000 people, which we normally call a "large town," taken as the significant lower level of urbanization and considered comparable to one of the 25 largest population centers in the world, which range from 2,600,000 people

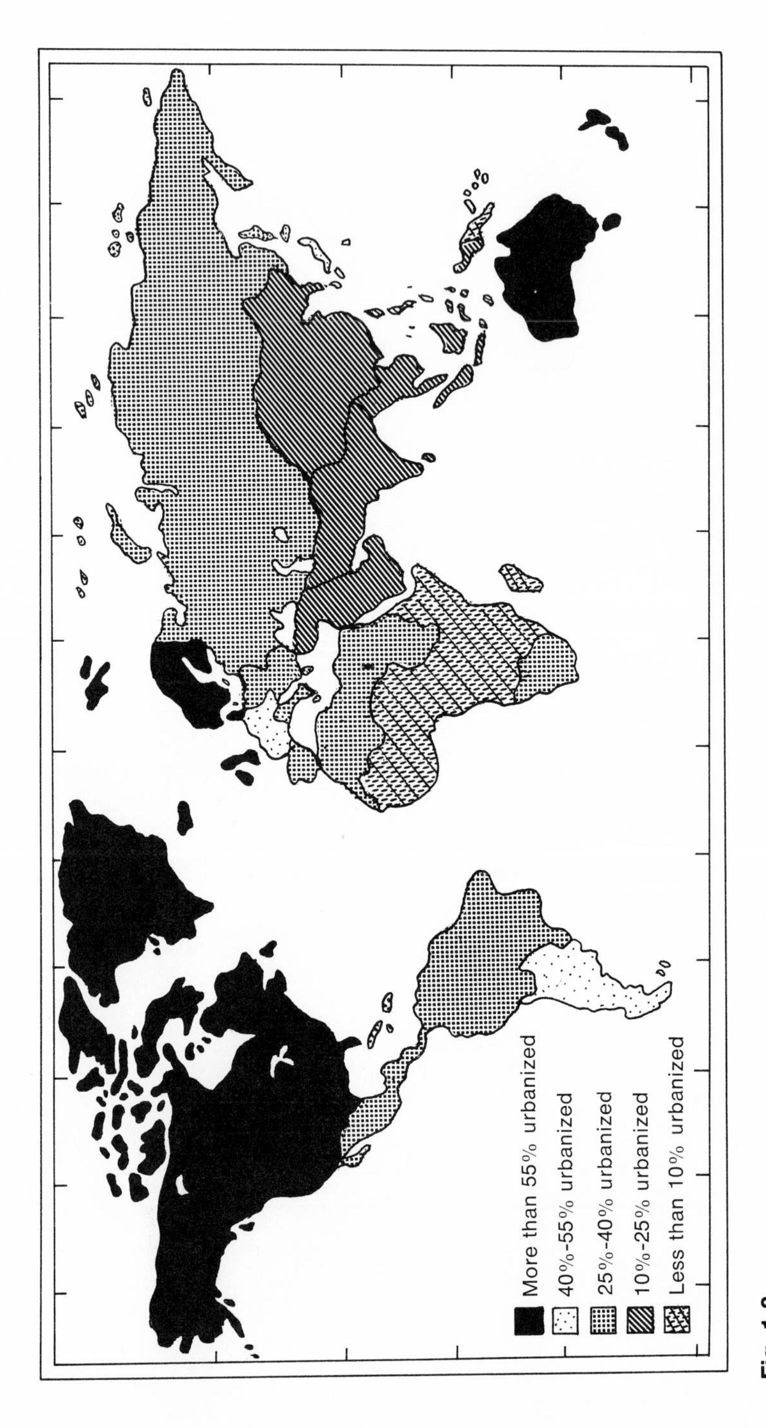

Fig. 1-8
Urbanization levels in major regions of the world expressed as percentage of total population in communities of 20,000 or more inhabitants as of 1960.

(Boston) to 16,000,000 people (New York)? The criterion of 20,000 people is based on the kinds of goods, services, and problems that the community respectively provides and experiences. Although, to be sure, cities with millions of people differ significantly from large towns, all communities of 20,000 or more inhabitants tend to have a common pattern of life. It should also be remembered that more than 70% of the world's total population still lives either in rural communities or in small towns.

Our modern society has become increasingly urban. This can be seen from the data in Table 1-6, which compares the number of people living in variously sized urban communities throughout the world for the years 1920 and 1960. These data show that although the number of people living in large towns, cities, and big cities increased by a factor of 2½ to 3 during these 40 years, the number of people living in multimillion cities increased by a factor of almost five. Over the same period of time, the world population increased by a factor of only one half, indicating the trend toward urbanization.

A further analysis of the distribution of the world's population among the variously sized urban communities is given in Table 1-7, in which the various regions of the earth are compared. The two extreme patterns of population distribution can be seen in Africa and North America. Generally, however, the majority of the urban

Table 1-6
Number of people (millions) living in various sized urban communities throughout the world

	Size of urban community			
Year	*Large town* (20,000-99,999)	*City* (100,000-499,999)	*Big city* (500,000-2,499,999)	*Multimillion city* (2,500,000 or more)
1920*	97	64	71	35
1960†	224	184	211	169

*Estimated world population, 1.9 billion.
†Estimated world population, 2.8 billion.

Table 1-7
Percentage of urban population living in variously sized urban communities (1960)

	Size of urban community			
Region	*Large town* (20,000-99,999)	*City* (100,000-499,999)	*Big city* (500,000-2,499,999)	*Multimillion city* (2,500,000 or more)
Africa	32	32	27	9
Asia*	31	23	26	20
Europe†	34	25	29	12
Latin America	30	19	21	30
North America	12	21	28	39
World	29%	24%	26%	21%

*Includes Australia and New Zealand.
†Includes Soviet Union.

populations of all regions of the world reside in "cities" of various sizes (100,000 or more inhabitants). The majority of the urban populations of North America and Latin America live in big cities and multimillion cities, about one third living in multimillion cities. The rise of the multimillion city is comparatively recent. In 1920 there were seven such communities in the world, with the following distribution: Asia, 1; Europe, 4; and North America, 2. In 1960 there were 26 such communities, distributed as follows: Africa, 1; Asia, 9; Europe, 6; Latin America, 4; and North America, 6.

Although the rise of urban communities, especially those of city size, has brought many benefits in the form of goods and services, it has also created many problems that threaten the well-being of man. One group of problems relates to **pollution.** The residues of our industrial economy pollute our land areas with garbage, our water resources with sewage, and our air with gases (Figs. 1-9 to 1-11). In addition, there has been a tremendous infusion of chemicals into our environment through the extensive use of synthetic compounds, for example, detergents, pesticides, and herbicides, in everyday life. We even have an environmental hazard in the form of "noise," a definite characteristic of our urban communities.

Another group of problems threatening the well-being of city dwellers results from the crowding together of people at densities that far exceed what urban planners have considered the "optimum viable density," which is 4000 residents per square mile. In New York City the population density exceeds an average of 26,000 inhabi-

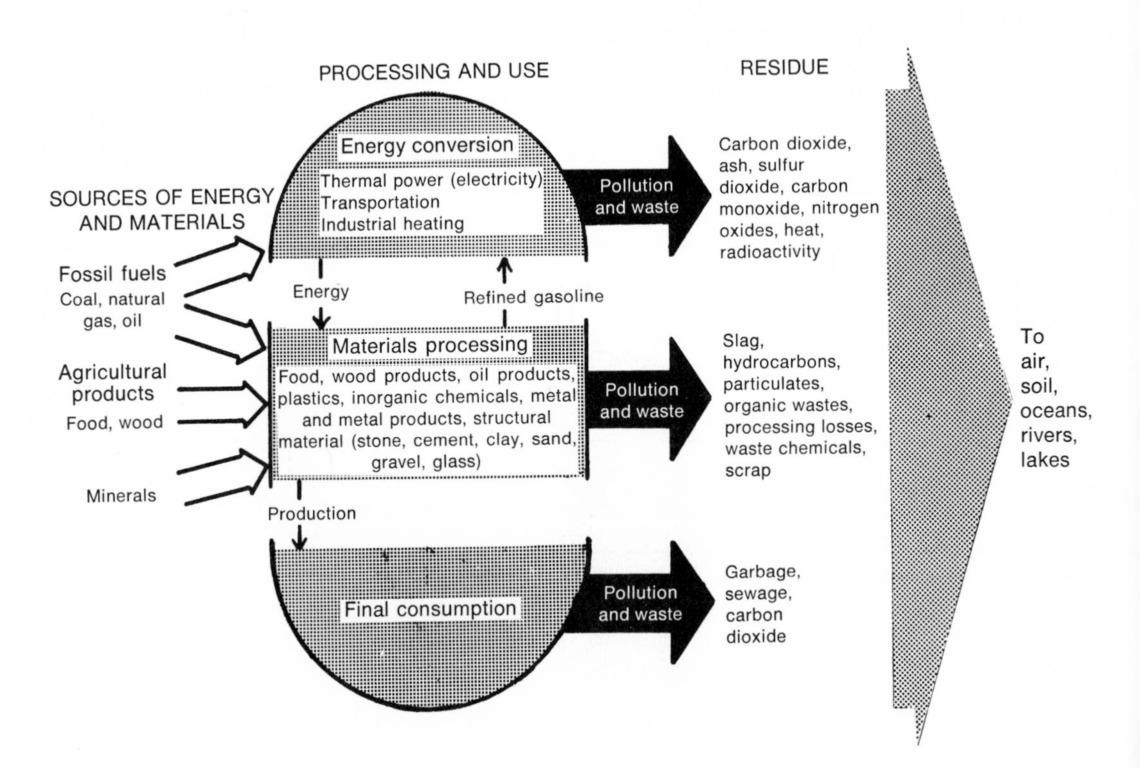

Fig. 1-9
Types and sources of pollution in our modern society.

Fig. 1-10
Contents of automobile ashtray discarded at the side of a road. (Copyright © 1976 by Theodore R. Lane.)

Fig. 1-11
Industrial and automobile pollution pouring smoke and particulate poisons into the air. (Copyright © 1976 by Theodore R. Lane.)

Fig. 1-12
Rapid growth of our metropolitan areas has led to a strain on the quality of city life and hence to a lessening of the city's attraction for migrants. City people are more likely than rural people to suffer from noise pollution, air pollution (which affects health and also makes it harder to keep clothes and homes clean), and crowding (which makes difficulties both for movement along the streets and for relationships within the family). (From U.S. Agency for International Development. 1976. World population growth and response. Washington, D.C.)

tants per square mile, with the density of Manhattan Island exceeding 75,000 people per square mile. This crowding has resulted in the appearance of a generalized type of aggressive behavior not found in less densely populated communities and in the appearance of mental illnesses of previously unknown frequency and severity. Such biological factors must be taken into account as society formulates policies to either promote or discourage the urbanization of a population (Fig. 1-12).

SUMMARY

It is important to realize that although the human species has achieved an unprecedented level of biological success, both in numbers and in distribution over the earth, its very success now threatens its own well-being. As we proceed with our study of biology, we shall return to and analyze in depth the problems of our modern society, both from an individual and a species point of view. Possible solutions to man's problems will be pointed out where such solutions are known. Where solutions are not presently available, we shall indicate the most promising areas of research. Our aim in the study of biology will be to find out how man can achieve a level of biological success that will be optimum for him, both as an individual and as a species. As the first step in our study, the next chapter deals with the relationship between man and his food supply.

SUGGESTED READINGS

Bouvier, L. F., and van der Tak, J. 1976. Infant mortality—progress and problems. Population Bull. **31**:1-33. Review of infant mortality trends in both developed and developing countries during the present century.

Coale, A. J. 1974. The history of the human population. Sci. Am. **231**:40-51. Analysis of the roles of birth rates and death rates in determining population size.

Demeny, P. 1974. The populations of the underdeveloped countries. Sci. Am. **231**:148-159. Study of the populations that account for three fourths of the human species and appear to be continuing their rapid growth rate for the rest of the century.

Ehrlich, P. R., and Ehrlich, A. H. 1970. Population, resources, environment. W. H. Freeman & Co., Publishers, San Francisco. Comprehensive sourcebook for the study of questions related to population growth and the ability of our environment to sustain such large numbers of people.

Freedman, R., and Berelson, B. 1974. The human population. Sci. Am. **231**:30-39. Discussion of the age structure of various populations and predictions of their future sizes.

United Nations Department of Economic and Social Affairs. 1969. Growth of the world's urban and rural population, 1920-2000. Population Stud. No. 44 ST/SOA/Series A/44. Comparison of population trends of the more developed and less developed areas of the world.

United Nations Department of Economic and Social Affairs. 1974. Concise report on the world population situation in 1970-1975 and its long-range implications. Population Stud. No. 56 ST/ESA/Series A/56. Detailed discussion of the factors that contribute to population growth and the long-range projections for each of the major geographical areas.

United States Agency for International Development. 1976. World population growth and response. Population Reference Bureau, Inc., Washington, D.C. Overview of major worldwide population developments between 1965 and 1975 and the responses of various governments to the problems caused by the current population explosion.

Westoff, C. F. 1974. The populations of the developed countries. Sci. Am. **231**:108-121. Study of the populations that constitute about one fourth of the human species and appear to be approaching a stable size.

In our modern society, the supermarket is the immediate source of our food. The need for adequate amounts of high-quality food is a worldwide problem of ever increasing importance. (Copyright © 1976 by Theodore R. Lane.)

CHAPTER 2 Man and his food supply

LEARNING OBJECTIVES

- What is a food cycle and of what types of organisms does it consist?
- What do we obtain from the food we eat?
- Why is it necessary that the molecules contained in the food we eat be continuously recycled?
- Why must the energy contained in the food we eat be continuously replenished by the sun?
- Which types of organic compounds are found in living material, and what is the function of each?
- What are the relative differences in chemical composition between the bodies of plants and animals?
- What should a well-balanced diet provide each day?
- From a biological point of view, what constitutes food quality?
- From a biological point of view, what constitutes food quantity?
- What are the relative differences in plant and animal foods with regard to quality and quantity?
- On which continents do we find those human populations that suffer from the poorest nutrition?
- What are the two types of nutritional diseases that affect 40% of the world's human population?
- What are the three main approaches used to solve the world's food problems?
- Over the past 20 years, what has been the relationship between world food production and human population growth?

In the previous chapter, we discussed the criteria of biological success for the individual and the species and found that in both cases mankind has done remarkably well. We know that man's achievements in the fields of science and technology have led to our present high level of well-being. However, we have the natural desire to further improve our situation. In order to make intelligent decisions in formulating public policy that will optimize our biological success, we need to understand just what the human organism is and how it functions in the world. We shall begin our study by analyzing man's body composition. This analysis requires that we take into account the fact that the material used to make up our bodies and the energy we use for carrying out all of our activities come from the food we eat. The availability of food sufficient to feed the world's populations has become critical in our modern world and threatens to cause even more problems in the future.

■ FOOD CYCLE

With relation to food production and use, a given organism belongs to one of three categories: producer, consumer, or decomposer. The primary **producers** of food are the green plants, both microscopic and macroscopic. Producers consume

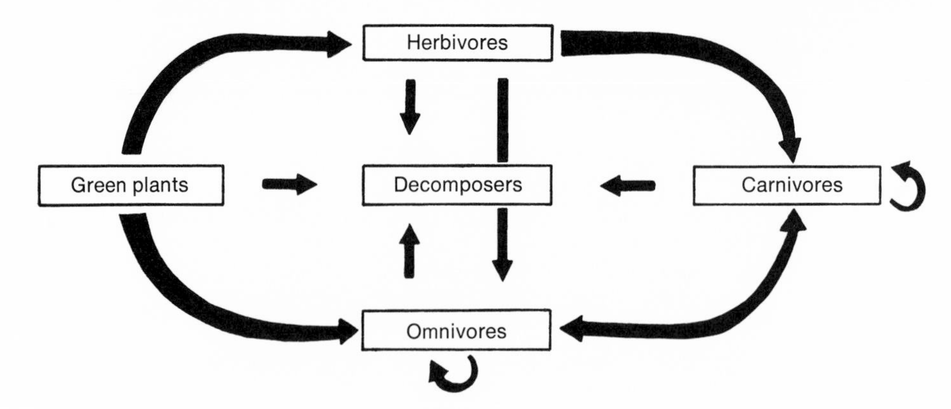

Fig. 2-1
Interrelationships among producers, consumers, and decomposers in a food cycle.

part of the food they manufacture, but they normally make more food than they need. This excess then becomes available to other consumers either as by-products, for example, seeds and fruits, of the plants or as parts of the plants themselves, for example, carrots, potatoes, and lettuce.

The principal **consumers** of food are the animals, both microscopic and macroscopic. Some animals eat only the by-products of producers, others eat the producers themselves, and still others eat both the by-products and the producers. Animals that eat only plant material are called **herbivores,** those whose diet consists solely of other animals are called **carnivores,** and those that eat both plant and animal material are called **omnivores.** Man belongs to the omnivore group of consumers.

Completing the food cycle are the **decomposers.** They are the organisms of decay that obtain their food through the chemical breakdown of the bodies of both producers and consumers. The primary decomposers are bacteria, yeasts, and molds. The interrelationships of the various types of organisms involved in food production and use are shown in Fig. 2-1.

■ FOOD AS A SOURCE OF BUILDING MATERIALS

If food is to be of use as a source of building materials in the growth of organisms, it must supply the kinds of substances an organism can use in constructing its body. In order to understand the problems involved in producing the proper kinds and amounts of food, it is necessary for us to examine the chemical composition of organisms and the availability of the same chemicals on our planet.

The earth consists of four interrelated and interacting components (Fig. 2-2): (1) the **lithosphere,** the earth's crust; (2) the **hydrosphere,** the oceans, ice caps, glaciers, lakes and rivers, and moisture; (3) the **atmosphere,** the envelope of air surrounding the earth; and (4) the **biosphere,** the earth's thin film of living matter. The organisms making up the biosphere, of which man is only one, necessarily live in the other three components of our world. An apparent exception to this statement may be raised, since today organisms can exist in space capsules far outside the earth's atmosphere. Such situations may best be regarded as small portions of the earth nipped off and projected temporarily into space.

Fig. 2-2
The four major components of the earth are visible in this beach scene. The lithosphere is seen in the sand and rocks, the hydrosphere in the ocean, the atmosphere in the sky, and the biosphere in the dog, grass, and trees. (Copyright © 1976 by Theodore R. Lane.)

All forms of matter, wherever found, consist of small units called **atoms.** There are at least 106 different kinds of atoms. A substance that is composed of only one kind of atom is called an **element.** Table 2-1 lists the elements that make up more than 99.99% of the human body and the availability of these same elements in the lithosphere.

The four most common elements in the human body are oxygen, carbon, hydrogen, and nitrogen. The same is true for all organisms in the biosphere. Of the four most common elements in the human body, Table 2-1 indicates that only oxygen is readily available from the lithosphere. When we compare the organisms living in the hydrosphere with the chemical elements available to them, we find that only oxygen and hydrogen are present in large amounts. For organisms living in the atmosphere, only oxygen and nitrogen are readily available. It is apparent that there is little correlation between the chemical elements needed by living organisms and the elements available on earth. This lack of availability of needed elements indicates the

Table 2-1
Chemical elements found in the human body and their availability in the earth's crust*

Name	Symbol	Average % of human body	Average % of earth's crust
Oxygen	O	65.0	46.6
Carbon	C	18.5	0.03
Hydrogen	H	9.5	0.1
Nitrogen	N	3.3	Trace
Calcium	Ca	1.5	3.6
Phosphorus	P	1.0	0.1
Potassium	K	0.4	2.6
Sulfur	S	0.3	0.05
Chlorine	Cl	0.2	0.05
Sodium	Na	0.2	2.9
Magnesium	Mg	0.07	2.1
Iodine	I	0.01	Trace
Iron	Fe	0.01	5.0

*Copper (Cu), cobalt (Co), manganese (Mn), zinc (Zn), boron (B), silicon (Si), and aluminum (Al) have been found in trace amounts in various organisms.

Fig. 2-3
Over 15 million tons of scrapped autos require disposal annually. There are now over 100 machines in the country that shred them into metal chips for the steel furnaces. (From Turner, C. E. 1971. Personal and community health, ed. 14. The C. V. Mosby Co., St. Louis.)

Table 2-2
Chemical compounds found in living material

Substance	Average % of plants	Average % of animals
Water	75.0	67.0
Inorganic salts	2.0	4.0
Carbohydrates	20.0	0.8
Lipids (fats)	0.1	13.0
Proteins	2.0	15.0
Nucleic acids and others	0.9	0.2

crucial role played by decomposers in the food cycle. Without decomposers, scarce elements would be locked up in the dead bodies of the producers and consumers of food, and all organisms would eventually die. It is only because of the recycling of scarce elements for use by later generations that organisms are able to survive on this planet. This fact takes on tremendous meaning when we realize that our modern society has only recently become aware of the need to decompose and recycle the materials contained in the "dead" bodies of its manufactured products, for example, automobiles and refrigerators (Fig. 2-3). Decomposition and reuse of elements is apparently a law of nature that must be observed if living organisms, including man, are to continue to exist on earth.

The elements listed in Table 2-1 are not found in a free form in an organism's body. They normally occur in compounds of two types, inorganic and organic. **Inorganic** compounds are found in the nonliving world. Examples of inorganic compounds include water (H_2O), ordinary table salt (NaCl), carbon dioxide (CO_2), and calcium carbonate ($CaCO_3$). **Organic** compounds are compounds of carbon that contain hydrogen. It will be noted that carbon dioxide and calcium carbonate do not contain both carbon and hydrogen and, hence, are classified as inorganic. Organic compounds are of many types; the four most important kinds found in living material are carbohydrates, lipids, proteins, and nucleic acids. **Carbohydrates** and **lipids** are used mainly as sources of energy, but are also used in the formation of certain parts of the organism. **Proteins** are used primarily for the construction and repair of the individual, but may be used as energy sources, if the need arises. **Nucleic acids** are the hereditary material of the organism. The average amounts of the various compounds found in living plants and animals are listed in Table 2-2.

An examination of Table 2-2 reveals both similarities and differences in the chemical composition of plants and animals. One of the similarities, namely, the large amount of water found in both types of organisms, is a reflection of the importance of water in the chemical reactions of all living material. With respect to differences, we find that plants are composed of a great deal more carbohydrate and considerably less protein and lipid than animals. These differences in body composition reflect alternative methods of meeting various needs. For example, the material that makes up the bulk of the organism's body differs between plants and animals. The wood of plants is mainly cellulose, which is carbohydrate. In the case of animals, the skeleton and muscles are mainly protein. Another difference lies in the type of compound used in storing excess energy. Here, too, plants utilize carbohydrates for this purpose, whereas animals store excess energy mainly in lipids.

■ FOOD AS A SOURCE OF ENERGY

Energy is the capacity to do work and always involves motion. There are a number of different forms of energy: heat, light, electricity, chemical bonds, etc. Organisms perform their various activities with the energy of chemical bonds, namely, the energy that binds the constituent atoms of food together. For a detailed description of atoms and how they combine with one another, see Appendix A.

□ Types of chemical bonds

We shall consider three types of chemical bonds: ionic, covalent, and hydrogen bonds.

In compounds formed through **ionic bonding,** there is a transfer of one or more electrons from one atom to another. This transfer process results in the atoms of the compound becoming oppositely charged; the particular atom that acted as an electron donor becomes positively charged, whereas the atom that acted as a electron recipient becomes negatively charged. All electrically charged atoms are called **ions.** Although ions are charged, the molecule they form is usually electrically neutral, because the oppositely charged atoms generally attract each other strongly enough to function electrically as a single entity. Important exceptions to this situation exist. Compounds such as sodium chloride (NaCl) have a strong tendency to dissociate into separate ions when placed in a liquid medium; in the case of NaCl, a sodium ion (Na^+) and a chlorine ion (Cl^-) are formed. As we shall see in later chapters, the charge on a particle greatly affects its ability to enter and leave an organism's body and also affects the functioning of the organism.

In **covalent bonding,** there is a sharing of electrons between atoms. Since there is no actual transfer of electrons, the participating atoms are not charged. A most important group of compounds is formed through the covalent bonding of their constituent atoms. These substances are the organic compounds, which, as mentioned before, always contain both carbon and hydrogen and may, in addition, contain the atoms of other elements. The element of central importance in organic compounds is **carbon.** Carbon forms compounds only by sharing electrons and can do so even with other carbon atoms. This ability of carbon atoms to bond covalently to one another permits the formation of molecules of tremendous size, with many different combinations of other atoms as well. As noted earlier in this chapter, four groups of organic compounds are extremely important for living organisms: carbohydrates, lipids, proteins, and nucleic acids. These types of compounds will be considered in greater detail later.

Hydrogen bonding is a special type of chemical linkage that involves the element hydrogen. This type of bond formation occurs when a hydrogen atom that is covalently bound to one atom forms a weak linkage with a second atom. Hydrogen bonds usually form between hydrogen and either oxygen or nitrogen. The formation of a hydrogen bond results in the hydrogen atom being shared simultaneously between two atoms and creating a bridge between them. Hydrogen bonding is particularly important in maintaining the structure of an organism's genetic material, as will be discussed later.

☐ Bond energy

Every bond in every molecule represents a quantity of chemical energy equal to the amount of energy that was necessary to link the atoms together. Each of the three types of bonds (ionic, covalent, hydrogen) requires a different amount of energy for its formation. This energy is measured in units of heat called calories. A **calorie** is the amount of heat required to raise the temperature of 1 gram (gm) of water from 14.5° to 15.5° C at a pressure of 1 atmosphere. One thousand calories is designated as a **kilocalorie** (kcal). The calorie is sometimes referred to as a "small calorie" and the kilocalorie (the calorie found in diet tables) as a "large calorie."

The amount of energy required to break a particular chemical bond is usually given in terms of "per mole" of the compound. A **mole** (gram molecular weight) is the sum of the mass numbers of the constituent atoms of a compound expressed in grams. The number of molecules in a mole of a compound is fantastically large and is the same for all compounds. It is useful to designate chemical and physical characteristics of compounds on a molar basis (per mole), since this indicates the involvement of a fixed number of molecules.

The breaking of hydrogen bonds requires 1 to 6 kcal/mole, depending on the particular compound; ionic bonds have an energy content of 10 to 30 kcal/mole, whereas covalent bond energies vary from 50 to 110 kcal/mole. These data indicate that hydrogen and ionic bonds are relatively weak and easily broken, whereas covalent bonds are strong and very stable. Therefore it is not surprising to find that the compounds making up most of the substance of a living organism are **covalent compounds.**

☐ Energy transfer among organisms

As a result of the great number of chemical bonds involved in the molecules of an organism, each individual is a vast storehouse of potential energy that can be used to do work. However, as this energy is depleted in carrying out various activities, it must be replenished. For strictly consumer-type organisms (herbivores, carnivores, and omnivores), the replenishment of energy is accomplished by eating the bodies of other organisms and obtaining some of their storehouses of potential chemical energy. Decomposers (bacteria, yeasts, and molds) replenish their energy through the chemical breakdown of the dead bodies of other organisms.

At first glance, it might seem that consumers and decomposers could form a self-sustaining cycle in which energy is passed continuously from organism to organism. Under such an arrangement, no source of energy outside the system composed of living things would be required. However, such a cycle is not possible, because as energy is converted from one form to another, a certain amount of it is lost to the nonliving world. When you turn this page or when the heat from your body warms the air around you, the energy consumed by these actions is lost to you. The energy lost to a living system must be replaced from an outside source.

It is quite evident that the producer-type organisms (mainly green plants) were omitted from the hypothetical consumer-decomposer system discussed previously. The green plants are not usually able to obtain energy from other living forms. Producers, as their name implies, manufacture their own food. In order to do this, they require an outside, nonliving source of energy, such as the sun. Green plants are able to use the energy of sunlight to bind atoms together into organic molecules.

The energy of the chemical bonds is then available for use by the other organisms of the food cycle.

An important point emerges from our discussion of food as a source of both building materials and energy for organisms. The basic building materials available in food are never destroyed. They can and, in fact, must be recycled for use by future generations. The energy available in food is eventually lost to the living world and must be continuously replenished from the nonliving world.

WORLD FOOD PROBLEMS

We have discussed food both as a source of building materials for the body and as a source of energy for life processes. To understand the basis of the present world food crisis and the even greater dilemma we face in the future, we must analyze the daily food requirements of humans and the availability of food throughout the earth. In establishing whether a diet is providing sufficient building materials for an individual, the criterion is the amount of **protein** consumed. The number of **calories** consumed by a person is the criterion for determining whether a diet is providing sufficient energy to carry on life's activities. Obviously, a well-balanced diet is one that provides sufficient amounts of both protein and calories. An individual whose diet seriously lacks protein is said to be **malnourished.** (As will be discussed in a later chapter, a lack of any other essential nutrient also results in malnourishment.) An **undernourished** person is one who does not receive enough calories per day.

Although the need for protein and calories varies with the individual and his stage of development, it has been estimated that for adults a proper diet should provide about 57 grams (gm) of protein and 2350 kilocalories (kcal)/day. The worldwide distribution of various types of diets is shown in Fig. 2-4. An examination of this map shows that people in large areas of Africa and Asia are suffering both malnutrition and undernutrition. As many as a billion and a half people throughout the world fall into one of these categories. The situation has resulted in the recognition of specific diseases caused by diets deficient in proteins and calories. Two such diseases are kwashiorkor and marasmus.

The importance of protein in world nutrition has been emphasized in the last two decades with the identification of the condition called **kwashiorkor** as a protein-deficiency disease. Kwashiorkor is a West African word that means "the sickness a child develops when another baby is born." The disease occurs when children are weaned from their mother's milk to a diet of starchy cereal, which is practically devoid of protein. This shift in diet usually occurs at the time of the birth of another child and manifests itself even if calories are abundantly provided. The affected child suffers from a failure to grow both in weight and in length and experiences vomiting, diarrhea, and a loss of appetite. The ability of the child to combat infection is diminished considerably, and death is usually attributed to an infection, such as measles or pneumonia, that normally would not be fatal.

A second nutritional disease that has become widespread in areas whose inhabitants suffer from inadequate diets is **marasmus.** This disease also occurs in children and reflects a state of undernutrition. The child with marasmus is very thin and wasted and has wrinkled skin and enormous eyes. It is not unusual for both malnutrition and undernutrition to occur simultaneously, and the tendency has been to con-

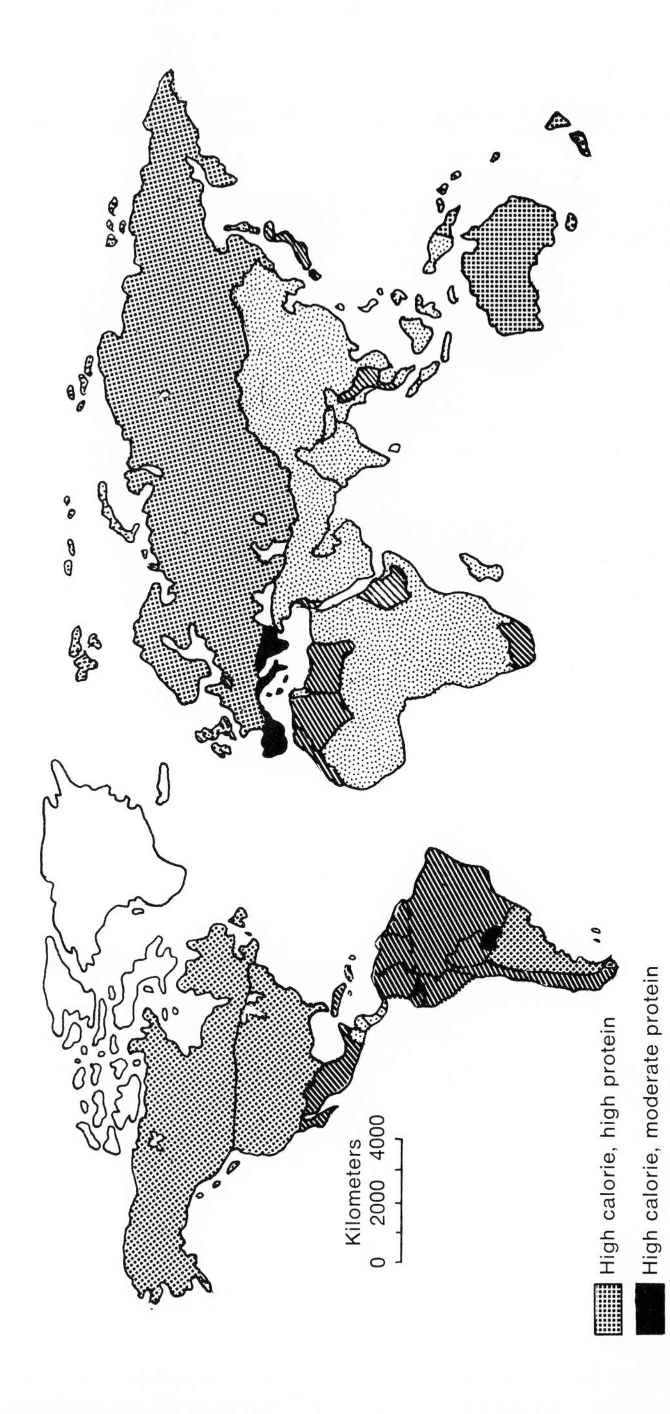

Fig. 2-4
Distribution of various types of diets throughout the world.

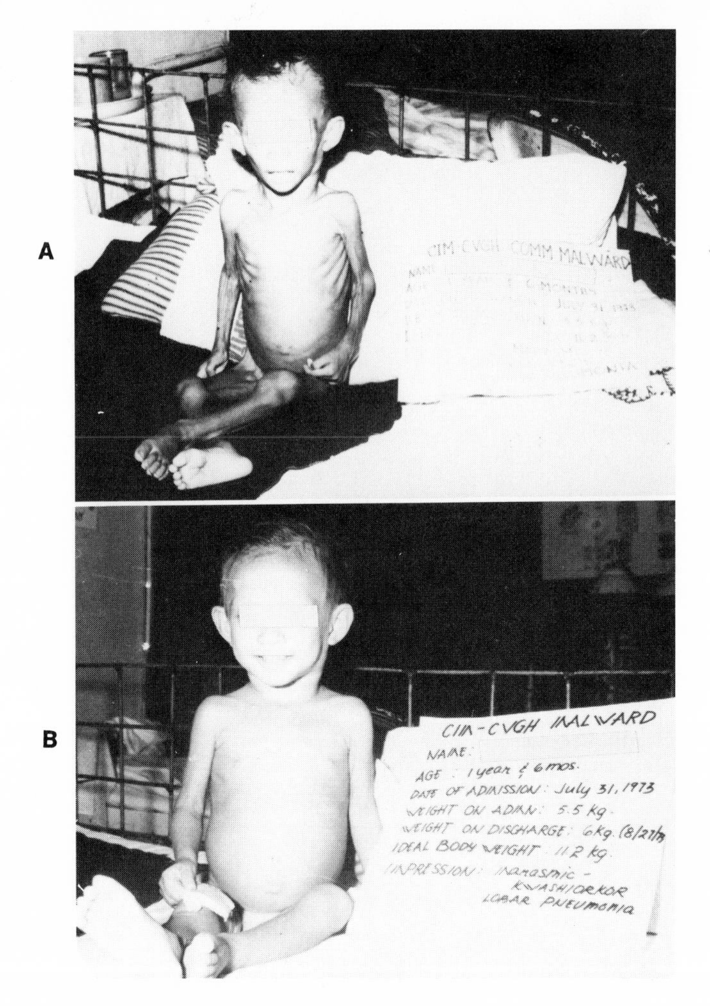

Fig. 2-5
A, Child with marasmic kwashiorkor on admission to hospital. **B,** Child on discharge 1 month later. (Photographs courtesy Dr. F. Solon, Cebu Institute of Medicine, Philippines; from Guthrie, H. A. 1975. Introductory nutrition, ed. 3. The C. V. Mosby Co., St. Louis.)

sider the resultant condition as a single disease, referring to it as "protein-calorie deficiency." Here, too, death results from some infectious or parasitic disease that in most cases would not be fatal in properly nourished children (Fig. 2-5).

□ Food pyramid

Having identified the types of food problems that are presently being suffered by more than 40% of the world's population, we might ask, "What are the prospects for alleviating this situation?" Many factors complicate the problem. Some relate to the qualitative aspects of food, whereas others deal with quantitative factors. For an understanding of this problem, we must again consider the food cycle that was discussed earlier in this chapter and shown in Fig. 2-1. As noted before, green plants

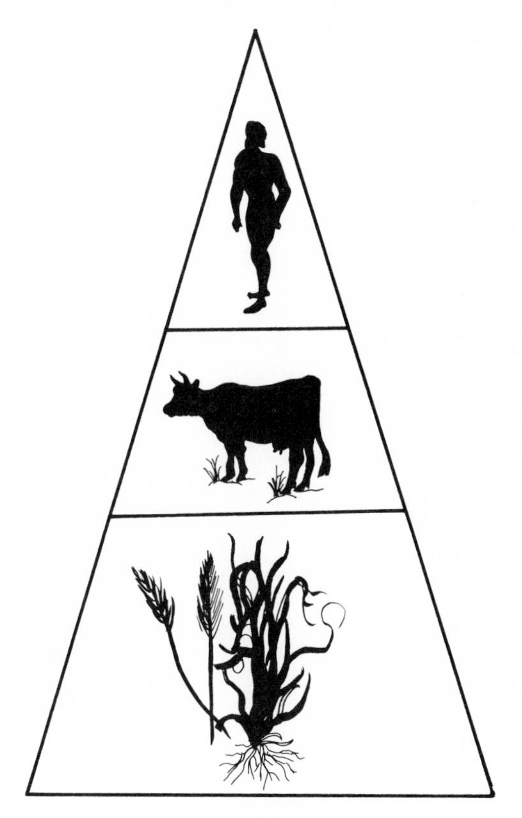

Fig. 2-6
Food pyramid involving green plants, herbivores, and man.

are the ultimate producers of food on earth. They, or their by-products, can be eaten by herbivores or omnivores, both of which can subsequently be eaten by carnivores or omnivores. It is obvious that in our modern society there is virtually no predation on man by carnivores or other omnivores. We shall therefore limit our consideration to green plants, herbivores, and human beings. These three types of organisms can be considered as forming a food pyramid, as shown in Fig. 2-6. The pyramid shape reflects the efficiency of the system in making usable energy from green plants available to man (Fig. 2-7).

Transfer of energy. Energy enters the food cycle in the form of radiation from the sun, which green plants use to bind together small molecules to form carbohydrates. In this process, called **photosynthesis,** solar energy is converted into chemical energy. In subsequent chemical reactions, the green plants use the energy of the manufactured carbohydrates to produce all the various other chemical compounds of which they are composed. The plant body and its by-products, in turn, may be eaten by herbivores and omnivores who chemically break down the plant compounds into their component parts. Through this process, these animals can obtain some of the plant's chemical energy for their own activities. The animals also use some of the parts of the plant molecules to build and repair their own bodies. When herbivores are eaten by carnivores and omnivores, the chemical compounds of the herbivores'

Fig. 2-7
One type of food pyramid involving producers (hay), herbivores (cattle), and omnivore (man). (From Kaluger, G., and Kaluger, M. 1976. Profiles in human development. The C. V. Mosby Co., St. Louis.)

bodies are broken down, and there is again a reuse of some of the energy and molecular components.

Thus there is a transfer of both **energy** and **molecules** through the food pyramid. However, the efficiency of the transfer of energy in the system just discussed is extremely low. In photosynthesis, only about 1% of the sunlight falling on green plants is converted to chemical bond energy. At the next level of the pyramid, roughly 10% of the energy present in plants becomes incorporated into the bodies of the animals that eat the plants. The same rate of efficiency occurs at the next stage; about 10% of the energy in the bodies of herbivores becomes incorporated into carnivores and omnivores. Thus at each transfer of energy in the food pyramid, about 90% of the chemical energy stored in the organisms of a lower level is **unavailable** to those at the next higher level. This means that there is considerably more usable energy available to organisms lower in the food chain than to those in higher positions.

The rates of energy transfer just noted permit us to make the following admittedly oversimplified calculations. It takes the energy of about 10,000 pounds of wheat to

Fig. 2-8
Corn plant, a major mechanism for trapping the sun's energy and converting it to chemical energy. The midwestern United States is blanketed with millions of acres of corn plants, as well as soybeans and other major crops, and heavy tropical vegetation and the tremendous amount of algae in the oceans trap even more energy. Even so, less than 1% of the solar energy that reaches the earth ever appears as photosynthetic products. (From Lane, T. R., ed. 1976. Life the individual the species. The C. V. Mosby Co., St. Louis.)

produce 1000 pounds of cattle. This amount of meat can then provide enough energy to produce about 100 pounds of human being. It would appear that if man moved down one step in the food pyramid, 10 times as much energy would be directly available, and 10,000 pounds of wheat could then be utilized to produce 1000 pounds of human being.

Unfortunately, the problem of undernutrition of much of the world's population cannot be solved in this simple manner for the following reason. It will be recalled that man is an omnivore and can utilize both plants and other animals as food. In actual fact most of the diet of the world's human population is plant material. Thus man is, in most instances, already at the lowest position possible in the food pyramid.

Therefore we must look to increased supplies of food rather than any change of diet to prevent and alleviate undernutrition (Fig. 2-8).

Transfer of molecules. In addition to the transfer of energy, the food pyramid in Fig. 2-6 also represents the transfer of atoms and molecules from plants to herbivores to man. The critical building material, as we noted previously, is protein, and the lack of it in the diet results in malnourished individuals. Unfortunately, not all proteins are equal in acting as sources of building materials, an important fact that greatly complicates the world food crisis. To understand the problem of protein in our diets, we must take a closer look at these highly complex organic compounds.

Proteins are composed of **amino acids** joined together in long chains. There are 20 different amino acids that are important in making the proteins of organisms. The characteristics of a protein are determined not only by which amino acids are used and the number of times they are repeated, but also by the sequence in which they are joined together. Since each amino acid may be used any number of times in any relation to other amino acids, the potential for the formation of different proteins is enormous. The human body contains hundreds, if not thousands, of different proteins either as enzymes or as structural molecules.

It is important to realize that each species builds proteins that are characteristic of itself. For example, a type of protein, such as hemoglobin, that is similar in the dog, duck, and human cannot be interchanged among the different species. This species specificity of proteins affects the transfer of molecules in the food pyramid in that proteins found in plants cannot be used directly by herbivores, nor can the proteins found in herbivores be used directly by carnivores or omnivores. At each stage of the food pyramid, the proteins of the consumed organism must first be broken down to their amino acids, and these amino acids must then be built up into the proteins characteristic of the consumer. Thus it is clear that *the ultimate value of a food protein lies in its amino acid composition.* It also follows that if the protein of a particular food is deficient in an amino acid required for the proteins of a consumer, the consumer will not be able to manufacture the necessary proteins and will suffer malnourishment. This last point is extremely important in considering the world food crisis.

Plants are able to build the amino acids they need for their proteins. Animals, however, have only a limited capacity to manufacture amino acids. As an example, the adult human being can manufacture 12 of the amino acids required for protein production, but must be supplied with sufficient quantities of the eight others through food. The amino acids that must be supplied in man's diet are called **essential** amino acids. Proteins that contain all the essential amino acids are called **complete** proteins. They normally contain about 50% essential and 50% nonessential amino acids.

All animal proteins, except gelatin, are complete proteins. However, as we have seen in the food pyramid, using herbivores as a source of protein is essentially a wasteful process, since we must first grow the plants on which to feed these animals. We also know that there is a 90% loss of energy in passing through each stage of the pyramid. The tremendous expense in producing animal protein for human consumption is reflected in the fact that meat, fish, poultry, and dairy products provide less than one third of the world's proteins. Fig. 2-9 illustrates the amounts of plant and

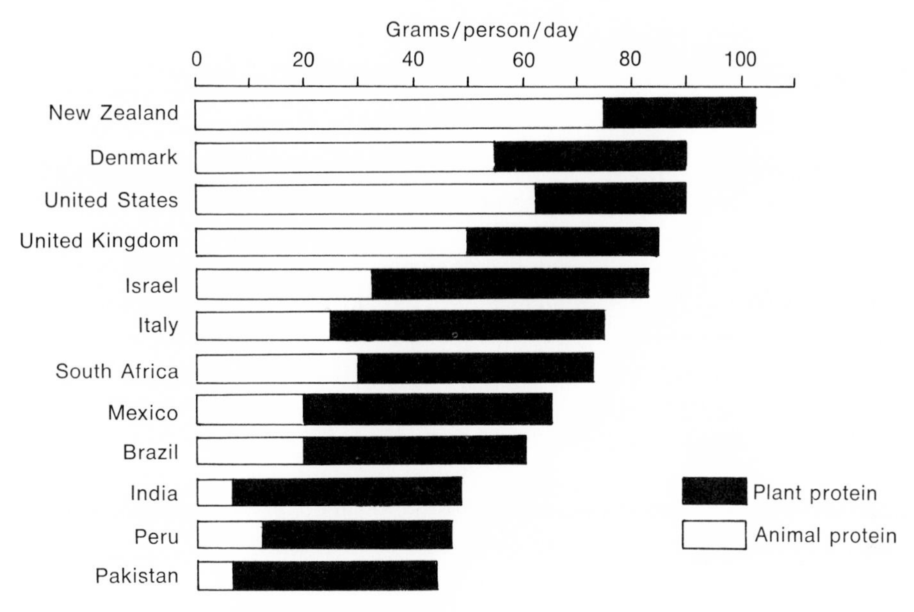

Fig. 2-9
Average amounts of plant and animal proteins in the daily diets of some populations.

Fig. 2-10
These foods all contain protein but of varying quantity and quality. Beef is about 25% protein and contains all the essential amino acids. Gelatin is almost pure protein, but lacks tryptophan, valine, and isoleucine. Rice contains about 5% protein, but lacks threonine and lysine. (Copyright © 1976 by Theodore R. Lane.)

animal protein in the daily diets of some of the world's human populations. It is quite clear that the use of animal protein as the main source of amino acids is restricted to relatively few countries. Most of the world's human beings rely on plant protein for their major source of amino acids.

Unfortunately, the use of plants as the sole, or even major, source of amino acids has its drawbacks. All vegetable proteins, except those found in nuts, are **incomplete** proteins. The most frequently used source of plant proteins consists of the seeds of such cultivated grasses as rice, wheat, corn, sorghum, oats, barley, rye, and other cereals. Collectively, these plants and their seeds are called **grain.** The seeds of the various bean and pea plants constitute another important but lesser used source of plant protein. Beans and peas belong to the group of plants called **legumes.**

In general, cereal grains are deficient in the essential amino acids **lysine** and **tryptophan,** whereas legumes are deficient in **methionine.** Of course, it is possible to achieve a balanced protein diet by consuming both types of plant foods, and some human populations have developed such eating patterns, which provide all the essential amino acids. An example of this combination of foods is found in the diet of many Mexicans, which includes large amounts of tortillas (a type of bread made from corn) and frijoles (made from beans). However, it is not simple or even feasible to change the diets of large portions of the world's human population, many of whom eat largely unbalanced diets (Fig. 2-10).

□ Solving the world's food problems

Three approaches have been taken to solve the world's food problems: (1) improving the quality of the food produced, (2) increasing crop production, and (3) obtaining new sources of food. Singly and collectively, they have met with only slight success in meeting the needs of our rapidly increasing world population.

Improving the quality of food produced. The solution to the problem of essential amino acid deficiencies in various plant proteins may in time be solved by modern biological research in the field of plant genetics. Recent studies have revealed that certain varieties of rice, wheat, corn, etc. do have all the essential amino acids in the amounts that man requires, that is, the seeds from these varieties contain complete proteins. Unfortunately in many cases, the varieties do not grow well in the areas of the world where they would be most useful. In other cases, there are differences in taste and color that make them unacceptable to the people who would benefit most by eating them. Hopefully, further research and educational programs will eventually solve these problems (Fig. 2-11).

Another drawback to the use of plants as the sole or major source of amino acids is the relatively poor percentage of protein content in plants. As shown in Table 2-2, part of this problem stems from the fact that, relative to animals, very little of the total plant body is composed of protein. The seeds that man uses as food actually have a much higher protein content than the total plant, but even this higher percentage is relatively low compared to animal tissues. Table 2-3 lists the various protein sources used by man and the percentage protein content of each. It is obvious that meat, poultry, and fish are much richer sources of protein than cereal grains, potatoes, and even milk. The advantage of milk lies in the completeness of its nutritional content, making it truly the perfect food for developing infants.

Fig. 2-11
New high-lysine corn, discovered by Purdue University scientists, known as opaque-2. This corn has twice as much lysine, a protein building block, as normal corn and therefore greater food value than other corn hybrids. (Courtesy Purdue University Agriculture Experiment Station Information Service, Lafayette, Ind.; from Arnett, R. H., Jr., and Bazinet, G. F., Jr. 1977. An introduction to plant biology, ed. 4. The C. V. Mosby Co., St. Louis.)

Table 2-3
Protein sources and their percentage of protein content

Protein source	Protein content (%)
Animals	20-30
Cereal grains	5-13
Legume seeds	10-40
Potatoes	1-4
Milk	3-4

According to Table 2-3, the protein content of some legumes is actually greater than that of animals. This fact raises the question of why legumes do not play a larger role in the diets of all people. The answer is that yields of legumes per acre are much smaller than the yields of cereal grains. The result has been that, although more than 1 billion tons of cereal grains are produced annually over the world, only about 80 million tons of legumes are produced. The need for increased production of all types of food is widely recognized, and a good deal of biological research has been directed to developing means to improve food production.

Increasing crop production. It was pointed out in Chapter 1 that the success of the Industrial Revolution was in large part the result of changes in agricultural methods that led to an increased production of food by a smaller farm labor force. These changes included the mechanization of the farm, the use of fertilizers, and the development of high-yielding genetic varieties of crops.

Mechanization of the farm was made possible by the invention of the internal combustion engine and the tractor, which rapidly replaced the horses and other draft animals that had been used previously to augment man's own limited muscle power. The use of machines powered by diesel fuel or gasoline meant that the energy present in the chemical bonds of the petroleum molecules was being used as a substitute for the chemical bond energy of the oats, corn, and hay that were grown as feed for the draft animals. The replacement of these animals by the tractor also resulted in the release of millions of acres that had been devoted to raising feed and the use of these acres to produce food for man. In the highly mechanized farms of today, the expenditure of energy from nonrenewable fossil fuel (petroleum) is sometimes greater than the energy obtained from the food produced. This deficit will be of little consequence for our modern society until the time is reached when the nonrenewable fossil fuels become really scarce. At the present time, the continuing need to produce more food each year for the ever increasing human population has prevented a proper consideration of this problem.

The use of fertilizers revolves around the lack of proper nutrients in the soil for maximum plant growth and seed production. The critical elements, in most cases, are nitrogen, phosphorus, and potassium (Fig. 2-12). Where fertilizers have been used intensively, yields of grains have doubled, tripled, and even quadrupled. The worldwide use of fertilizers is shown in Table 2-4. It is clear that the farmers of Africa, Asia, and Latin America use far less fertilizer than the world average, whereas the

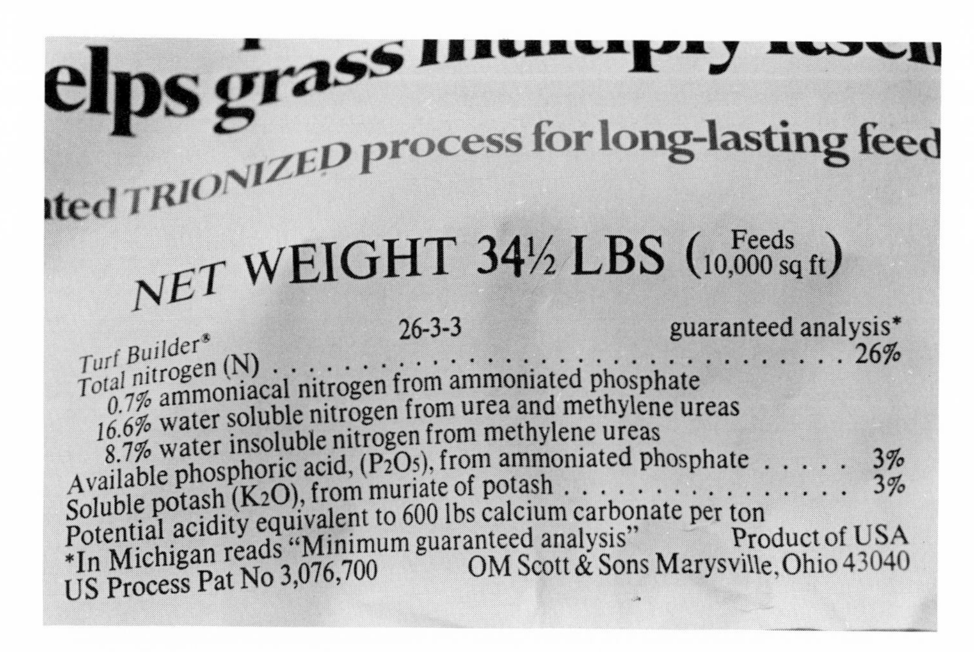

Fig. 2-12
Composition of a common fertilizer. (Copyright © 1976 by Theodore R. Lane.)

Table 2-4
World use of fertilizers (1967 to 1968)*

Region	Acres cultivated (millions)	Fertilizer used (millions of tons)	Fertilizer per acre (pounds)
Africa	575	2.0	6
Asia†	975	9.0	18
Europe‡	1000	30.0	60
Latin America	200	1.5	14
North America	635	20.0	60
Total	3385	62.5	37 (average)

*From United Nations. 1969. The state of food and agriculture, 1969. Report by the Food and Agriculture Organization of the United Nations. Rome.
†Includes Australia and New Zealand.
‡Includes Soviet Union.

Table 2-5
Effects of use of mechanization, fertilizers, and plant breeding (hybrids) on maize (corn) production in the United States (1933 to 1963)*

Year	Use of fertilizers (based on 1933 = 1)	% Crop as hybrids	Yield (bushels/acre)
1933	1.0	0.1	22.6
1943	1.9	52.4	32.2
1953	4.2	86.5	39.9
1963	6.5	95.0+	67.6

*From U.S. Department of Agriculture. 1966. Agricultural statistics. Washington, D.C.

farmers of Europe and North America engage in an extensive use of plant nutrients. If more fertilizer were used in regions where population size is greatest and the rate of population growth is highest (see Table 1-4), there would be more food for the people who need it most.

An increasingly important method of improving food production is the use of high-yielding genetic varieties of cereal grains, legumes, etc. These new varieties, in most cases, have been produced by scientists working in experimental projects designed to raise crop yield. In one experimental approach, strains of a plant from different parts of the world are studied for desirable characteristics. Those strains with different desirable qualities are then crossed (hybridized), and the hybrid offspring are crossed among themselves until a new strain is developed containing the desired traits. Perhaps the best known example of this type of plant breeding is the recent development of the so-called miracle seed of the "Green Revolution." These are semidwarf varieties of wheat and rice that can double the yield of grain per acre. The man who was instrumental in promoting the development of these strains, Dr. Norman Borlaug, received the Nobel Peace Prize in 1970 in recognition of the importance of the present world food crisis in shaping the policies of nations.

The best example of the effects of the combined use of mechanization, fertilizers, and genetic varieties (hybrids) in raising the yield of a crop is corn production in the

United States. From the data shown in Table 2-5, it is quite clear that a combination of the agricultural techniques just discussed has resulted in a tripling of the yield of corn over the 30-year period from 1933 to 1963. It is difficult to speculate as to whether further improvements in crop yield are possible and if so what their magnitude might be.

We have examined several methods for increasing food production that have resulted in significant increases of available food in some regions of the world. But how do these increases in food production relate to world population increases? The data in Fig. 2-13 indicate world food production, world population size, and per capita food production for the years 1950 to 1973. The values are relative and based on 1954 as the "standard year." It is quite apparent that world food production has barely kept pace with population growth. This situation is the result in part of the continuing rise of world population and in part of an increase in meat consumption in Europe, the Soviet Union, and Japan.

Increased meat production, as noted earlier, requires that an increasing part of the grain produced be diverted from human consumption to use as animal feed. This means that the gain in acreage available for human food production, achieved through the Industrial Revolution, is now being offset by a change in diet of a considerable portion of the world's population.

Obtaining new sources of food. In a continuing search for solutions to the world's food problem, biologists have explored a number of possibilities for increasing the amount of available food. It should be remembered that the problem of food is twofold: the requirement *to supply sufficient amounts of all the essential amino acids* and the requirement *to supply a sufficient quantity of calories (energy).*

One approach to the problem of providing complete plant proteins other than

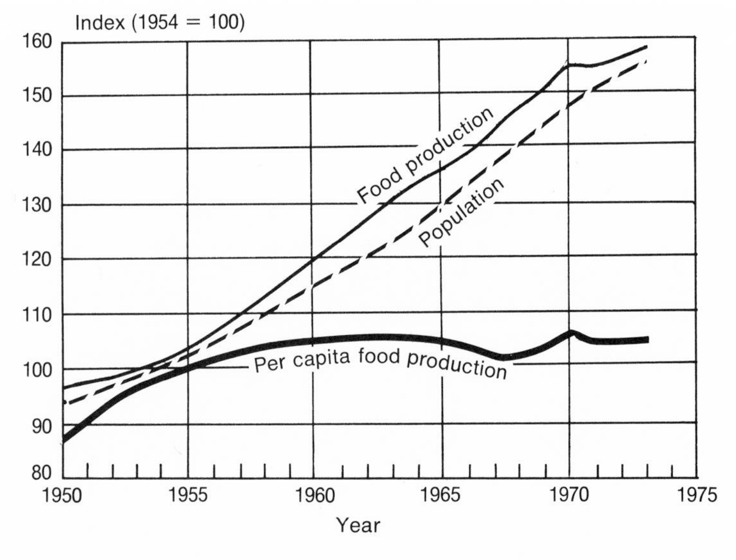

Fig. 2-13
Relative changes in world human population and food production for the period 1950 to 1973.

through the genetic improvement of crops has been to add the missing essential amino acids to the food products that are made from plants. Examples of this type of project, called **amino acid fortification,** include the addition of the amino acid lysine to wheat flour, the fortification of rice with synthetic granules of lysine and threonine, and the addition of lysine to corn flour. This approach to the problem of food improvement has great potential, but it is too early to evaluate its success.

Complete animal protein can also be added to foods through the use of a protein concentrate called **fish flour,** which is made from types of fish that are ordinarily not used as food. Whole fish are ground up, and their lipids and water are extracted, leaving a heavy concentrate of protein material. The resulting product is a completely tasteless and odorless grayish powder containing 90% fish protein. This concentrate can be added to processed foods or baked into bread, cookies, and crackers. Unfortunately, the cost of producing the fish flour is so high that it is not competitive with plant and milk proteins, even though the plant proteins do not contain some of the essential amino acids. At present, it is difficult to speculate whether fish flour will play a significant role in improving the world's critical food situation (Fig. 2-14).

Among still other ideas for providing new sources of food has been the use of microorganisms as food supplements. Both single-celled green algae and bacteria have been used for this purpose. The organisms are cultured in tremendous numbers, ground up, and their protein contents extracted. The protein concentrate called **single-cell protein** is made into a powder that can be added to flour or other foods. A great advantage of the use of microorganisms in food production is their ability to grow on varied types of substrates. These include such waste materials as manure, garbage and sewage, and even petroleum. The ability of microorganisms to utilize waste materials results in the direct recycling of otherwise scarce elements and

Fig. 2-14
Fish flour preparation served to children in Rangoon day nursery to combat protein deficiency. FAO nutritionist helps the Burmese nutrition staff with this program. (FAO photograph by S. Bunnag; from Williams, S. R. 1973. Nutrition and diet therapy, ed. 2. The C. V. Mosby Co., St. Louis.)

molecules in the food cycle. The direct use of petroleum would permit a most efficient method of eventual transfer of the chemical bond energy in the petroleum molecules to man. It has been estimated that we could produce 100% of the protein required by the world's inhabitants through the use of between 15% and 20% of the world's production of petroleum.

The widespread use of single-cell proteins is prevented by the tastes and dietary customs of the different groups of the world's population. The utilization of this product as animal feed, thereby freeing more of the world's cereal grain production for human consumption, is hampered by the lower cost of the plant protein supplements now being used for cattle feed. Although it seems ironic that cost analysis influences such basic biological problems as food production for human beings, it is nonetheless a factor in our modern society. It requires not only that we develop more methods for obtaining food, but that these methods be more efficient and less expensive than those now available.

■ SUMMARY

In this chapter we examined the food cycle and its twin roles of providing both the building materials and the energy necessary for sustaining living organisms on the earth. We also considered the world's food problems as related to providing both essential amino acids (food quality) and calories (food quantity) for human beings. We found that the present world food crisis is complicated by two factors of human origin: (1) the unequal distribution of presently available food among the various segments of the world's population and (2) the presence of an ever increasing number of human beings on the earth.

The problem of feeding the people of the world is further rooted in differences in food quality and food quantity between plant and animal materials. Plant materials are 10 times more efficient to produce than animal meat, but most plant proteins are incomplete and must somehow be supplemented.

An analysis of attempts to solve the problems of food quality as well as food quantity indicates that we have not yet been successful in improving the food situation for most of the people of Africa, Asia, and Latin America. Even the development of new sources of food has not solved the present world food crisis. It is difficult to be certain, but mankind appears to have just about reached the limit of the food material he can extract from the earth's resources. Further expansion of the world's population will undoubtedly result in even more widespread starvation than we are witnessing today. In the next chapter, we shall expand our discussion of food requirements and man's needs for specific types and amounts of food as determined by varying conditions of development and work.

SUGGESTED READINGS

Dumont, R., and Rosier, B. 1969. The hungry future. Praeger Publishers, Inc. New York. Detailed study of the number of undernourished and malnourished people in the world.

Food and Agriculture Organization of the United Nations. 1975. The state of food and agriculture. 1974, Rome. World review, by regions, of population, food supply, and agricultural development.

Myrdal, G. 1974. The transfer of technology to underdeveloped countries. Sci. Am. **231**:173-182. Analysis of the effects of improved agricultural productivity on fertility control in these countries.

Poleman, T. T. 1975. World food: a perspective. Science **188**:510-518. Comparison of the increases in food production and human population during the past 20 years.

Revelle, R. 1974. Food and population. Sci. Am. **231**:160-171. Discussion of the maximum size of the human population that our modern agricultural technology could possibly support.

Sanchez, P. A., and Buol, S. W. 1975. Soils of the tropics and the world food crisis. Science **188**:598-603. Discussion of the role that agricultural programs in tropical countries can play in meeting the world's food needs.

Trowell, H. C. 1954. Kwashiorkor. Sci. Am. **191**: 46-50. In-depth discussion of this widespread nutritional disease.

Wittwer, S. H. 1975. Food production: technology and the resource base. Science **188**:579-584. Analysis of the need to institute a massive program in agricultural research and technology to improve food production in order to feed the growing world population.

Woodwell, G. M. 1970. The energy cycle of the biosphere. Sci. Am. **223**:64-97. Discussion of the food cycle as it relates to energy loss and its replenishment by the sun.

Four food groups, each of which contributes to a balanced diet. **A,** Milk group. **B,** Meat and fish group. **C,** Vegetables and fruit. **D,** Bread and cereals. (Courtesy National Dairy Council; from Turner, C. E. 1971. Personal and community health, ed. 14. The C. V. Mosby Co., St. Louis.)

CHAPTER 3 Essential nutrients for energy and growth

LEARNING OBJECTIVES

- What is a nutrient?
- Which bodily functions are performed by nutrients?
- What is an essential nutrient?
- What is a carbohydrate?
- Why are carbohydrates considered essential nutrients?
- What are the different types of carbohydrates, and what are their food sources?
- What is the relationship between the carbohydrate glucose and diabetes?
- What is a lipid?
- Why are lipids considered essential nutrients?
- What is the relationship between lipids and heart disease?
- What is a protein?
- What is an amino acid?
- Why are proteins, and hence amino acids, considered essential nutrients?
- What differences are there in the daily protein requirements of men versus women and children versus adults?
- What is the relationship between the amino acid phenylalanine and the hereditary disease phenylketonuria?

Our discussions on the size, distribution, and food problems of the world's human population pointed out the unfortunate fact that about 40% of the human beings on this earth are either undernourished or malnourished. Even more disturbing is the estimate that about 400 million of these people are experiencing absolute starvation, and some 50% of those starving are children. Although most of the world's critical food problems are centered in Africa, Asia, and parts of Latin America (Fig. 2-4), significant instances of malnutrition and undernutrition also occur in Europe and North America. Either out of ignorance or as a result of poor eating habits, people can suffer from inadequate nutrition even in the midst of ample food supplies. Unfortunately, most people are not aware of their own daily food requirements, nor do they understand that these food requirements change with their stage of development and level of activity.

Our aim here and in Chapter 4 will be to analyze the food requirements of human beings and the consequences that result if these requirements are not met. It is clear that if we are to make intelligent decisions to meet the needs both of ourselves and our fellow men, we must understand the basic food requirements of the individual and the type of diet that will meet these requirements.

HUMAN FOOD REQUIREMENTS

An individual's diet is defined as what the person usually eats and drinks. In order for a diet to be balanced, it must provide all the nutrients essential for the normal functioning of the body. An **essential nutrient** is one that cannot be synthesized by the body and must be provided to the organism through food. However, as was pointed out in an earlier chapter, nutrients essential for one species may not be essential for another.

In the body, nutrients serve several functions: (1) to supply energy, (2) to provide the molecules for growth and repair of body tissues, and (3) to regulate body processes (metabolism). We have already indicated that carbohydrates and lipids are the main sources of energy for the body, with proteins acting as secondary possibilities. We have also pointed out that proteins, more specifically their amino acids, act as the main source of molecules for the growth and maintenance of body tissues. However, in addition to proteins, a rather large number of so-called "mineral elements," for example, calcium and iron, are required for the production of different types of tissues in the body. Finally, the proper regulation of the body's metabolism requires that the nutrients present in foods contain ample amounts of proteins, mineral elements, vitamins, and water.

Some nutrients are present in a wide variety of foods, and there is little likelihood of a deficiency occurring. On the other hand, some nutrients are found only in a restricted number of foods and, as such, will be available in less than optimal amounts if the **variety** of foods in a diet is limited. We shall now examine each of the various essential nutrients, indicating the amounts required each day, their availability in different foods, and the diseases that result from their absence or improper metabolism.

Carbohydrates

Carbohydrates, commonly referred to as sugars and starches, are classified as essential nutrients despite the fact that they can be replaced by fats and proteins as sources of energy. The reason for considering carbohydrates as essential is the occurrence of body malfunctions in those individuals whose diets are completely lacking in this type of nutrient. These malfunctions include the breakdown of a considerable portion of the body's protein, the loss from the body of very large amounts of the essential mineral element sodium, the loss of energy, and the experience of fatigue. All these effects of a diet devoid of carbohydrates are reversed by the addition of carbohydrates to the diet.

The various malfunctions experienced by those on a carbohydrate-free diet are not caused by any lack of energy, but are the result of disturbances of other biochemical processes, especially those in which carbohydrates serve as vital components, such as in nucleic acid synthesis.

Carbohydrates are organic compounds composed of carbon, hydrogen, and oxygen. In simple carbohydrates, the proportions of the three elements are $C_nH_{2n}O_n$, or $C_n(H_2O)_n$. In compounds formed by the union of two or more simple carbohydrates, the proportions of the three elements are $C_n(H_2O)_{n-1}$. It is clear that the ratio of hydrogen to oxygen atoms in carbohydrates is the same as that found in water. Carbohydrate molecules vary greatly in their degree of complexity. Our appreciation of their important roles in body functions depends on our understanding their composition and characteristics.

Monosaccharides. The basic type of carbohydrate molecule is the monosaccharide, also called simple sugar. Some sugars contain as few as three carbons (trioses), others contain five carbons (pentoses), still others contain six carbons (hexoses), and some even more. Although both the trioses and pentoses play important biological roles, it is the hexoses, the six-carbon sugars, that are considered the most important, because the hexoses serve as the basic molecules from which both simpler and more complex carbohydrates can be formed. The hexoses can also be converted, through appropriate chemical reactions, into lipids and proteins.

Three hexoses are most important in human nutrition: glucose (also called dextrose), fructose (also called levulose), and galactose. They have the same overall chemical formula, but differ from one another in the pattern in which the hydrogen and oxygen atoms are attached to the chain of carbon atoms (Fig. 3-1). The different structural formula of each monosaccharide results in its characteristic sweetening power, solubility, and other properties. Of the three hexoses, only glucose and fructose are found in their simple forms in foods (mainly fruits and some vegetables). In most cases, the hexoses required for body processes are derived from the breakdown of more complex carbohydrates (starches, etc.) in the process of digestion.

One important aspect of human well-being concerns **glucose,** which is the only carbohydrate found in the general circulation of the body. It occurs both in the blood plasma and the red blood cells, where it is generally referred to as **blood sugar.** The blood of a normal person who is hungry usually contains about 100 milligrams (mg) of glucose/100 milliliters (ml) of blood. This concentration of glucose in the blood is called the **normal fasting blood sugar level.** The level rises after a meal and then falls gradually until it reaches the fasting level, which is accompanied by the onset of hunger. People whose blood sugar levels rise above 160 mg/100 ml are said to be **hyperglycemic;** those whose levels fall below 80 mg are called **hypoglycemic.**

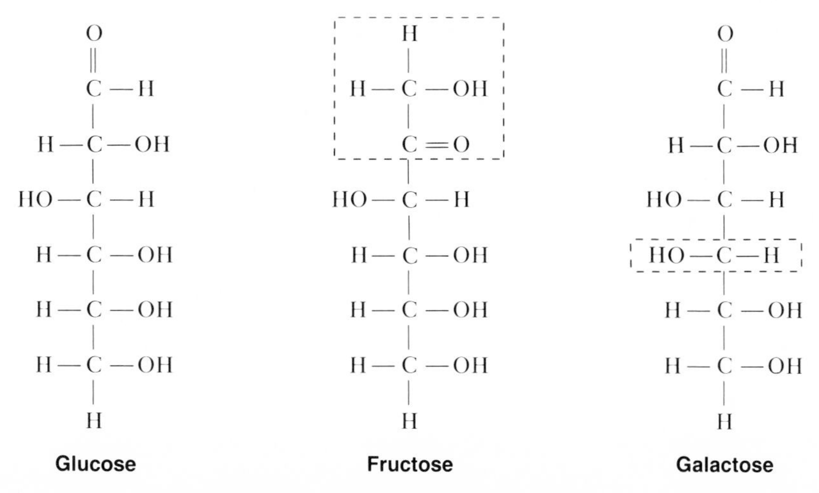

Fig. 3-1
Hexoses important in human nutrition.

If an individual's blood sugar rises so high that some of it passes through the kidney and appears in the urine, the individual suffers from **diabetes.** The prevalence of this disease over the entire world is not known, but, based on the data for the United States, it is apparently widespread. In the United States, there are 2.8 million known diabetics, and it is estimated that another 1.6 million Americans have diabetes, but are not aware of it. Combining these two numbers, we arrive at a frequen-

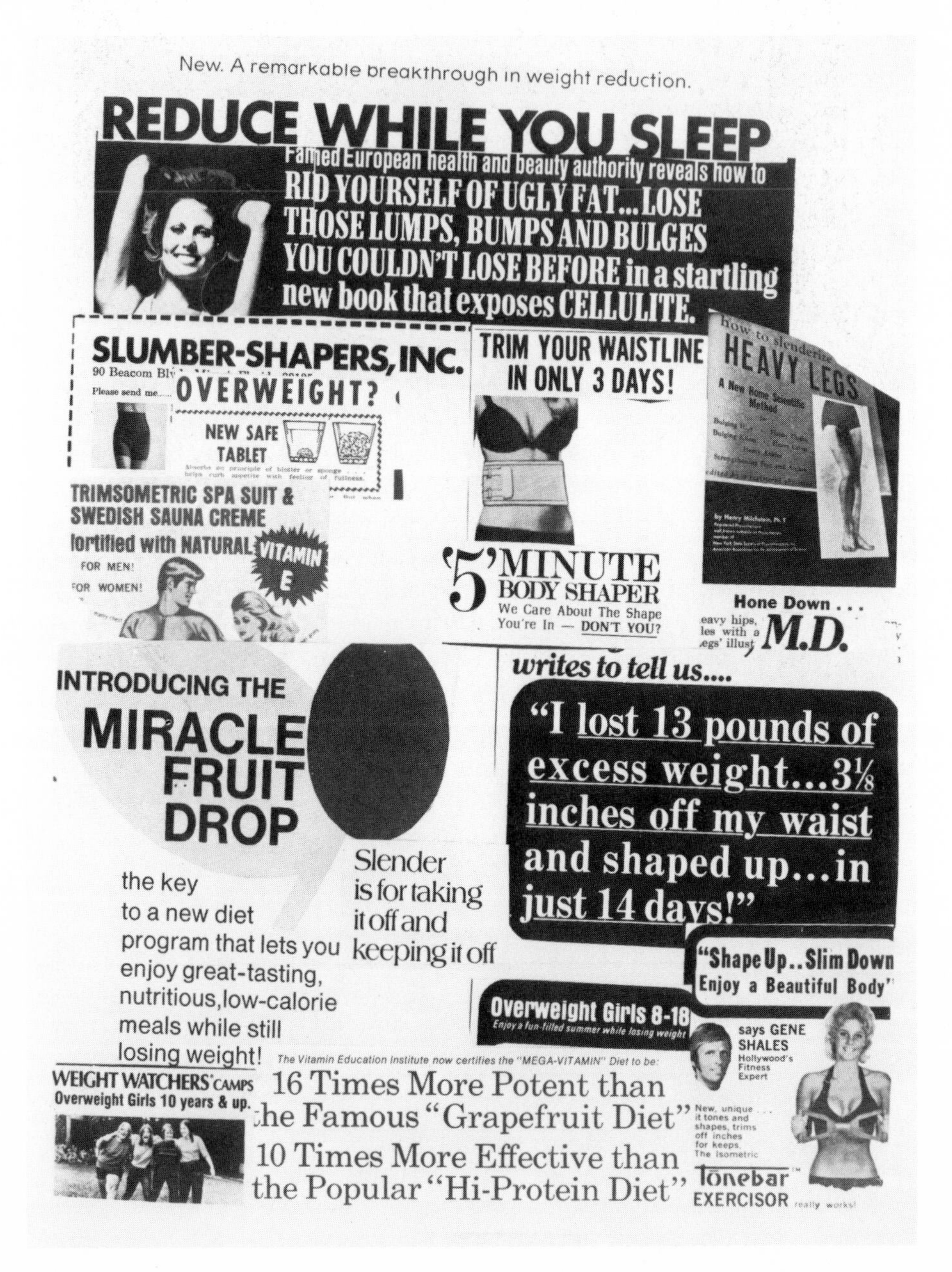

Fig. 3-2
Examples of weight control aids advertised in popular magazines. (From Guthrie, H. 1975. Introductory nutrition, ed. 3. The C. V. Mosby Co., St. Louis.)

cy of 2% for the occurrence of diabetes in the general population. This is an unusually high overall incidence for any disease.

Diabetes is about 40 times more prevalent in persons age 65 to 75 years than in those under age 25 years. Most cases of diabetes become apparent in adults between the ages of 50 and 75 years. For this older group of diabetics, diet is the primary therapy in controlling the disease. These people are instructed to limit themselves to a daily caloric intake of no more than 1500 kilocalories (kcal) (normal average is 2500 kcal), which can only be achieved by a severe restriction of the consumption of carbohydrates, especially in the pure sugar form of candy, ice cream, etc. The importance of diet and weight in diabetes can be seen in the fact that two thirds of the female diabetics and one half of the male diabetics in this country are more than 20% overweight. Although obesity does not cause diabetes, it does tend to bring out the disease in those who are susceptible to it. It is quite clear that carbohydrates not only provide energy for the individual, but also affect his well-being, sometimes negatively, in many other ways (Fig. 3-2).

Disaccharides. It was pointed out earlier that monosaccharides, as such, are not commonly found in foods. More common are the disaccharides, also called double sugars, that are formed when two monosaccharides are joined together chemically. In this process, one molecule of water is split off from between the two sugars (H from one and OH from the other), and the resulting chemical bonds join them into one disaccharide molecule (Fig. 3-3). The reverse of this process occurs in digestion, as we shall discuss in a later chapter.

All disaccharides have the same overall chemical formula ($C_{12}H_{22}O_{11}$), but differ from one another in the arrangement of their constituent atoms. The most common disaccharide is **sucrose,** ordinary table sugar, which is formed by the fusion of glucose and fructose. Granulated sugar is 100% sucrose. The world consumes about 30 million tons of sucrose a year, two thirds of it coming from sugar cane and one third of it from beet sugar (Fig. 3-4).

Another very important disaccharide is **lactose,** a combination of glucose and galactose. Lactose ("milk sugar") is found only in milk and is the major source of the monosaccharide galactose. Although lactose is the major energy-providing compound for human infants, lactose cannot be utilized by most adults, because, as most human beings grow older, they lose their ability to produce the enzyme that breaks down lactose into its component monosaccharides, and the body's cells can only utilize carbohydrates when they are in monosaccharide form.

A serious problem arises in infants who cannot utilize the monosaccharide galactose that they obtain from the breakdown of lactose molecules. The galactose ac-

$$C_6H_{12}O_6 \quad + \quad C_6H_{12}O_6 \quad \overset{(1)}{\rightarrow} \quad C_{12}H_{22}O_{11} \; + \; H_2O$$

$$\text{Monosaccharide + Monosaccharide} \quad \underset{(2)}{\leftarrow} \quad \text{Disaccharide + Water}$$

Fig. 3-3
Diagram showing synthesis of a disaccharide from two monosaccharides, *(1)*, and hydrolysis, *(2)*, the reverse process.

Fig. 3-4
The major sources of disaccharides in our diets are refined sugar, which is 100% sucrose, and milk, whose solid material consists of 40% to 50% lactose. (Copyright © 1976 by Theodore R. Lane.)

cumulates in the blood and causes a wide range of problems. The infant fails to thrive, experiences vomiting and diarrhea, becomes severely retarded mentally, and develops cataracts. If untreated, the infant usually dies. However, it has now become relatively simple to treat those affected by eliminating from their diets all carbohydrates that contain galactose (including milk, of course). When such infants are raised on galactose-free diets, all adverse symptoms disappear. Fortunately, only 1 child in about 50,000 suffers from this condition, which is called **galactosemia.** The condition is genetically determined and occurs only in individuals who carry two genes for this trait.

There are many other disaccharides. One of these is **maltose,** which is formed from two glucose molecules. Maltose, produced in large amounts by germinating grain, is important in the brewing of beer, during which process it is fermented by yeasts and converted to alcohol.

Polysaccharides. The monosaccharides and disaccharides are the types of carbohydrates commonly referred to as **sugars.** A third group, the polysaccharides, are much more complex and include, among others, starches, glycogen, and cellulose. Polysaccharides are usually composed solely of glucose molecules that are linked together in long chains, which may be branched or unbranched. A polysaccharide molecule may be composed of as many as 2000 glucose molecules. The number of glucose subunits and their arrangement within the polysaccharide determine its particular characteristics.

Starch is the principal carbohydrate storage compound of higher plants. Each

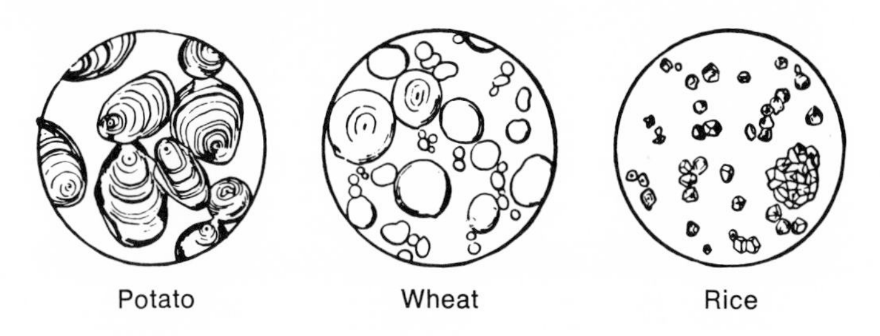

Fig. 3-5
Starch granules of different sizes from different sources. (From Fearon, W. R. 1940. An introduction to biochemistry, ed. 2. The C. V. Mosby Co., St. Louis.)

plant species deposits long, **unbranched** chains of starch of various lengths in its own characteristic size starch grain. Thus the starch grains produced by potato, rice, wheat, and corn plants can be distinguished from one another (Fig. 3-5). The starches themselves, however, are chemically identical, except in chain lengths, and our bodies are able to digest them all, breaking them down into their constituent glucose molecules for use by the individual cells.

A great number of animals, man included, store a limited amount of carbohydrate in the form of another polysaccharide called **glycogen,** sometimes referred to as "animal starch." It takes the form of long, **branched** chains of glucose units. Glycogen is stored in liver and muscle cells, which are the only two animal tissues, other than blood and milk, that normally contain carbohydrates in any appreciable amount.

Another important plant polysaccharide is **cellulose.** Cellulose, like starch and glycogen, is composed of many glucose units. They are arranged in **unbranched** chains, but are linked in a slightly different manner from those of starch. Table 2-2 reveals that 80% of the solid material of a plant is carbohydrate. Most of this carbohydrate is in the form of cellulose that makes up the **wall** around every plant cell. Unfortunately, man does not possess the enzyme necessary to break the chemical bonds between the glucose subunits of cellulose. If he did, there would be less of a food crisis today, at least from an energy standpoint. The indigestibility of cellulose does have a good side; cellulose contributes **bulk** to our diets, which is important in exercising the muscles of the intestinal tract.

Although necessary as a precursor of certain body compounds, the major function of carbohydrates is to provide a source of energy. However, because carbohydrates are essential nutrients, the Food and Nutrition Board of the U.S. National Research Council recommends that 100 grams (gm) of carbohydrate be included in everyone's diet each day. When considering carbohydrate as an energy source, it is interesting to note that the amount of energy provided by carbohydrates is almost constant regardless of the type consumed. One gram of carbohydrate provides about 4 kcal of energy regardless of whether it is monosaccharide, disaccharide, or polysaccharide. Table 3-1 lists the percentages of carbohydrates found in various foods. Since every gram of carbohydrate yields the same number of calories, we can easily calculate the number of calories provided by various quantities of the different foods.*

*28 gm = 1 ounce.

Table 3-1
Carbohydrate content of various foods*

Food	Carbohydrates (%)	Food	Carbohydrates (%)
Sugar, granulated	100.0	Ice cream	20.6
Cornstarch	87.6	Lima beans	19.8
All-purpose flour	76.1	Corn	18.8
Maple syrup	65.0	Grapes	15.7
Enriched white bread	50.5	Apple	14.5
Whole wheat bread	47.7	Beans, green	7.1
Muffins	42.3	Cabbage	5.4
Rice	24.2	Beef liver	5.3
Macaroni	23.0	Whole milk	4.9
Bananas	22.2	Oysters, raw	3.4
Potatoes, baked	21.1	Pears, cooked	2.0

*From U.S. Department of Agriculture, 1963. Handbook No. 8. Washington, D.C.

In looking at world food production, we find that carbohydrates have an important place. Carbohydrate-rich plants are easy to grow, have a high energy yield per unit of land, and, as a result, are relatively inexpensive sources of energy. We therefore find that when the amount of money available for food is reduced, the proportion of carbohydrate foods in a diet increases. As was pointed out in Chapter 2, when this type of situation becomes extreme, individuals will suffer from malnutrition.

Lipids

Lipids, like carbohydrates, are composed principally of carbon, hydrogen, and oxygen. However, lipids have a much smaller proportion of oxygen in their molecules than do carbohydrates. The chemical structure of lipids is complicated by the fact that they are actually composed of two different types of compounds, namely, glycerol and fatty acids (Fig. 3-6). A **glycerol** molecule is a three-carbon chain in which a hydroxyl group (OH) is attached to each carbon atom. A **fatty acid** is a chain of carbon atoms with a carboxyl (COOH) group at one end.

The number of carbon atoms in a fatty acid chain varies with the particular acid. Most of the fatty acids involved in edible lipids contain from 4 to 24 carbon atoms. A molecule of lipid is formed when one molecule of glycerol combines with one, two, or three molecules of fatty acid. The general term used to describe the combination of glycerol and fatty acid is "glyceride," and lipids may be monoglycerides, diglycerides, or triglycerides.* The fatty acids involved in the formation of a particular diglyceride or triglyceride need not be identical, and in most cases they are different. Over 98% of the lipids found in food are triglycerides. The formation of a simple triglyceride lipid is shown in Fig. 3-6.

Saturated and unsaturated fatty acids. From the preceding discussion, it should

*A newer term, **acetylglycerol,** is technically more correct than **glyceride,** but the latter will be used in this text because it is currently the more familiar term.

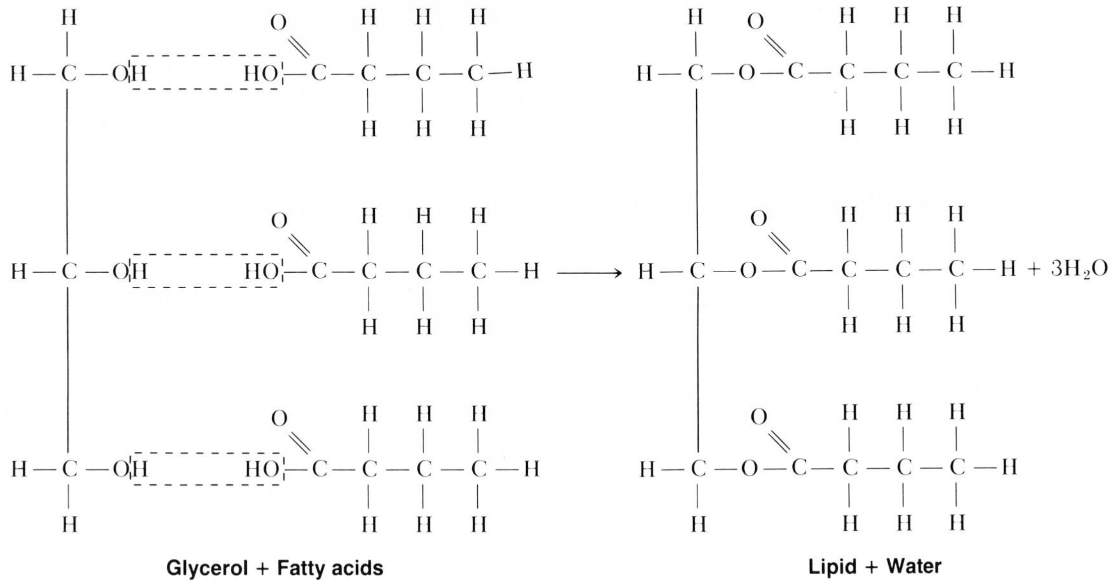

$\rightarrow + 3H_2O$

Glycerol + Fatty acids **Lipid + Water**

Fig. 3-6
Synthesis of a lipid.

be apparent that the differences among lipids result from differences in their fatty acid composition. The lipids found in foods can exist either in a solid form at room temperature, in which case we call them **fats,** or as liquids, in which case we call them **oils.** Whether a lipid is a solid or a liquid at room temperature is a function of its fatty acid structure. If every one of its internal carbon atoms has two hydrogen atoms attached to it, so that all the carbon-to-carbon bonds are single bonds, as in Fig. 3-6, then the fatty acid is said to be **saturated** with hydrogen atoms. If, however, two adjacent carbon atoms are connected by a double bond (because each lacks one hydrogen atom), the fatty acid is said to be **unsaturated,** because it contains less than the maximum number of hydrogen atoms. If a double bond occurs in two or more places in the chain, the fatty acid is called **polyunsaturated.** Examples of a saturated and an unsaturated triglyceride are shown in Fig. 3-7.

If saturated fatty acids predominate in a lipid, the lipid will have a relatively high melting point; it will be solid at room temperature and will be called a fat. If unsaturated fatty acids predominate, the lipid will have the reverse characteristics and will be an oil. Linoleic acid is the most commonly occurring polyunsaturated fatty acid. It represents 75% of the fatty acids in safflower oil and 54% in corn oil, but only 2% in butterfat.

Although the same basic lipid components (glycerol and the various fatty acids) are available to most organisms, each species produces its own characteristic types of lipids. Most animal fats are relatively high in saturated fatty acids, whereas most vegetable oils are relatively high in unsaturated fatty acids. Exceptions to this generalization include poultry and fish fats, which contain highly unsaturated fatty acids, and coconut oil, which is composed of saturated fatty acids. Although it is saturated, coconut oil is liquid because it has a preponderance of short-chain fatty acids. Mar-

```
H    O     H   H   H   H   H   H   H   H   H   H   H   H   H   H   H
|     \\   |   |   |   |   |   |   |   |   |   |   |   |   |   |   |
H—C—O—C—C—C—C—C—C—C—C—C—C—C—C—C—C—C—C—H
  |        |   |   |   |   |   |   |   |   |   |   |   |   |   |   |
  |        H   H   H   H   H   H   H   H   H   H   H   H   H   H   H
  |
  |  O     H   H   H   H   H   H   H   H   H   H   H   H   H   H   H
  |   \\   |   |   |   |   |   |   |   |   |   |   |   |   |   |   |
H—C—O—C—C—C—C—C—C—C—C—C—C—C—C—C—C—C—C—H
  |        |   |   |   |   |   |   |   |   |   |   |   |   |   |   |
  |        H   H   H   H   H   H   H   H   H   H   H   H   H   H   H
  |
  |  O     H   H   H   H   H   H   H   H   H   H   H   H   H   H   H
  |   \\   |   |   |   |   |   |   |   |   |   |   |   |   |   |   |
H—C—O—C—C—C—C—C—C—C—C—C—C—C—C—C—C—C—C—H
  |        |   |   |   |   |   |   |   |   |   |   |   |   |   |   |
  H        H   H   H   H   H   H   H   H   H   H   H   H   H   H   H
```

Saturated lipid

```
H    O     H   H   H   H   H   H   H   H   H   H   H   H   H   H   H   H   H
|     \\   |   |   |   |   |   |   |   |   |   |   |   |   |   |   |   |   |
H—C—O—C—C—C—C—C—C—C—C—C=C—C—C=C—C—C—C—C—C—H
  |        |   |   |   |   |   |   |           |           |   |   |   |   |
  |        H   H   H   H   H   H   H           H           H   H   H   H   H
  |
  |  O     H   H   H   H   H   H   H   H   H   H   H   H   H   H   H   H   H
  |   \\   |   |   |   |   |   |   |   |   |   |   |   |   |   |   |   |   |
H—C—O—C—C—C—C—C—C—C—C—C=C—C—C=C—C—C—C—C—C—H
  |        |   |   |   |   |   |   |           |           |   |   |
  |        H   H   H   H   H   H   H           H           H   H   H   H   H
  |
  |  O     H   H   H   H   H   H   H   H   H   H   H   H   H   H   H   H   H
  |   \\   |   |   |   |   |   |   |   |   |   |   |   |   |   |   |   |   |
H—C—O—C—C—C—C—C—C—C—C—C=C—C—C=C—C—C—C—C—C—H
  |        |   |   |   |   |   |   |           |           |   |   |   |   |
  H        H   H   H   H   H   H   H           H           H   H   H   H   H
```

Unsaturated lipid

Fig. 3-7
Comparison of saturated and unsaturated lipids.

garine and vegetable shortenings are examples of vegetable oils that have been intentionally changed by saturating at least some of the carbon-to-carbon double bonds with hydrogen to produce fats with certain desired physical characteristics. These "hydrogenated" materials are harder and more resistant to spoilage (Fig. 3-8). Table 3-2 is a listing of the percentages of lipid and saturated and unsaturated fatty acids found in some representative foods.

Fig. 3-8
Almost all margarines and shortenings are now made from vegetable oils, which are generally more unsaturated than animal fats. (Copyright © 1976 by Theodore R. Lane.)

Table 3-2
Lipid and fatty acid composition of 100 gm of selected foods*

Food	Total lipids (%)	Saturated fatty acids (%)	Unsaturated fatty acids (%)
Coconut oil	100	86	7
Corn oil	100	10	81
Soybean oil	100	15	72
Olive oil	100	11	83
Safflower oil	100	8	87
Butter	81	46	29
Mayonnaise	79.9	14	57
Bacon	52	17	30
Peanut butter	50.6	9	39
Cream cheese	37.7	21	13
Beef, rib	37.4	18	17
Ham	23	4	6
Chicken	17.1	5	9
Egg, cooked	11.5	4	6
Tuna, canned	8.2	3	4
Milk, whole	3.7	2	1

*From U.S. Department of Agriculture. 1963. Handbook No. 8. Washington, D.C.

Nutritional roles. Lipids serve as a concentrated source of energy. Each gram of lipid, whether of animal or plant origin, oil or fat, provides 9 kcal of energy. This is 2¼ times as much energy as can be obtained from a gram of carbohydrate. Lipids, specifically fats, are the primary form in which an animal stores excess energy, and the amount of body fat in an animal is determined by the animal's energy balance. In humans, a certain amount of body fat, about 20% of body weight for women and 15%

for men, is considered normal and desirable. More than this results in overweight and, in extreme cases, obesity, with all the health problems that beset such individuals (Fig. 3-2).

Lipids are also involved in the absorption of vitamins A, D, E, and K from the intestinal tract. The importance of vitamins is discussed in Chapter 4. For our present consideration, it is important to note that vitamins A, D, E, and K can pass into the cells of the intestinal tract only if they are attached to molecules of fat. The molecules of fat used in this process must be present in the food consumed by the individual. These four vitamins are called **fat-soluble vitamins.** The complete elimination of fat from a person's diet leads to a reduced intake of these nutrients, even though they may be present in adequate amounts in the food consumed. When diets are marginal in their vitamin content, fats can become essential nutrients in preventing the occurrence of vitamin-deficient diseases.

The reason for listing lipids as an essential nutrient is best seen in the case of one of the fatty acids, **linoleic acid.** This polyunsaturated fatty acid cannot be produced by the body and must be obtained from the diet. The absence of linoleic acid has been shown to result in a reduced growth rate of experimental animals; it also affects the structure and some of the enzyme functions of the animals' cells. A complete lack of linoleic acid in the diet will result in malnutrition. Although the percentage of linoleic acid is more than 50% in such foods as corn oil, soybean oil, and safflower oil, lesser and variable amounts are present in hydrogenated fats or spreads prepared from these oils. However, with the increasing use of vegetable oils rather than animal fats in the American diet, the daily consumption of this essential fatty acid is increasing.

Lipids and heart disease. It has been known for quite some time that certain human diseases are either caused by or associated with dietary factors, for example, the association of diabetes and carbohydrates. One type of heart disease appears to be associated with **cholesterol,** a fatlike substance, and **blood triglycerides,** fats used for energy storage. This form of heart disease, called **atherosclerosis,** involves a thickening of the walls of the arteries, which results in a narrowing of the arterial passageways. Atherosclerosis is caused by the accumulation of large amounts of cholesterol just inside the innermost layer of the arterial wall. If a clot forms in the artery, it can block completely the already narrowed blood vessel, withholding oxygen and food material from the tissues normally served by the artery. Such an occurrence in the arteries that serve the muscles of the heart will quickly prevent the heart from continuing its normal function; the individual will experience a "heart attack" because of a "coronary thrombosis" (thrombosis meaning clot). It has been found that persons who suffer heart attacks almost always have above-normal levels of blood cholesterol and triglycerides. In addition, populations that derive most of their calories from saturated lipids exhibit a much higher incidence of heart disease than do populations that depend more on unsaturated lipids and carbohydrates (Fig. 3-9).

Efforts have been made to reduce the incidence of heart attacks by dietary manipulation in so-called "high-risk groups," that is, middle-aged men with family histories of heart disease who are working under high emotional tension. Unfortunately, this approach has not met with the hoped-for success. A restriction on the

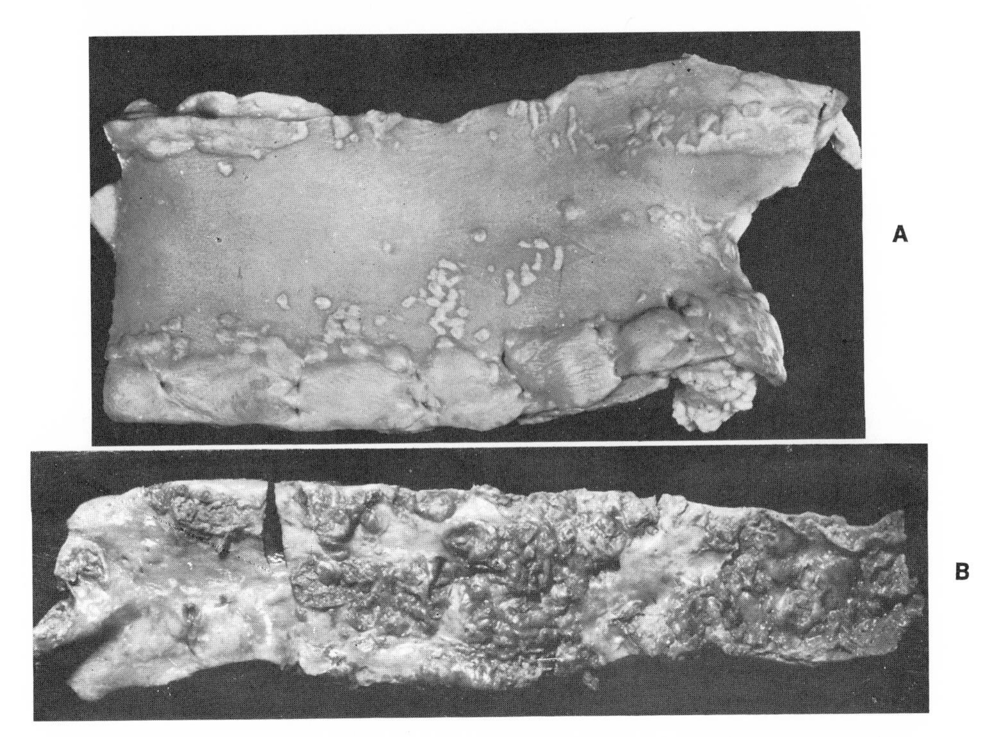

Fig. 3-9
Atherosclerosis. Two sections of aorta opened lengthwise to show early, **A**, and advanced, **B**, "takeover" of an otherwise smooth lining by atheromatous plaques. (From Anderson, W. A. D., and Kissane, J. M., eds. 1977. Pathology, ed. 7. The C. V. Mosby Co., St. Louis.)

consumption of such cholesterol-containing foods as eggs, meat, and liver has not resulted in a lowering of blood cholesterol levels or in a reduction of heart attacks. Apparently, this is because the body itself synthesizes cholesterol from lipids, carbohydrates, and proteins whenever their amounts in the diet exceed the body's need for energy.

A second dietary approach is for high-risk individuals to use more vegetable oils and less animal fats. This dietary restriction does lead to a reduction in blood cholesterol levels, but not to a reduction in the incidence of heart attacks.

A third approach is to limit the body's synthesis of triglycerides. These fats are produced either directly from dietary lipids or from lipids formed by the body using excess carbohydrates or proteins. For unknown reasons, the production of triglycerides, and also cholesterol, by the body is greater when excess carbohydrates occur in the form of sugars rather than the more complex starches. As a consequence of these findings, it is usually recommended that high-risk individuals modify their diets to consume polyunsaturated rather than saturated lipids and limit their total caloric intake to the amount necessary to maintain optimum body weight. Unfortunately, even a rigid adherence to these dietary restrictions has **not** been found to be universally effective in eliminating atherosclerosis and subsequent heart attacks. More research is needed into the causes of heart diseases, especially concerning the factors that predispose particular individuals and families to this problem.

□ Proteins

Although one nutrient may not actually be more important than another for the survival and well-being of an individual (since they are all essential), we find that proteins do play a major role in the development and maintenance of all organisms. Proteins are found in every cell and in all cellular constituents, that is, bone, muscle, etc., of the body. More than half of the dry weight of a human being is protein. All enzymes and many hormones are proteins; these substances are vital components in the regulation of body processes. In the absence of proteins, there is a failure in body growth that is followed by a loss of already established functional body tissues. Should protein deprivation continue for any extended period of time, the affected individual will die.

Amino acids. In an earlier discussion, it was pointed out that proteins are linear combinations of amino acids, some of which are essential nutrients for human beings. There are 20 different kinds of amino acids that are found in naturally occurring proteins. These are listed in Table 3-3. It should be noted that one of the amino acids, **histidine,** cannot be manufactured by infants, but can be produced by adults. Although it is not understood why this occurs, changes of this type during development are quite common. All amino acids consist of carbon, hydrogen, oxygen, and nitrogen atoms; some also contain sulfur. In addition, all amino acids contain both a carboxyl group (COOH) and an amino group (NH_2). Fig. 3-10 shows the simplest of the amino acids, namely, **glycine.**

Proteins are formed when amino acids are joined together in a long chain. In the chemical union of two amino acids, a molecule of water is formed using the OH from the carboxyl group of one amino acid and an H atom from the amino group of an adjacent amino acid. This process is depicted in Fig. 3-10. The chemical linkage between the two amino acids is called a **peptide bond,** and a linear chain of amino acids built up by this process is called a **polypeptide.**

As was mentioned earlier, each species builds its own characteristic proteins, and the unique qualities of each type of protein are determined not only by which amino acids are used and the number of times they are repeated, but also by the **order** in which they are joined together—the amino acid sequence. It should also be remem-

Table 3-3
Naturally occurring amino acids in food and body tissue

Classification	Amino acid	Classification	Amino acid
Essential dietary components for all human beings	Isoleucine Leucine Lysine Methionine Phenylalanine Threonine Tryptophan Valine	Nonessential	Glycine Glutamic acid Arginine Aspartic acid Proline Alanine Serine Tyrosine Cysteine Asparagine Glutamine
Essential for infants	Histidine		

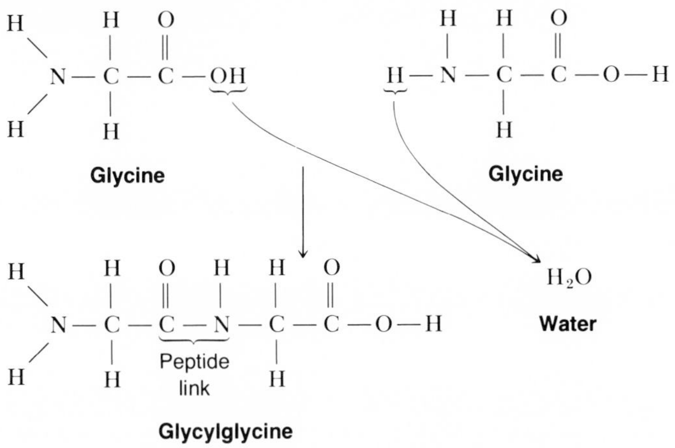

Glycine

Glycine

H_2O

Water

Peptide link

Glycylglycine

Fig. 3-10
An amino acid combination. (From Levine, L. 1973. Biology of the gene, ed. 2. The C. V. Mosby Co., St. Louis.)

bered that proteins in food are classified either as **complete,** meaning they contain all the essential amino acids required for growth, or as **incomplete,** meaning they lack one or more of the essential amino acids. Should a single incomplete protein become the sole dietary source of protein, no new tissue can be formed, nor can worn-out tissue be replaced.

Nutritional roles. Proteins, through their amino acids, function in various ways in the body. The primary function of amino acids is to form the proteins used in the growth of the individual and in the replacement of worn-out tissues. However, whether the amino acids are used in this fashion depends on the person's caloric intake. If the caloric intake of the individual is adequate, the amino acids are used for the synthesis of the body's proteins. However, if caloric intake is insufficient to meet the individual's energy needs, the amino group is removed (the amino acid is **deaminated**), and the nonnitrogenous fraction of the amino acid is used as a source of energy. The removed amino group is eventually excreted by the body as urea in the urine formed by the kidneys.

On the other hand, the diet may provide a sufficient caloric intake, and, in addition, there may be more amino acids available than are needed for protein synthesis. Under these conditions, the excess amino acids will be deaminated, the nitrogen excreted, and the nonnitrogenous portion of the amino acids will be converted into glycogen or fat and stored in the body. Unfortunately, unlike carbohydrates and lipids, amino acids and proteins are not stored by the body and must be provided by the diet each day.

Another very important function of amino acids is their use in the formation of essential body compounds such as hormones and enzymes. Some hormones and all enzymes are composed of proteins. Hormones and enzymes will be discussed in some detail later in this text; for the present, it is important to note that these substances are so vital to the body that their formation and replacement is reduced only when severe protein deprivation occurs. Proteins function in a number of other ways in the body, relating mainly to the excretory and circulatory systems. These will be considered when we discuss these systems of the body.

As stated earlier, proteins may be used as a source of energy. When used as fuel, each gram of protein yields 4 kcal of energy, the same amount of energy that is obtained from 1 gm of carbohydrate. The possible use of amino acids as energy sources by the body has been taken into account in setting up national standards for the daily intake of proteins. Table 3-4 lists the amounts of protein that ought to be consumed daily by individuals in various age groups. The difference in daily protein requirement between adult men and women is a reflection of the difference in their average weights. Specific conditions also influence protein requirements. In the case of women, the additional need for protein to provide for the growth and development of both the embryo and the nursing child is reflected in the recommendation in Table 3-4 that pregnant and nursing individuals consume 10 and 20 more gm of protein each day, respectively. Growing children need increasing amounts of protein in their diets to provide the materials needed for skeleton and muscle production. The continued daily need for large amounts of protein in adult life results from the constant wearing down and repair of body tissues. It has been estimated that every molecule of protein in the body is replaced every 160 days.

The minimal daily needs for essential amino acids are given in Table 3-5. Histidine is shown as an essential amino acid only for infants, adults being able to synthesize histidine from simpler compounds. Here we have an instance of increased biochemical function with age. As with most such physiological changes during the lifetime of an individual, we have yet to discover its significance.

Although we have spoken in a general way of the protein content of foods, we have not looked at the percentage of protein found in some of our commonly eaten foods. These are shown in Table 3-6. In assessing the value of a diet, we need to look at the distribution of dietary protein throughout the day's meals. It is recommended that at least one third of the daily consumption of protein be from animal sources in

Table 3-4
Recommended daily amounts of protein for selected age groups*

Group	Protein (gm)
Adult men (70 kg)	65
Adult women (58 kg)	55
During pregnancy	+10
During lactation	+20
Children	
1-3 yr	25
3-6 yr	30
6-8 yr	35
8-10 yr	40
Adolescents	
14-18 yr (boy)	60
14-18 yr (girl)	55

*From Food and Nutrition Board. 1968. Publication No. 1694. National Academy of Sciences–National Research Council, Washington, D.C.

order to ensure an ample supply of all essential amino acids. Our traditional food combinations tend to link a complete protein with an incomplete protein. The use of milk with cereal, cheese with macaroni, meat with rice, and peanut butter with bread are examples of this type of combination. It is interesting to note that our customary eating habits reflect our biological needs for essential amino acids. This fact is all the more remarkable if we consider that a combination such as peanut butter and bread cannot be an extension of a food habit of primitive man, as these foods were not available at that time.

Amino acid enzyme diseases. Earlier we discussed the protein-deficiency disease **kwashiorkor.** The lack of any essential amino acid in a diet will prevent the formation of proteins by an individual, no matter how plentiful the supply of all other amino acids. The lack of a single essential amino acid will result in the individual suffering kwashiorkor just as if there was a general lack of protein in the diet.

There are amino acid–related diseases that involve the ability of the body to transform one amino acid into another. An example of these is the condition known as **phenylketonuria.** This disease, which occurs in one out of 15,000 children, is inher-

Table 3-5
Minimal daily requirements of essential amino acids*

Amino acid	Infants (mg/kg)	Adults	
		Female (mg)	*Male (mg)*
Isoleucine	126	450	700
Leucine	150	620	1100
Lysine	103	500	800
Methionine	45	350	200
Phenylalanine	90	220	300
Threonine	87	305	500
Tryptophan	22	157	250
Valine	150	650	800
Histidine	34	—	—

*From Food and Nutrition Board. 1959. Publication No. 711. National Academy of Sciences–National Research Council, Washington, D.C.

Table 3-6
Protein contribution of 100 gm portions of commonly eaten foods*

Food	Protein (gm)†	Food	Protein (gm)
Chicken	32	Haddock	20
Pork	29	Eggs	13
Beef	29	Waffles	9
Lamb	28	Lima beans	8
Peanut butter	28	Peas	5
Sardines	21	Milk	4
Salmon	20		

*From U.S. Department of Agriculture. 1963. Handbook No. 8. Washington, D.C.
†28 gm = 1 ounce.

Fig. 3-11
PKU. This child is a delightful, perfectly developed 2 year old. Screened and diagnosed at birth, she has eaten a carefully controlled low phenylalanine diet and is growing normally. (From Williams, S. R. 1973. Nutrition and diet therapy, ed. 2. The C. V. Mosby Co., St. Louis.)

ited and occurs in individuals who carry two genes for this trait. Phenylketonuria involves the essential amino acid **phenylalanine,** a small amount of which is needed for protein synthesis. Most diets contain much more phenylalanine than is needed for the production of proteins. In normal individuals, this excess phenylalanine is converted to the nonessential amino acid **tyrosine,** which is then used in other biochemical pathways of the body. Those individuals who suffer from phenylketonuria lack the enzyme that converts phenylalanine to tyrosine. As a result of this enzyme deficiency, phenylalanine and some of its derivatives, especially **phenylpyruvic acid,** accumulate in the blood and urine. Affected children suffer from mental retardation, and this characteristic of the disease has led to its being called "phenylpyruvic idiocy." It is quite fortunate for these affected individuals that phenylketonuria can be controlled by diet. If affected children are given a low phenylalanine diet starting before 6 months of age, they will develop with normal mental abilities. Quite clearly the low phenylalanine diet must provide enough of the amino acid to permit a sufficient amount of protein synthesis to meet the growth requirements of the developing child (Fig. 3-11).

It is unfortunate that for every amino acid, both essential and nonessential, at least one enzyme deficiency disease has been found. In some cases, dietary restrictions of the particular amino acid serve to avoid the effects of the disease. However, in others, the biochemical pathways are so complex that it is not possible, as yet, to control the disease.

■ SUMMARY

A balanced diet must provide all the nutrients required for energy production, growth and repair of body tissues, and regulation of body processes. In this chapter, we examined the role of carbohydrates as energy suppliers, their availability in the different foods we eat, and the diseases that result from malfunctions in their metabolism. We also reviewed the role of lipids as suppliers of energy, their function as transporters of certain vitamins into the cells of the intestinal tract, and the association of lipids with heart disease. Finally, we considered proteins both as energy sources and as providers of the building components, amino acids, for body tissues. We examined the daily requirements for essential amino acids in different groups of people, the proportion of protein in commonly eaten foods, and an example of an amino acid–enzyme disease that can be cured by dietary control.

The availability of energy and amino acids alone is not sufficient to ensure the proper functioning of the body. Other essential nutrients, namely, vitamins, minerals, and water, are also required. In the next chapter, we shall consider these other essential nutrients, and we shall also examine the overall daily energy needs of the individual under various conditions and the effects on the body of prolonged starvation.

SUGGESTED READINGS

Food and Agriculture Organization of the United Nations. 1973. Amino acid content of foods and biological data on proteins. FAO Nutr. Stud. No. 24. Comprehensive review of the quality of proteins found in different foods.

Food and Nutrition Board. 1960. Evaluation of protein nutrition, Publication No. 711. National Academy of Sciences–National Research Council, Washington, D.C. Analysis of the role of proteins in health and disease.

Food and Nutrition Board. 1966. Dietary fat and human health, Publication No. 1147. National Academy of Sciences–National Research Council, Washington, D.C. Thorough review of the role of lipids in various diseases.

Guthrie, H. A. 1975. Introductory nutrition, ed. 3, The C. V. Mosby Co., St. Louis. Extremely readable and authoritative book on both basic and applied nutrition.

Jelliffe, D. B., and Jelliffe, E. F. P. 1975. Human milk, nutrition, and the world resource crisis. Science **188**:557-561. Plea for the return to breast feeding as a means of partially solving the food crisis in developing countries.

Krehl, W. A. 1955. The nutritional significance of the carbohydrates. Borden Rev. Nutr. Res. **16**:85-101. Very fine review of the importance of carbohydrates in our diet.

Latham, M. C. 1975. Nutrition and infection in national development. Science **188**:561-565. Discussion of the role of malnutrition in causing high fatality rates from normally nonfatal diseases.

Spain, D. M. 1966. Atherosclerosis. Sci. Am. **215**:48-56. In-depth discussion of the disease and how dietary factors contribute to it.

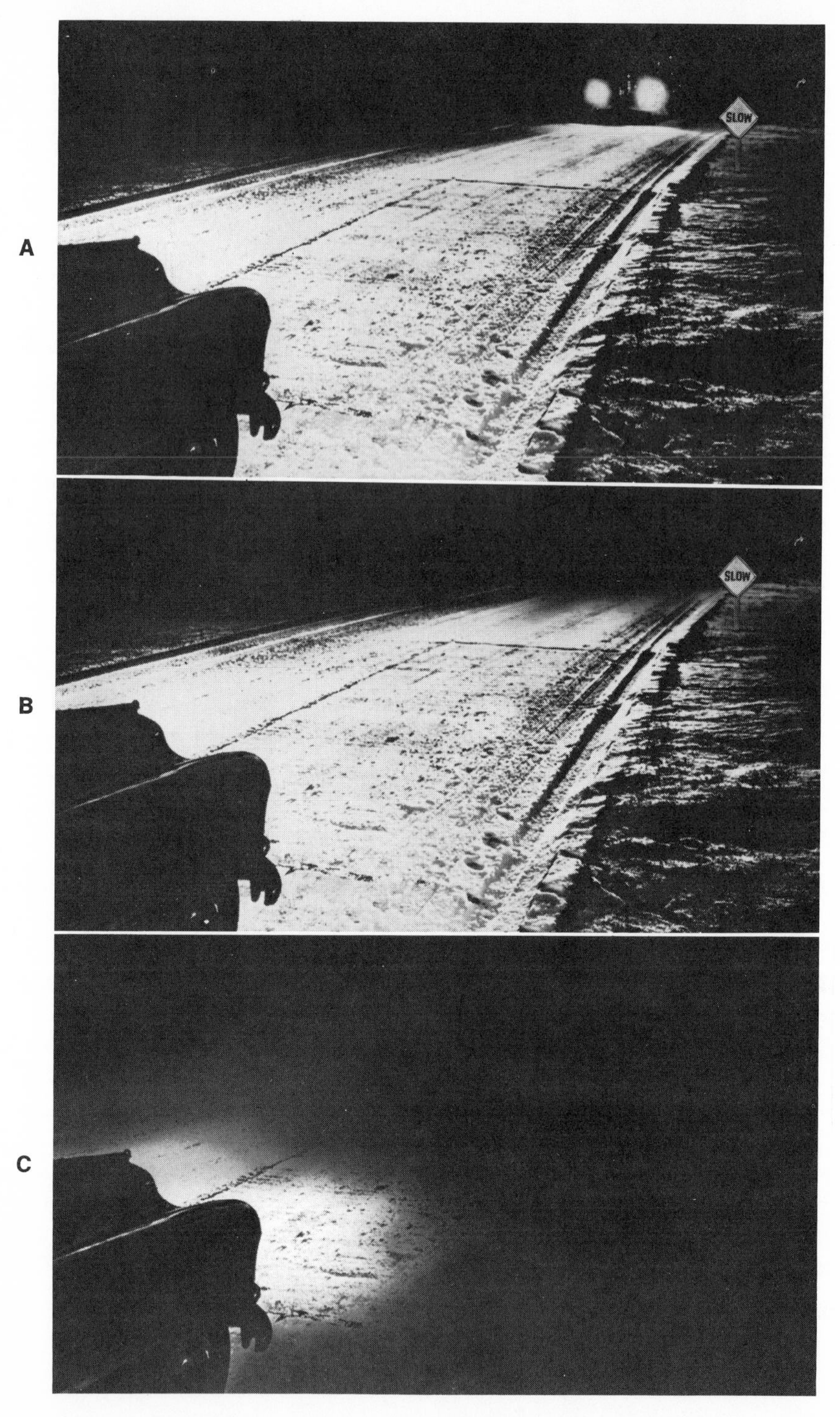

Night blindness. The loss of visual acuity in dim light after exposure to bright light is illustrated here. **A**, Both normal individual and vitamin A–deficient subject see headlights of an approaching car. **B**, After car has passed, normal individual sees a wide stretch of road. **C**, Vitamin A–deficient subject can barely see a few feet ahead and cannot see road sign at all. (Courtesy The Upjohn Co., Kalamazoo, Mich.; from Guthrie, H. A. 1975. Introductory nutrition, ed. 3. The C. V. Mosby Co., St. Louis.)

CHAPTER 4 Other nutrients and energy requirements

LEARNING OBJECTIVES

- What is a vitamin?
- How are vitamins classified?
- Which diseases are caused by vitamin deficiency and which by vitamin excess?
- Which mineral elements are essential for human beings?
- What are the nutritional roles of mineral elements?
- Which diseases are caused by mineral element deficiencies?
- Why is water considered an essential nutrient?
- Which functions does water perform in the body?
- Which factors determine the individual's daily water requirement?
- Which factors determine an individual's daily energy requirements?
- What are the daily energy requirements for individuals in various occupations?
- What is the sequence of metabolic events that occurs during starvation?

Carbohydrates, lipids, and proteins are not the only essential nutrients that we need to obtain from our food. Others include vitamins, various mineral elements, and water. With rare exceptions, no one food provides all the factors necessary for the proper functioning of the body, and thus we must develop food habits that provide a balanced diet. It is important to understand that an individual's energy requirement depends on such factors as physical activity, age, body weight, etc., and that changes in energy needs occur rather continuously throughout life. Although we tend to stress the role of food in providing for our well-being, starvation still occurs in our modern world, unfortunately, and there is still need to study its effects on the body, while working to eliminate it from our society.

■ VITAMINS

Vitamins are essential nutrients, needed in very small amounts, that are necessary for the normal functioning of the body. Vitamins as a class have **no** particular chemical structure in common. Each vitamin performs a different and specific biochemical task. As was the case with amino acids, a vitamin can be "essential" for one species, but not for another. Plants manufacture all the vitamins they need from carbon dioxide, water, and the elements available to them in the soil.

□ Fat-soluble and water-soluble vitamins

Vitamins are classified either as fat soluble or water soluble. Each term indicates the compound that can dissolve the particular type of vitamin. Fat-soluble vitamins

(vitamins A, D, E, and K) are composed solely of carbon, hydrogen, and oxygen. Any excess of these vitamins over the daily needs of the body can be stored, and as a result, they are not absolutely necessary components of the daily diet. Water-soluble vitamins (vitamin C and eight members of the vitamin B complex) are also composed of carbon, hydrogen, and oxygen. However, in addition, most of them contain nitrogen, and some contain other elements, such as cobalt or sulfur. These vitamins are not stored by the body and must be supplied in the diet every day.

□ Nutritional role of vitamins

Unlike the essential nutrients discussed thus far, vitamins are not used as sources of energy, nor are they used as structural components of the body. Many vitamins are used as necessary adjuncts of enzyme activity. Enzymes often work with smaller organic molecules called **coenzymes.** The specific compound being acted on (substrate) attaches to the surface of the protein catalyst (the enzyme), and the coenzyme acts as a donor or acceptor of those atoms that are added to or removed from the substrate during the chemical reaction. Most vitamins or their derivatives function as coenzymes or are necessary in the formation of coenzymes. (However, there are many coenzymes that are **not** vitamins.)

Although vitamins are needed in only very small amounts, the complete lack of a vitamin in a diet inevitably leads to a characteristic disease. The recommended daily amount of a particular vitamin is often given in terms of **international units** (IU). These are the standardized doses of a vitamin that will cause a specific effect either in promoting growth or in the prevention of disease. This method of describing doses is complicated by the fact that 1 IU of one vitamin does not necessarily correspond to the same amount of another vitamin (1 IU of vitamin A is equal to 0.3 micrograms [μg], whereas 1 IU of vitamin E is equal to 1 milligram [mg]). Table 4-1 lists the daily requirements for each vitamin. It will be noted that the quantities involved vary from vitamin to vitamin, although compared to carbohydrate, lipid, and protein requirements, the amounts of vitamins needed are indeed small.

□ Diseases caused by vitamin deficiency and vitamin excess

Vitamins were first discovered in studies of diseases that developed as a result of the absence of particular foods from people's diets. In our modern society, the knowledge of the beneficial effects of an adequate intake of vitamins and the detrimental effects associated with their deficiency has resulted in a tremendous concern over daily vitamin intake. Consequently, there has been a tremendous rise in the production and sale of vitamin supplements. Unfortunately, the vitamin supplements available provide far greater amounts of vitamins than an individual can conceivably need. There is, of course, a tremendous economic waste involved in purchasing unneeded vitamins, but even more importantly, excessive intake of most of the fat-soluble vitamins (vitamins A, D, and K) can have harmful effects. Included in Table 4-1 is a list of the diseases caused by an insufficiency of the various vitamins (Fig. 4-1), as well as the harmful effects caused by the excessive consumption of some. At the present time, we do not understand why harmful effects result from an excess of only some of the fat-soluble vitamins.

Table 4-1
Vitamins: daily requirements, effects, and food sources

Vitamin	Daily requirements*	Diseases caused by deficiency	Diseases caused by excess	Important food sources
Fat-soluble				
Retinol (A)	5000 IU	Night blindness, corneal degeneration	Loss of hair, resorption of bone	Fruits and vegetables, liver, fish
Calciferol (D)	400 IU	Rickets (defective bone formation—failure to calcify)	Hypercalcemia (high blood calcium levels)	Cod-liver oil, eggs, milk, and butter
Tocopherol (E)	30 IU	Hemolysis (destruction) of red blood cells	—	Vegetable oils, eggs, fish, and butter
Phylloquinone (K)	2 mg	Prolonged blood clotting time	Hemolytic anemia (decreased life span of red blood cells)	Vegetables and fruits, eggs
Water-soluble				
Ascorbic acid (C)	60 mg	Scurvy (anemia, bleeding of gums, hardening of muscles of legs)	—	Citrus fruits, vegetables
Thiamin (B_1)	1.5 mg	Beriberi (gradual loss of body tissue, mental confusion, heart failure)	—	Cereals, meat, legumes, fish
Riboflavin (B_2)	1.5 mg	Cheilosis (cracks in lips, corneal degeneration)	—	Milk, liver, eggs, cereals, vegetables
Niacin	30 mg	Pellagra (digestive disturbances, skin eruptions, mental disturbances)	—	Meat, milk, vegetables, legumes
Pyriodoxine (B_6)	2 mg	Convulsions, anemia	—	Meat, liver, eggs, vegetables, cereals
Pantothenic acid	10 mg	Impaired muscular coordination, mental depression	—	Meat, eggs, cereals, vegetables
Folic acid	0.05 mg	Anemia, gastrointestinal disturbances	—	Vegetables, liver
Cobalamin (B_{12})	5 μg	Pernicious anemia	—	Meat, liver, milk
Biotin	150 μg	Dermatitis (skin inflammation), gastrointestinal disturbances	—	Liver, vegetables, milk, eggs

*The abbreviations used are as follows: IU = international unit; mg = milligram ($^1/_{1000}$ gram); μg = microgram ($^1/_{1,000,000}$ gram).

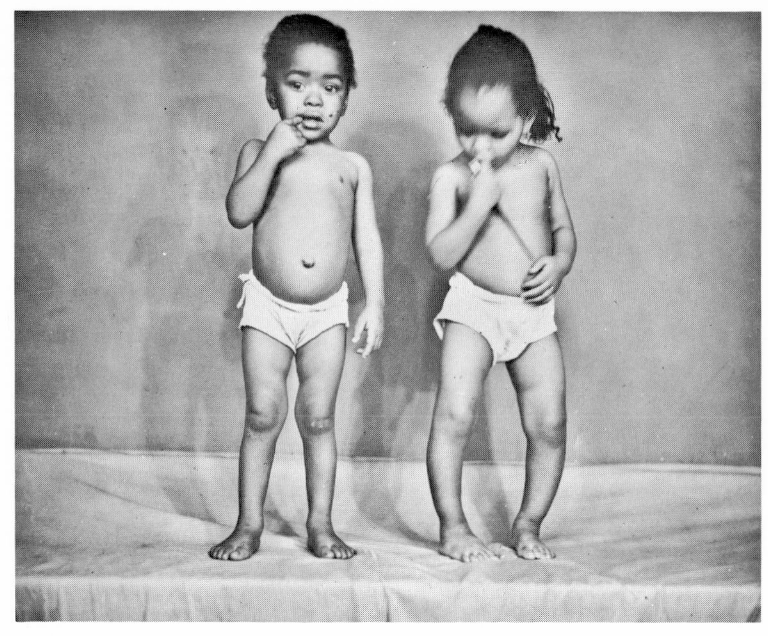

Fig. 4-1
Children suffering from rickets as a result of a vitamin D deficiency. Defective bone formation has led to knock-knees of child on left and bowlegs of child on right. (From files of Parke, Davis & Co. Therapeutic notes. Detroit; courtesy Dr. Tom Spies and Dr. Orson D. Bird; from Williams, S. R. 1973. Nutrition and diet therapy, ed. 2. The C. V. Mosby Co., St. Louis.)

■ MINERAL ELEMENTS

Approximately 4% of the human body—about 6 pounds (2.7 kilograms [kg]) of an adult's total body weight—consists of inorganic salts (see Table 2-2). These mineral elements are necessary for the proper functioning of the individual and are essential in that if they are not supplied in proper amounts in the diet, illness and death will result. Up to now, 17 chemical elements have been proved necessary for human beings. However, chemical analysis may reveal an additional 20 to 30 elements present in the human body. Some of these may be as yet unproved essential elements of the body, but most of them are undoubtedly contaminants from the soil, air, or water of our environment.

□ Macronutrients and micronutrients

Minerals can occur in the body in combination with organic compounds, such as, for example, the iron in the hemoglobin of blood, with other inorganic ions, such as the calcium phosphate of bone, or as free ions, such as the calcium in the fluid between cells. The essential mineral elements are sometimes grouped into two categories based on their relative amounts in the body: **macronutrient elements,** each of which constitutes more than 0.005% of total body weight and hence is found in a concentration greater than 50 parts per million (ppm), and **micronutrient elements,** each of which is found in a concentration of less than 0.005% of the body. It must be stressed that the quantity of a particular mineral element in the body is not

Table 4-2
Classification and concentrations of mineral elements in the body

Classification	Elements	% Body weight
Macronutrient elements (>0.005% body weight)	Calcium	1.5-2.2
	Phosphorus	0.8-1.2
	Potassium	0.35
	Sulfur	0.25
	Sodium	0.15
	Chlorine	0.15
	Magnesium	0.05
Micronutrient elements (<0.005% body weight)	Iron	0.004
	Zinc	0.002
	Selenium	0.0003
	Manganese	0.0002
	Copper	0.00015
	Iodine	0.00004
	Molybdenum	—
	Cobalt	—
	Chromium	—
	Fluorine	—

necessarily a reflection of its importance to the well-being of the person. In Table 4-2, the percentages of body weight for those essential mineral elements that have been accurately determined are listed.

☐ Nutritional roles of mineral elements

Mineral elements are used in many diverse ways by the body, depending on the particular element, and in some cases, a given element has more than a single function. Some elements are constituents of body tissues, such as bones and teeth; others are components of functioning body compounds, such as vitamins, hormones, and enzymes. Still other mineral elements function in the maintenance of the acid-base balance of body fluids or as necessary factors in specific body processes.

Calcium and phosphorus occur in large concentrations in bones and teeth, together forming a significant constituent of these tissues. The amount of calcium needed to meet the body's requirement for bone growth varies with the rate of skeletal development. At birth, there are about 28 grams (gm) of calcium in the infant's body, whereas an adult's body contains about 1200 gm, representing an average daily increase of 0.17 gm throughout the individual's development. A good deal of the calcium consumed by an individual is not absorbed by the body, but passes out with the feces. The percentage of loss of dietary calcium is about 75%. Therefore a great deal more calcium must be ingested than is actually needed for body functions. The recommended daily intake of calcium varies from 800 milligrams (mg) for children (1 to 8 years of age) to 1400 mg for adolescents (13 to 19 years of age) to 800 mg for adults (20 years of age and older). The body's changing need for calcium reflects the individual's growth in height rather than weight.

Phosphorus makes up about 650 gm of a 70 kg adult body, and 85% to 90% is found in the bones and teeth, where it is in the form of calcium phosphate. The remaining 10% to 15% is distributed throughout all the cells of the body. In every cell, phosphorus is a constituent element of the genetic material (DNA) and is also vital in any reaction that involves the uptake or release of energy. For infants under 6 months of age, the recommended daily intake of phosphorus is 400 mg; for all older individuals, 800 mg is suggested. In contrast to calcium, about 70% of dietary phosphorus is absorbed by the body.

Mineral elements, as mentioned earlier, play a vital role in body functions as component parts of vitamins, hormones, and enzymes. Thiamin (vitamin B_1) contains sulfur as an integral part of its structure, whereas cobalamin (vitamin B_{12}) contains cobalt. The hormone thyroxin, which regulates the overall energy level of the body, contains the element iodine, and hemoglobin, the blood compound that transports oxygen to the cells, is an iron-containing molecule. In the case of enzymes, we find, for example, that both copper and iron are part of the enzyme cytochrome oxidase, which is involved in the cell's energy-producing biochemical reactions; molybdenum is part of xanthine oxidase, which causes the liver to release its stores of iron for use by other tissues; zinc is part of the pancreatic juice enzyme carboxypeptidase, which splits proteins into their component amino acids during digestion.

Still other mineral elements function in the maintenance of the acid-base balance of body fluids, especially that of the blood and the fluids surrounding the cells. The acidity level (pH) of the body is not affected by the organic acids present in foods, since these compounds are broken down to carbon dioxide, water, and energy; however, these acids do affect the taste of the food. On the other hand, the mineral elements contained in food do affect acid-base balance. Some minerals form acids when in solution: the acid-forming elements found in food are chlorine, sulfur, and phosphorus. These are found in high concentration in foods containing proteins, such as meat, fish, poultry, eggs, and cereals. Such foods are therefore designated as acid-forming foods. Mineral elements that are basic, or alkaline, in solution include calcium, sodium, potassium, and magnesium. These elements tend to predominate in fruits and vegetables.

It is interesting to note that even citrus fruits, such as grapefruits, lemons, and oranges, which have an acid taste because of their organic acids, are alkaline-forming foods, because the organic acids in the fruits are used for energy production, and the mineral elements are predominantly basic or alkaline elements. We can easily see that the acidity level of the bodies of people on extremely high protein diets will tend to increase, whereas the acidity levels of strict vegetarians will tend to do the opposite. The body, through various devices, some of which will be discussed in later chapters, can normally rid itself of excess amounts of either acids or bases. Death as a result of an acid-base imbalance is seldom caused by dietary factors alone.

In addition to the functions just mentioned, mineral elements also act as necessary factors in specific body processes. For example, the speed of clotting of a person's blood depends in part on the amount of calcium present in the blood; the absorption of nutrients from the digestive tract is often a mineral-dependent process. The absorption of carbohydrates from the digestive tract is facilitated by magnesium and sodium, and the absorption of the vitamin cobalamin (vitamin B_{12}) is aided by calcium.

We shall discuss nerve impulse transmission in detail in a later chapter. However, for our present consideration of the nutritional needs of the individual, it is important to indicate the vital roles played by mineral elements in the functioning of the nervous system. A nerve impulse is essentially an electrical stimulus that passes through a nerve fiber. This process is brought about by a change in the electrical charge that exists on the membrane of the nerve cell. The change in electric charge is caused by the flow of sodium ions from the surrounding body fluids into the nerve cell and the flow of potassium ions out of the nerve cell. This flow of ions is initiated by whatever factor in the environment has served as the stimulus for nerve excitation. A return of the nerve to its "resting state" follows a reversal of the flow of these ions into and out of the cell. The overall process of nerve impulse transmission depends on the exchange of sodium and potassium ions across the nerve cell membrane.

The nutritional roles of the remaining essential mineral elements are quite varied. Selenium, in a manner not well understood, is necessary for normal growth. This can be seen best in those people who suffer from both the protein deficiency disease kwashiorkor and selenium deficiency. In these cases, treatment with adequate protein is ineffective until sufficient selenium is also provided. One of the other essential mineral elements, manganese, is necessary for normal skeletal development and is also a component of the enzymes involved in the synthesis of fatty acids.

A number of body processes suffer when the dietary intake of chromium is low: there is a reduced tolerance for glucose, a retardation of growth, and a disturbance of those chemical reactions involving amino acids.

Table 4-3
Daily adult requirements of mineral elements and important food sources

Mineral element	Daily requirements*	Important food sources
Calcium	800 mg	Milk and milk products, fish, flour, green leafy vegetables
Phosphorus	800 mg	Meat, fish, poultry, eggs, and cereals
Potassium	1 gm	Fish, vegetables, meat
Sulfur	?	Meat, fish, vegetables
Sodium	0.5 gm	Meat, fish, poultry, salt
Chlorine	?	Salt
Magnesium	350 mg	Cereals, seafood, vegetables
Iron	15 mg	Liver, spinach, fish, eggs
Iodine	150 μg	Drinking water, spinach, dairy products
Zinc	10 mg	Meat, cereals
Selenium	?	Cereals, vegetables
Manganese	4 mg	Cereals, vegetables
Copper	2 mg	Meats, shellfish, cereals
Molybdenum	?	Legumes, cereals, vegetables
Chromium	?	Vegetables, cereals, fruit
Cobalt	0.6 μg	Meat, shellfish, poultry, milk
Fluorine	?	Drinking water, tea, coffee, legumes, vegetables

*The abbreviations used are as follows: mg = milligram ($^1/_{1000}$ gram); gm = gram; μg = microgram ($^1/_{1,000,000}$ gram).

The last essential mineral element that we shall consider is fluorine. It attaches to bones and teeth, forming crystals that serve to add strength and rigidity to these structures. The presence of the fluorine-containing crystals tends to make teeth more resistant to the corrosive effect of acids in the mouth and thus reduces the amount of tooth decay.

The daily adult requirements for each of the 17 essential mineral elements are listed in Table 4-3, along with important food sources. In some cases, the daily requirements for human beings are not known, because it has not been possible to study the effects of limited amounts of some elements in the daily diet of a large number of individuals. In addition to those mineral elements that have been established as essential for the proper functioning of the body, there is a rather large group of elements for which essentiality has not yet been established, although there is evidence of their participation in certain biochemical reactions. These elements are vanadium, barium, arsenic, bromine, strontium, cadmium, and nickel. Finally, there are those elements that are found in the body, but for which no role has been elucidated. They are presumed to be contaminants from the environment. These include gold, silver, aluminum, tin, bismuth, gallium, and lead.

□ Diseases caused by mineral element deficiencies

A deficiency of any one of the essential mineral elements will result in a disorder of the body. In this study, we shall concentrate on those diseases that are currently playing an important role in the health and well-being of the members of our modern society. We find that people whose diets as children lacked a sufficient amount of calcium are of short stature and, in addition, are susceptible to tooth decay. However, since diets that are low in calcium are frequently also low in protein, it is difficult to be certain that a lack of calcium is the primary cause of these effects, although it is undoubtedly a contributing factor. The situation related to calcium applies to phosphorus as well, since both elements are intimately associated with bone and tooth structure.

It is important to note that suboptimal dietary intake is not the only cause of mineral deficiencies. Severe cases of diarrhea can result in a mineral deficiency in the body because the passage of food through the intestine may be so rapid that proper absorption of a particular mineral element cannot occur. Potassium deficiency can develop by this means, and if prolonged, may cause heart abnormalities and overall muscle weakness.

Another mineral element, magnesium, can also be reduced below its necessary level in the body as a result of improper absorption from the intestinal tract. When this occurs, the individual suffers from uncontrolled muscle contractions. Initially they occur as tremors. However, under prolonged magnesium deficiency, the uncontrolled contractions become more severe until convulsive seizures occur. One factor that can produce this condition is excessive alcohol consumption. Alcohol increases the rate of magnesium excretion from the body; this at least in part accounts for the uncontrolled trembling so characteristic of chronic alcoholics.

A rather prevalent disease in our modern society is **anemia,** a condition in which there is a reduction in either the number of red blood cells or the amount of hemoglobin per red blood cell. In either case, the individual suffers fatigue after only mild exertion and a decreased resistance to infection. In severe cases of anemia, death can result from infections that would cause only relatively mild illnesses in healthy individuals. Although many factors can produce anemia, a lack of dietary iron is the most prevalent cause of the disease in infants. This problem is brought about because of the relatively small amount of iron in milk and is compounded in our society by the fact that cow's milk contains only about half as much iron as human milk. One study showed that whereas 42% of breast-fed infants were anemic, 70% of bottle-fed infants were similarly afflicted. Therefore it is generally recommended that the diet of infants include enriched cereals in order to ensure the intake of an amount of iron sufficient to meet their needs.

The last mineral element that we shall consider in this discussion of mineral deficiency diseases is iodine. It is found in every cell of the body in minute amounts, but most of it (70% to 80%) is concentrated in the thyroid gland, where it is used in the formation of the hormone thyroxin. Thyroxin is secreted by the thyroid gland into the circulatory system and distributed to all cells of the body. As mentioned earlier in this chapter, thyroxin regulates the overall energy level of the body. But the amount of thyroxin produced by the cells of the thyroid gland is limited by the availability of iodine. A drastic reduction in the daily intake of dietary iodine, that is, below 20 μg, causes an increase in both the size and number of cells of the thyroid gland in an apparent effort by the body to maximize thyroxin production. This growth and division of the thyroid gland cells results in an increase in size of the thyroid gland, a condition known as **simple goiter** (Fig. 4-2).

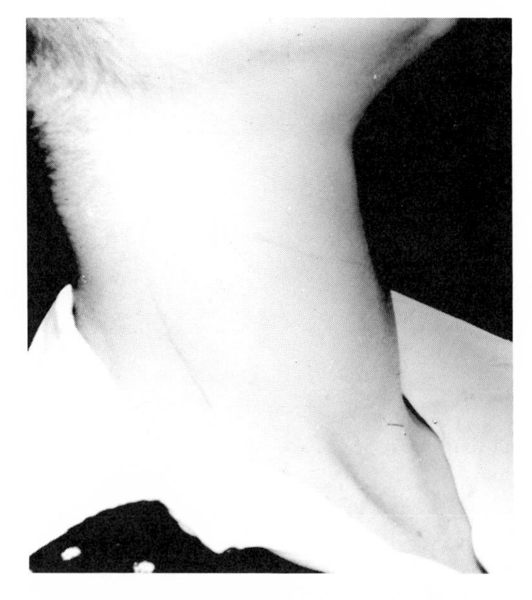

Fig. 4-2
Typical case of adult goiter identified in the 1970 *National Nutrition Survey*. (Courtesy Dr. W. J. McGanity, University of Texas, Galveston, Texas; from Guthrie, H. A. 1975. Introductory nutrition, ed. 3. The C. V. Mosby Co., St. Louis.)

Fig. 4-3
Areas of the world in which goiters are common.

Of the estimated 200 million cases of simple goiter in the world today, the majority are caused by a dietary lack of iodine. This problem tends to occur in those geographical regions where the soil is of glacial origin or where heavy annual rainfall and flooding have removed the iodine from the soil. This disease does not occur in areas near coastlines, where iodine-containing vapors, formed from the oceans, deposit this element directly onto the soil. In the United States, the areas of highest incidence of simple goiter are those that border the Great Lakes and the Rocky Mountains. The areas of the world where goiter is endemic are shown in Fig. 4-3.

A number of methods have been devised to add iodine to the diets of people in areas where goiter is endemic. These have included the addition of iodine to the area's water supply, feeding iodized salt to cows to increase the iodine content of their milk, and the iodination of table salt. Of these methods, the iodination of table salt has proved most effective in reaching a majority of the people in the affected areas. An example of the success of this approach can be seen in the state of Michigan. A study in 1921 revealed that 47% of the schoolchildren suffered from this type of simple goiter. An intensive educational program was instituted to promote the use of iodized salt. By 1925 the incidence of this disease had dropped to 32%, and by 1951 only 1% of the population had the disease. Having achieved this dramatic success, the health authorities made the mistake of terminating the educational campaign for the continued use of iodized salt on the assumption that a permanent dietary pattern had been established. Unfortunately, the incidence of simple goiter began to increase, and the campaign had to be reinstituted. This experience and others like it have shown that it is necessary to maintain educational campaigns to ensure proper dietary patterns and thus prevent nutritional diseases.

WATER

Water is without doubt a crucial ingredient of our daily diet. Although a human being can live for weeks and even years without the intake of some essential vitamins and minerals, the complete absence of water for more than 3 or 4 days is usually fatal. Water comprises 60% of our total body weight and is a constituent of every cell of the body. It is present in widely varying concentrations in different tissues: 72% of muscle, 25% of adipose tissue, and 10% of bone and cartilage. Human urine is 97% water, whereas blood is 80% water.

The human body contains about 45 liters* of water. Of this amount, 30 liters is

*One liter = 1.06 quarts.

Table 4-4
Daily water requirements under various conditions

Category	Daily water requirement (ml/kg body weight)
Infant	110
10-year-old child	40
Adult whose environmental temperature is	
72° F	22
100° F	38

found within the cells of the body, and 15 liters is either in the blood (3 liters) or in the spaces (12 liters) between the cells. Any given molecule of water usually shifts from one position in the body to another many times before it leaves the body.

The daily water requirement of an individual depends not only on body weight, but also varies with age and environmental temperature. The younger the person, the greater is the water requirement per unit of body weight; the higher the environmental temperature, the greater is the need for water. Table 4-4 gives some average requirements for daily water consumption under various conditions.

□ Functions of body water

Water performs various functions, depending on what region of the body it occupies. The water that constitutes the fluid portion of the blood serves to carry in solution various nutrients, hormones, and waste products from one part of the body to another. When it is in the spaces between cells, water carries the nutrients and hormones from adjacent blood vessels to the cell surfaces. It also collects waste products from the cells and acts as the vehicle by which the waste products reach and enter the blood vessels. There is a constant interchange of water molecules between the fluid portion of the blood and the fluid in the spaces between the cells.

Within each cell, water is used in many ways. It serves to transport materials within the cell and also acts as a medium in which biochemical reactions take place. In addition, it is a necessary reactant in a number of reactions and is incorporated into many of the compounds that form the structural components of a cell.

In the digestive system, the water of saliva serves as a lubricant, facilitating the passage of food down the esophagus. The water of digestive juices in the stomach and small intestine is necessary as a reaction medium and is also used in the biochemical reactions that break complex nutrients, for example, polysaccharides and proteins, into their simpler component parts, in this case, monosaccharides and amino acids.

Another important role of water is in the regulation of body temperature. As a result of many of its biochemical reactions, the body produces a large amount of heat. Some of this heat serves to maintain the body's temperature at 98.6° F. However, usually more heat is produced than is necessary for the maintenance of normal temperature. The most effective method of ridding the body of excess heat is through the evaporation of water from the surface of the skin. The water that is evaporated from the body surface is secreted as **perspiration** by glands in the skin. Evaporation, the changing of a substance from a liquid to a gaseous state, requires energy in the form of heat, and the body is constantly losing heat and thus cooling itself through this process. The loss of body heat through evaporation amounts to about 25% of an individual's total caloric expenditure.

□ Sources of body water

We obtain our body water from several sources: the beverages we drink, the solid foods we eat, and those biochemical processes within the cells of our bodies that involve the breakdown of carbohydrates, lipids, and proteins. The typical pattern of daily water consumption and loss by the body is shown in Table 4-5. Of the 1100 milliliters (ml) of water consumed in liquid form each day, about 40% is drinking water, whereas the rest comes from milk and other beverages. It may be surprising to

Table 4-5
Typical pattern of average daily consumption and loss of water

Components	Amount of water (ml)
Sources of water	
Liquids	1100
Solid food	750
Biochemical reactions	350
Total	2200
Losses of water	
Urine	1200
Evaporation (skin and lungs)	900
Feces	100
Total	2200

Table 4-6
Water content of representative foods*

Food	Water (%)	Food	Water (%)
Lettuce	96	Chicken	63
Asparagus	92	Beef	47
Milk	87	Bread	36
Oranges	86	Butter	15
Potatoes	80	Gelatin	13

*From U.S. Department of Agriculture. 1963. Handbook No. 8. Washington, D.C.

learn that solid foods supply about one third of our daily water needs. However, as indicated in Table 4-6, so-called solid foods contain appreciable amounts of water.

☐ Effects of excessive water loss or intake

Severe biochemical difficulties result from either excessive water loss or excessive water intake. Abnormal water loss can accompany diarrhea, vomiting, or fever. Abnormal water retention, called **water intoxication,** can result from many causes, including reduced kidney function. A reduction of body water by 10% represents severe dehydration, and a 20% reduction is fatal. An increase of body water by 10% above normal levels produces **edema** (swelling caused by the accumulation of excessive fluid in the intercellular spaces of the body). Further increase results in the excess water entering the cells of the body, causing them to swell. When this occurs in the brain cells, convulsions, coma, and death can follow.

One further aspect of water loss needs to be considered. At high temperatures, there is a significant amount of water lost through perspiration. Along with the loss of water, there is also an appreciable loss of salt, sodium chloride, of which the sodium is most critical. The reduced amount of body water will cause the individual to feel thirsty and will lead to an increased consumption of liquids. However, if the lost sodium is not also replaced, its concentration in the body fluids will become very

Fig. 4-4
Steelworker, example of an individual whose water and salt consumption must be carefully coordinated. (From Kaluger, G., and Kaluger, M. F. 1976. Profiles in human development. The C. V. Mosby Co., St. Louis.)

diluted, and the individual will suffer from water intoxication, with the threat of the severe consequences noted previously. Water intoxication is a serious problem in some of our modern industries (iron and steel) in which men work at high temperatures. Under such conditions, it is necessary that workers take salt tablets when they drink water (Fig. 4-4).

■ ENERGY REQUIREMENTS OF THE INDIVIDUAL

An individual's energy requirements depend on four interacting variables: physical activity, body weight, age, and climate. In childhood and adolescence, there are additional energy needs for growth, as is also true during pregnancy and lactation (Fig. 4-5).

All the calories provided by the diet are obtained from carbohydrates, lipids, and proteins. In the typical American diet, carbohydrate supplies about 50% of the energy; lipid, about 35%; and protein, about 15%. The relative amounts of these

Fig. 4-5
A, Sheep in the Armsby calorimeter as part of a metabolism experiment in animal nutrition, for which the chamber was originally designed. **B,** Two men entering the same calorimeter to participate in a 72-hour metabolism study after the unit's conversion for use with human subjects in the 1950's. In both photos, note the thickness of the walls and the intervening air spaces. (Photographs by R. Beese; courtesy Dr. G. Barron, The Pennsylvania State University, University Park, Pa.; from Guthrie, H. A. 1975. Introductory nutrition, ed. 3. The C. V. Mosby Co., St. Louis.)

three types of food in a diet will depend on economic, social, and cultural factors. In economically depressed communities, carbohydrates form a larger portion of the diet than they do in financially well-to-do groups. We also find that individuals will often retain the food habits of their families and community regardless of changes in financial position. The general field of study called **social nutrition** deals with why people eat what they do and how eating habits can be changed. Social nutrition will become increasingly important as we attempt to improve the health of an ever greater population (Fig. 4-6).

□ Physical activity

In every population, individuals vary greatly in the extent of their physical activity. Although there is no precise way to categorize occupations, a generally accepted classification is as follows:

1. Involved in light activity
   a. Men: office workers, most professional men (doctors, lawyers, teachers, etc.)
   b. Women: office workers, most professional women, housewives in houses with mechanical household appliances

SOYBEAN LECITHIN
Here is natural, organic soy lecithin. Rich in natural choline and inositol. Available in granules for use with water, juice or sprinkled on cereals, salads, etc., and in convenient easy to use capsule form.
CAPSULES—Fortified with Vitamin D
100—1.00 500—4.50 1000—6.90
*Postage Paid**

SEA-PRO CERTIFIED* PROTEIN
*Clean Protein from Gardens-of-Sea!

ROYAL JELLY
50 MG. CAPSULES
Here is the sensational but controversial Royal Jelly, offered as a new food for human consumption in this convenient, easy to take form.

DOLO-ZYME tablets
Original Dolomite plus Sturdegest!

IRISH SEA MOSS
A rich bounty from the sea in its natural unadulterated form.
Unbleached 12 oz.—50c
Add Postage—See Back Page

CITRÒN-X TABLETS
LEMON BIOFLAVONOIDS
Research workers have been very hopeful regarding the value of some of the ingredients contained in this exciting formula as a dietary supplement.
Three Tablets Contain:
Vitamin C (From Rose Hips Extract) ....100 mg.
Lemon Bioflavonoid Complex ....100 mg.
Rutin (From Buckwheat) .... 30 mg.
Hesperidin Complex .... 30 mg.

GARLIC PARSLEY
TABLETS
Millions of these Garlic Parsley tablets have been sold to satisfied, happy users. Thousands can't be wrong. If

Vegetable Broth
NATURAL BRAND
A combination of natural dehydrated vegetables for making a tantalizing delicious vegetable broth.
8-oz.—1.00 1-lb.—1.65
Add Postage—See Back Page

Alfalfa Seeds
Here are finest quality Alfalfa Seeds money can buy. Carefully selected to meet Natural Sales Company's high quality standards.
1-lb.—70c 5-lbs.—3.25
Add Postage—See Back Page

VEGETABLE CONCENTRATES
ORGANICALLY GROWN
Here you can get the added nutritional benefits of these natural vegetables in handy, convenient tablets at a low, economical price. Add postage on items below. Shipping weight 500 tablets 1 lb., 1000 tablets 2 lbs.

CABBAGE TABLETS	500—1.95 — 1000—3.70
ALFALFA TABLETS	500—1.50 — 1000—2.15
CELERY TABLETS	500—1.90 — 1000—3.25
DULSE TABLETS	500—2.25 — 1000—3.50
RICE POLISH TABLETS	500—1.70 — 1000—2.69
WATERCRESS TABLETS	500—2.80 — 1000—4.90
OKRA TABLETS	500—1.95 — 1000—3.70
11 VEGETABLE TABLETS	500—2.80 — 1000—4.90

Containing Beets, Carrots, Celery, Endive, Kale, Lettuce, Alfalfa, Parsley, Spinach, Tomato, Kelp and excipients.

C-RUTIN
Three tablets contain 100 mg. of Vitamin C derived from Rose Hips and 50 mg. Rutin derived from Buckwheat which contains subnutritional amounts of factors called by some investigators Vitamin P. Plus excipients.

SUNFLOWER SEEDS
HULLED — ORGANICALLY GROWN
If you have not already enjoyed this wonderful food, order today. Rich in bone forming Calcium. Excellent source of B-1 and Niacin. High in protein. Delicious as a cereal or added to other foods.
1-lb.—1.10 — 5-lbs.—5.25
Add Postage—See Back Page

'FROM-THE-SEA' No. A-134
KELP·ALL
Derived entirely from natural rich Pacific Kelp. Provides a fine all natural source of Iodine and minerals in trace amounts to supplement daily diet. Small tabs, easy-to-take. Note extra special buy!

AMO TABS
AMINO ACID TABLETS
A source of essential Amino Acids obtained from the enzymatic predigested protein of yeast.

CAROB POWDER
An alkaline food to substitute for cocoa. High in natural sugar. Ideal for chocolate-like drinks, pies, cakes and other confections.

Unsaturated Fatty Acids
(VITAMIN F)
Now Natural Sales Company's low cost purchase plan makes it possible for you to obtain these valuable Natural, Unrefined Protective Food Oils at a sensible price.
Each capsule contains 168 mg. of Unsaturated Fatty Acids (Vitamin F) from Linseed Oil. Principally Lenoleic and Linolenic Acids.

PAPAYA B-PAPS
Contains Papaya, Papain and Natural Vitamin B-1, Vitamin B-2 and Niacin. An aid in the digestion of protein.

ROSE HIPS
Here is the same Rose Hips you have read and heard so much about. An excellent natural source of Vitamin C. Specially selected and picked for Natural Sales Co. these Rose Hips are sun-dried and cleaned without artificial processing. Once you try Natural brand Rose Hips you'll prefer them always, just as many thousands of others do.
SUPREME | ROSE HIPS REGULAR | TEA | POWDER

MORVITE
Fabulous New Food Supplement. 20 Food Factor Supplying 34 Natural Vitamins and Minerals.

DESICCATED LIVER
This is pure — very highest quality, defatted Liver.
Dried at low heat to retain all the beneficial factors. Finest

WHEY
Superior quality. Rich in milk minerals and lactose — the intestinal floral sugar that combats putrefaction and excess protein decomposition.

Sensational New Nutritional Development
MIRACLE WAFERS
Delicious Candy-Like Wafer, chuck full of Delightful Ingredients:
YOGHURT LICORICE
WHEY MOLASSES
BANANA FLAKES ROYAL JELLY

NATURAL-ORGANIC
PAPAYA ENZYME

ACIDOPHILUS PLUS
Exclusive With Natural Sales!

Fig. 4-6
Composite of items offered for sale in the catalogue of a health food store. (From Guthrie, H. A. 1975. Introductory nutrition, ed. 3. The C. V. Mosby Co., St. Louis.)

Table 4-7
Effect of occupation on average daily expenditure of energy (kcal/kg* body weight)†

Group	Occupation			
	Involved in light activity	*Moderately active*	*Very active*	*Exceptionally active*
Men	42	46	54	62
Women	36	40	47	55

*One kg = 2.2 pounds.
†From World Health Organization. 1973. WHO Tech. Rep. Ser. No. 522.

Fig. 4-7
Miner, example of an individual whose daily energy requirement is 54 kcal/kg body weight. (Courtesy World Health Organization; from Turner, C. E. 1971. Personal and community health, ed. 14. The C. V. Mosby Co., St. Louis.)

2. Moderately active
    a. Men: workers in light industry, students, farm workers on mechanized farms
    b. Women: workers in light industry, students, department store workers, housewives in houses without mechanical household appliances
3. Very active
    a. Men: unskilled laborers, athletes, mineworkers, steel workers, farm workers on nonmechanized farms
    b. Women: athletes, dancers, farm workers on nonmechanized farms
4. Exceptionally active
    a. Men: lumberjacks, blacksmiths, rickshaw pullers
    b. Women: construction workers

The daily energy requirements of the various types of occupations are listed in Table 4-7. The data are given in terms of "kcal/kg body weight" in order to permit the calculation of any individual's caloric needs (Fig. 4-7).

□ Body weight

Body weight affects energy expenditure in a number of ways. The greater the individual's weight, the more physical work is required to move the entire body or any part of it. In addition, the number of calories expended in maintaining posture while standing is determined by weight. It should be noted that the energy requirements per unit of body weight for men are higher than for women who are of the same total weight and in the same occupational categories as the men. This sex difference in energy requirements is a result of the fact that energy expenditure is measured in terms of fat-free body mass, and the female body generally contains a larger proportion of fat.

□ Aging

The energy requirements of adults decline with age even when there is no change of occupation. This reduction in caloric needs is the result of a general decline in level of physical activity, a change in body weight, and an increased occurrence of diseases and disabilities. There is no evidence that energy expenditure declines significantly, either at work or at leisure, between the ages of 20 and 39 years. However, from the

Table 4-8
Average daily energy requirements (kcal/kg body weight) of moderately active adults at different ages*

Age (years)	Men	Women
20-39	46	40
40-49	44	38
50-59	42	36
60-69	37	32
70-79	32	28

*From World Health Organization. 1973. WHO Tech. Rep. Ser. No. 522.

Table 4-9
Daily energy requirements of children and adolescents by age groups*

Age (years)	Daily energy requirements (kcal/kg body weight)
Children	
<1	112
1-3	101
4-6	91
7-9	78
10-12 (males)	71
10-12 (females)	62
Male adolescents	
13-15	57
16-19	49
Female adolescents	
13-15	50
16-19	43
Moderately active adults	
Males	46
Females	40

*From World Health Organization. 1973. WHO Tech. Rep. Ser. No. 522.

age of 40 years onward, there is a general reduction in the amount and speed of movement, both on the job and during nonoccupational activities.

There is an even sharper decline in general activity after the age of 60 years. In people 60 to 69 years of age, the limitation of physical activity attributable to disease or disability increases sharply and becomes even greater after the age of 70 years. Table 4-8 shows the daily energy requirements of **moderately active** adults at different ages. The data are given in terms of kilocalories per kilogram of body weight so that the daily energy requirements can be calculated, taking into account both age and body weight.

The energy requirements of infants, children, and adolescents include those calories needed for growth as well as those needed for general activities. As shown in Table 4-9, the energy requirements of young people, per unit of body weight, far exceed those of adults regardless of occupation. A moderately active adult needs proportionally less than half the calories of an infant child.

□ Climate

Man, like other mammals, has the capacity to increase his body's production of heat in response to a cold environment. However, man is also able to use clothing and temperature-controlled housing to provide himself with a generally comfortable skin temperature of 91.4° F (33° C).

The energy expended in standard tasks is the same in different climates, except possibly in extreme and unusual environments. However, climate may have a marked effect on the amount and the timing of physical work and recreation. The social adaptations to climate do affect the caloric needs of different communities and must be taken into account in estimating the food requirements of groups living in different environments.

In setting up a climatic standard for judging energy requirements, it is assumed that people living in the temperate zone (United States, France, Russia, China, Japan, Chile, Argentina, South Africa) are exposed to an average annual temperature of 50° F (10° C). It has been estimated that the caloric needs of an individual living in the temperate zone are **increased** by 3% for every **decrease** of 18° F (10° C) below the reference temperature. On the other hand, the energy requirements of an individual are decreased by 5% for every increase of 18° F (10° C) of the average annual external temperature above the reference temperature.

□ Pregnancy and lactation

During pregnancy, extra energy is needed both for the growth of the fetus and for the activities of the heavier mother. The total caloric cost of a pregnancy is about 80,000 kcal. This needed extra energy allowance requires an average increase of about 150 kcal/day in the first trimester and 350 kcal/day in the second and third trimesters.

About half of the 80,000 kcal required for a pregnancy is provided by an increased lipid storage on the part of the mother. Fat deposition accounts for about 4 kg of a total weight gain of 12.5 kg during pregnancy in a well-nourished woman. Fat deposition begins early in pregnancy and represents an increase in the woman's energy reserve. This reserve provides protection against a possible food shortage at a later stage of pregnancy, when the needs of the fetus are increasing rapidly, or it may be utilized during lactation, when maternal energy needs are high.

There are very good physiological, hygienic, and psychological reasons for breast-feeding infants. Breast milk contains appropriate amounts of carbohydrates, lipids, proteins, and minerals and seems to offer young infants protection against disease as well. A study made in Chile revealed that, in 1954, 95% of the 1-year-old children were breast-fed, but by 1974 breast-feeding was restricted to only 20% of the 2-month-old infants. The same study found that Chilean infants who were bottle-fed during their first 3 months of life suffered a mortality rate three times that of breast-fed infants.

The cause of this tragedy in Chile and its occurrence in many other developing countries can be traced to two basic problems: (1) the use of bottle-feeding by mothers incapable of reading the instructions for the proper preparation of the powdered milk and (2) the kitchens where there is no clean water for mixing and sterilizing. It is clear that safe bottle-feeding can only occur in a community with a relatively high level of education and sanitation and a standard of living that permits the purchase of adequate equipment for proper sterilization.

The average daily amount of milk produced by a nursing mother is 850 ml. Human milk has an energy content of about 0.72 kcal/milliliter (ml), for a total of 600 kcal. The human body produces milk with an efficiency of about 80%. This means that a lactating female needs 750 kcal/day above that required by her occupation, age, body weight, and climate. During a 6-month period of lactation, the total energy requirement is about 135,000 kcal.

• • •

We have considered the energy requirements of the individual under various conditions. We now want to analyze that unfortunate phenomenon that continues to haunt relatively large segments of the world population, namely, starvation.

■ STARVATION

Starvation, the prolonged deprivation of food, has been experienced by human beings throughout history. To a varying extent, famine appears to accompany most disasters, whether they are of natural origin, such as droughts or floods, or man-made, such as wars. The deleterious effects of starvation are both severe and permanent in young children. A child who has suffered food deprivation very early in life and for an appreciable length of time will never reach normal size for his age, even though he is later fed well enough to restore a normal rate of growth. In addition to reduced stature, a permanent impairment of brain function may occur, leading to below-average mental ability. Prolonged starvation will, of course, cause a child's early death.

In the case of adults, starvation leads to a reduced level of activity, loss of weight, increased susceptibility to disease, and premature death. However, the human body can survive without food for long periods. A well-authenticated case is Terrence MacSwiney, the Irish revolutionist. In his famous hunger strike in a British prison in 1920, MacSwiney survived for 74 days before dying of starvation. A typical pattern of weight loss during food deprivation is shown in Fig. 4-8. It should be noted that the loss of weight is not linear, but gradually decreases with time.

How does the body accommodate itself to prolonged starvation? To explain the body's ability to withstand food deprivation for long periods of time, we must begin with an examination of the body's chemical needs. The primary need of the body is for energy with which to perform vital functions. The need for molecules for growth

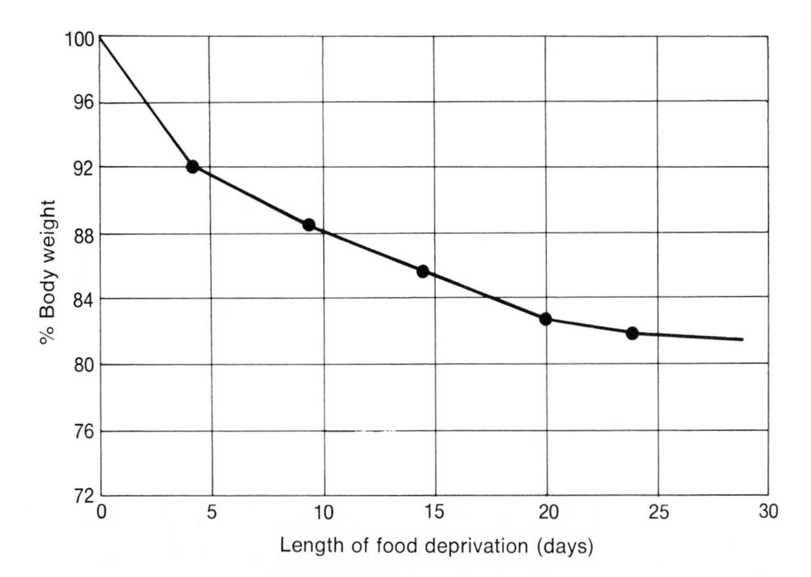

Fig. 4-8
Weight loss during starvation.

Table 4-10
Fuel composition of an average 70 kg man

Fuel	Weight (kg)	Energy (kcal)
Tissues		
Fat (adipose triglyceride)	15	135,000
Protein (mainly muscle)	6	24,000
Glycogen		
In muscle	0.150	600
In liver	0.075	300
Total		159,900
Circulating		
Glucose (in extracellular fluid)	0.020	80
Free fatty acids (in blood plasma)	0.0003	3
Triglycerides (in plasma)	0.003	30
Total		113

and repair is secondary to the need for energy. Table 4-10 is a listing of the "fuel composition" of an average 70-kilogram (kg) man. The data indicate that body fat is the great storehouse of energy for man, with protein representing a secondary reserve. However, as we shall see, the use of these energy sources by the various parts of the body does not follow a simple pattern.

The complicating factor in the use of the body's energy sources is the nervous system, especially the brain. Normally only glucose can supply energy to the brain, which requires 125 grams (gm) of glucose each day. Strangely, that much glucose is never available in the body at any one time. The body's main reserve of glucose is in the form of glycogen in the liver and amounts to only 75 gm. Moreover, not all of this 75 gm of glycogen is available for use by the brain, because the liver tends to retain a fair amount of glycogen for stressful emergencies. In fact, in the absence of a continued intake of glucose, the available glycogen from the liver can supply the brain's need for energy for only a few hours. This situation presents a problem because there are relatively long periods of time each day, especially between dinner and breakfast, during which virtually no food is consumed. (This is one reason never to skip breakfast.)

In the absence of a continued intake of carbohydrates, the body begins to break down some of the proteins of its muscles, forming amino acids. A number of these amino acids can be transformed chemically by the liver into glucose for use by the brain. At the same time, some of the triglycerides in the adipose tissue of the body are broken down to provide glycerol and fatty acids. The glycerol can be used by the liver to make additional glucose, whereas the fatty acids can be used as an energy source by cells other than those of the nervous system. These processes are diagrammed in Fig. 4-9.

The first stage of prolonged starvation resembles, in its biochemistry, the overnight fast of everyday life. Based on an initial expenditure of 2000 kcal/day, we can calculate that this energy is supplied through the breakdown of 85 gm of muscle protein and 180 gm of adipose tissue triglyceride. However, this rate of protein

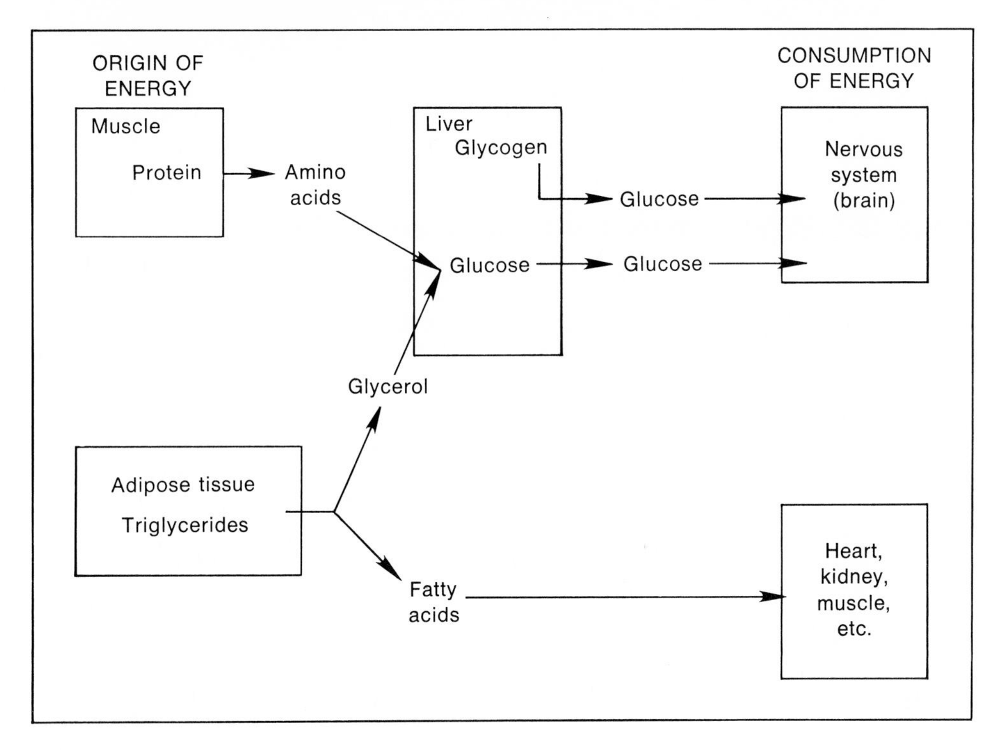

Fig. 4-9
Energy metabolism under overnight fasting conditions.

breakdown, if continued, would result in the loss of 10% of the body's muscle protein each week in order to supply glucose for the brain and other parts of the nervous system. Human beings cannot survive this degree of protein depletion for very long. Therefore some other way must be available to supply energy to the brain under conditions of extended food deprivation.

It has been found that on prolonged fasting, a change occurs in the body's biochemistry with relation to fatty acids. Some of these are transformed by the liver into two other compounds, namely, beta hydroxybutyrate and acetoacetic acid. These compounds are members of the group of molecules known as **ketones.** Ordinarily, the body does not form ketones from fatty acids. The importance of this change in body biochemistry is the fact that the brain can utilize ketone as an energy source, thus eliminating the need to produce glucose from body protein. Unfortunately, the breakdown of body protein is not completely eliminated, even in prolonged starvation, and a reduced amount of destruction of body proteins continues.

One of the consequences of the body's conservation of protein during starvation is that urination (for the excretion of nitrogen) is reduced. This reduction results in a decreased need for water intake. If the individual's loss of water through perspiration is minimal, a cup of water a day will be sufficient to maintain the body's water balance.

The overall picture of loss of the various chemical constituents of the body during prolonged starvation is shown in Fig. 4-10. The initial reduction in weight is caused

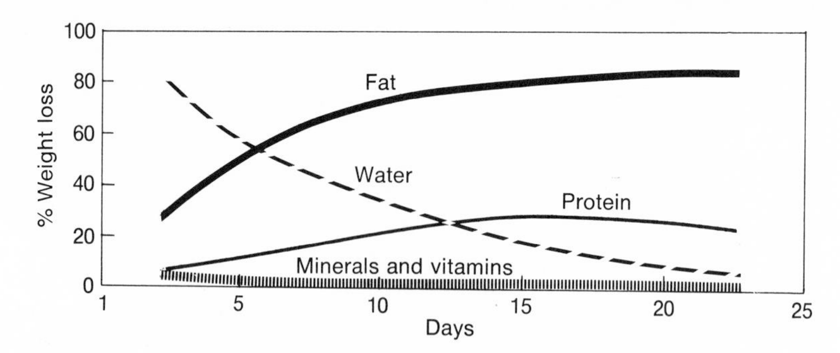

Fig. 4-10
Changes in loss of various chemical constituents of the body during starvation.

mainly by a loss of water and secondarily of fat and protein. As starvation continues, however, a progressively greater proportion of the weight loss is accounted for by the consumption of body fat. Although it is generally believed that the pancreatic hormones insulin and glucagon play an important role in bringing about the changes in body biochemistry that occur during starvation, the mechanisms by which these hormones accomplish their tasks have not been worked out.

■ SUMMARY

In this chapter, we examined the role of vitamins as coenzymes in body metabolism, the diseases caused by a deficiency of the various vitamins, and the diseases caused by excessive intake of some vitamins. We then considered the different mineral elements with regard to their various functions in the body, their availability in different foods, and the diseases caused by their deficiencies. The last discussion in this area centered on the essential nutrient water, its many functions in the body, its daily requirements, and the effects of its excessive intake or loss.

Completing our study of essential nutrients, we examined the energy requirements of the individual as affected by occupation, body weight, aging, climate, pregnancy, and lactation. The caloric needs of a person reflect all these interacting variables. We concluded our study of the types and functions of food with an analysis of the effects of prolonged starvation on the body.

The first four chapters of this book examined the concept of biological success both for the individual and the species, the general need for food in the functioning of the human body, the availability of food on our planet, and the specific constituents of food that contribute to our well-being. In the next section, we will examine how the needed nutrients are obtained from the food we eat, how we distribute the nutrients to the cells of our body, and how we utilize the nutrients for our well-being.

SUGGESTED READINGS

Dwyer, J. T., and Mayer, J. 1975. Beyond economics and nutrition: the complex basis of food policy. Science **188**:566-570. Thoughtful discussion of the many factors that go into establishing a national food policy.

Loomis, W. F. 1970. Rickets. Sci. Am. **223**:77-91. Fascinating discussion of the relationship between solar ultraviolet radiation and calciferol, vitamin D, in preventing rickets.

Mayer, J. 1975. Management of famine relief. Science **188:**571-577. Very fine analysis of the causes, physiological and psychological effects, and treatment of starvation.

Walsh, J. 1973. Vitamin B_{12}: after 25 years, the first synthesis. Science **179:**266-267. Well-written account of the discovery, characteristics, and synthesis of the only vitamin containing a metal ion, that is, cobalt.

Williams, S. R. 1973. Nutrition and diet therapy, ed. 2. The C. V. Mosby Co., St. Louis. Up-to-date, comprehensive, and authoritative book on nutrition and health.

World Health Organization. 1973. Energy and protein requirements. WHO Tech. Rep. Ser. No. 522. General discussion of the energy requirements of the individual and the role played by proteins in meeting these requirements.

World Health Organization. 1974. Handbook of human nutritional requirements. WHO Monogr. Ser. No. 61. In-depth discussion of the roles of vitamins and mineral elements in the body and the daily requirements of each.

Young, V. R., and Scrimshaw, N. S. 1971. The physiology of starvation. Sci. Am. **225:**14-21. In-depth discussion of the effects of starvation on body functions.

Contrast size and shape of first and second set of teeth. (Courtesy H. Armstrong Roberts, Philadelphia; from Turner, C. E. 1971. Personal and community health, ed. 14. The C. V. Mosby Co., St. Louis.)

CHAPTER 5 Making our food usable

LEARNING OBJECTIVES

- What must be done to food to make it usable by the cells of the body?
- Which aspects of digestion are accomplished by the mouth?
- What diseases affect our teeth and gums?
- Which aspects of digestion are accomplished by the stomach?
- Which aspects of digestion are accomplished by the small intestine?
- Which aspects of digestion are accomplished by pancreatic enzymes?
- Which aspect of digestion is accomplished by bile?
- Which aspects of digestion are accomplished by the large intestine?
- What are the disorders of the large intestine?

The food that we consume is usually not in the chemical form that can be used by our bodies. The various compounds first must be broken down into their basic molecules (polysaccharides into monosaccharides, lipids into glycerol and fatty acids, and proteins into amino acids). The process by which this is accomplished is called **digestion.** The type of chemical reaction involved in the digestion of food is related to that by which polysaccharides, lipids, and proteins are formed. The formation of a disaccharide from two monosaccharides was described in Chapter 3 (see Fig. 3-3). It is important to note that in this process, a molecule of water is formed by contributions of atoms from both monosaccharides. This mechanism of combining molecules, with the accompanying production of water, is called **condensation.** Fig. 3-6 shows that the formation of a lipid is also the result of condensation, and the same is true of the formation of an amino acid chain (a protein), as shown in Fig. 3-10.

When polysaccharides, lipids, and proteins are digested, the process of condensation is reversed. Water molecules are added to these large compounds at the sites of chemical bonding between adjacent molecules (H to one molecule, and OH to its neighbor). These bonds are then split, releasing molecules of monosaccharides, glycerol, fatty acids, or amino acids. The use of water to split complex molecules into simpler ones is called **hydrolysis.** The chemistry of digestion is basically simple, because this same process, hydrolysis, is responsible for the digestion of all three major types of food.

After digestion is completed, the usable molecules must be distributed to the cells of the body where the molecules are used as sources of energy, as component parts of cell products (enzymes, hormones, etc.), or for the growth and division of the cells themselves. In addition, the unusable constituents of food must be eliminated from our bodies.

The tasks involved in food consumption and its chemical alteration are accomplished by the digestive system. This body system is a continuous series of connecting cavities and tubes that receive food into the body, reduce the nutritive components of the food to their usable forms, and evacuate the residues. We shall now examine the different parts of the digestive system (mouth, pharynx, esophagus, stomach, small intestine, large intestine, rectum) and see how each performs its specific functions. An overall diagram of the human digestive system is shown in Fig. 5-1.

MOUTH

The mouth is the point of ingestion of food; it consists of the lips, cheeks, jaws, tongue, and palate. These structures combine with other associated organs to perform the following functions: (1) testing food to determine its suitability (temperature, taste, texture); (2) chewing and grinding food into particles small enough to be

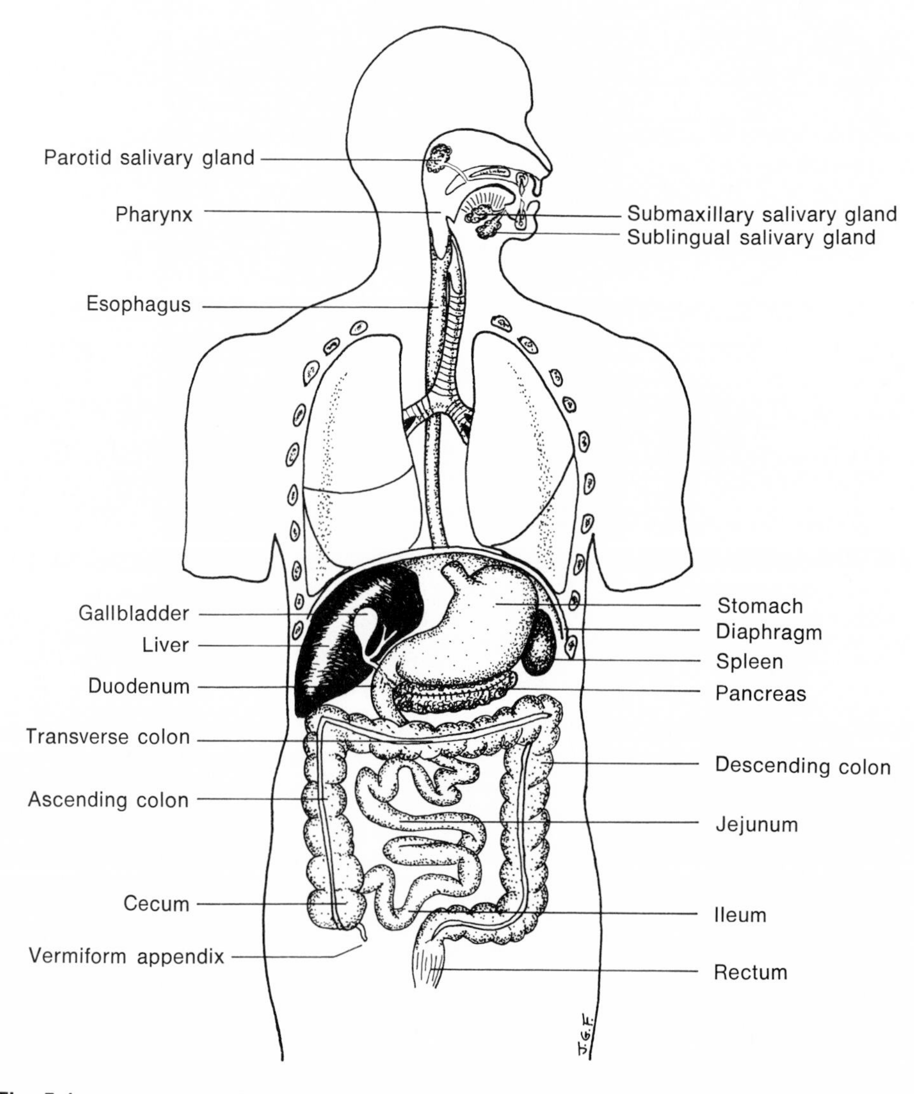

Fig. 5-1
Human digestive system. Liver has been folded back to reveal stomach, duodenum, and gallbladder. Note that spleen is part of circulatory system.

swallowed; (3) adding enzymes to initiate the chemical process of digestion; and (4) shaping the chewed food into a lump, called a **bolus,** for swallowing.

□ Testing food

As food passes through our lips and enters the oral cavity, a large number of sensory receptors are stimulated to provide information about the food. Coarse, irregular, or sharp substances are detected by pain, touch, and pressure receptors in the skin of the lips, gums, cheeks, roof and floor of the mouth, and tongue. Other receptors yield information on the temperature of the food. The tongue has special sensory receptors, called **taste buds,** that are sensitive to substances that are sweet,

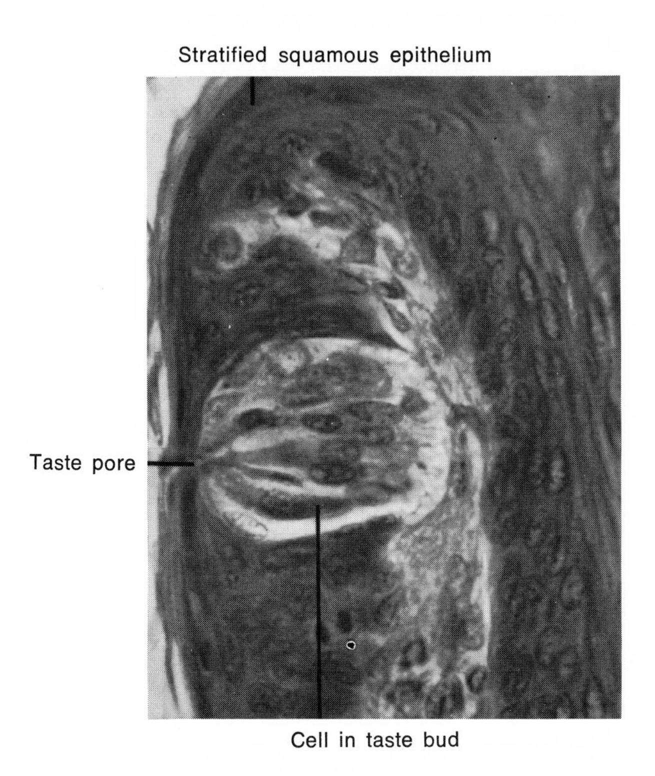

Fig. 5-2
Taste bud showing pore. (×640.) (From Bevelander, G., and Ramaley, J. 1974. Essentials of histology, ed. 7. The C. V. Mosby Co., St. Louis.)

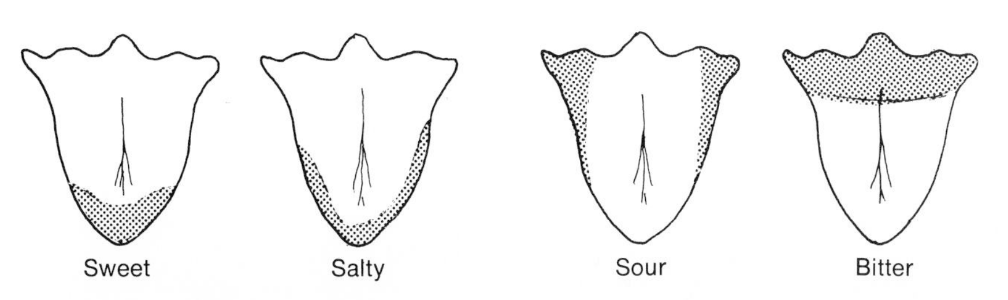

Fig. 5-3
Distribution of different types of taste buds in tongue.

sour, bitter, or salty (Fig. 5-2). The receptors of these four basic tastes tend to be concentrated on different parts of the tongue, as shown in Fig. 5-3.

The bitter taste, when it occurs in high intensity, usually causes a person to reject the particular food. This reaction to bitter-tasting substances is undoubtedly of great benefit, because many of the deadly chemical compounds found in poisonous plants are intensely bitter in taste.

A person's ability to taste varies with age; as people grow older, their taste sensitivity decreases. This change results in large measure from a decrease in the number of taste buds in the tongue. A 75-year-old man has only 36% of the number of taste buds in his tongue that he had at age 30 years.

Chewing and grinding food

Chewing and grinding food into small particles is the function of the teeth. Humans have four different types of teeth, as illustrated in Fig. 5-4. Each type is adapted for a different function. In the front of each jaw are chisel-shaped **incisors** that are specialized for biting. Behind the incisors, on each side in each jaw, is a pointed **canine** tooth adapted for tearing food. Next in order, in adults, are **premolars** and **molars.** These last two types of teeth are wide and have flattened surfaces that function in crushing and grinding food.

An adult normally has 32 so-called **permanent** teeth, 16 in each jaw. Starting in the front and midline of the mouth, on each side of each jaw, there are two incisors, one canine tooth, two premolars, and three molars. The third molars are the "wisdom" teeth, which do not develop in all individuals. Human beings, and many other

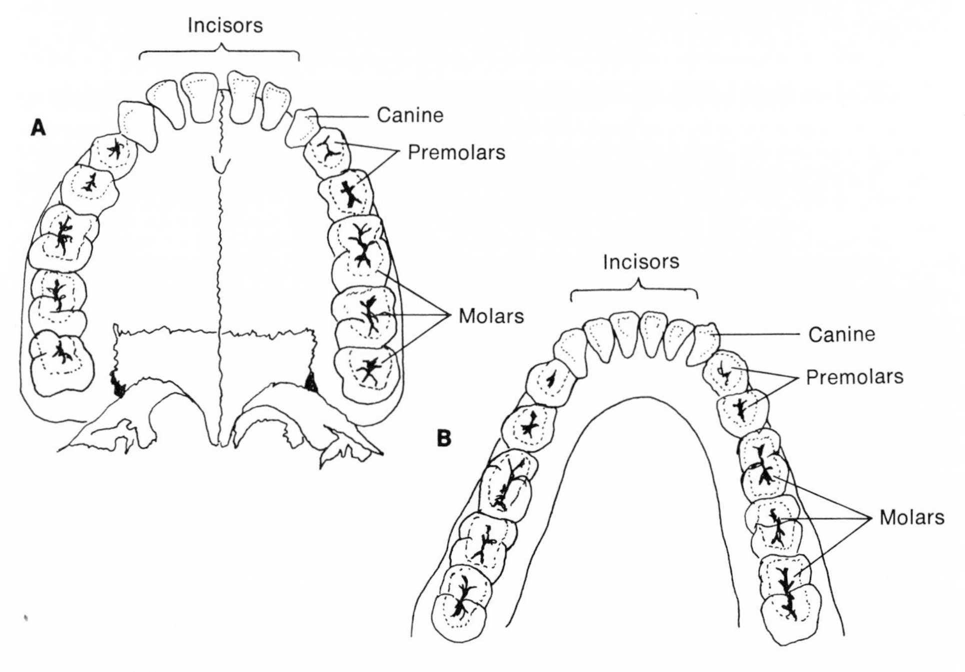

Fig. 5-4
Types and distribution of human teeth. **A,** Upper jaw. **B,** Lower jaw.

animals, have two sets of teeth. A child's first set of teeth, called **milk** teeth, are 20 in number; the 12 molars are lacking. Fig. 5-5 shows the parts and internal structure of a human molar tooth.

All teeth consist of three parts. That part of the tooth located above the gum is called the **crown;** the portion surrounded by gum is called the **neck;** and the part of the tooth that is embedded in the socket of the jaw is the **root.**

The internal structure of a tooth consists of four sections, enamel, dentine, pulp and cement, each with a different function. The outer surface is covered by a layer of **enamel.** The enamel covers only the crown of a tooth and the upper part of its neck. Enamel is the hardest substance in the body; it is 97% inorganic matter, mainly small crystals of calcium phosphate, and 3% organic material, mainly protein. Beneath the enamel is found a layer of **dentine.** It composes the main body of the tooth, and although it is stronger than the bones of the skeleton, dentine is not as strong as enamel. Dentine consists of 72% inorganic and 28% organic material. The innermost section of a tooth is called the **pulp.** In it are found the nerves and blood vessels that function to maintain the tooth in the jawbone. The tooth is fastened to the jawbone by a substance called **cement,** which is secreted by the cells of a membrane, called the **periodontal membrane,** that lines the tooth socket.

The decay and infection of our teeth and gums represent a considerable problem in our modern society. It has been estimated that Americans now spend about $2 billion annually repairing diseased teeth and gums. But the complete repair of the damaged teeth and gums of all the people in the United States would cost $8 billion more annually than we now spend. At the present time, the average child reaching school age has three decayed teeth. Of the children older than 14 years of age, 80% have an average of 11 cavities in their teeth, and 75% to 80% have some form of gum disease. When we examine older individuals, we find that cavities tend to occur less

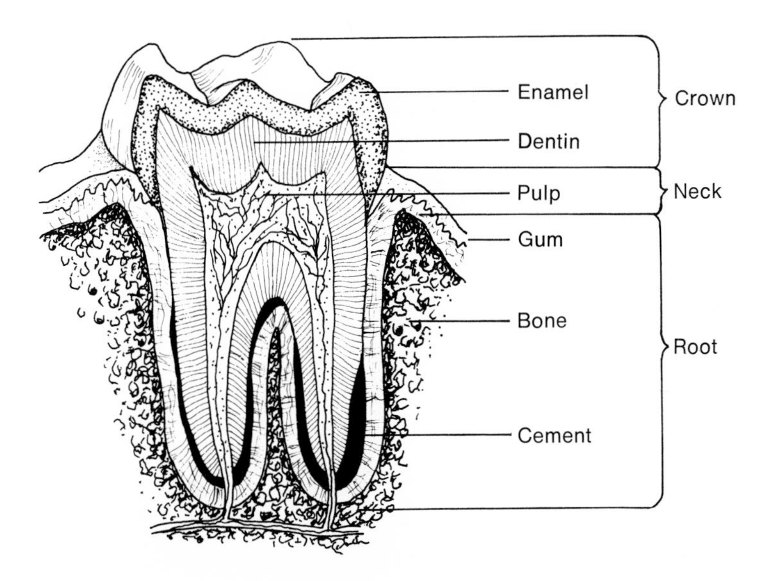

Fig. 5-5
Parts of molar tooth.

frequently, although gum disease becomes more prevalent. Two thirds of the people 35 years of age or older have serious gum problems.

What causes the formation of dental cavities, also called **caries?** It is generally believed that acids produced by acid-producing bacteria in the mouth cause erosion first of the enamel and secondarily of the dentine. If untreated, the destruction of the tooth continues, often leading to infections of the surrounding tissues. The acids that destroy the teeth are produced as by-products of the bacterial metabolism of dietary carbohydrates. Studies of the diets of different human populations and the incidence of dental caries in these same populations indicate that the disaccharide sucrose, our ordinary table sugar, is the greatest single cause of tooth decay (Fig. 5-6). It will be recalled that sucrose is formed as a result of the fusion of glucose and fructose. Of these two monosaccharides, fructose appears to be mainly responsible for the high

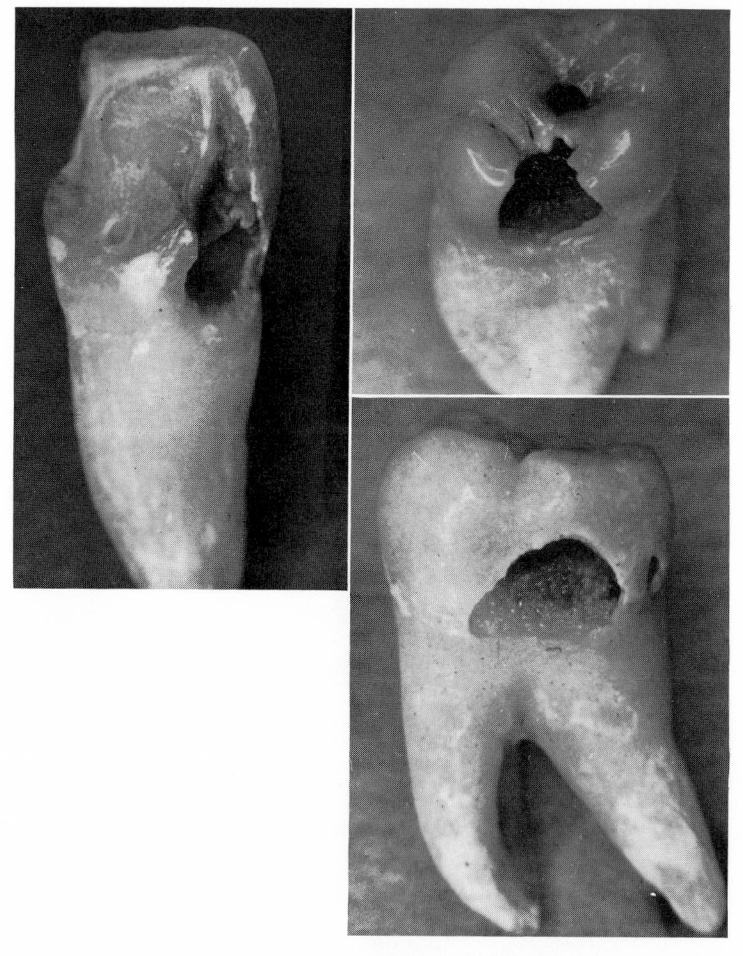

Fig. 5-6
Three examples of caries caused by neglect that commonly result in loss of the tooth because of pulp degeneration. (From Gilmore, H. W., and Lund, M. R. 1973. Operative dentistry, ed. 2. The C. V. Mosby Co., St. Louis.)

incidence of dental caries. The evidence for this statement comes from a study of people suffering from **hereditary fructose intolerance,** which makes them extremely sensitive to fructose. These individuals become violently ill if they eat even small amounts of sucrose or fruits containing fructose. Consequently, they tend to avoid sweet foods of all kinds and obtain their carbohydrates from starchy foods instead. (Starches are polysaccharides composed solely of glucose molecules.) It has been found that people suffering from this hereditary intolerance have far fewer dental caries than the rest of the population.

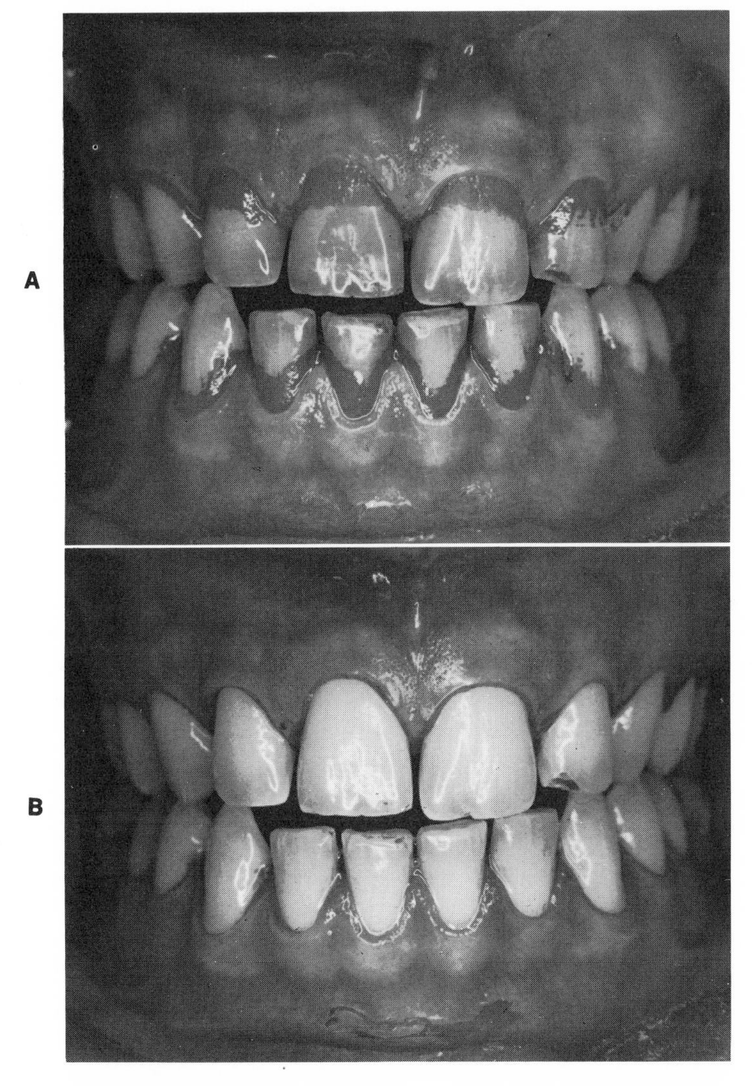

Fig. 5-7
A, Rinsing with 6% basic fuchsin reveals plaque not visible on unstained teeth. **B,** After cleansing teeth, mouth rinsed again with 6% basic fuchsin indicates that plaque has been completely removed. (From Bernier, J. L., and Muhler, J. C., eds. 1970. Improving dental practice through preventive measures, ed. 2. The C. V. Mosby Co., St. Louis.)

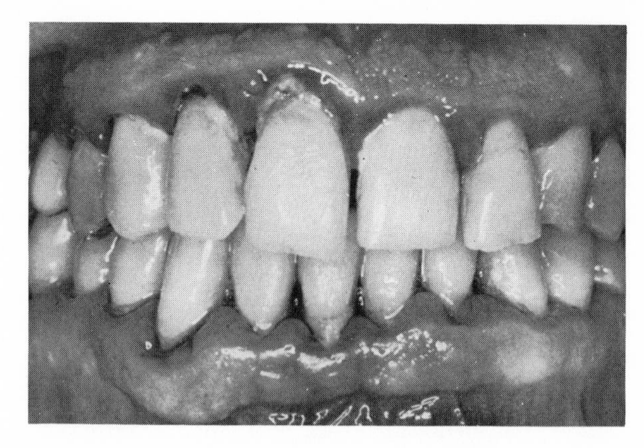

Fig. 5-8
Typical case of periodontitis. (From Grant, D. A., Stern, I. B., and Everett, F. G. 1972. Orban's periodontics—a concept—theory and practice, ed. 4. The C. V. Mosby Co., St. Louis.)

In our discussion of dental caries, we must examine how the bacteria are able to remain on the teeth long enough for cavities to form. Bacteria produce a sticky substance that binds both themselves and food particles to the teeth. This sticky, almost invisible film is called **plaque.** Plaque is formed continuously and should be removed each day through the use of dental floss and vigorous brushing of the teeth (Fig. 5-7).

In adults over the age of 35 years, **periodontal disease,** not dental caries, is the principal cause of tooth loss. This disease is caused by bacterial enzymes and toxins that attack and destroy the cells of the gums. In its earliest stage, periodontal disease is known as **gingivitis,** or inflammation of the gums. If the disease is allowed to persist, the gum will become separated from the tooth, and a crevice will be formed between gum and tooth. The destruction of the gum is progressive and will include the area around the root of the tooth and may even involve the bone of the jaw. If the disease is not treated, loss of the teeth is inevitable. Treatment of periodontal disease usually involves cutting away loose and inflamed gum tissue around the teeth and instituting rigorous home dental care (Fig. 5-8).

□ Adding enzymes

The chemical breakdown of food begins in the mouth. However, only the polysaccharides are involved, and these are broken down by the enzyme **salivary amylase** (also called **ptyalin**) into the disaccharide maltose. In addition to exposure to salivary amylase, food in the mouth is mixed with **mucus,** which helps to moisten the food and lubricate its passage through the pharynx and esophagus. Salivary amylase and mucus, together with a large amount of water, make up the secretion we call **saliva.**

Saliva is produced by three pairs of salivary glands that open into the mouth (Fig. 5-1). The largest salivary gland, the **parotid gland,** produces a type of saliva that is composed mainly of water and amylase. The same type of secretion is produced by the **submaxillary gland.** However, the **sublingual gland** secretion is mainly mucus.

Saliva is normally slightly acid (pH 6.0 to 7.0). Salivary amylase works well in this pH range, but the enzyme is inactivated by the extremely acid condition of the stomach. As a result, its action on starch is of short duration.

□ Shaping food for swallowing

After the ingested food has been chewed into particles of proper size and mixed with enzyme, mucus, and water, it is shaped into a bolus and swallowed. Much of the shaping of a bolus is done by the tongue, using the hard palate portion of the roof of the mouth to help do the job.

During the act of swallowing, the tip of the tongue is raised and pressed against the hard palate just behind the upper front teeth. The upper surface of the tongue forms a trough into which the food or water is channeled. The posterior end of the tongue is lowered while the middle portion of the tongue is raised, forcing the food or water backward through the mouth and downward into the pharynx.

■ PHARYNX

The pharynx is the cavity behind the mouth where the digestive and respiratory passages cross. In addition to the mouth cavity, six other structures connect with the pharynx. These include (1) two internal openings (**nares**) from the nasal cavity, (2) two **eustachian tubes** leading to the cavities of the middle ears and functioning to equalize pressure on the ear drums, (3) the **glottis** opening into the larynx (the voice box, located at the upper end of the trachea, the main respiratory tube), and (4) the **esophagus** leading to the stomach.

When food is swallowed (Fig. 5-9), it is first pushed into the pharynx by movements of the tongue. In passing through the pharynx, the food initially crosses the pathway for air, which extends from the nasal cavity to the glottis. Food is prevented from returning to the mouth by the raised middle portion of the tongue. The soft palate, including the **uvula** (Fig. 5-9), is elevated, thereby closing the passage between the nose cavity and the pharynx. Food is prevented from entering the glottis by raising the larynx so that the glottis is pressed against both the base of the tongue and a flap of tissue called the **epiglottis.** The raising of the larynx can be observed in the movement of the "Adam's apple" (the outer surface of the larynx) each time we swallow. A person cannot swallow and inhale at the same time.

It takes 1 to 2 seconds for the food bolus to pass through the pharynx and into the esophagus. Any chemical digestion of the food that occurs during this short time interval is a continuation of the action of the salivary amylase that was mixed with the food in the mouth.

■ ESOPHAGUS

Up to this point, the digestive tract structures that we have studied (mouth and pharynx) have been "cavities." Beginning with the esophagus, the tubular nature of the alimentary canal becomes apparent. The esophagus conducts food from the pharynx to the stomach. This is accomplished by a series of waves of contractions and relaxations, called **peristalsis,** as shown in Fig. 5-10. It takes approximately 5 to 10 seconds for a wave of peristalsis to pass from the pharynx to the stomach, the length of the esophagus being about 25 centimeters (cm) (10 inches).

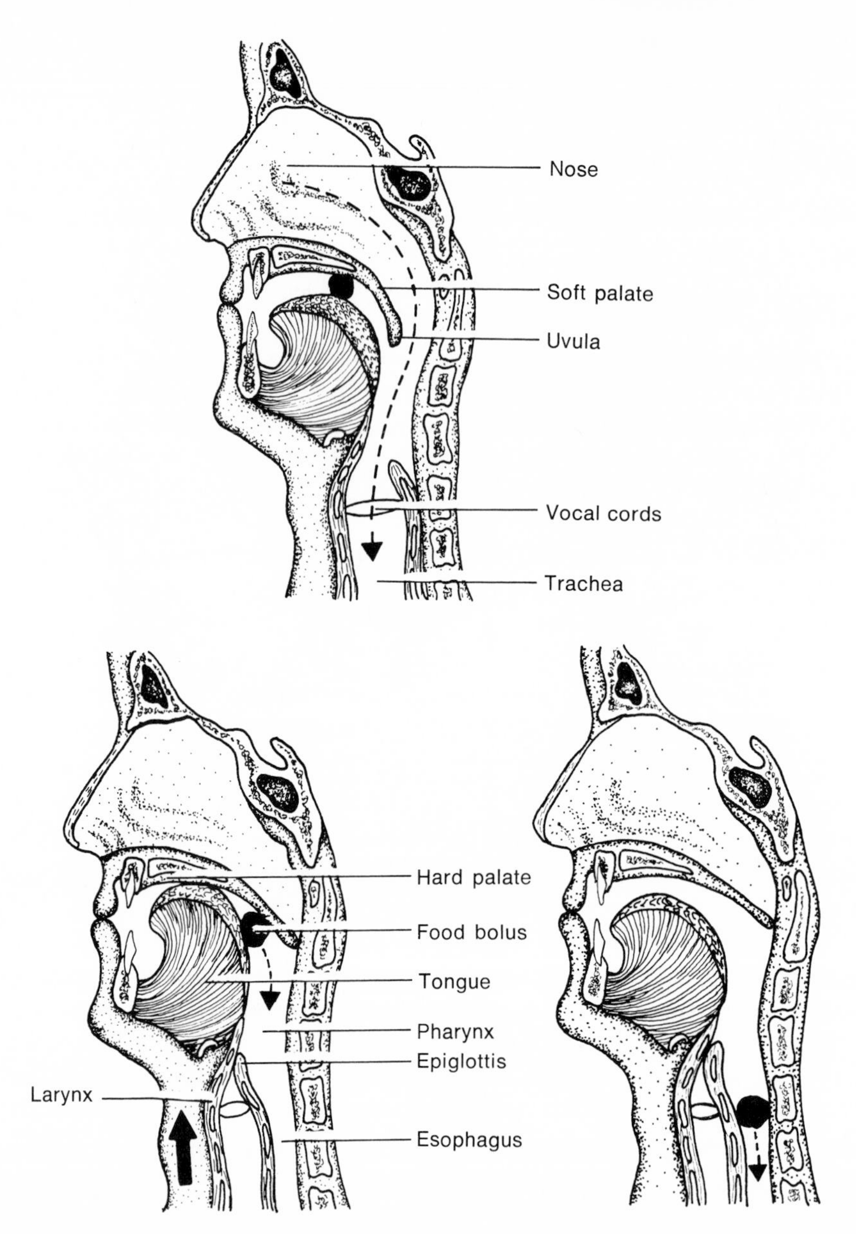

Fig. 5-9
Position of various parts of mouth and pharynx while swallowing a food bolus.

At the lower end of the esophagus, the muscles of the esophagus are formed into a ring called the **cardiac sphincter.** The cardiac sphincter controls the opening of the esophagus into the stomach and opens only when food is in the lower end of the esophagus. When closed, the cardiac sphincter serves to prevent food in the stomach from entering the esophagus. When food is not being swallowed, the esophagus is a collapsed tube.

As stated previously, in the esophagus, as in the pharynx, the only chemical digestion is a continuation of salivary amylase enzyme activity. The only esophageal

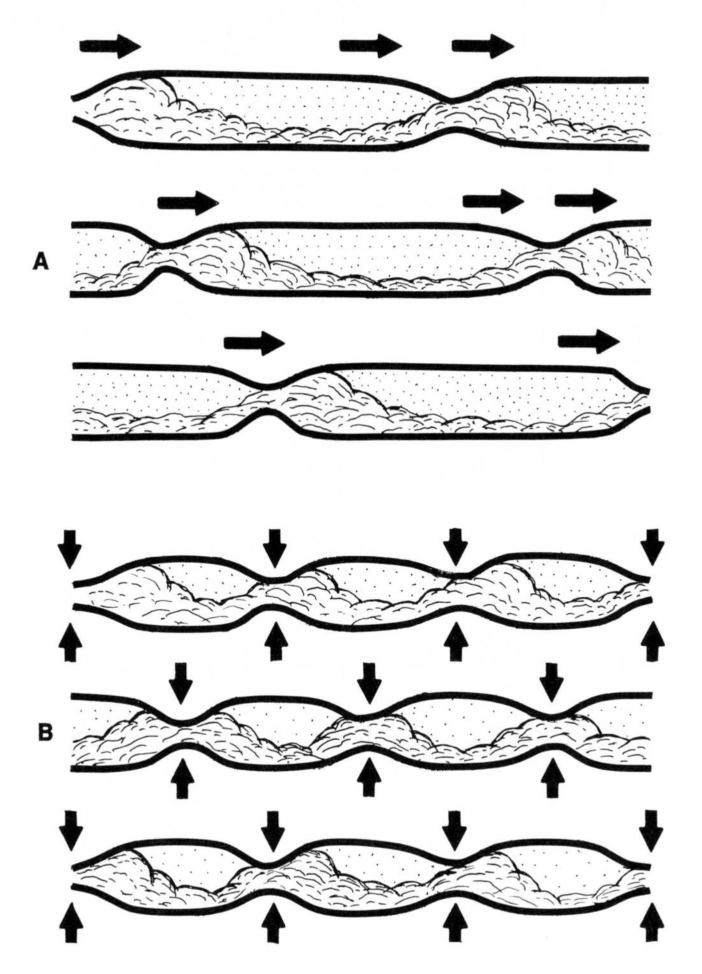

Fig. 5-10
Sequence of muscle contractions during, **A**, peristalsis and, **B**, segmentation contractions.

secretion is mucus. It provides lubrication for swallowing and protection for the lower end of the esophagus against the corrosive effect of gastric juice that periodically enters from the stomach, causing "heartburn."

■ STOMACH

The stomach performs many functions in the digestive process. The stomach (1) receives swallowed food from the esophagus; (2) stores it until it can be accommodated by the lower portions of the digestive tract; (3) produces "gastric juice," which consists mainly of mucus, hydrochloric acid, and the enzyme pepsin; (4) mixes the food with the gastric juice to form a milky, semifluid mixture called **chyme;** (5) performs a limited amount of absorption of water and certain drugs into the bloodstream; and (6) moves the chyme into the small intestine at a rate suitable for continued digestion and absorption by the small intestine.

□ Parts of the stomach

Based on structural and functional characteristics, the stomach may be considered as consisting of two sections (Fig. 5-11), although there are no sharp lines of demarca-

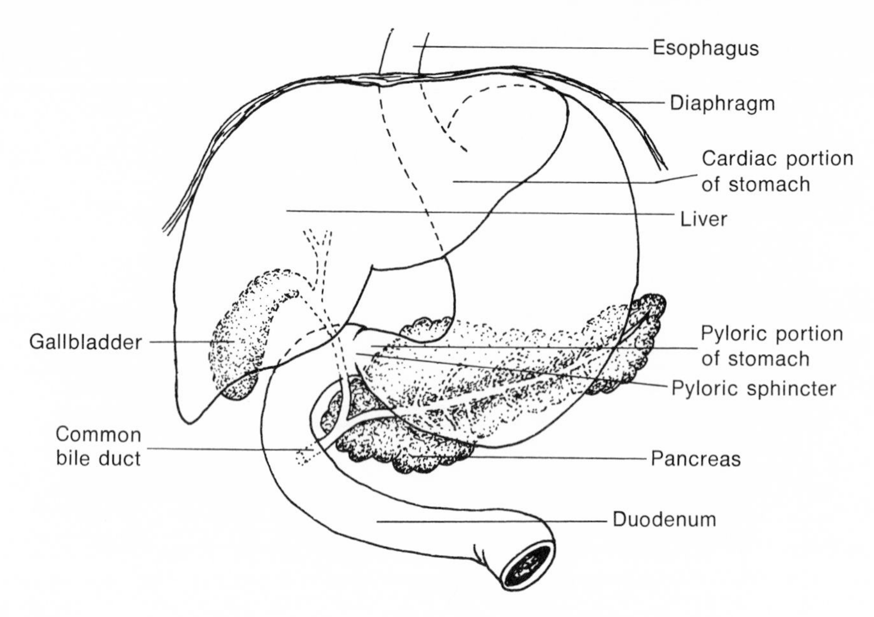

Fig. 5-11
Stomach, duodenum, and associated digestive glands. Dotted lines indicate position of structures located underneath other organs.

tion separating these regions. The region nearest the esophagus is called the **cardiac** portion. Food entering the stomach forms concentric circles in the cardiac portion. The contraction of the muscles in this region tends to mix the food with the gastric juice and to move it down into the lower portion of the stomach, called the **pylorus.** In the pylorus, the further mixing of food and gastric juice occurs until chyme is formed and moved into the small intestine.

□ Chemical digestion of food

Each of the components of gastric juice is secreted by a different type of cell in the inner lining of the stomach. One type of cell produces **hydrochloric acid** (HCl). HCl has a very low pH and, even in a mixture of water, food, etc., gives the stomach contents a very acid pH of 2.0 to 3.0. This pH is much too low for the continued activity of salivary amylase, which therefore ceases when the salivary amylase reaches the stomach. However, the acid contents of the stomach are optimum for the action of the gastric juice enzyme **pepsin.**

Pepsin is the principal enzyme produced by the cells of the stomach. The enzyme is synthesized inside the cells in the form of **pepsinogen,** which has no digestive activity. However, once pepsinogen is released from a cell and comes into contact with HCl, it is immediately transformed into pepsin. This enzyme breaks down proteins into variously sized peptide chains (polypeptides). The further digestion of peptides into amino acids occurs later in the small intestine.

Ever since it was demonstrated that pepsin could digest protein, biologists have raised the question: Why does the stomach not digest itself? One answer is that sometimes it does undergo autodigestion. Under certain circumstances, the gastric

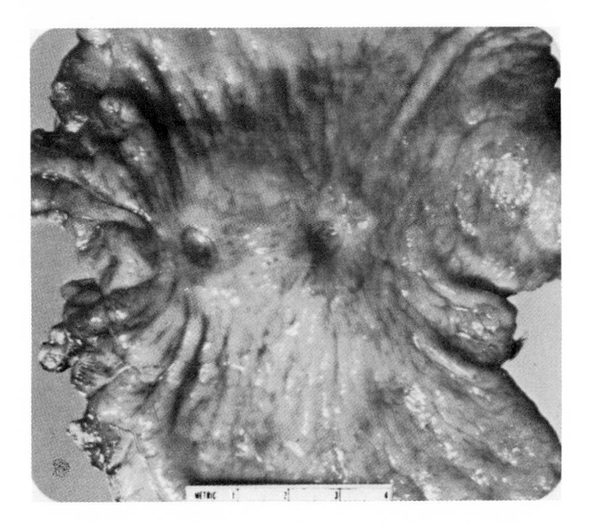

Fig. 5-12
Inner lining of stomach, showing two circular ulcers on otherwise smooth surface. (From Anderson, W. A. D., and Kissane, J. M., eds. 1977. Pathology, ed. 7. The C. V. Mosby Co., St. Louis.)

juice can destroy cells of the inner lining of the stomach and produce an **ulcer** (Fig. 5-12). In our modern society, about 1 person in 100 has a stomach ulcer. Ulcers are more frequent among white-collar workers than in laborers. People who are subjected to extreme anxiety for a long time seem particularly prone to ulcers. It has been found that ulcer patients have a high rate of gastric secretion during the period between meals, when the stomach is empty. These people tend to secrete 15 to 20 times more gastric juice than occurs normally. This huge amount of gastric juice overwhelms the stomach's normal protective mechanisms and produces the equivalent of open sores on the inner lining of the stomach.

What are the normal protective mechanisms of the stomach against self-digestion? There are two protective devices. One is preventive in nature and involves the secretion of **mucus** by a large number of cells in the stomach lining. The mucus forms a continuous layer over the stomach cells and serves to prevent the HCl and pepsin from coming in contact with the cells. The other protective device is remedial; it involves the rapid and continuous shedding of the uppermost cells of the stomach lining and their replacement by new cells produced within the stomach wall. The human stomach normally sheds about half a million cells per minute, thereby renewing the entire inner lining every 3 days. About 25% of the dry weight of each bowel movement is composed of dead stomach and intestinal cells. Through this continuous renewal of its inner surface lining, the stomach in effect automatically repairs any minor damage that may occur to its cells as a result of the digestion process.

□ Absorption from the stomach

Relatively few ingested substances can be absorbed directly from the stomach into the circulatory system for distribution to the rest of the body. Those that can include water and, unfortunately, certain drugs. The most important of these drugs is **alcohol.** When alcohol is taken into the body, about one fifth of it passes directly into the circulatory system from the walls of the stomach, and the remainder is absorbed from the small intestine. Alcohol is not broken down by digestion, but is transported, as alcohol, to the various tissues of the body. The chief effect of alcohol is that of a narcotic on the central nervous system. The seriousness of this problem is obvious from the fact that at the current time, about 60,000 people are killed and about 500,000 people severely injured yearly in automobile accidents in the United States, more than half of which involve drunken drivers.

A problem of lesser magnitude is **aspirin** (acetylsalicylic acid), which also passes through the cells of the inner lining of the stomach and enters the circulatory system. In passing through the stomach cells, aspirin destroys them and in so doing permits HCl to penetrate the area and cause bleeding. The blood loss after taking two aspirin tablets usually amounts to between 0.5 and 2.0 ml. This amount of blood loss is quite negligible in relation to the overall well-being of the individual. However, there have been cases in which habitual aspirin users have developed anemia by losing blood at a rate higher than the body's production of red blood cells.

Emptying of the stomach

After the food bolus has been thoroughly transformed into chyme by the action of HCl and pepsin, it is moved by peristalsis into the lower end of the pyloric region of the stomach. At the point of juncture of the stomach and small intestine is a ring of muscle called the **pyloric sphincter.** As the waves of peristalsis increase in intensity, they drive the chyme to the pyloric sphincter, which opens periodically for short increments of time, permitting gradual passage of the stomach's contents into the small intestine.

The rate at which food is emptied from the stomach is controlled by the small intestine through the production of the hormone **enterogastrone,** which regulates the contractions of the stomach. The rate of emptying the stomach is slowed when too much food is already in the small intestine, when the chyme is excessively acid, or when there is too much lipid in the food. Among the different foods that affect the rate of emptying the stomach, lipids have the greatest inhibitory effect, sometimes delaying the passage of a lipid-rich meal for as long as 3 to 6 hours. Proteins have an intermediate effect, and carbohydrates have only a mild delaying effect.

On leaving the stomach, the chyme consists of undigested carbohydrates, lipids, and proteins and, in addition, maltose and peptide molecules, which result, respectively, from the partial digestion of starches by salivary amylase and proteins by pepsin.

SMALL INTESTINE

The small intestine (Fig. 5-1) is a coiled tube about 20 feet in length. It consists of three regions. The first section, called the **duodenum,** is 1 foot long and functions mainly in digestion. Although some absorption of digested material occurs in the duodenum, most absorption occurs in the remainder of the small intestine. The second segment, the **jejunum,** measures approximately 8 feet in length. In it, the digestive process continues, and an increased amount of absorption takes place. The last region, called the **ileum,** is 11 feet long. The digestive and absorptive processes are completed here, except for the absorption of some water and vitamins.

Our discussion of digestion up to this point has revealed that the enzymatic processes that occur in the mouth and stomach do not result in the formation of the basic molecules (monosaccharides, glycerol, fatty acids, and amino acids) that are usable by the body cells. It is only in the small intestine that these usable molecules are formed.

The enzymes whose actions complete the digestive process come from the pancreas, aided in part by a secretion called bile from the liver, and from the cells of the intestine itself. In contrast to pepsin, all the enzymes that function in the intestine require an alkaline medium for their activity. The alkaline medium is provided mainly by the juice secreted by the pancreas, which has a high sodium carbonate content and a pH of 8.5. Pancreatic juice is the major means by which the acidity of stomach chyme is neutralized and made alkaline.

□ Pancreatic juice enzymes

Pancreatic fluid is a complete digestive juice because it contains enzymes that split carbohydrates, lipids, and proteins. The carbohydrate-splitting enzyme is **pancreatic amylase.** It catalyzes the cleavage of polysaccharides into maltose. As we shall see later, the production of monosaccharides is the job of enzymes secreted by the intestinal wall cells.

The lipid-splitting enzyme is **pancreatic lipase.** It breaks down lipids into glycerol and fatty acids. The work of pancreatic lipase is facilitated by the **liver** secretion **bile.** Bile is produced in all parts of the liver, and most of it is collected by a system of ducts that lead into the **gallbladder,** where it is stored until needed after a meal. From the gallbladder, the bile is carried through the **common bile duct** to the duodenum. A small amount of bile is continuously transported directly from the liver to the duodenum through secondary ducts.

In the duodenum, the bile breaks up large lipid globules into very small ones, thus providing an overall greater surface area on which the enzyme pancreatic lipase can act. Bile, similar to pancreatic juice, is alkaline and helps to neutralize the acid chyme coming from the stomach. Most of the pancreatic juice enters the duodenum through ducts that join the common bile duct (Fig. 5-11). However, as in the case of the liver, there are a number of secondary ducts that carry a small portion of the digestive juice directly from the pancreas to the duodenum.

The pancreas produces three protein-splitting enzymes. Two of them, **trypsin** and **chymotrypsin,** break down proteins into peptide chains of various lengths. In their action, they resemble the stomach enzyme pepsin. The third pancreatic protein-splitting enzyme is called **carboxypeptidase.** This enzyme starts at one end of a peptide chain (the end with a free COOH group) and splits off one amino acid at a time. In doing so, carboxypeptidase completes the process of protein digestion, making amino acids available for use by the body.

One of the problems that can arise in the digestive process is a blocking of the common bile duct by **gallstones** (Fig. 5-13). These "stones" are hard little pellets formed by the fusion of varying amounts of the following constituents of bile: calcium salts, bile pigments, and cholesterol. When the common bile duct is blocked, large

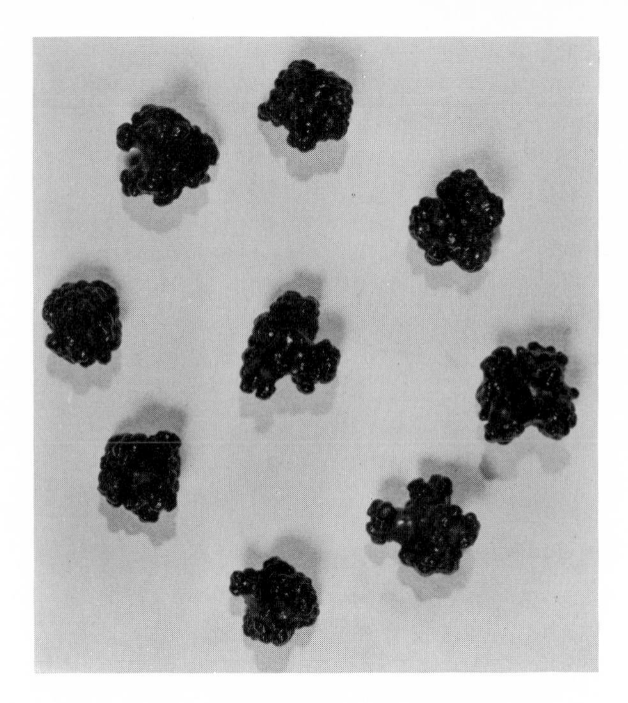

Fig. 5-13
Gallstones removed from obstructed gallbladder. (From Anderson, W. A. D., and Kissane, J. M., eds. 1977. Pathology, ed. 7. The C. V. Mosby Co., St. Louis.)

amounts of both bile and pancreatic juice are prevented from entering the duodenum. Without a sufficient flow of bile and pancreatic juice, as much as three quarters of the lipid material entering the small intestine remains undigested, as does a third to a half of the proteins and polysaccharides. As a result, large portions of the ingested food are not absorbed, and copious, fatty feces are excreted.

When the common bile duct is blocked by one or more gallstones, pressure builds up on the bile in the liver and gallbladder, forcing the bile into the bloodstream. The bile is then distributed throughout the various body tissues and fluids. In the skin and eyeballs of an affected individual, the pigments of the bile produce a yellow color, a condition known as **jaundice.** If the blockage of the common bile duct is not correctable through medication, it is necessary to operate and remove both the bile duct and the gallbladder (Fig. 5-14). After this type of operation, bile and pancreatic juice enter the duodenum only through the secondary ducts, mentioned earlier in this discussion, which become enlarged in response to their greater use.

Thus far we have described the digestive processes that take place in the duodenum as a result of the actions of the enzymes of pancreatic juice. The activities of these enzymes result in the formation of the basic molecules of fat digestion (glycerol and fatty acids) and protein digestion (amino acids). However, there still remains the completion of carbohydrate digestion plus, as we shall see, an auxiliary pathway that aids in protein digestion. These latter aspects of the overall digestive process are accomplished by enzymes of the intestinal juice.

Fig. 5-14
Surgically removed gallbladder that has been opened to show complete filling of organ by gallstones. (From Anderson, W. A. D., and Kissane, J. M., eds. 1977. Pathology, ed. 7. The C. V. Mosby Co., St. Louis.)

Intestinal juice enzymes

The inner lining of the duodenum contains cells that secrete digestive enzymes. These cells produce three disaccharide-splitting enzymes: (1) **maltase,** which hydrolyzes maltose to form glucose molecules; (2) **sucrase,** which divides sucrose into its component simple sugars glucose and fructose; and (3) **lactase,** which splits lactose into the monosaccharides glucose and galactose. Maltase completes the digestive process begun by the salivary and pancreatic amylases, and sucrase and lactase break down the two most common disaccharides contained in our diets.

Another enzyme produced by cells of the intestinal lining is the peptide-splitting enzyme **aminopeptidase.** As suggested by its name, this is an enzyme that splits off amino acids located at the amino (NH_2) end of a peptide chain. It acts in a manner similar to carboxypeptidase from the pancreas, but at the other end of a polypeptide. Aminopeptidase and carboxypeptidase enzyme molecules, working at their respective ends of the peptide chains, speed the protein digestion process to completion. A summary of the various digestive enzymes, their substrates, and products is given in Table 5-1.

After the digestion of ingested food into its basic molecules, there still remain two tasks for the digestive system: the absorption of the now available nutrients and the evacuation of unusable residues. These functions are performed by the remainder of the small intestine, the large intestine (colon), the rectum, and the anus.

Absorption of digested nutrients

Most of the absorption of digested nutrients occurs in the jejunum and ileum of the small intestine, although some does occur in the duodenum. The absorptive surface of the intestine has many folds, which serve to increase the surface area. In addition, literally millions of small fingerlike projections, called **villi,** are located over the entire surface of the small intestine (Fig. 5-15). The villi project about 1 millime-

Table 5-1
Digestion

Site	Secretion	Enzymes	Substrates	Products
Mouth	Saliva	Salivary amylase	Polysaccharides	Maltose
Stomach	Gastric juice	Pepsin	Proteins	Polypeptides
Small intestine	Pancreatic juice	Pancreatic amylase	Polysaccharides	Maltose
		Lipase	Lipids	Glycerol and fatty acids
		Trypsin	Proteins	Polypeptides
		Chymotrypsin	Proteins	Polypeptides
		Carboxypeptidase	Polypeptides	Amino acids
	Intestinal juice	Aminopeptidase	Polypeptides	Amino acids
		Maltase	Maltose	Glucose
		Sucrase	Sucrose	Glucose and fructose
		Lactase	Lactose	Glucose and galactose

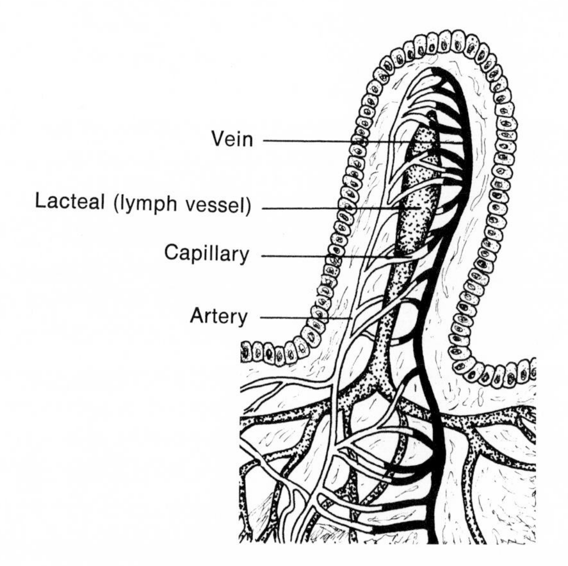

Fig. 5-15
Structure of a villus.

ter (mm)* from the surface of the inner lining and are the main organs of absorption of the small intestine.

Absorption involves the passage of digested nutrients into and through the cells of the inner lining of the intestine. On leaving the cells, the food passes into vessels of the circulatory system for distribution throughout the body. The mechanism by which nutrients leave the intestinal tract and enter its cells depends on the relative concentrations (number of molecules per unit volume) of the various types of molecules in both locations.

---

*1 mm = 1/25 inch.

Table 5-2
Daily secretion of digestive juices

Source	Amount (ml)	pH
Saliva	1200	6.0-7.0
Stomach secretions	2000	1.0-3.5
Small intestine secretions	3000	7.8-8.0
Pancreatic secretions	1200	8.0-8.3
Bile	700	7.8
Large intestine secretions	60	7.5-8.0
Total	8160	

When the concentration of a particular substance is greater in the intestinal cavity than in the surrounding cells, the substance passes into the cell without any expenditure of energy on the part of the cell. This process is called **diffusion,** and the difference in concentration of a particular substance in the intestinal contents and in the cells is called a **diffusion gradient.** However, certain molecules are absorbed even when their concentration is greater in the cells than in the intestinal contents. Under these conditions, the cell expends energy and does work in moving these molecules against a diffusion gradient. This process is called **active transport.** In fact, most of the nutrients are absorbed from the intestinal cavity by active transport and thus do not depend on the diffusion gradient. Water and some mineral elements pass into the cells of the intestinal lining by diffusion.

After leaving the intestinal cells, the nutrients enter the circulatory system. Most nutrients enter the capillaries of the villus (Fig. 5-15). Glycerol and fatty acids, however, follow a different path. These substances are resynthesized into lipids by the cells of the villus and are released in the form of droplets, which are too large to enter the capillaries. The lipids are taken up by the **lacteals,** which are lymph vessels. The lymph vessels eventually join one of the large veins of the body. By this mechanism, lipids are transported throughout the body along with all other nutrients.

The total quantity of liquid that must be absorbed each day from the digestive tract is equal to the ingested liquid plus that contained in the various digestive system secretions. A normal adult consumes about 1.5 liters of liquid and secretes about 8.1 liters of digestive juices every day (Table 5-2). Of this total of 9.6 liters of fluid in the small intestine, about 95% is absorbed in the jejunum and ileum, leaving only 5% to pass into the large intestine.

□ Emptying of the small intestine

Food in the small intestine is moved along by **peristalsis,** as shown in Fig. 5-10. However, there is a second type of movement, called **segmentation contractions,** that serves to mix the food with the digestive secretions present in the small intestine. During segmentation contraction (Fig. 5-10), the muscles of a small section of the intestine contract simultaneously at a number of different, equally spaced points. The net effect of this type of muscle action is to temporarily divide the small intestine

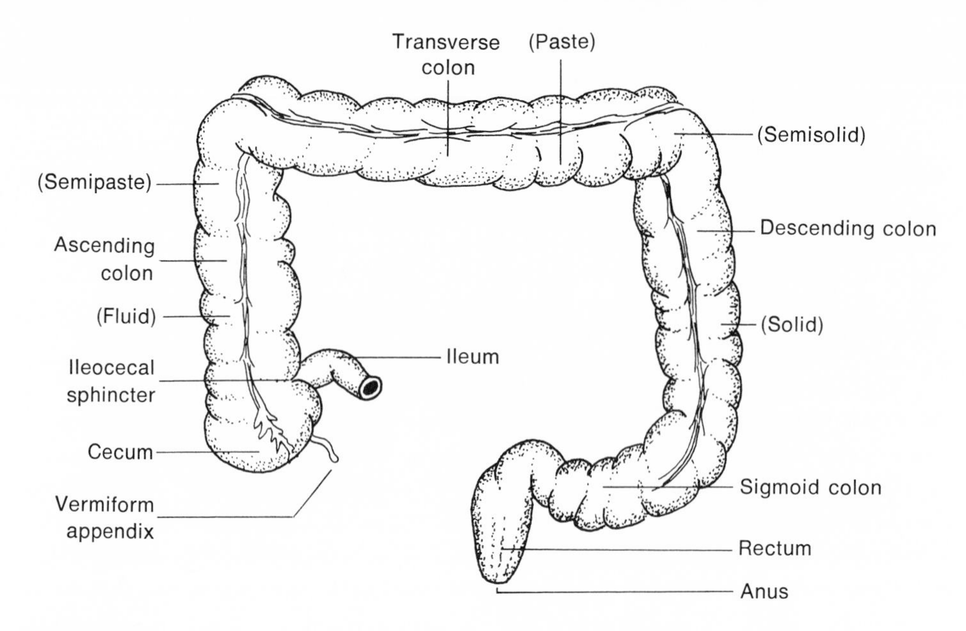

Fig. 5-16
Large intestine, rectum, and anus. Words in parentheses refer to consistency of waste material in large intestine.

into regularly spaced segments. Subsequently, this action is repeated at a different set of contraction points each time. In a given section of the small intestine, these successive segmentation contractions may occur eight or nine times each minute, keeping the food and digestive juices thoroughly mixed together.

The rate of peristalsis of the small intestine is quite slow. A normal meal requires about 3 to 10 hours to pass through. At the point of junction of the small and large intestines, there is a thickening of the muscles of the intestinal wall to form the **ileocecal sphincter** (Fig. 5-16). This sphincter regulates the passage of material from the small to the large intestine. The degree of contraction of the ileocecal sphincter is controlled primarily by the **cecum,** the first section of the large intestine. When the cecum is empty, the ileocecal sphincter is relaxed, and the reverse occurs when the cecum is filled. The material entering the large intestine consists largely of indigestible foods (cellulose, etc.), large numbers of dead stomach and intestinal lining cells, bile pigments, and water.

LARGE INTESTINE, RECTUM, AND ANUS

The **large intestine,** or **colon,** is about 5 feet in length and has the general shape of an inverted U (Fig. 5-16). It is divided into the ascending, transverse, descending, and sigmoid colons. The terminal 6 inches of the digestive tract is a straight tube, called the **rectum,** that ends with the **anus.** The large intestine, as its name implies, is larger in diameter and has thicker walls than the small intestine. The small intestine empties into the side of the colon a short distance above the lower end of the ascending colon. Below the point of entrance of the small intestine, the ascending colon forms a blind sac, called the **cecum,** at the tip of which is a small projection, the appendix, which is about the size of the little finger. Man's appendix is of no known use to him. Unfortunately, it can become infected and must then be removed by a surgical operation (appendectomy).

Most of the digested nutrients have been removed from the material reaching the colon, which performs two additional functions: (1) absorption of water and mineral elements from the chyme and (2) storage of fecal material until it can be expelled. In general the ascending and transverse colons are principally involved in absorption, and the descending and sigmoid colons are mainly concerned with temporary storage.

□ Absorption of water and mineral elements

Absorption of materials by the cells of the inner lining of the large intestine is relatively slow. It is therefore not surprising to find that the colon has no villi. Approximately 500 milliliters (ml) of chyme pass through the ileocecal sphincter into the large intestine each day. Most of the water and mineral elements in the chyme are absorbed in the colon, leaving only about 100 ml of fluid to be evacuated in the feces.

Large numbers of bacteria are also found in the large intestine. These organisms live on the undigested food that reaches the colon. The significance of these bacteria in the life of an individual is not clearly understood. However, the presence of these organisms, called **coliform bacteria,** in the water used for drinking, bathing, etc. indicates fecal contamination.

The only secretion of the large intestine is mucus, which functions to protect the cells of the inner lining of the colon from the bacteria that live in this section of the digestive tract. Mucus also serves to hold the fecal material together, thus facilitating its evacuation.

Feces normally are about three-fourths water and one-fourth solid matter. The solid matter is composed of about 30% bacteria, 25% dead stomach and intestinal cells, 10% to 20% fat, 10% to 20% inorganic matter, 5% undigested food roughage, and 2% to 3% protein. The brown color of the feces comes from the bile that is added to the chyme in the duodenum. Human bile has a deep yellow-orange color, but in the intestine, the bile pigments undergo chemical alterations that change them to dark brown. When the bile duct is obstructed, the feces are whitish or clay colored.

□ Storage and evacuation of feces

Most of the feces are stored in the descending and sigmoid colons and moved into the rectum only just before or during defecation. The muscular movements of the large intestine somewhat resemble those of the small intestine. There are two types of movements: one serves to continuously mix the material in the colon, and the other moves the material along.

Mixing movements of the large intestine consist of a number of circular constrictions that occur simultaneously over a relatively short segment of the colon. This is followed by relaxation of the contracted muscles and a repetition of the process in an adjoining portion of the colon. The net effect of continuously mixing the contents of the large intestine is to expose the material to the cells of the inner lining, thereby ensuring maximum absorption of water.

Peristaltic movements consist of a series of consecutive muscle contractions, at adjoining points, that move the material in the large intestine toward the rectum and anus. These movements do not occur continuously throughout the day; they are most frequent for about 10 minutes during the first hour or so after eating breakfast and appear to a lesser extent after each meal.

The occurrence of peristaltic movements in the large intestine after a meal results from stimulation of the muscles of the colon by the nervous system. This stimulation takes place by means of a reflex called the **gastrocolic reflex.** Filling the stomach with food produces a reflex action that results in the emptying of the colon. When a mass of feces is moved into the **rectum,** resulting in its distention, an accompanying stimulation of the nerves in its walls brings about a feeling of fullness and a need to defecate. From 12 to 24 hours are required for digestive waste products to pass through the colon and rectum.

Defecation can be controlled to a large extent by the individual. If for some reason elimination of feces cannot occur, the rectum can store them for a considerable period of time. Children are usually trained to defecate only at certain times of the day and only in socially convenient situations. Adults also usually restrict their elimination of digestive wastes, either consciously or unconsciously, to a particular time in their daily pattern of activity.

The act of defecation depends partly on the contraction of muscles that the individual can control (voluntary muscles) and partly on some that he cannot (involuntary muscles). An individual consciously contracts the **voluntary muscles** of his abdominal wall to aid in the elimination of his feces and simultaneously relaxes the outer ring of muscles (sphincter) around the anal opening. Also involved in defecation is the contraction of the **involuntary muscles** of the large intestine and rectum.

□ Disorders of the large intestine

The most common disorders of the large intestine relate to the speed with which food passes through it. Slow passage of food results in excess absorption of water and the formation of hard feces, causing **constipation.** Rapid passage results in an insufficient absorption of water and the formation of loose feces, causing **diarrhea.**

Constipation is a very common phenomenon. Its most frequent cause is irregular bowel habits that develop through a lifetime of inhibition of normal defecation needs. Our modern, largely sedentary type of life tends to decrease intestinal motility and thus contributes to the high incidence of constipation. Also, our diets are usually low in indigestible plant fiber material (cellulose) and do not provide the **roughage** that is necessary to keep the muscles of our intestines active and strong.

Symptoms of constipation include a feeling of fullness, headache, and irritability. These effects appear to be caused by the presence of feces in the rectum. Experiments on human volunteers have shown that packing the rectum with cotton produces the same effects.

Constipation can usually be alleviated by combining the following: (1) proper diet to include a good deal of unprocessed cereals, such as wheat, corn, barley, and oats; (2) extra fluid intake; and (3) regular exercise, especially of the abdominal muscles.

Diarrhea is most commonly caused by an infection in the stomach or intestine or by nervous tension. In the case of infections, the lining of the infected portion of the digestive tract becomes irritated and secretes large quantities of water. This acts to dilute the irritating factors and to cause a rapid movement of the feces toward the anus, thereby washing the infectious agent out of the system.

In the case of nervous system–induced diarrhea, we find that during periods of greatly increased tension, such as at examination time, there is an increase both in peristaltic contractions and mucus secretion by the large intestine. In extreme cases, even the chyme in the small intestine is rapidly moved through the large intestine to the anus, causing the loss of extremely large quantities of water.

Depending on the cause of the diarrhea, the method of correcting the condition will vary. In the case of infections, medication may be required to rid the body of the infecting organism. Relief from nervous tension–induced diarrhea requires rest and freedom from worry and anxiety. In all cases of diarrhea, a bland diet of water, tea, broth, and nonirritating foods is recommended.

SUMMARY

Making our food usable requires that we reduce it in size to minute particles and break it down chemically to certain basic molecules. These two tasks are accomplished by our digestive system, which includes the digestive tract itself and various digestive glands.

The mouth functions mainly to test food and reduce it to manageable size. Although some chemical digestion does occur in the mouth, it is fragmentary and of short duration. The stomach functions mainly to store the relatively large quantities of food consumed at each meal. Here, too, chemical digestion is of secondary importance, although a considerable amount of protein digestion does occur. The digestive process is accomplished totally in the small intestine.

However, the food and water in our digestive tracts are of no value to us as long as they remain there. The digested nutrients must be absorbed from our intestines into the bloodstream and transported to the cells of the body if we are to use the nutrients for energy and growth. The function of food absorption is performed by the cells of the inner lining of the small intestine. The task of internal transportation is accomplished by our circulatory system, which we shall discuss in Chapter 6.

SUGGESTED READINGS

Brooks, F. P. 1970. Control of gastrointestinal function. Macmillan Publishing Co., Inc., New York. Short, well-illustrated discussion of human digestion.

Davenport, H. W. 1972. Why the stomach does not digest itself. Sci. Am. **226**:86-93. In-depth discussion of the safety mechanisms that protect the stomach lining from the actions of pepsin and hydrochloric acid.

Kalmus, H. 1958. The chemical senses. Sci. Am. **198**:97-106. Analysis of smell and taste as related to the testing of food.

Kmet, J., and Mahboubi, E. 1972. Esophageal cancer in the Caspian littoral of Iran. Science **175**:846-853. Study of the population with the highest known incidence of this disease and the factors that might be responsible for causing this type of cancer.

Kretchmer, N. 1972. Lactose and lactase. Sci. Am. **227**:70-78. Analysis of the loss of the ability to produce the enzyme that breaks down milk sugar as we grow older.

Neurath, H. 1964. Protein-digesting enzymes. Sci. Am. **211**:68-79. Discussion of enzyme structure and how these digestive enzymes function.

Scherp, H. W. 1971. Dental caries: prospects for prevention. Science **173**:1199-1205. Review of the role of certain foods in causing tooth cavities.

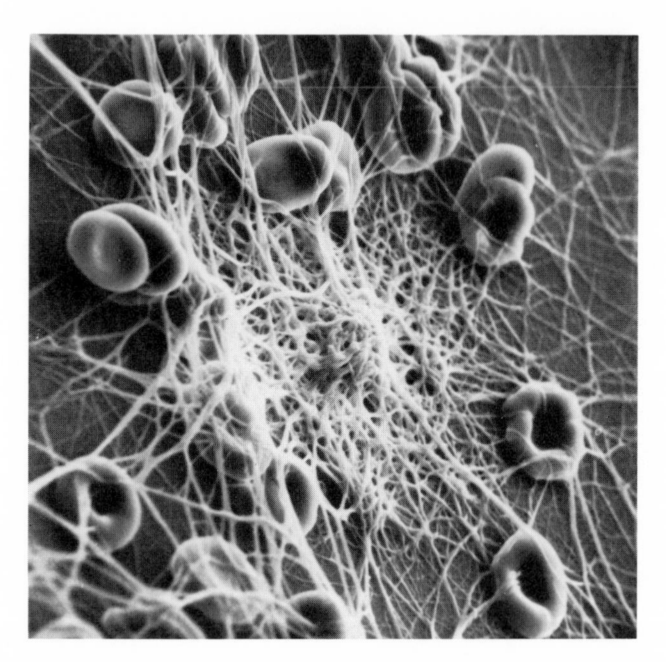

Scanning electron micrograph (×5180) of human red blood cells entrapped in fibrin clot. Clotting is initiated after tissue damage by disintegration of platelets in blood, leading to complex series of intravascular reactions that end with conversion of a plasma protein, fibrinogen, into long, tough, insoluble polymers of fibrin. Fibrin and entangled erythrocytes form blood clot, which arrests bleeding. Aggregation of platelets probably underlies raised mass of fibrin in center. (Courtesy N. F. Rodman, University of Iowa, Iowa City, Iowa; from Hickman, C. P., Hickman, F. M., and Hickman, C. P., Jr. 1974. Integrated principles of zoology, ed. 5. The C. V. Mosby Co., St. Louis.)

CHAPTER 6 Our internal transportation system

LEARNING OBJECTIVES

- What are the functions of our circulatory system?
- Of what three subdivisions does our circulatory system consist?
- What is the primary function of our red blood cells?
- What are the different types of anemia and the characteristics of each?
- What is the primary function of our white blood cells?
- What are the various types of white blood cells, and in what situation is each important?
- What are the different types of leukemia and the characteristics of each?
- What steps are involved in blood clotting?
- What three conditions can lead to uncontrolled bleeding from injuries?
- How do we develop immunity against infectious and poisonous agents?
- Which factor limits the successful transplantation of body organs from one person to another?
- In what way does the transfusion reaction involving ABO blood groups differ from that involving Rh antigens?
- What are the different types of chambers in the heart, and what does each do?
- What types of blood vessels do we have, and what does each do?
- What are the two main circulatory pathways in the body?
- Of what two phases does a heartbeat consist, and what controls it?
- How does exchange of materials occur between the circulatory system and the cells of the body?
- What is hypertension, and how does it contribute to cardiovascular disease?

Thus far we have studied the nutritional needs of man and the mechanisms by which we make our ingested food usable for energy and growth. This chapter deals with the way digested food is distributed to the cells of the body. Our internal transportation system not only distributes food, it also removes waste products from the cells, carries other materials (hormones, inactive enzymes) throughout the body, transports and exchanges substances, for example, oxygen and carbon dioxide, between the individual and the environment, and protects us from invasion by foreign particles and pathogenic organisms.

Our internal transportation system, called the **circulatory system,** consists of three subdivisions: (1) the pumping structure, or heart; (2) the conducting vessels, including arteries, capillaries, veins, and lymphatic vessels; and (3) the transportation medium, that is, blood and lymph. The transporting medium (blood and lymph) is the actual vehicle in which materials are carried throughout the body and will be discussed first. The heart and conducting vessels function to bring the blood and lymph from one group of cells to another.

BLOOD AND LYMPH

Blood consists of two separable components: a cellular portion and a liquid portion. The **cellular** portion is composed of red blood cells, white blood cells, and blood platelets. The liquid part, called **plasma,** consists of water, glucose, salts, and three types of proteins (albumins, globulins, and fibrinogen). The average human body contains about 5 liters of blood, of which 45% consists of cells and cell fragments, and 55% is plasma.

Lymph is the fluid that fills the spaces between cells. In order for a substance to reach the cells, it must pass from a blood vessel through the lymph. The reverse process must occur when wastes are discharged by cells. Lymph is very similar in composition to blood plasma, except that it contains much less protein.

We shall now examine the various types of cellular units found in blood (Fig. 6-1) and see how each performs its specific functions.

Red blood cells (erythrocytes)

In each cubic millimeter (mm^3) of blood, there are about 5.5 million red blood cells in males and about 5.0 million in females. Newborn infants have about 6 to 7 million erythrocytes/mm^3; this number decreases after birth, and the adult number is reached at about 3 months of age.

In the fetus, red blood cells are manufactured in the liver, spleen, and bone marrow. However, by the time the infant is born, red blood cell production is exclusively the job of the bone marrow. In its formation, each erythrocyte is initially a complete cell with a nucleus. (Cells and their nuclei will be discussed in detail in the next chapter.) But, as the erythrocyte completes its development, before it is re-

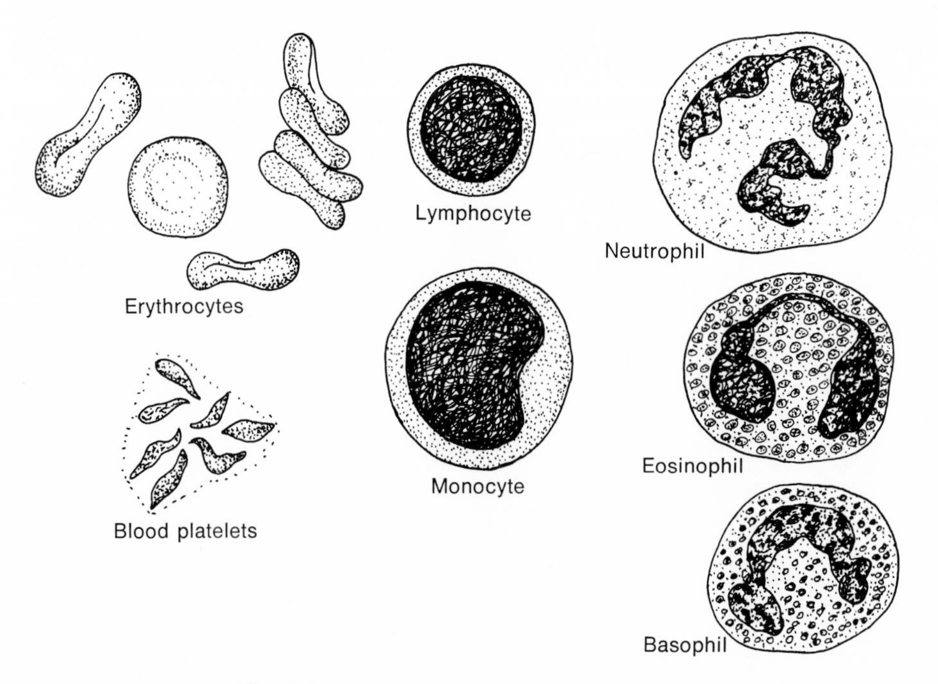

Fig. 6-1
Various types of cellular units found in blood.

leased into the bloodstream, the nucleus is extruded from the cell. *Thus all normal red blood cells lack nuclei.*

An erythrocyte has an average life span of 120 days. At the end of this time, it becomes fragile and breaks down. The fragments of the red blood cells are picked up by the spleen, liver, and bone marrow to be used either in the production of new erythrocytes or for other biochemical processes of the body. The limited life span of red blood cells requires that the body provide for the continuous production of these cells.

Erythrocyte function. The primary function of erythrocytes is the transportation of oxygen to the cells of the body. The availability of sufficient red blood cells for this important function is ensured by the fact that erythrocyte production is regulated by the amount of oxygen available to the cells of the body. When the amount of oxygen coming to the cells is low, the rate of red blood cell formation increases.

A reduced cellular level of oxygen occurs, for example, in persons living at very **high altitudes,** where the oxygen content of the air is low. Also, people engaged in **heavy physical activity,** especially athletes, tend to have insufficient oxygen for their needs. Thus it is not unusual to find that the red blood cell counts of people living at high altitudes and athletes living at any altitude range as high as 6.5 to 8.0 million/mm^3, 30% to 60% higher than normal.

Erythrocytes carry oxygen by means of the iron-containing protein **hemoglobin,** which makes up about 33% of each red blood cell. Every 100 milliliters (ml) of whole blood of a healthy individual contains between 14 and 16 grams (gm) of hemoglobin. In the lungs, where oxygen is abundant, hemoglobin of the circulating erythrocytes unites with oxygen to form **oxyhemoglobin.** In the body tissues, where oxygen is scarce, oxyhemoglobin releases its oxygen and becomes ordinary hemoglobin once more.

Anemias. Any deficiency in the oxygen-carrying capacity of the blood is called anemia. The symptoms of this condition include fatigue and loss of energy. The lack of oxygen causes the death of cells, an overworking of the heart, and, not infrequently, heart failure. Diminished oxygen-carrying capacity can stem from several causes: (1) a reduction in the number of red blood cells in the body (pernicious anemia), (2) a reduction in the amount of hemoglobin in each erythrocyte (thalassemia), or (3) a change in the chemical structure of the hemoglobin molecule that reduces its oxygen-carrying efficiency (sickle cell anemia).

Pernicious anemia. Pernicious anemia results from a reduced production of erythrocytes by the body caused by a deficiency of the essential nutrient **cobalamin,** vitamin B_{12} (see Table 4-1). This vitamin is required for normal cell division, and its shortage greatly diminishes the rate of red blood cell production.

A lack of cobalamin in the body can result from a dietary deficiency. More frequently, however, the problem lies in the faulty absorption of this essential nutrient from the intestinal tract. The defect in absorption is not caused by a malfunction of the cells of the intestinal tract, but rather by a lack of production of a mucopolysaccharide (amino group–containing polysaccharide), called the **intrinsic factor,** by the cells of the stomach. The mucus secreted by the stomach normally contains this mucopolysaccharide, which combines with cobalamin and makes the vitamin available for absorption. When the stomach cells do not secrete this substance, cobalamin is not absorbed from the intestinal tract. As a result, red blood cell production is reduced, and pernicious anemia develops. The disease can be cured by the injection or ingestion of large amounts of cobalamin.

Thalassemia. Thalassemia is caused by a reduction in the amount of hemoglobin in each red blood cell. Whereas there are 14 to 16 gm of hemoglobin/100 ml of whole blood in a normal person, the blood of those suffering from thalassemia have only about 5 gm of hemoglobin/100 ml. The disease is genetically determined and occurs in individuals who carry two genes for the trait.

Shortly after birth, infants with this disease show its clinical symptoms (enlarged liver and spleen and accumulation of fluids in body tissues) and must receive repeated transfusions to remain alive. Individuals suffering from thalassemia are unable to cope with infections; most die before the age of 13 years, and very few survive to age 20 years. Unfortunately, there is as yet no known cure for this severe form of anemia.

There are a number of different forms of the disease, all uniformly fatal. The incidence of thalassemia is not constant throughout the world. It occurs mainly in people of Mediterranean, Asian, and African origin. In these populations, the incidence of the disease is about 1 in 400 people.

Sickle cell anemia. In sickle cell anemia, a change in the chemical structure of the hemoglobin molecule results in a reduced oxygen-carrying capacity of the red blood cells despite the fact that the amount of hemoglobin per cell and the number of red blood cells per cubic millimeter are normal.

The protein portion of the hemoglobin molecule consists of four polypeptide chains. To each chain is attached an iron-containing (heme) compound. This heme group is the "working" portion of the hemoglobin molecule in that it contains a central iron atom to which oxygen atoms can be loosely bound. However, it is the protein portion of the hemoglobin molecule that actually determines the functional efficiency of the iron atom.

Most (98%) of the hemoglobin molecules of normal people consist of two identical polypeptide chains, called **alpha** chains, and two other identical polypeptide chains, called **beta** chains. In sickle cell anemia, the amino acid composition of the beta chain is the crucial factor. Of the 146 amino acids that make up each of the beta chains of both normal and sickle cell hemoglobin, there is only one difference between them. Amino acid 6 of normal hemoglobin is **glutamic acid,** whereas the comparable amino acid of sickle cell hemoglobin is **valine.** It is not presently known why a difference in one amino acid should cause such a large difference in oxygen-carrying capacity.

Sickle cell anemia is genetically determined and occurs in individuals carrying two genes for the trait. People who suffer from this disease normally show only a moderate anemia. However, the blood cells of these people tend to stick together and clog the small blood vessels of the body, preventing the normal flow of blood. This usually results in pain in the stomach region and at various joints of the body and can result in damage to internal organs. In addition, people with sickle cell anemia have difficulty coping with infections and may require periodic transfusions.

Unlike thalassemia, sickle cell anemia is **not** uniformly fatal, and people having the disease may achieve a normal life span. However, individuals with this disease must be careful to avoid infections, they should choose occupations that do not

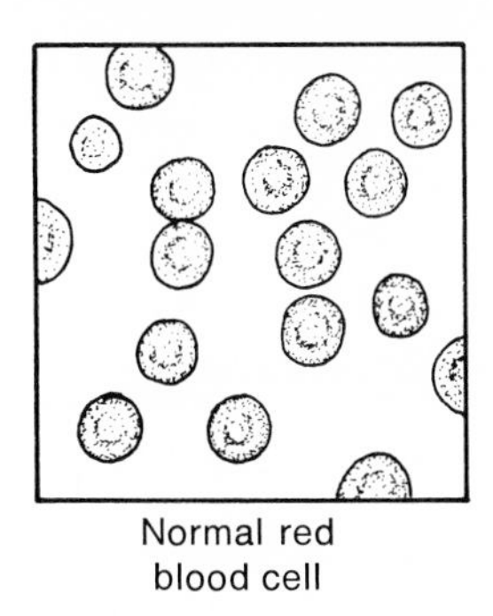

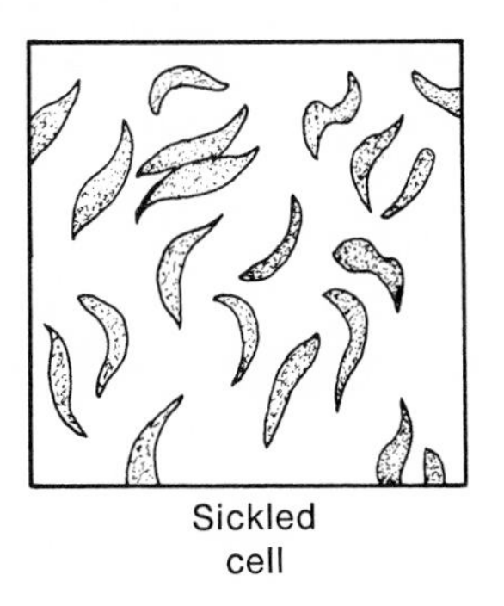

Fig. 6-2
Normal and sickled red blood cells.

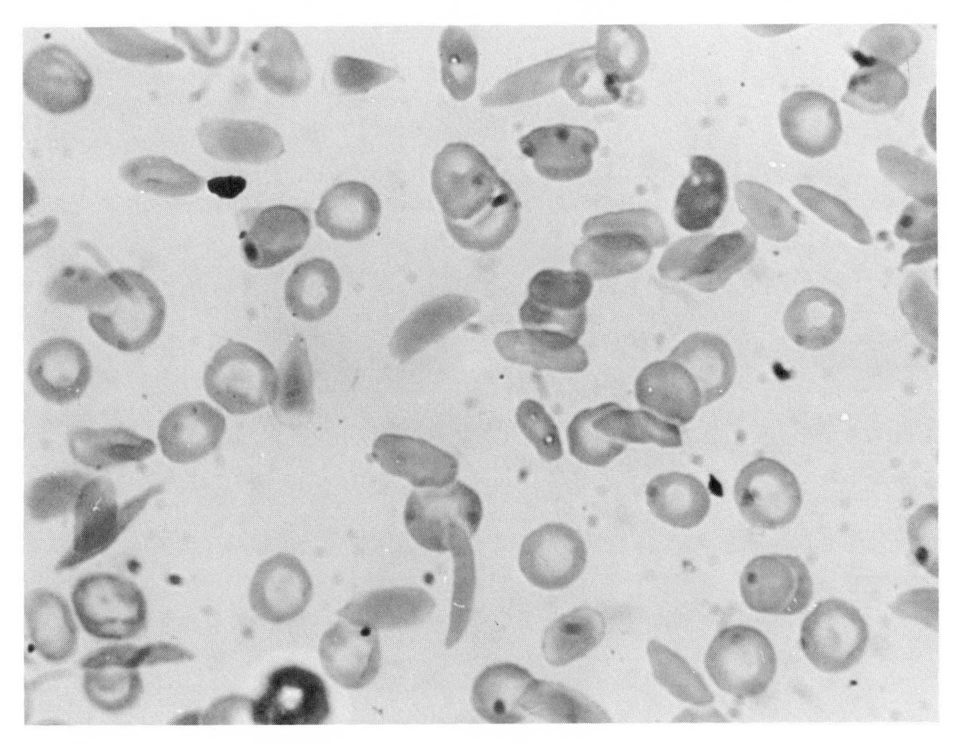

Fig. 6-3
Red blood cells from patient with sickle cell anemia crisis, showing crescent shape of sickled cells. (×970.) (From Dougherty, W. M. 1976. Introduction to hematology, ed. 2. The C. V. Mosby Co., St. Louis.)

involve strenuous physical labor, and they must maintain a well-balanced diet, rich in proteins, vitamins, and mineral elements.

Sickle cell anemia is diagnosed on the basis of the peculiar shape assumed by the red blood cells under laboratory conditions. Whereas normal erythrocytes are doughnut shaped, those of people with this disease tend to take on a sickle-shaped form under conditions of low oxygen, thus giving the disease its name (Figs. 6-2 and 6-3).

A less pronounced form of sickling is seen in the red blood cells of people who have only one gene for this disease. Such individuals are said to exhibit the **sickle cell trait.** Persons with the sickle cell trait lead normal lives and can engage in strenuous physical activity, including such sports as track, basketball, and football.

This disease is found mainly in Africa, the Mediterranean area, and southern Asia. In the United States, it has thus far been found only in Negroes. It is estimated that about 8% of Negro Americans have the sickle cell trait, whereas sickle cell anemia occurs in about 1 in every 500 births. As yet, there is no known cure for sickle cell anemia.

Lead poisoning anemia. Our modern technology has produced conditions that can result in a form of anemia involving (1) reduced amount of hemoglobin per red blood cell, (2) reduced number of erythrocytes per cubic millimeter of blood, and (3) reduced average life span of the red blood cells. This kind of anemia is caused by the ingestion of lead and is especially prevalent in children between 1 and 6 years of age.

In large, old cities in the United States, there is a marked concentration of lead poisoning cases in slum areas. Here the apartment walls often have many coats of paint on top of one another, the older coats sometimes dating back 20 to 50 years.

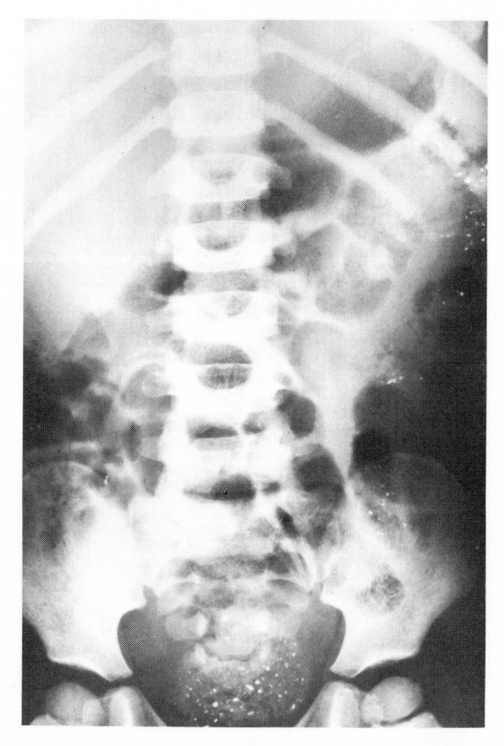

Fig. 6-4
Roentgenogram shows lead particles in a 3-year-old girl with pica (chewing on painted plaster). Small dots throughout the abdomen represent lead particles inside the bowels. (Courtesy X-ray Department of Children's Hospital of Michigan, Detroit; from Waldbott, G. L. 1973. Health effects of environmental pollutants. The C. V. Mosby Co., St. Louis.)

Until about 1940, paint containing lead was frequently used for painting apartments. As a result, in buildings built before 1940, paint peelings and loosened plaster provide a dangerous source of lead to children who tend to swallow such particles (Fig. 6-4).

Lead poisoning, associated with the ingestion of paint peelings, develops only after a considerable amount of paint is swallowed. From 3 to 6 months of fairly steady paint ingestion is necessary before clinical symptoms become apparent. Affected children usually exhibit a loss of appetite, abdominal pain, vomiting, and irritability. In addition to producing an anemia, lead poisoning has a severe effect on the kidneys and nervous system of the individual and may actually lead to death.

The effects of lead poisoning, including the anemia, can be treated and cured by modern medicine, providing it is not too far advanced. In addition, and most importantly, lead poisoning is preventable. Once children have been treated for lead poisoning, they should not be re-exposed to the conditions that produced the disease. There is great need to remove areas of high lead concentrations from our environment, whether this means removal of all old layers of paint from apartment walls or the complete clearing away of slums. Our modern society must recognize the medical and human costs involved in retaining environments that produce disease, debilitation, and even death in a segment of its population.

□ White blood cells (leukocytes)

White blood cells differ from red blood cells in three ways: (1) leukocytes retain their nuclei throughout their life span, (2) leukocytes lack hemoglobin (hence are colorless), and (3) leukocytes move actively into and out of blood vessels and tissue spaces.

The primary function of white blood cells is to combat infectious agents that invade the body. The body is constantly exposed to bacteria, viruses, and other microscopic organisms that enter the mouth, anus, respiratory passages, and even urinary tract. These organisms are capable of causing diseases if they invade the deeper tissues. Basically, the body protects itself in two ways: (1) by the action of its leukocytes, which engulf the invading agents, a process called **phagocytosis,** and (2) by the formation of **antibodies,** which attach to and immobilize the disease-causing organisms.

There are five types of white blood cells: neutrophils, eosinophils, basophils, monocytes, and lymphocytes. The normal percentage of each type of leukocyte in the

Table 6-1
Percentage of each type of leukocyte in blood

Leukocyte	%
Neutrophil	62.0
Eosinophil	2.3
Basophil	0.4
Monocyte	5.3
Lymphocyte	30.0

bloodstream is shown in Table 6-1. Each type of white blood cell has its own specialized characteristics, but the functions of some are not fully understood.

Neutrophils, eosinophils, and basophils are produced in the bone marrow, whereas lymphocytes and monocytes are formed in the spleen, thymus, and tonsils, a group of organs known collectively as **lymph glands.** Although the white blood cells contain nuclei, they do not undergo cell division. The average life span of leukocytes is very short, only 2 to 4 days, and their supply must constantly be replenished (Fig. 6-5).

Infection and leukocyte function. When an injury occurs to some body tissue, a series of events occurs that prevents the invasion and spread of disease-causing organisms. The blood supply to the injured tissues increases, and large quantities of fluid and protein leave the blood vessels and enter the intercellular spaces near the point of injury. This causes a swelling and reddening (**inflammation**) of the area, which in effect seals off the injured site. This activity prevents the spread of the organisms or any poisons (toxins) produced by them.

When tissues are damaged, there is a movement of white blood cells to the site of injury and also an increase in leukocyte production by the bone marrow and lymph glands. Normally, whole blood contains about 7000 white blood cells/mm^3. After an injury, the number of leukocytes may rise as high as 20,000 to 30,000 cells/mm^3.

White blood cells engulf not only invading microorganisms, but also pieces of

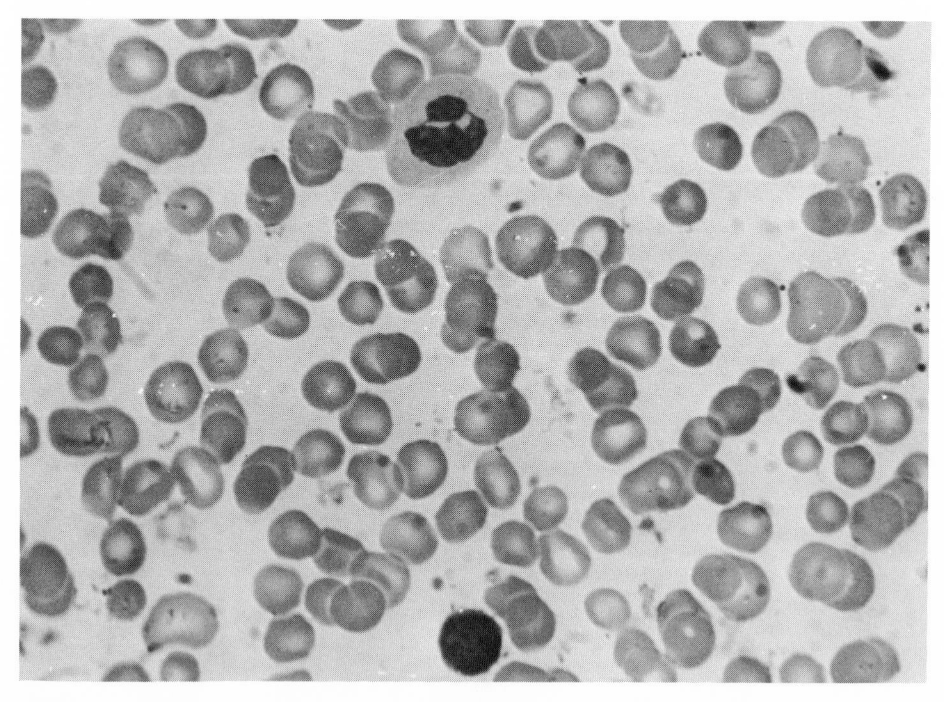

Fig. 6-5
Blood smear, showing erythrocytes whose thin central portion permits light to pass through. Note **neutrophil** (top) and **lymphocyte** (bottom). (×970.) (From Dougherty, W. M. 1976. Introduction to hematology, ed. 2. The C. V. Mosby Co., St. Louis.)

dead cells. The two most important phagocytic leukocytes are **neutrophils** and **monocytes.** They continue to ingest foreign and dead material until they themselves are killed by the accumulated dead and decaying material. After a few days, an aggregation of dead tissue cells, bacteria, leukocytes, and plasma forms at the site of the injury. This mixture is called **pus.**

Our discussion of infection and leukocyte function is not restricted to external body injuries. An identical series of events occurs when the appendix becomes infected with bacteria from the large intestine. In the case of **appendicitis,** however, surgery is required to remove the appendix and thus prevent the spread of the infection throughout the body.

Although the functions of some types of white blood cells are not known, we do know that they are associated with certain specific diseases. For example, people who suffer from hay fever, asthma, or other allergies have a greatly increased number of **eosinophils** in the bloodstream. Those with whooping cough or pernicious anemia show a marked increase in the number of **lymphocytes.**

The remaining white blood cell, the **basophil,** is not involved in phagocytosis, but it does produce certain chemicals that help the healing process after injury or infection.

Leukemias. Leukemia is a general term applied to all disease in which there is an *uncontrolled production of white blood cells.* This excess of leukocytes is the result of an abnormal increase in the growth and division of the leukocyte-forming cells of the bone marrow or lymph glands. Depending on whether the bone marrow or a lymph gland is involved, we find an increase in a particular type of white blood cell.

Leukemia cells, as is true of all types of cancerous cells, spread throughout the body and establish themselves at many different places. Wherever they are, the cells grow and divide so rapidly that tremendous demands are made on the body for nutrients, especially amino acids and vitamins. As a result, the energy of the individual is greatly depleted, and the person becomes weak and eventually dies of pneumonia or some other infection.

Leukemias are classified on the basis of (1) the speed of spread of the disease, being termed either **acute** or **chronic,** and (2) the type of leukocyte that is produced in excessively large numbers, for example, **myelocytic** indicating cells produced by the bone marrow, **lymphocytic** indicating lymphocytes, and **monocytic** indicating monocytes.

Each year in this country, leukemia develops in about 19,000 people and takes the lives of approximately 15,000 people. Of the various types of leukemia, **acute lymphocytic leukemia** is the one most frequently seen in children. Although the other acute and chronic leukemias are occasionally seen in children, they appear primarily in adults.

Medical research has discovered a number of chemical compounds that are effective in controlling, and at times even curing, some types of leukemia. Chemotherapy has been found most effective against the **acute leukemias,** bringing hope to the children afflicted with acute lymphocytic leukemia. Unfortunately, the chronic forms of leukemia appear to be resistant to chemotherapy.

The factors that cause leukemia to develop are unknown. There is evidence that would implicate a number of diverse agents. It is known that there is a higher incidence of leukemia in those individuals who have been exposed to high levels of radiation, for example, survivors of atomic bombings in Japan. However, even among people who received the largest radiation doses, only 1 in every 100 developed leukemia.

There is evidence that genetic factors may be involved. Identical twins have the same genetic makeup, and if one identical twin develops leukemia before the age of 5 or 6 years, the other twin has about a one in five chance of also developing the disease within a short time. This is not true of fraternal twins, who are no more similar in genetic makeup than any other brothers or sisters.

More recently, there has been a great deal of emphasis and research on viruses (minute parasites that live and reproduce within the cells of all organisms) as possible causative agents of leukemia. The stress on viral research has been generated by the discovery that leukemia in other animals, for example, mice, can be caused by viruses. Supportive evidence for a hypothesis concerning virally induced leukemia comes from studies of other types of human cancers, such as breast cancer. These studies show the presence of certain viruses in cancerous cells and their absence in noncancerous cells of the same individual. Although an association of virus and cancer does not prove that the virus has caused the cancer, this type of observation has stimulated a great deal of research to determine if viruses do, in fact, cause leukemia and other malignancies.

☐ Blood platelets

Blood platelets, unlike red blood cells and white blood cells, are not complete cells; they are fragments of certain giant cells produced in the bone marrow. There are about 300,000 platelets/mm^3 of whole blood. Platelets have a life span of about a week and are important in initiating the process of blood clotting, as will be discussed later in this chapter.

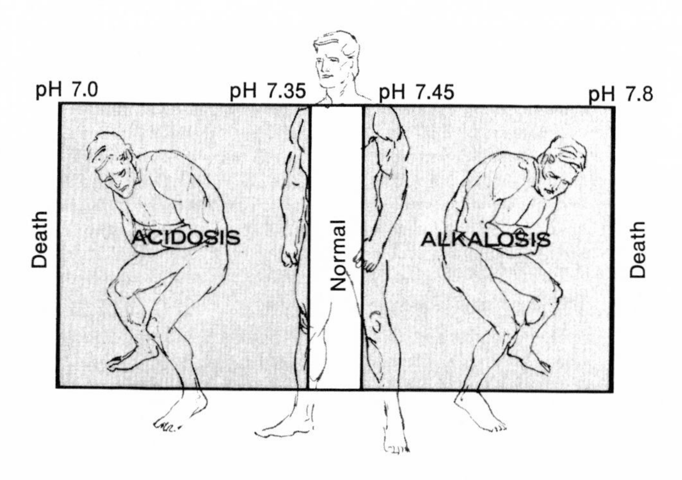

Fig. 6-6
pH as a matter of life and death. (From Cutter Laboratories, Inc. Parenteral solutions handbook. Berkeley, Calif.; from Brooks, S. M. 1972. Basic biology, a first course. The C. V. Mosby Co., St. Louis.)

□ Plasma

We began our discussion of blood by pointing out that it consists of a cellular portion (45%) and a liquid portion (55%), called **plasma.** Plasma is about 90% water, 7% proteins, 0.9% salts, and 0.1% glucose, with lesser amounts of other substances. Plasma is slightly alkaline, having a pH of 7.4 (Fig. 6-6). The proteins found in blood plasma are of three types: albumins (4%), globulins (2%), and fibrinogen (1%).

The blood plasma has a number of functions. It is the vehicle that transports almost all materials, except oxygen, throughout the body. The proteins present in plasma are involved in blood clotting, in the immunity responses of the individual to foreign organisms and substances, and in the maintenance of the water balance of the body. Some of the roles of plasma in body functions are examined here.

Blood clotting. In man, as in many other organisms, the breaking of a blood vessel can lead to a serious loss of blood and even death. This problem exists regardless of whether there is a rupture of a small internal blood vessel or an injury inflicted on the body from the outside. The seriousness of this problem is undoubtedly reflected in the body's elaborate mechanism that serves to prevent the accidental loss of any large quantity of blood.

Although blood clotting is a very complicated process, consisting of many biochemical steps, it basically involves the conversion of the plasma protein **fibrinogen** to **fibrin.** Fibrin forms a tangled mat of fibers, called a **clot,** that traps both red and white blood cells. After a while, the clot shrinks and squeezes out a straw-colored fluid, called **serum.** Serum is similar in chemical composition to plasma except that it lacks the protein fibrinogen. The overall clotting process is summarized in a highly simplified manner, in Fig. 6-7. There are many "clotting factors" and biochemical reactions other than those shown. However, the overall clotting process can be understood by studying the changes that take place in three proteins—thromboplastin, prothrombin, and fibrinogen. Whenever a break occurs in a blood vessel, the broken cells release an inactive form of the enzyme **thromboplastin** into the plasma. Simultaneously, blood platelets release a clotting factor, called **factor III.** Factor III combines with the inactive form of thromboplastin and in so doing transforms it into an

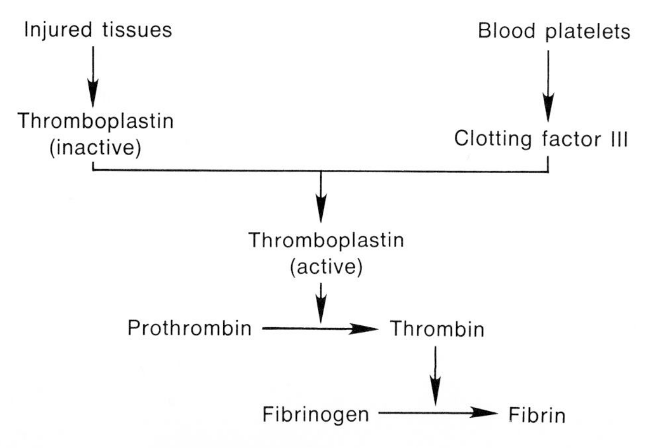

Fig. 6-7
Basic events in blood clotting.

active enzyme. Active thromboplastin then converts another inactive plasma protein, **prothrombin,** into the active enzyme **thrombin.** Thrombin, in turn, converts **fibrinogen** into fibrin.

Our interest in the process just described stems from the fact that there are human beings whose blood does not clot normally and who may suffer fatal hemorrhages after relatively minor external or internal injuries. There are three conditions that can lead to uncontrolled bleeding from injuries: (1) vitamin K deficiency, (2) hemophilia, and (3) platelet deficiency.

Vitamin K deficiency. Vitamin K, phylloquinone (see Table 4-1), is necessary for the formation of prothrombin. Vitamin K is one of the fat-soluble vitamins and cannot be absorbed from the intestinal tract if fat digestion has been inadequate. Thus a lack of prothrombin can result not only from a deficiency of vitamin K in the diet, but also from a failure of the liver to secrete bile or from an obstructed bile duct. The absence of prothrombin, in turn, will lead to uncontrolled bleeding after injury. This problem can be corrected by injecting vitamin K directly into the affected individual.

Hemophilia. Hemophilia is an hereditary disease characterized by excessive bleeding after even slight cuts of the body. The inability of the blood to clot is caused by the absence of a plasma factor not shown in Fig. 6-7. This condition is also correctable. Affected individuals must be injected periodically with the missing factor.

Blood platelet deficiency. Some people suffer from a diminished number of platelets in the blood. When the number of platelets falls below 70,000/mm^3, these people have a tendency to bleed, as do hemophiliacs. This situation can be successfully treated by periodic transfusions of platelets from blood donors.

Immunity. Earlier in this chapter, we discussed the role of white blood cells in protecting the body against invading organisms by engulfing them. Our blood plasma contains an additional form of defense against both infections (bacteria, viruses) and poisonous agents (toxins). This second defense mechanism is a chemical system involving the production of specific protein molecules, called **antibodies,** that attach themselves to and neutralize foreign organisms and poisons.

Antibody formation. When a foreign particle, called an **antigen,** invades the bloodstream, the body begins developing antibodies against it. The antibodies are produced in lymph glands or other lymphoid tissues distributed throughout the body. After they are formed, the antibodies are released into the blood plasma. Chemically, antibodies are **globulin**-type proteins, which form the **gamma globulin** portion of the plasma proteins. Once formed, antibodies give the individual an **immunity** from the deleterious effects that may have accompanied the initial exposure to the particular infectious or poisonous agent. We are all familiar with the fact that after once having chickenpox, measles, or whooping cough, we do not contract the disease again. By the same mechanism, we gain immunity through **vaccination** against a disease, that is, by receiving an injection of either dead or weakened organisms or their toxins (Fig. 6-8).

Antibody action. Each antibody neutralizes its particular antigen in a specific way. Some antibodies attach themselves to the antigens and cause them to form clumps (**agglutinate**). Others combine with bacteria and cause the bacteria cell to split apart (**lyse**). In most cases, after the antibody has neutralized the antigen, the

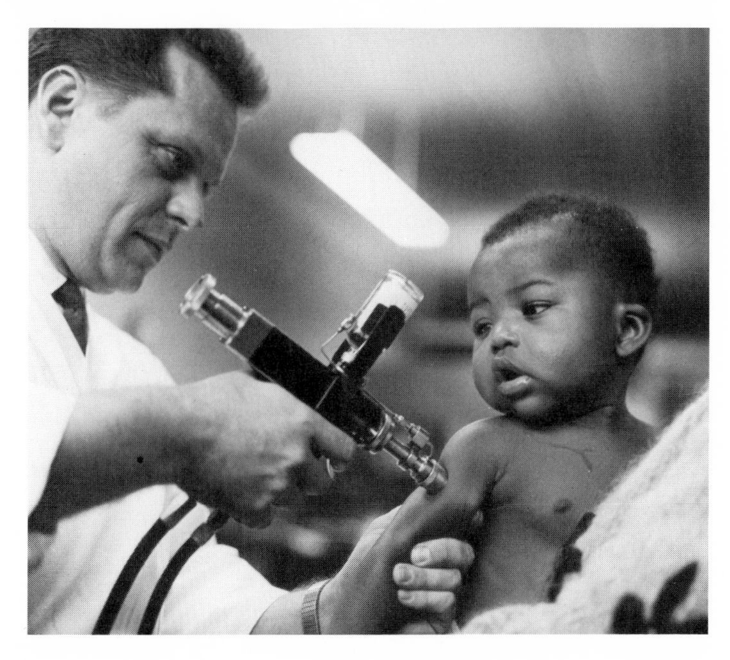

Fig. 6-8
Jet injector (air gun) delivery of measles vaccine to young Chicagoan. Procedure is instant and pain free. (From American Medical Association. 1966. Medical news. J.A.M.A. **196:**29; from Smith, A. L. 1977. Principles of microbiology, ed. 8. The C. V. Mosby Co., St. Louis.)

entire antigen-antibody complex is ingested either by leukocytes in the bloodstream or by large phagocytic cells in the lymph glands.

Allergies. Our ability to develop an immunity to specific infectious and poisonous agents is a powerful weapon in the fight against disease. However, there are circumstances in which this defense mechanism is not at all beneficial and may even result in serious damage to the body. The development of **allergies** (hypersensitivity) to dust, pollen, certain fabrics, some foods, and certain antibiotics are examples of deleterious antibody production.

Transplantation reaction. Modern medicine has developed a high degree of surgical proficiency in transplanting tissues and organs from one person to another. Unfortunately, any foreign cells transplanted into a recipient will cause the development of antibodies against the transplanted cells. The only exception to this situation involves transplants from one identical twin to another.

The severity of the **transplantation reaction,** when unrelated people are involved, can be seen in the fact that kidney transplants generally last about 5 years, liver and heart transplants 1 to 2 years, and lung transplants only about 1 month. In an attempt to reduce graft rejections, people receiving transplanted organs are given rather large doses of **cortisone,** a chemical compound that suppresses antibody formation. Unfortunately, such treatment leaves the patient without defenses against infection, which is why so many transplant patients eventually die of pneumonia.

Autoimmunity. Another instance in which antibody production becomes dangerous to the individual is when a person develops antibodies that react against his own proteins. This occurs when an individual is infected with an organism whose protein products are antigenically similar to those of the person's own tissues. The development of an immunity against the body's own tissue proteins is called **autoimmunity.** A number of diseases that occur in middle and late life are believed to be autoimmune diseases, including rheumatoid arthritis and rheumatic fever.

Blood transfusion reactions. Whenever an individual has lost a great deal of blood, either through accident or disease, it is necessary to replace the lost blood as soon as possible. The most efficient way to do this is to transfer blood from a healthy individual (**donor**) to the person requiring it (**recipient**). Unfortunately, prior to 1910, there was a rather high incidence of death among the recipients.

AB antigens. In 1910 Karl Landsteiner, an Austrian physician, discovered that the red blood cells of some people can be agglutinated and destroyed by antibodies in the plasma of other individuals. He further discovered that there are two antigens associated with human erythrocytes, and he called them A and B. He then referred to the corresponding antibodies as anti-A and anti-B. These antibodies are not developed in response to any known infection; rather they appear shortly after birth.

Every human being belongs to one of four blood groups, depending on the presence, absence, or combination of these antigens and antibodies. The four blood groups are A, B, AB, and O. People who have type A blood cells have anti-B antibodies in the plasma; people with blood type B have anti-A antibodies; people with blood type AB have neither type of antibody; and people with blood type O have both anti-A and anti-B antibodies. Table 6-2 shows the possible donors and recipients for each blood group. Individuals who are blood type O are called **universal donors,** because they can give their red blood cells to people of all other blood groups. Individuals who are blood type AB are called **universal recipients,** because they can receive erythrocytes from people of any other blood group.

The discovery of the various blood group types and their corresponding antibodies explained the previously observed death of some recipients during blood transfusions. For example, if a blood transfusion was performed between a donor of blood type B and a recipient of blood type A, the cells of the donor were agglutinated and destroyed by the anti-B antibodies in the plasma of the recipient. The donor's clumped blood cells would clog the blood vessels of the recipient, causing death.

Transfusion reaction problems occur only when red blood cells are being transfused. If blood plasma alone is transfused, no antigen-antibody reaction occurs. Thus despite the fact that blood plasma from a type B individual contains anti-A antibodies, it may be safely given to a person of blood type A. The reason for this is the slowness

Table 6-2
Human blood groups

Blood group	Antigen in red blood cell	Antibodies in plasma	Groups person can give blood to	Groups person can receive blood from
A	A	Anti-B	A, AB	A, O
B	B	Anti-A	B, AB	B, O
AB	A and B	None	AB	AB, A, B, O
O	None	Anti-A and anti-B	O, A, B, AB	O

with which the transfusion is administered. In the transfusion, the antibodies of the donor are diluted in the recipient's plasma and are not able to agglutinate the recipient's erythrocytes, although they may attach themselves to and destroy a few individual cells.

Rh antigens. In addition to being type A, B, AB, or O, human red blood cells are either Rh positive or Rh negative. However, in the case of the Rh antigen, the plasma normally contains no antibody against it. If an Rh-negative individual receives a transfusion containing red blood cells from an Rh-positive person, the lymph tissues of the recipient will **develop** antibodies against Rh-positive cells. A subsequent transfusion from the same or another Rh-positive donor will then result in a transfusion reaction and death of the recipient. For reasons not well understood thus far, a transfusion of Rh-negative erythrocytes into an Rh-positive individual does **not** stimulate antibody production against Rh-negative cells. Rh-negative erythrocytes apparently contain no Rh antigen.

A transfusion reaction involving the Rh antigen can occur between a woman and her unborn child despite the fact that maternal and fetal bloodstreams do **not** mix. If a woman is Rh negative and her husband Rh positive, the fetus may be Rh positive. (The inheritance of ABO and Rh blood factors will be discussed in Chapter 9.) In a normal pregnancy, some red blood cells from the blood of the fetus usually enter the maternal bloodstream, especially as membranes and vessels are ruptured during the birth process. This is the equivalent of a small blood transfusion from the unborn child to the mother. If the mother is Rh negative and the fetus is Rh positive, the red blood cells of the fetus will stimulate the mother's lymph tissues to produce anti–Rh-positive antibodies.

If a subsequent pregnancy also involves an Rh-positive fetus, some of these Rh antibodies will pass from the mother's plasma into the unborn child's bloodstream and cause the clumping of erythrocytes. This condition is called **erythroblastosis fetalis.** In extreme cases, the fetus dies before birth. More frequently, the child is born alive, but dies shortly thereafter.

Modern medicine has developed a technique that prevents the formation of anti–Rh-positive antibodies by the mother. Shortly after the birth of an Rh-positive child to an Rh-negative mother, the mother is given an injection of anti–Rh-positive antibodies obtained from another person. The anti–Rh-positive antibodies attach themselves to and destroy the fetal red blood cells that are dispersed throughout the maternal bloodstream. The destruction of the fetal erythrocytes prevents their acting as antigens for the production of anti–Rh-positive antibodies.

Because the small number of fetal erythrocytes are scattered widely throughout the mother's bloodstream, there is virtually no danger of the formation of dangerous clumps of fetal cells in the blood vessels of the mother. It should also be noted that this technique can be used successfully only because most of the transfer of fetal cells occurs during the birth process. If large numbers of fetal erythrocytes were transferred early in pregnancy, there would be no way to prevent the formation of anti–Rh-positive antibodies.

The injected anti–Rh-positive antibodies are themselves gradually destroyed in the maternal bloodstream. However, they are not replaced by the female's lymph tissue, as the potential antigens (fetal Rh-positive erythrocytes) are no longer present.

HEART AND CONDUCTING VESSELS

We shall now turn our attention to the heart and conducting vessels. These structures function to bring the blood and lymph from one group of cells to another. Human beings, just as many other animals, have a **closed circulatory system.** In such a system, the blood is always contained within the heart and blood vessels, and nutrients, wastes, hormones, etc. must be able to pass through the walls of the blood vessels to travel to or from the cells of the body.

Our closed circulatory system consists of a heart and three types of vessels: arteries, capillaries, and veins. In addition, there are lymphatic vessels that serve to channel excess tissue fluid to one of the large veins, where the fluid flows into the blood and becomes part of the plasma. The lymphatic vessels, as was discussed in Chapter 5, also function in lipid absorption from the small intestines.

Heart

The human heart (Fig. 6-9) is a powerful muscular pumping organ composed of four chambers: two **atria** (right and left) and two **ventricles** (right and left). Atria receive blood from the various organs of the body, whereas ventricles push the blood to the various organs. The vessels that carry blood **to the atria** are called **veins;** those that carry blood **away from the ventricles** are called **arteries.**

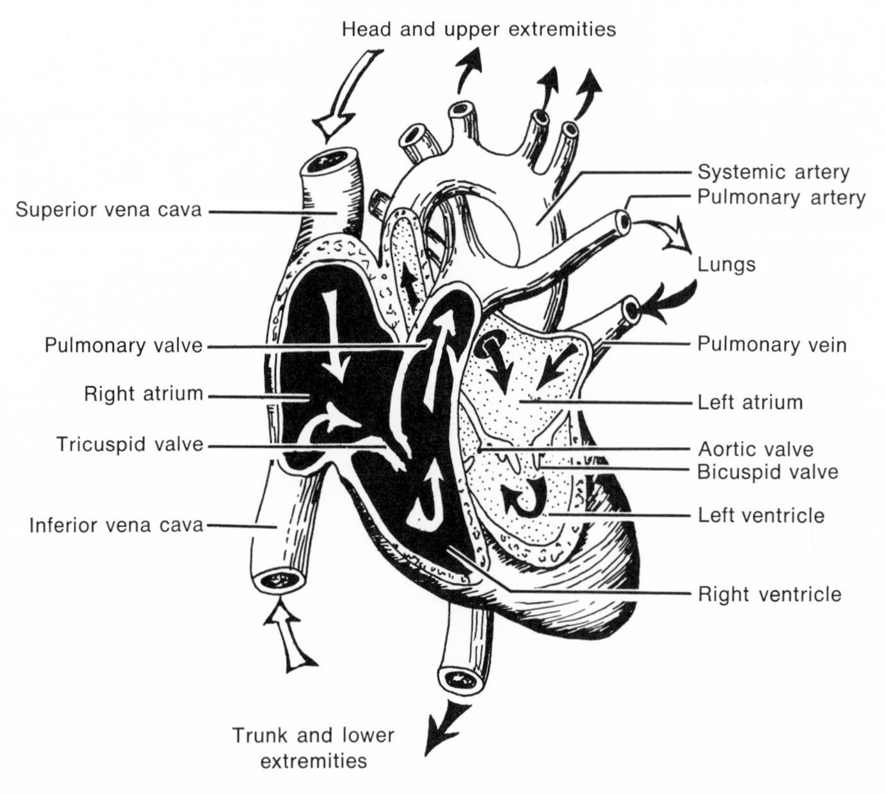

Fig. 6-9
Human heart and associated blood vessels.

Overall circulatory pathway. A diagrammatic representation of the overall circulatory pathway of humans is shown in Fig. 6-10. The **right atrium** receives blood from all parts of the body, except the lungs. When the right atrium contracts, blood is forced into the right ventricle. A **tricuspid valve** between the two chambers prevents the blood from returning to the right atrium. Contraction of the right ventricle forces blood into the pulmonary artery and on to the capillaries of the **lungs.** At the beginning of the pulmonary artery is a **semilunar (pulmonary) valve** that prevents blood from returning to the right ventricle.

After passing through the capillaries of the lungs, the blood flows into the **pulmonary veins** and is brought to the **left atrium.** There are **no** valves at the openings of the pulmonary veins to the left atrium. However, rings of muscle tissue around the veins in that region prevent blood from returning to the veins.

When the left atrium contracts, blood is pushed into the **left ventricle.** There is a **bicuspid valve** between the left atrium and the left ventricle that prevents the blood from returning to the left atrium. Contraction of the left ventricle forces blood into the **systemic artery** and on to the capillaries of all parts of the **body,** except the lungs. At the beginning of the systemic artery, there is another semilunar valve, the **aortic valve,** that prevents the blood from flowing back into the left ventricle.

The presence of valves at strategically located places around and within the heart ensures the flow of blood in a continuous circuit. From the capillaries of the body's various organs and tissues, the blood flows into large veins, called the **venae cavae,** and back to the **right atrium,** thus completing the cycle. Blood from the head, arms, and upper part of the body enters the right atrium through the **superior vena cava;**

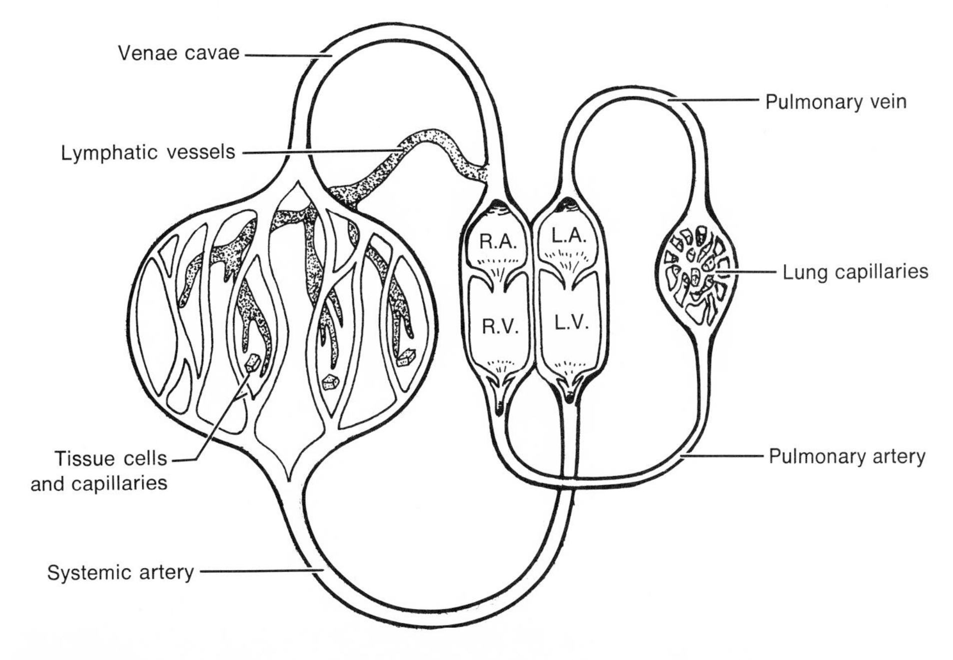

Fig. 6-10
Overall circulatory pathway. Note all blood goes to lungs for oxygenation before it goes out to body tissues.

blood from the lower part of the body returns to the heart through the **inferior vena cava.** As occurred between the pulmonary veins and the left atrium, in the present situation, we find that rings of muscle tissue prevent the blood from returning to the venae cavae from the right atrium.

The overall passage of blood through the body ensures a maximum supply of oxygen for all our cells. This is accomplished by first sending every drop of blood to the lungs each time it is returned to the heart from some other region of the body.

Heartbeat. The heart normally contracts about 72 times each minute, during which time it pumps about 5 liters of blood. This quantity of blood is the equivalent of the total amount of blood contained in the body. However, not every drop of blood passes through the heart every minute, because some circuits through the body are longer than others.

Each heartbeat consists of a wave of contraction involving both the atrial and ventricular muscles. The contraction of the heart muscles is called **systole;** it is followed by a period of relaxation of these muscles called **diastole.** The heartbeat that the physician hears through the stethoscope is divisible into two sounds that have been likened to the syllables "lub dup." The first sound, "lub," is low pitched, relatively soft, and of long duration. It is caused by the snapping shut of the bicuspid and tricuspid valves when the ventricles begin to contract. The second sound, "dup," is high pitched, sharper, and of short duration. This sound is caused by the closing of

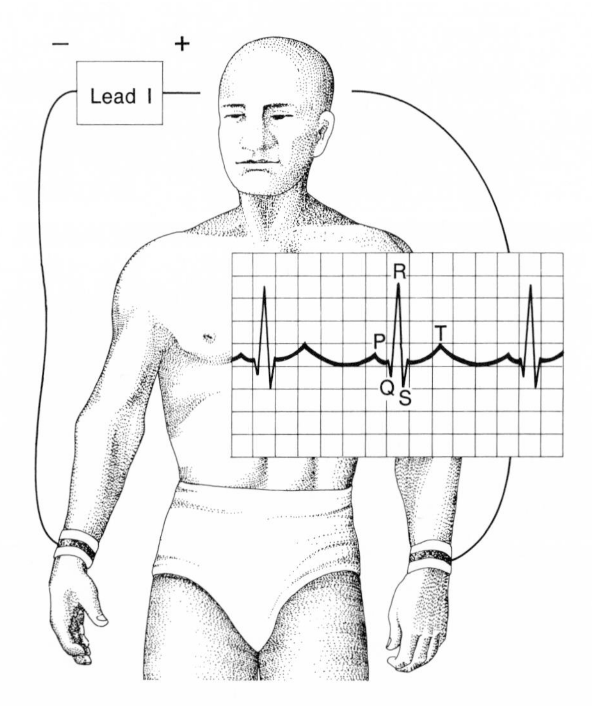

Fig. 6-11
Electrocardiogram (ECG) of normal heart. P corresponds to the passage of nerve impulses through the atria, QRS occur when nerve impulses pass through the ventricles, and T represents the return of the ventricular muscle cells to a resting state. (From Brooks, S. M. 1975. Basic science and the human body: anatomy and physiology. The C. V. Mosby Co., St. Louis.)

the semilunar (aortic and pulmonary) valves when the ventricles begin to relax (Figs. 6-11 and 6-12).

Any damage to the bicuspid or tricuspid valves because of disease or aging will affect the quality of the first heart sound (lub). Correspondingly, any damage to the semilunar valves affects the quality of the second heart sound (dup). Such damage may occur in individuals suffering from syphilis, rheumatic fever, or other diseases. Fortunately, modern surgery has progressed to the point where heart valves can be repaired and replaced relatively safely despite the fact that the operation is quite complicated and requires cutting into the heart itself (**open heart surgery**).

The beating of the heart is controlled by special cells that are located in the heart tissue itself. In the wall of the right atrium is a mass of cells, called the **sinoatrial node** or **pacemaker**, that initiates the heartbeat and regulates its overall rate. A second region of heartbeat control is located between the atria and just above the ventricles. It is called the **atrioventricular node.** The atrioventricular node controls the contraction of the ventricles and acts as a pacemaker if the sinoatrial node is destroyed by injury or disease. In cases where the regulation of the heart's contractions has become erratic, modern medicine can provide the affected individuals with artificial pacemakers and thereby extend the lives of these individuals for many years.

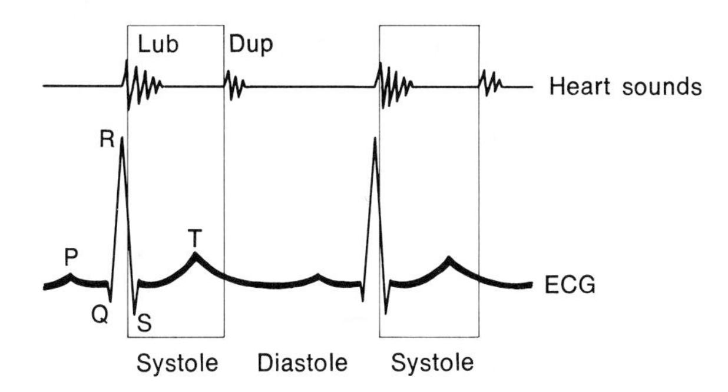

Fig. 6-12
The two heart sounds and their relation to the ECG. (From Brooks, S. M. 1975. Basic science and the human body: anatomy and physiology. The C. V. Mosby Co., St. Louis.)

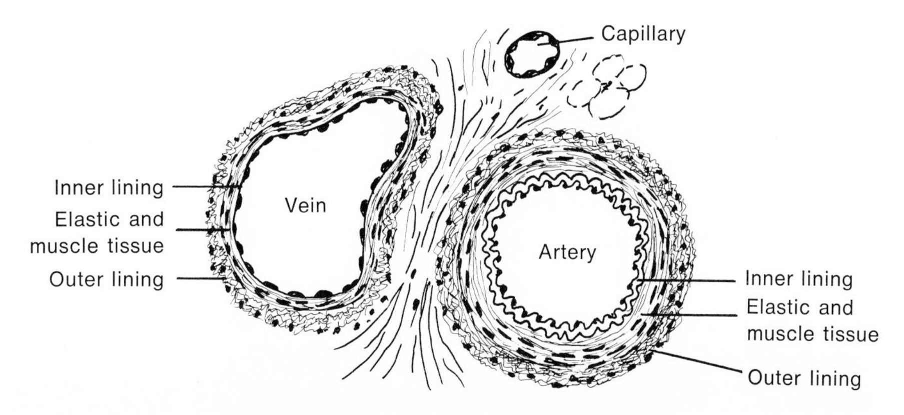

Fig. 6-13
Cross section of artery, capillary, and vein.

Conducting vessels

Blood vessels. As already noted, there are three main types of blood vessels (Fig. 6-13): (1) **arteries,** which carry blood from the heart to the body organs; (2) **veins,** which bring blood from the body organs back to the heart; and (3) **capillaries,** extremely thin tubes that connect the arteries and veins. It is only through the capillaries that substances can pass into and out of the circulatory system.

Arteries and veins. Arteries and veins differ structurally in two ways. First, arteries are thicker then veins, because the arteries have a great deal more elastic and muscle tissue in their walls. These tissues provide arteries with both the flexibility and power needed to move blood throughout the body. Second, veins, but not arteries, are equipped with valves along their length to prevent the backflow of blood. The power to move the blood through the veins does **not** come from the contractions of the ventricles. That force is dissipated by the time the blood passes through the capillaries and reaches the veins. Blood is moved along in the veins through pressure put on these vessels by body movements, with the valves maintaining the direction of flow.

Capillaries. Capillaries are extremely thin walled, and it is through their walls that materials are exchanged between the circulatory system and the body cells (Fig. 6-14). In all blood vessels, blood is under a pressure that is generated by the heart and the walls of the blood vessels themselves. This force is called **blood pressure.** In the capillaries, however, in addition to blood pressure, a second force is generated by the presence of proteins in the blood plasma. This second force is called **osmotic pressure.** Its mode of action will be discussed in Chapter 7. These two forces tend to counteract one another; blood pressure forces materials out of the capillaries, whereas osmotic pressure pulls materials into the capillaries. Because the walls of arteries and veins are too thick to permit passage of materials, osmotic pressure does not operate in these large blood vessels, and blood pressure serves only to push the blood along.

Exchange of materials. At the arterial end of the capillary, the force of the blood pressure (measured in millimeters of mercury [mm Hg]) exceeds that of the osmotic

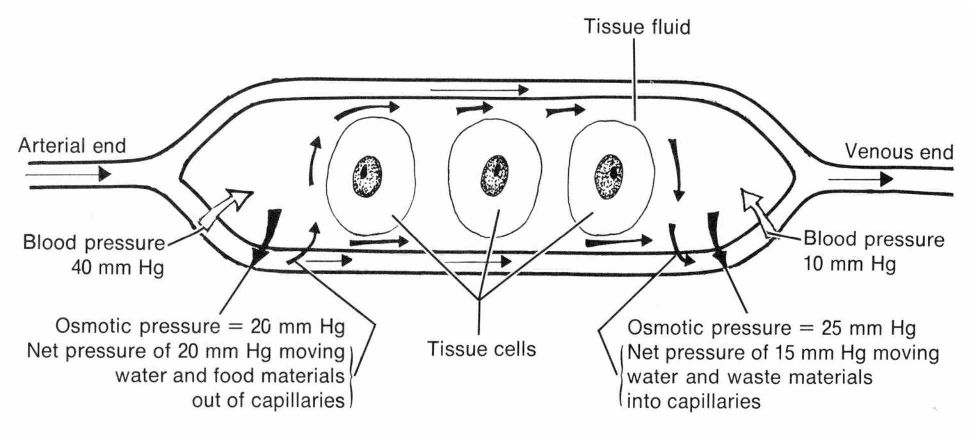

Fig. 6-14
Transfer of water and materials between capillaries and tissue fluid.

pressure. As a result, water and many substances dissolved in the blood plasma (monosaccharides, lipids, amino acids, salts, etc.) are forced out through the capillary walls and into the **tissue fluid** that surrounds the capillaries and the cells. However, the proteins in the blood plasma are large and **cannot** pass through the walls. Furthermore, because of the loss of water from the capillaries (just noted), the **concentration** of the proteins in the blood plasma increases, as does the osmotic pressure of the blood.

At the venous end of the capillary, the force of the osmotic pressure exceeds that of the blood pressure. As a result, most of the water and all of the waste materials in the tissue fluid pass **into** the capillary. About 10% of the liquid that leaves the capillary because of blood pressure at the arterial end is not pulled back. This excess liquid is removed from the tissue fluid and returned to the circulatory system through the lymphatic vessels.

This change in relative magnitudes of blood pressure and osmotic pressure results in the transfer of nutrients to the cells and the removal of waste materials from the cells. It is important to stress that this exchange of materials can occur only through the thin walls of the capillaries.

Lymphatic vessels. The lymphatic vessels constitute a special set of tubes that have two functions: (1) return of excess tissue fluid to the circulatory system and (2) absorption of digested lipids from the small intestine. The lymph system differs from

Fig. 6-15
Woman with elephantiasis in a South American city. Cause is blockage of lymphatic channels by microfilariae of genus *Wuchereria*, a nematode aschelminth. Worms are transmitted by several genera of mosquitoes. Name comes from the elephant-like legs and feet of victims. Fortunately, disease can be prevented by killing mosquitoes in their breeding places; also worms can be killed by drugs. (Courtesy E. S. Beneke; from Pettit, L. C. 1962. Introductory zoology. The C. V. Mosby Co., St. Louis.)

the blood system in that the lymph vessels serve only to return fluid to the heart. There are **no** arteries in the lymph system, only the equivalent of capillaries and veins (Fig. 6-10). The end point of the lymph system is the **superior vena cava.**

At some points, where lymph vessels unite, there are aggregations of cells known as **lymph nodes.** The lymph nodes, along with the lymph glands, are sites for the manufacture of lymphocytes and for the filtering out of bacteria and other foreign agents that have entered the lymph system.

The chemical composition of lymph varies depending on its location in the body. In the parts of the body from which the lymph begins its flow to the superior vena cava, the lymph is almost identical in composition with the tissue fluid surrounding it. However, in areas where organs, such as the liver, release proteins into the lymph, the chemical composition of the lymph will vary greatly from tissue fluid. Lymph contains less protein than blood, and it has **no** red blood cells.

If lymph is not returned to the veins about as rapidly as it is formed, it will accumulate in tissue spaces and cause swelling. This condition is known as **edema.** In tropical countries, humans may be infected with a parasitic worm, called *Wuchereria bancrofti,* that invades the lymph vessels and nodes. The presence of these worms in the lymph system prevents the normal flow of lymph to the superior vena cava. Consequently, lymph accumulates, especially in the legs and arms. Often there is a tremendous swelling of these parts of the body. The disease has been called **elephantiasis** because of the elephant-like appearance of the arms and legs of the affected individuals. Medicines are available to alleviate the condition. However, the most beneficial approach is to prevent the occurrence of the disease through the elimination of mosquitoes that transfer the worm from person to person (Fig. 6-15).

□ Specialized pathways

We have discussed the overall transit of blood through the body (Fig. 6-10), and the major routes are shown in Fig. 6-16. Most of the nonpulmonary pathways follow the expected pattern, that is, artery, capillaries, vein, and vena cava. However, there are two unusual pathways that we shall examine: (1) the hepatic-portal route and (2) the coronary circulation.

Hepatic-portal route. One exception to the rule that all veins carry blood directly to the heart is the hepatic-portal route. In this pathway, the capillaries from the spleen, stomach, pancreas, and intestines unite to form the **hepatic-portal vein.** This vein does **not** carry blood to the heart, but carries it to the **liver.** In the liver, the hepatic-portal vein breaks up into capillaries. Thus the liver is able to act on and store digested food as soon as the food has been absorbed into the circulatory system (from the intestine). The capillaries of the liver then unite to form the **hepatic vein,** which carries the blood from the liver to the **inferior vena cava.**

Coronary circulation. The muscles of the heart are **not** nourished by the blood within its chambers. The walls of the heart are much too thick to permit passage of materials to its cells. Instead, the heart is supplied with blood by a **coronary artery** that branches off the systemic artery at the point where that vessel leaves the heart. The coronary artery breaks up into capillaries, which re-form into a few small **coronary veins** that enter the right atrium. The coronary circulation has the most profuse network of capillaries in the body. Each muscle fiber of the heart is serviced by a capillary.

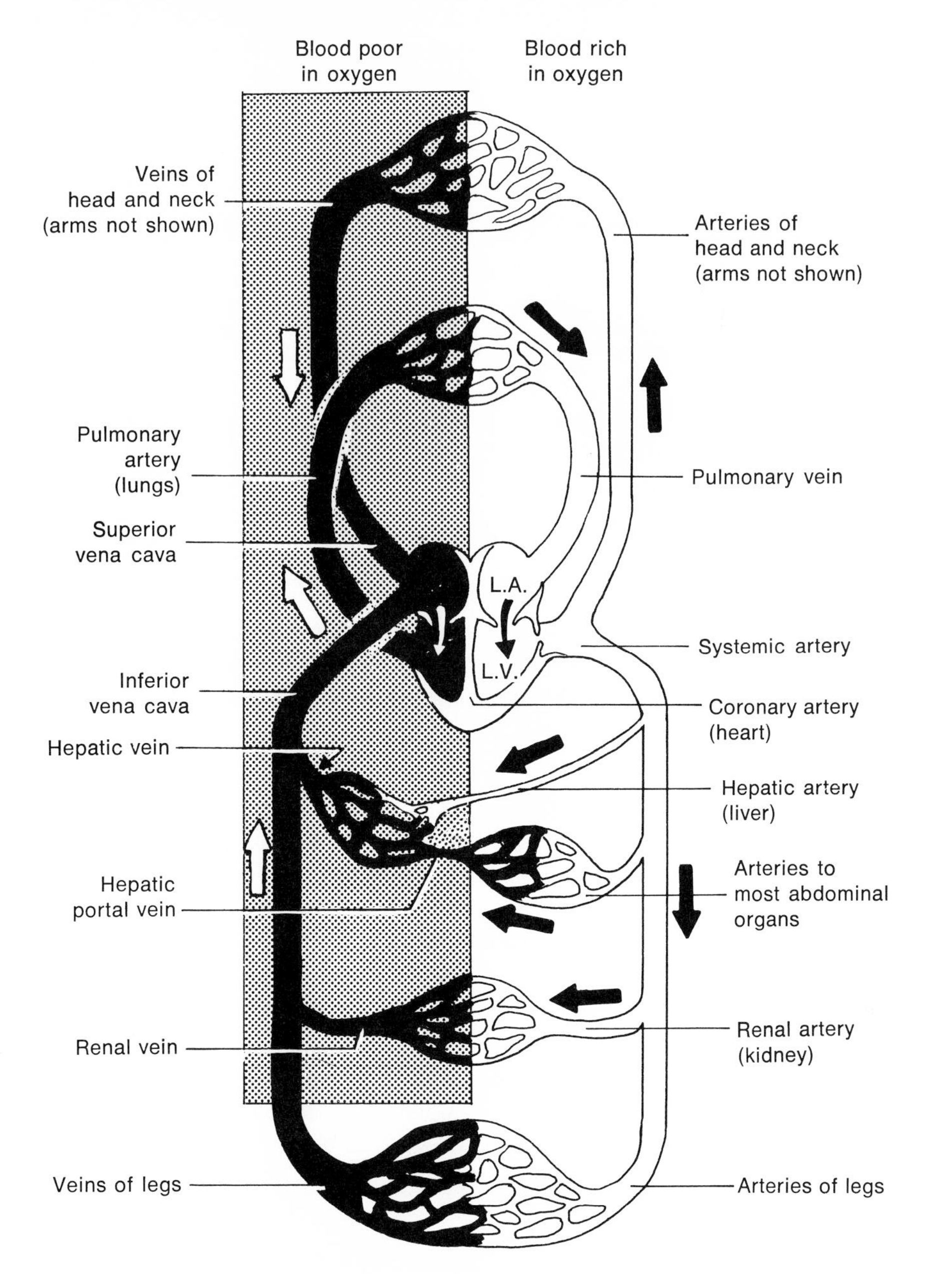

Fig. 6-16
Major routes of blood through body.

☐ Cardiovascular diseases

There are approximately 220 million people in the United States. Our annual death rate is 9.4/1000 people, resulting in approximately 2 million deaths per year from all causes and including all ages. Of these 2 million deaths, 1 million result from cardiovascular diseases, making these the nation's number one cause of death. Of these deaths, 675,000 are caused by diseases involving the heart itself, and 200,000 because of the breaking or clogging of blood vessels in the brain (stroke). The remaining 125,000 deaths caused by cardiovascular disease involve the blood vessels of such organs as the lungs, kidneys, spleen, and liver.

Although a number of diseases of the heart can cause death, the most frequent cause of death is **heart attack.** In a fatal heart attack, the heart stops beating because of the death of an appreciable number of its muscle cells. A mild heart attack occurs when a relatively small fraction of the heart's muscle cells die.

The death of heart muscle cells, called **infarction,** is usually caused by a blocking of one of the branches of the coronary artery, which carries blood to these cells. The cells die because of lack of oxygen. If the blood vessel is only partially occluded, the cells will receive a deficient amount of oxygen and will suffer some damage, a situation called **ischemia.** A short period of relatively mild ischemia results in the individual suffering **angina pectoris** (severe chest pain, often accompanied by a feeling of pressure or suffocation under the breastbone), whereas severe and prolonged ischemia leads to infarction and heart attack (excruciating chest pain, often accompanied by shortness of breath and heavy perspiration).

□ Hypertension (high blood pressure)

Many diseases increase the probability that a person will suffer a heart attack. One of these, **atherosclerosis** (the narrowing of arterial walls as a result of the deposition of cholesterol), was discussed in Chapter 3. Another disease that appears to be a contributing factor in most cardiovascular diseases is **hypertension** (high blood pressure).

What is blood pressure? As stated previously, blood pressure is the force exerted on the blood by the heart's pumping action. With each heart contraction (systole), blood pressure increases. This increased blood pressure is called **systolic pressure.** With each relaxation of the heart muscles (diastole), blood pressure decreases. This decreased blood pressure is called **diastolic pressure.** In normal young adults, the systolic pressure is about 120 mm Hg, and the diastolic pressure is about 80 mm Hg. These measures of blood pressure are commonly expressed as 120/80 mm Hg.

How is blood pressure measured? Blood pressure is measured by a special instrument called a **sphygmomanometer.** An inflatable band, connected to a mercury column, is placed on the upper arm and pumped up by means of a rubber bulb to cut off the blood supply to the lower arm. The pressure of the inflated band on the arm is measured by the height of the mercury column in the sphygmomanometer.

A stethoscope is placed on the inside surface of the arm just below the inflated band, which is slowly deflated. At the instant the physician or nurse hears the first thumping sounds, the height of the mercury column is recorded; this is the person's systolic pressure. The thumping sound is caused by spurts of blood passing through the arm when the contraction of the heart muscles produces a force just greater than that in the partially deflated band. The pressure in the band is further reduced until the sounds suddenly become dull and muffled. At this point, the height of the mercury column is again recorded; this value is the individual's diastolic pressure. The sounds become dull and muffled when the blood flow is no longer impeded by the band (Fig. 6-17).

How serious is hypertension? The seriousness of having high blood pressure can be seen in the estimates of life expectancy for people with normal and elevated blood

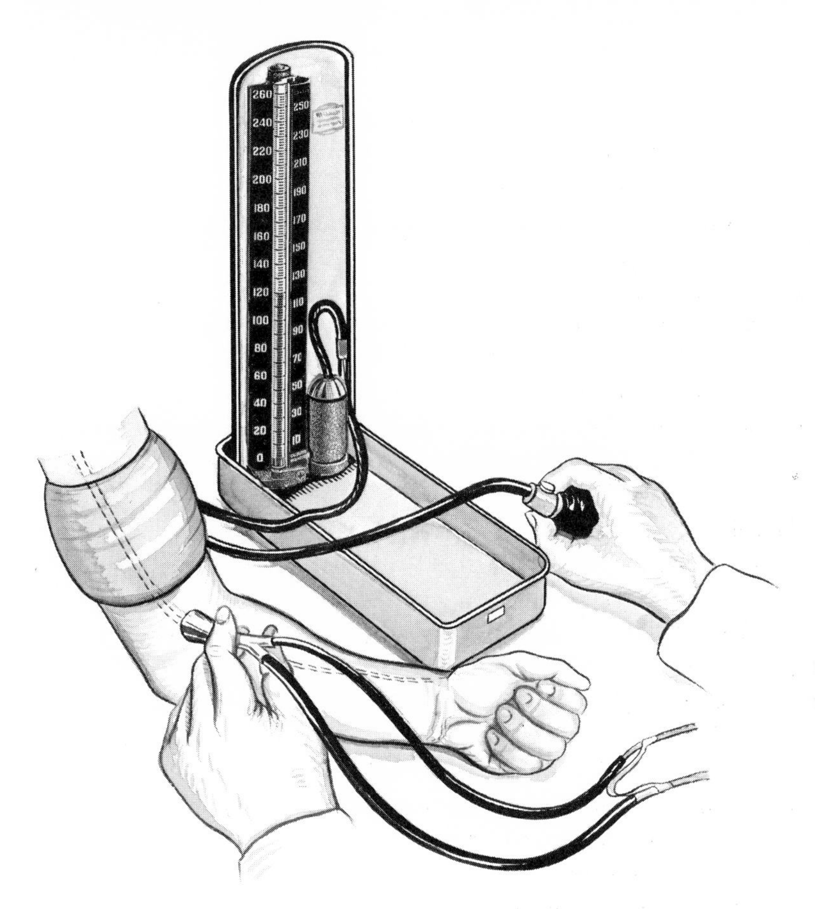

Fig. 6-17
Indirect measurement of blood pressure with a sphygmomanometer. Note relative positions of instrument, subject, and operator. (From Schottelius, B. A., and Schottelius, D. D. 1973. Textbook of physiology, ed. 17. The C. V. Mosby Co., St. Louis.)

pressures. In the United States, a male with normal blood pressure (120/80 mm Hg) can expect to live an average of 77 years. However, a male with 150/100 mm Hg blood pressure can expect to live only to age 65 years. For females, the comparable life expectancies are 82 and 73 years of age, respectively. The decrease in life expectancy of 12 years for males and 9 years for females is all the more frightening when we consider that 150/100 mm Hg is only a relatively moderate level of hypertension.

Two other facts are clear from the preceding data: (1) women have a significantly longer life expectancy than men and (2) women are better able to withstand the deleterious effects of hypertension than are men. The reason for these differences between the sexes is not at all well established, and much research still needs to be done in this field.

In the United States, high blood pressure is listed as the primary cause of 60,000 deaths a year. However, prolonged hypertension can result in heart attack, brain stroke, or kidney damage, one of which is then listed as the primary cause of death. Compared to an individual with normal blood pressure, someone with untreated high blood pressure is four times as likely to have either a heart attack or a stroke and twice as likely to develop kidney disease.

The tragedy of the just described effects of hypertension is that, in most cases, they are avoidable. Modern medicine can now treat virtually every patient with high blood pressure effectively and easily.

How prevalent is hypertension? Twenty-three million people in the United States alone suffer from high blood pressure (approximately 1 in 10 people). However, the disease is not equally distributed through all age groups. Hypertension is more prevalent among older people than among younger ones. Over the age of 21 years, one person in five has high blood pressure (systolic of 150 mm Hg or higher; diastolic of 95 mm Hg or higher). Unfortunately, one person in two will go undiagnosed because there are no distinct symptoms of the condition until it causes some damage to heart, brain, kidneys, lungs, etc.

Males and females are equally prone to hypertension, although, as discussed earlier, females seem better able to withstand its damaging effects.

Different racial groups in the United States show different frequencies of hypertension. One of every four adult Negro Americans has high blood pressure, compared with one of every seven adult Caucasian Americans. The reason for this racial difference is not clear and requires investigation.

What causes hypertension? Despite the increasing ability of medical science to control high blood pressure, **no** single cause has been found. A number of known diseases do produce high blood pressure, but these account for only 5% of hypertension cases. For the remaining 95% of the cases, called **primary** or **essential hypertension,** no known cause has been found, but there are a number of theories as to what produces high blood pressure.

Diet, especially as related to salt consumption, has been implicated in some situations. In a northern province of Japan, the people consume about 50 grams (gm) of salt each day (0.5 gm is the daily adult requirement [see Table 4-3]). Half of that particular Japanese population dies of stroke, a common complication of high blood pressure. In contrast, some of the populations of Africa eat almost no salt and rarely, if ever, develop hypertension.

Excess weight also appears to be linked to increased blood pressure. Each pound of fatty tissue requires about a mile of capillaries to nourish it. This means that the heart must work harder to pump blood through the extra network of blood vessels of an overweight person.

The role of heredity in hypertension also has supporting evidence. Those whose parents have high blood pressure are more likely to be hypertensive than are those whose parents have normal blood pressure.

In all of the instances just mentioned, the association of a particular factor with hypertension does not prove a cause-and-effect relationship. However, the contributing role of high blood pressure in cardiovascular disease makes this a most important field for research.

SUMMARY

We have studied our internal transportation system and have seen how materials are carried from one part of the body to another. We have also found that our blood functions not only in transportation, but also to protect us from invasion by foreign particles and pathogenic organisms.

Also considered were the diseases of the heart and its blood vessels (cardiovascular disease) and the diseases of blood (anemias and leukemias). The chapter closed with a discussion of high blood pressure, the most important contributing factor to cardiovascular disease.

The basic function of both the digestive and circulatory systems is to provide the cells of our body with the basic molecules they need for their activities, growth, and reproduction. In Chapter 7, we will consider the structure and function of cells in some detail, so that we may better understand ourselves.

SUGGESTED READINGS

Adolph, E. F. 1967. The heart's pacemaker. Sci. Am. **216:**32-37. Description of the importance of the heart's contractions in the maintenance of heartbeat.

Cerami, A., and Peterson, C. M. 1975. Cyanate and sickle-cell disease. Sci. Am. **232:**45-50. Description of a chemical compound that can make sickle-cell hemoglobin act like normal hemoglobin in a test-tube experiment. The feasibility of using this compound in human beings has not yet been demonstrated.

Freis, E. D. 1973. The modern management of hypertension, Veterans Administration Publication IB 11-35. U.S. Government Printing Office, Washington, D.C. In-depth review of the causes and treatment of high blood pressure.

Langer, W. L. 1976. Immunization against smallpox before Jenner. Sci. Am. **234:**112-117. Historical review of the two centuries of smallpox vaccination that preceded the modern technique.

Mayerson, H. S. 1963. The lymphatic system. Sci. Am. **208:**80-90. In-depth consideration of the physiological role of this important portion of our circulatory system.

McKusick, V. A. 1965. The royal hemophilia. Sci. Am. **213:**88-95. Discussion of this genetic disease and its effects on three generations of European royalty.

Nossal, G. J. V. 1969. Antibodies and immunity. Basic Books, Inc., Publishers, New York. Easily understood discussion of the body's protective reaction to foreign particles.

Phibbs, B. 1971. The human heart. The C. V. Mosby Co., St. Louis. Excellent guide to an understanding of the heart and its diseases.

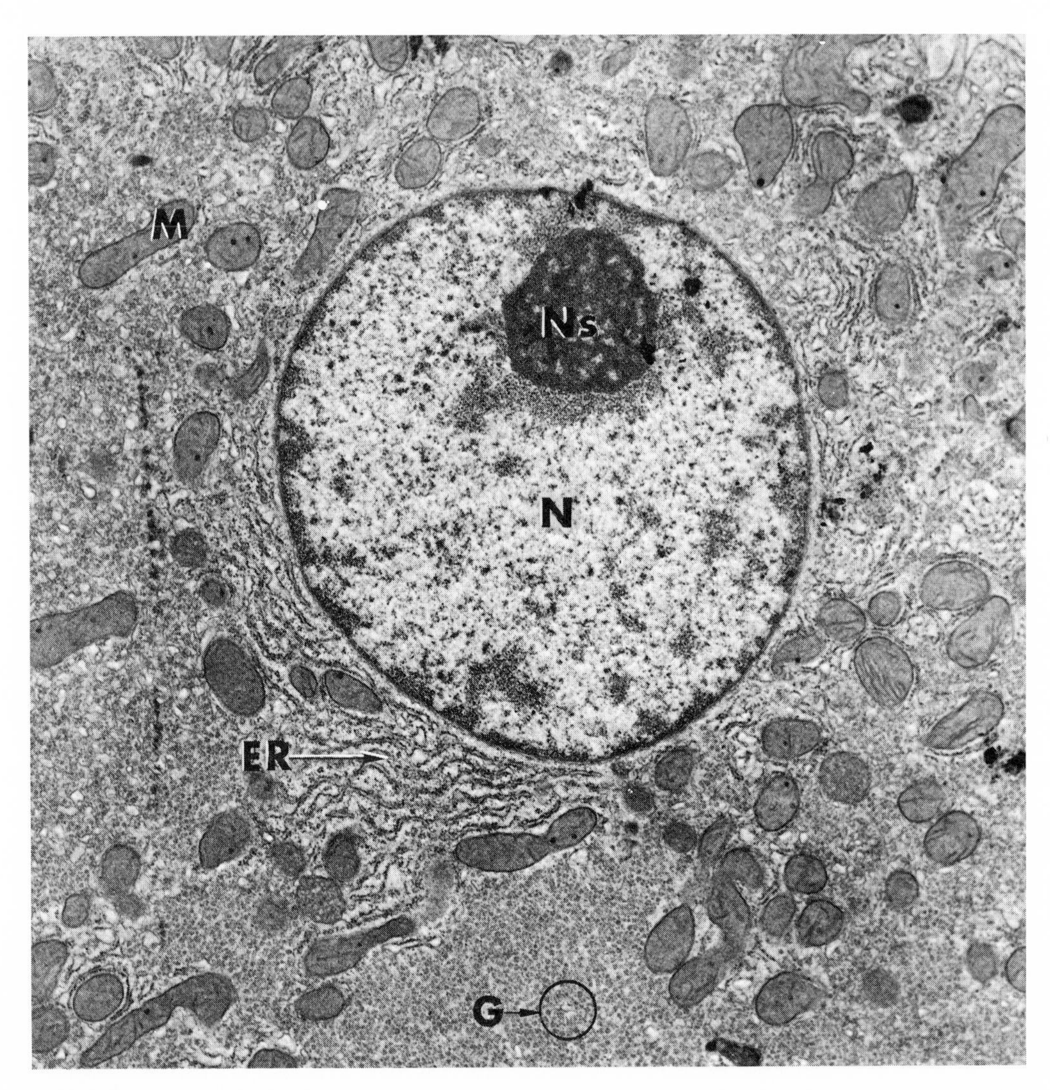

Electron micrograph of portion of human liver cell. *G*, Glycogen granules; *M*, mitochondrion; *ER*, endoplasmic reticulum; *N*, nucleus; *Ns*, nucleolus. Note double-walled "unit" nuclear membrane. (×15,000.) (Courtesy Dr. J. M. Layton; from Schottelius, B. A., and Schottelius, D. D. 1973. Textbook of physiology, ed. 17. The C. V. Mosby Co., St. Louis.)

CHAPTER 7 Our cells and their functions

LEARNING OBJECTIVES

- Which three important bodily functions take place in our cells?
- What are the various parts of a cell and the function of each?
- What are the three component parts of all body tissues?
- What are the four basic types of tissues in our bodies and the functions of each?
- Which disease results from a thickening of the mucus secreted by many of the body's cells, and what are the characteristics of the disease?
- What are the characteristics and cause of Hurler's syndrome?
- What are the characteristics of the muscles associated with the stomach and intestines?
- What are the characteristics of the muscles associated with the skeleton?
- What are the characteristics of the muscles that compose the heart?
- What are the characteristics and causes of muscular dystrophy?
- Which diseases affect nerve cells, and what are the characteristics and causes of each disease?

Our discussions of human populations, the world's food supplies, the nutritional needs of the individual, and the processes of digestion and circulation bring us logically to an examination of our cells and their functions. Most of the important bodily functions take place in our cells: energy is extracted from food, more living material is produced, and the process of reproduction is begun.

The human body is composed of millions of cells that are organized into tissues (muscle, bone, etc.). These tissues in turn combine in a variety of ways to form specific organs (heart, skull, etc.), and different organs unite to form organ systems (circulatory, skeletal, etc.). Although the total individual can do many things that no one cell can accomplish, it is actually the combined activities of many different kinds of cells that are responsible for what is achieved. Thus the cell is the ultimate unit of both structure and function.

A cell is a highly complex unit, consisting of many component parts, a number of which are physically interconnected. It is customary to consider a cell as composed of three main units: (1) plasma membrane, (2) nucleus, and (3) cytoplasm. However, this classification is somewhat misleading in implying nonoverlapping roles for these parts of the cell. As we shall see, this is not the case, and there is much interaction among these cell components. The living material of the cell is called **protoplasm.** This term applies equally to the plasma membrane, nucleus, and cytoplasm.

Because of their small size, the discovery of the existence of cells was possible only after the development of the microscope. It was Robert Hooke in 1665, using

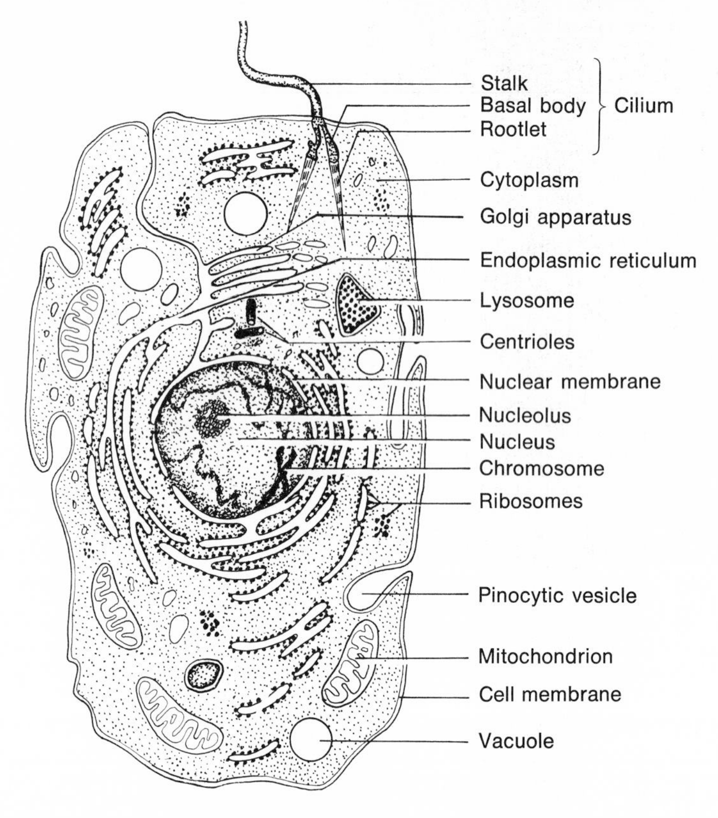

Fig. 7-1
Generalized cell. (From Levine, L. 1973. Biology of the gene. ed. 2, The C. V. Mosby Co., St. Louis.)

one of the then newly invented microscopes, who discovered and reported that a piece of cork consisted of tiny, boxlike cavities, which he called "cells." Hooke's observations were extended by other scientists to other types of plant tissues and also to animals. The subsequent widespread reports of the occurrence of cells in both plant and animal tissues led to one of the most important generalizations of biology, namely, the **cell theory,** which states that *all living organisms are composed of cells.*

A diagram of a generalized cell is shown in Fig. 7-1. There is no such thing as a "typical" cell, because cells are as varied in their organization as they are in their functions. The general characteristics of all cells will be discussed in this chapter.

CELL SIZE

As will be true of all their other characteristics, cells are extremely varied in size. The ostrich egg, despite its large size, is a single cell when it is first formed. In contrast, a human egg cell is right on the edge of visibility for the unaided eye. To appreciate what this means in terms of absolute sizes, let us consider a few basic points. Length may be measured in terms of inches, feet, yards, etc. or in terms of angstroms, microns, millimeters, etc. Although scientists prefer to utilize the metric

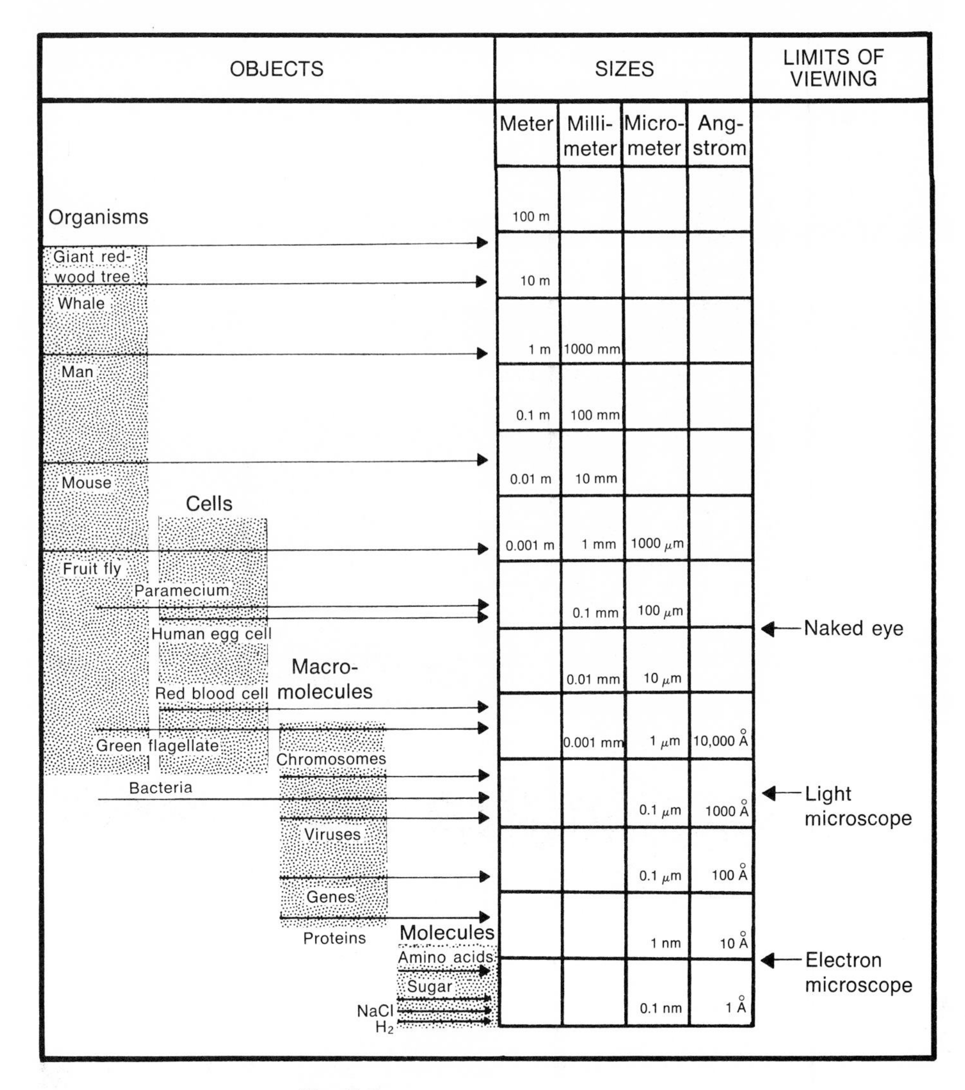

Fig. 7-2
Objects, sizes, and limits of viewing.

system (angstroms, etc.), the two systems of measurement are equatable, as shown here and in Fig. 7-2:

1 inch = 25.4 millimeters (mm)
1 mm = 1000 micrometers (μm)
1 μm = 1000 nanometers (nm)
1 nm = 10 angstroms (Å)

(For a detailed description of measurements, see Appendix C.)

The unaided human eye can identify a dot that is as small as 1/10 mm (1/254 inch) in diameter. A human egg cell is about this size. The best light microscope permits the identification of a dot that is as small as 0.2 μm across and so improves on the naked eye by about 500 times. The electron microscope increases our ability to see objects almost 400 times more. The electron microscope makes it possible to distin-

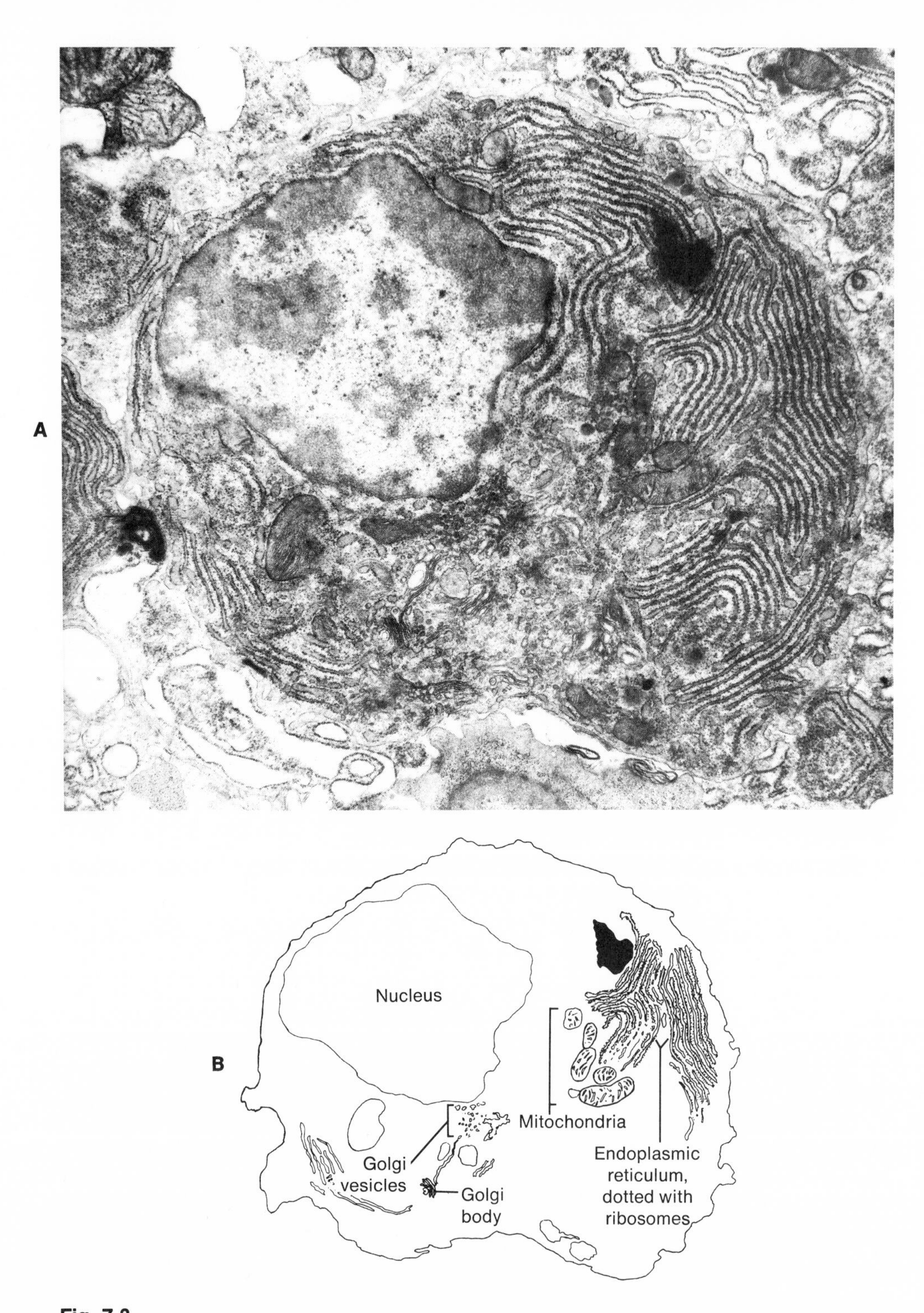

Fig. 7-3
A, Electron micrograph of human lymphocyte (Chapter 6). **B,** Sketch of cell, indicating various internal structures. (**A,** ×15,000.) (Courtesy J. Barrett, University of Missouri, Columbia, Mo.; from Lane, T., ed. 1976. Life the individual the species. The C. V. Mosby Co. St. Louis.)

guish dots about 5 Å in diameter, an improvement of roughly 200,000 times over the human eye.

While a human egg cell is approximately 100 μm in diameter, a human red blood cell is only about 7 μm across and hence well below the level of visibility for the human eye. The same is true of all cells of the body, most of which are about 15 μm in diameter. This size can be contrasted with the largest bacterial cells, which are about 5 μm long. The smallest bacteria and viruses are far beyond the limits of the light microscope and can only be seen through the use of the electron microscope. As our discussion of the cell proceeds, it will also become apparent that many of the structures inside a human cell are so small that they, too, can only be seen with the electron microscope (Fig. 7-3).

PLASMA MEMBRANE

At the outer edge of all cells is a membrane that controls the entry and exit of materials into and out of the cell. This is the **plasma membrane.** It has a thickness of approximately 75 Å, and, hence, its internal structure is not visible when viewed under the light microscope. However, with the electron microscope, the cell membrane is seen as two thin dark lines, each 20 Å thick, separated by a clearer region 35 Å thick. Based on biochemical evidence, the two thin lines are thought to consist of protein macromolecules, and the clearer area is thought to be filled with lipid molecules, thus forming a **lipoprotein** membrane.

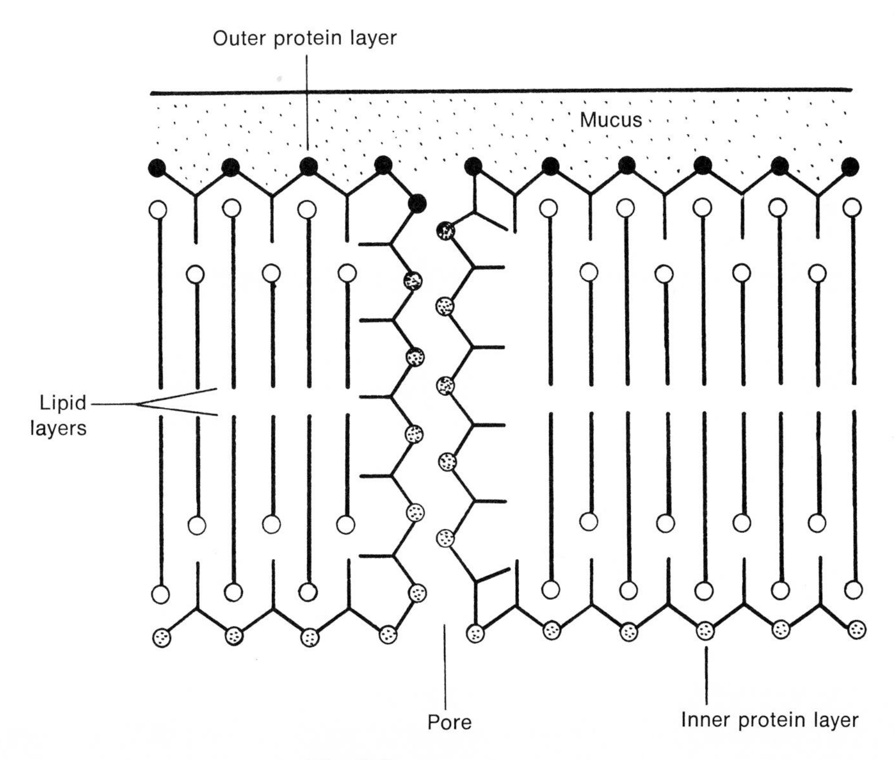

Fig. 7-4
Model of plasma membrane.

A proposed model of the cell membrane is diagrammed in Fig. 7-4. The cell membranes of neighboring cells are usually not in actual physical contact with one another. Between them, we find a cementlike substance that appears to hold the cells together. When cells are organized in sheets of tissue, this cementlike substance forms a supporting **basement membrane** that fuses the cells together. As can be seen in Fig. 7-4, the cell membrane is thought to contain pores.

Materials can only enter or leave a cell when they are contained in a liquid medium. This means that any portion of a plasma membrane that is to function in the passage of substances must be bathed by water. Clearly, the large number of cells that cover the outside of our bodies are without a surrounding film of water and therefore do not function in exchanging materials with the outside environment. This function is restricted to cells that are located within our bodies and in contact with a liquid medium. The cells lining the digestive tract, the lungs, and most other cavities of the body are such cells; their outer surfaces are covered by a thin layer of liquid.

The plasma membrane regulates the flow of substances into and out of the cell in a number of different ways, some of which are not well understood. However, every case involves the movement of molecules in a liquid medium. We shall now consider a number of situations that determine the passage of various materials through the plasma membrane and note how they are related to molecular motion.

□ Diffusion

The phenomenon of **diffusion** occurs when the movement of particles in a system is determined solely by the concentration of these particles in any one portion of the system. This phenomenon may be readily demonstrated by dropping a lump of sugar into a beaker of water. The sugar dissolves, and the individual sugar molecules move from their original positions in the beaker and eventually become dispersed equally throughout the liquid. The movement of sugar molecules is always from a region of high concentration to one of lower concentration. If a lump of sugar and a lump of salt are dropped into two separate parts of the beaker, both substances will distribute themselves equally in the beaker.

Diffusion is a very important mechanism by which the cell exchanges some materials with the environment, especially gases such as oxygen and carbon dioxide. Oxygen, which is normally higher in concentration outside the cells than inside, is constantly entering the cells. Carbon dioxide, on the other hand, is constantly leaving the cells for the opposite reason. These gases dissolve readily in the film of water covering the cell membrane.

Gases are able to enter and leave cells solely under the action of diffusion because their molecules are so small that they pass completely unhindered between and among the molecules of the cell's lipoprotein membrane. This is why poisonous gases, such as those contained in the exhaust fumes from automobiles, prove to be so dangerous to living organisms, including human beings.

□ Osmosis

Sometimes the cell membrane prevents the passage of most particles through itself, while letting water pass freely. A membrane that acts in this differential fashion

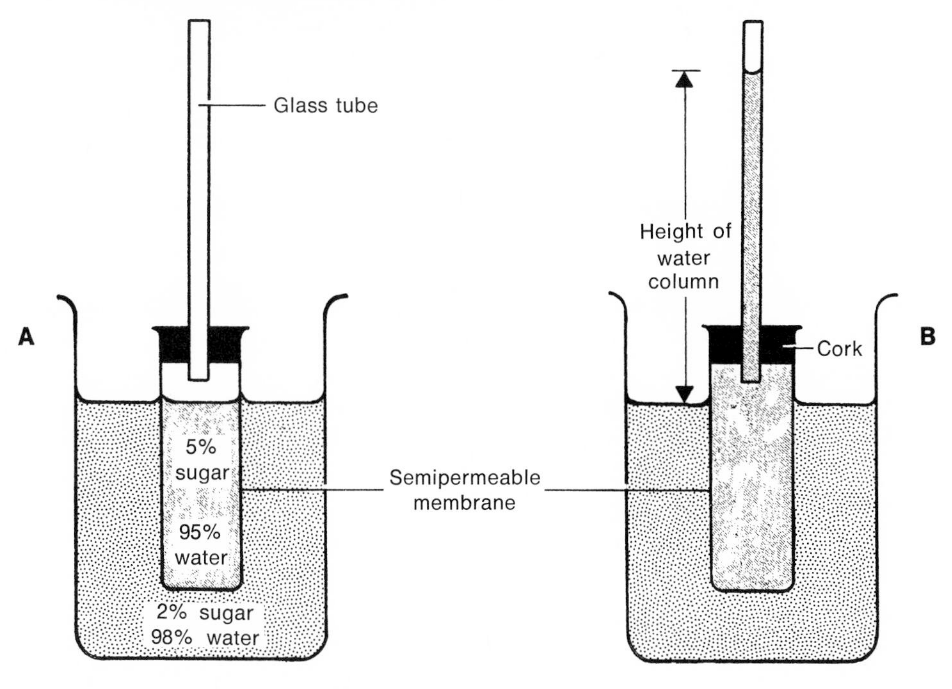

Fig. 7-5
Diagram illustrating osmosis.

is called a **semipermeable membrane.** A simple mechanical model of this type of system is diagrammed in Fig. 7-5 and illustrates the following experiment.

A semipermeable membrane, for example, one made of collodion, is formed into a sac. The sac is filled with a 5% sugar solution and is then fitted with a cork through which passes a glass tube. The sac is then placed in a beaker containing a 2% sugar solution. In this system, the sugar molecules, because of the semipermeable membrane, are unable to diffuse from an area where they are in high concentration (inside the sac) to one in which they are in low concentration (in the outside solution). However, the water molecules are free to do so, and they, in fact, move from the beaker into the membrane sac and up the glass tube. Such diffusion of water through a semipermeable membrane is called **osmosis.** The force operating to push the solution up the glass tube in such a system is called **osmotic pressure.**

The strength of the osmotic pressure will depend on the differences in concentration of particles on the two sides of the semipermeable membrane. Theoretically, the flow of water into the glass tube should continue until the concentrations of sugar in the sac and in the beaker are equal. In actual practice, however, we find that as the height of the water in the glass tube increases, its own weight pushes the molecules of water from the sac back into the beaker. This counterforce is called **hydrostatic pressure.** Eventually, the effect of hydrostatic pressure pushing water molecules out of the sac into the beaker will equal the effect of osmotic pressure operating in the opposite direction. When this situation is reached, the column of water in the tube will stop rising and will remain at that particular level.

Chapter 6 described the interaction between hydrostatic pressure (blood pressure) and osmotic pressure (caused by proteins in the blood) in the exchange of water and materials between the capillaries and the tissue fluid.

Active transport

Up to this point, we have discussed the role of the plasma membrane in gas exchange and water balance. However, solid materials, for example, food and waste materials, must also be capable of entering and leaving the cell, respectively. Therefore the plasma membrane of cells cannot be strictly semipermeable, but must permit passage of at least some solid materials. In other words, it is a **selectively permeable membrane.**

Fig. 7-4 shows that the plasma membrane consists of lipoprotein. The presence of lipids in the plasma membrane makes possible the relatively easy passage of many lipid-soluble substances through the membrane. These lipid-soluble substances enter and leave the cell under the force of diffusion. However, other types of molecules are also able to enter and leave a cell. The mechanism by which they do so involves the phenomenon of **active transport.**

Active transport is the movement of substances through the cell membrane as a result of the expenditure of energy by the cell. This is quite a different process than diffusion and osmosis, which do not require energy. Active transport involves either materials that cannot readily cross the membrane or ones that are present in lower concentration **outside** the cell (and thus tend to diffuse out rather than in). Although the actual mechanism of active transport is not well understood, it is assumed that some sort of "carrier molecule" is involved. The carrier molecule reacts chemically with the molecule to be transported, forming a compound that is soluble in the lipid portion of the plasma membrane. This new lipid-soluble compound then moves into the cell by diffusion, since it is in higher concentration at the periphery of the cell membrane than on its inside surface.

Once the compound reaches the inner surface of the cell membrane, it is broken apart by a chemical reaction, and the carrier molecule moves back through the cell membrane by diffusion, since the "free" carrier molecule is then present in higher concentration at the inside of the cell membrane than at its outside. Under this arrangement, a continuous two-way stream of molecules can provide the cell with its needs for various materials. It is the chemical reactions that initially join and later dissociate the carrier molecule and the transported molecule that require the expenditure of energy.

Membrane selectivity

All cell membranes are selectively permeable. Unfortunately, there is no general rule for the types of molecules that will be either admitted or excluded. For example, sodium and potassium ions are of similar size and charge, but they are handled differently by some cells. As illustrations, we can consider the red blood cells and the nerve cells of the human body. These cells are in constant contact with the body fluids, which contain both these ions, yet potassium ions are actively transported into these cells and concentrated there in a much greater proportion than found in the body fluids. Conversely, sodium ions are kept at a lower concentration in the cell than in the body fluids by actively transporting out most of the ions that may have diffused in. As will be discussed later, this ability of the cell membrane to distinguish between potassium and sodium is of great importance in the functioning of nerve cells.

But the cell membrane cannot completely regulate the exchange of materials between the cell and its surrounding environment. Some poisons can move freely into the cell and kill it, and some beneficial substances may be lost because the membrane cannot prevent them from diffusing out of the cell.

Earlier in this discussion of cell membranes, it was mentioned that they were thought to contain pores. These pores would presumably lead into membrane-bound channels, extending deep into the cell. The existence of pores in the membrane has not been conclusively demonstrated. They are estimated to be only 8 to 10 Å in diameter, and, as indicated in Fig. 7-2, this is the limit of visibility for the electron microscope. If they exist, the function of pores and their channels must be to bring certain substances toward the center of the cell without their having to traverse the cell's protoplasm. This kind of arrangement could be highly important if, in transit through the cell's cytoplasm, a particular substance might become involved in chemical reactions that would be detrimental to the cell. All substances present in the pores and their channels would be separated by the cell membrane from the rest of the cell's protoplasm and would eventually have to pass through the membrane if they are to become incorporated into the cell.

Although we have spoken of the plasma membrane as a regulatory device at the periphery of a cell, it must be remembered that it is an integral component of the cell itself. The plasma membrane is a portion of the cell's protoplasm and is essential for the proper organization and function of a cell.

ENDOPLASMIC RETICULUM

Our discussion of the cell's plasma membrane, up to this point, has stressed its role as the outer, limiting component of the cell. To this we must now add the fact that the cell's membrane folds inward in many places and forms small sacs, called **vesicles.** Some of these invaginations lead into the center of the cell. There channels connect with a complex set of vesicles that interlace the interior of the cell. The existence of this network of channels and vesicles was reported in 1945 by K. R. Porter, who named it **endoplasmic reticulum.** This component of the cell is shown in Fig. 7-1.

The endoplasmic reticulum connects with two other cell structures. It is continuous with the outer layer of a double membrane that surrounds the **nucleus** of the cell, and it is also continuous with the system of tightly packed, smooth-surfaced vesicles called the **Golgi apparatus.** Both these cell components will be discussed later in this chapter.

The membranes comprising the endoplasmic reticulum sometimes appear to be rough (irregular) and sometimes smooth. The difference in appearance is caused by the presence or absence, respectively, of small bodies called **ribosomes** on the protoplasmic side of the membrane, that is, a rough-appearing membrane has ribosomes attached to it (Fig. 7-6). As will be discussed in Chapter 8, the ribosomes are where proteins are synthesized by the cell. The association of ribosomes with the endoplas-

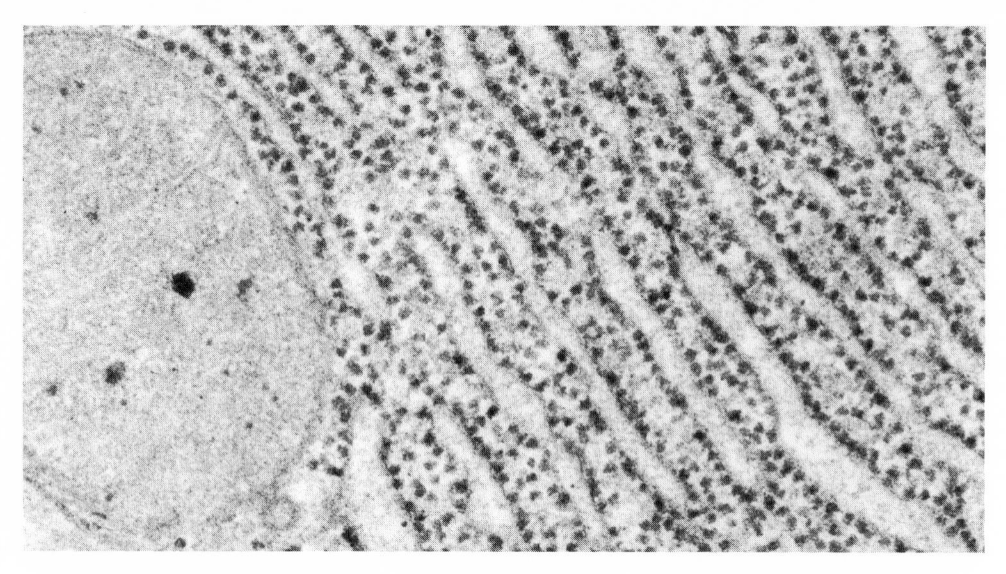

Fig. 7-6
Electron micrograph of portion of pancreatic cell showing endoplasmic reticulum with ribosomes (small dark granules). Oval body at left is mitochondrion. (×66,000.) (Courtesy G. E. Palade, The Rockefeller University, New York; from Hickman, C. P., Hickman, F., and Hickman, C. P., Jr. 1974. Integrated principles of zoology, ed. 5. The C. V. Mosby Co., St. Louis.)

mic reticulum suggests a function for this network of channels and vesicles. Through the use of radioactive chemicals as "tags," it has been found that products synthesized by the cell are distributed both within the cell and to the outside **through the endoplasmic reticulum.** There is also evidence that materials from the outside environment are moved directly to various parts of the cell, including the nucleus, through this system. All substances in the endoplasmic reticulum would, of course, have to pass through the selectively permeable membrane lining its channels before actually entering the cell's protoplasm. The continuity of nuclear membrane, endoplasmic reticulum, and plasma membrane indicates the close interrelationship of these components of the cell.

■ GOLGI APPARATUS

As long ago as 1898, Camillo Golgi discovered and described a system of vesicles (small sacs) in the cytoplasm of the brain cells of barn owls. Further study showed this type of structure to be present in virtually all cells. This system of vesicles was called a "reticular apparatus" by its discoverer, but the name was later to become **Golgi apparatus,** or **Golgi bodies** (Figs. 7-1 and 7-7). More recently, with the help of the electron microscope, the Golgi apparatus has been found to consist of a series of membrane-lined, flattened sacs arranged approximately parallel to one another. The membranes of the Golgi apparatus appear to be identical in construction to the cell's plasma membrane and are found to be continuous, at certain points, with the membrane of the endoplasmic reticulum. Thus the Golgi apparatus is linked to the internal transportation system of the cell.

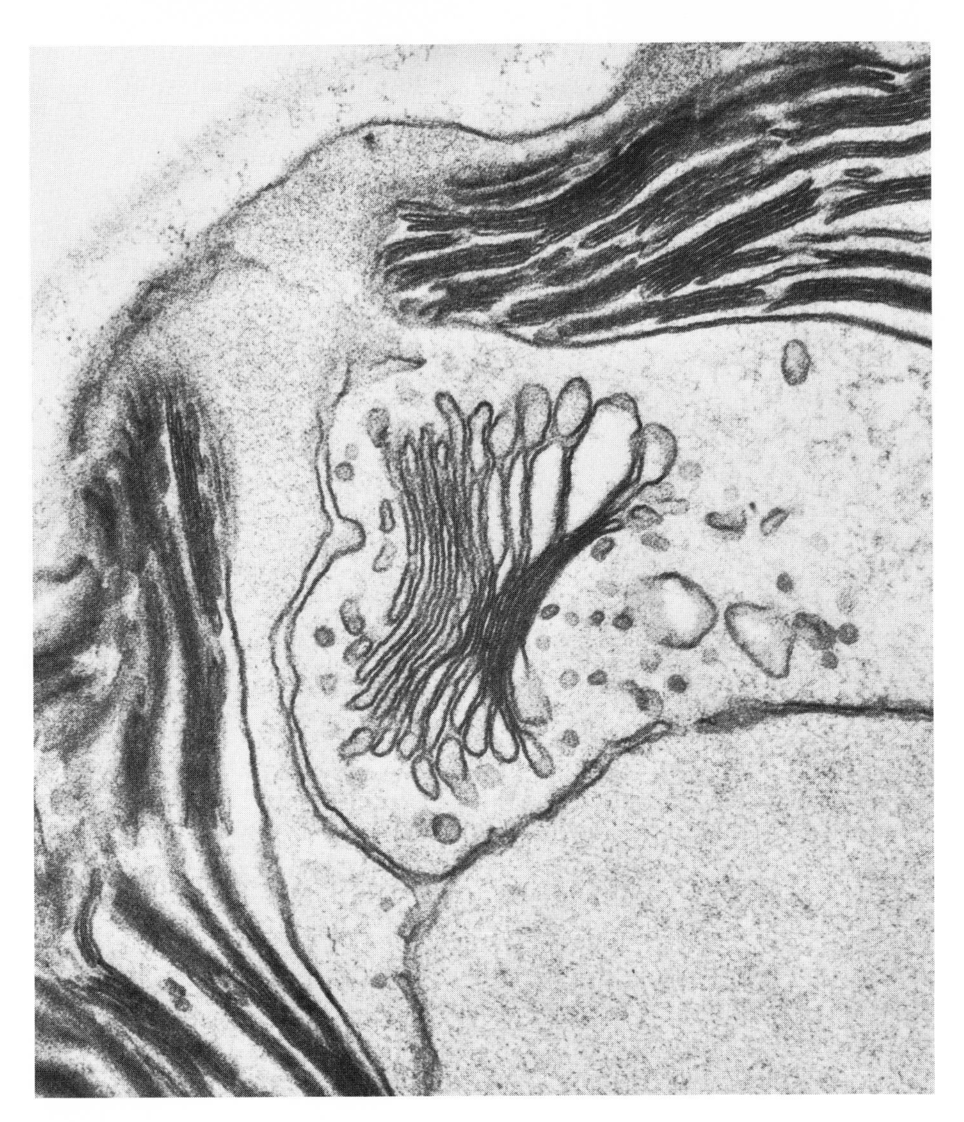

Fig. 7-7
Large Golgi apparatus showing stacked membrane structures and numerous vesicles. (×90,000.) (Courtesy Dr. R. M. Brown, University of Texas Cell Research Institute, Austin, Texas; from Brown, W. V., and Bertke, E. M. 1974. Textbook of cytology, ed. 2. The C. V. Mosby Co., St. Louis.)

Experimental evidence indicates that the Golgi apparatus functions as a storage depot. Proteins manufactured by the cell are stored there until secreted to the outside or distributed to other parts of the cytoplasm. Cells that are active in producing compounds for secretion, for example, cells of the pancreas, liver, and other glands, have an especially prominent Golgi apparatus.

Materials stored in the Golgi apparatus are organized into variously sized units that are then surrounded by lipoprotein membranes. These "units" of stored materials pinch off from the vesicles and, depending on their contents, travel either to other parts of the cytoplasm or to the plasma membrane. When reaching another part of the cytoplasm, the small veiscles open, and their stored materials are distributed to the surrounding protoplasm. When reaching the plasma membrane, the unit fuses with the cell membrane and discharges its contents to the outside.

NUCLEUS

In almost all cells, we find a large distinct body called the nucleus (Fig. 7-1). It was recognized as a regular feature of cells and given its name by Robert Brown in 1831. The periphery of the nucleus is marked by the presence of a double **nuclear membrane.** Each of the two membranes constituting the nuclear membrane is thought to consist of two layers of protein with lipid between them, thus resembling the cell's plasma membrane. As mentioned earlier in this discussion, the outer nuclear membrane is continuous with that of the endoplasmic reticulum. This means that the space between the two nuclear membranes must be continuous with that of the endoplasmic reticulum.

The presence of nuclear membranes makes it possible for the protoplasm of the nucleus to be quite different from that of the surrounding cytoplasm. In fact, the nuclear membranes are found to be highly selective in accepting materials from both the rest of the cell and from the endoplasmic reticulum.

There are points at which the two nuclear membranes appear to fuse. It is thought that some form of pore exists at these junctures and increases the efficiency of diffusion of materials from the cytoplasm to the nucleus and vice versa. However, even these nuclear pores are believed to be highly selective as to which materials will pass through (Fig. 7-8).

Within the nucleus are several types of distinct bodies: the chromosomes and one or more nucleoli. **Chromosomes** are in the shape of either rods or spheres and are composed of nucleic acid (DNA) and protein bound together in a complex called nucleoprotein. The DNA of the chromosomes is the hereditary material of the organism. It is divisible into basic units called **genes,** which are arranged in linear order within the chromosomes.

As will be discussed in Chapter 8, genes determine cell function through the production of proteins. The specific functions of various cells are the result of the different kinds of proteins produced by their genes. Not only do genes function in the daily activities of living cells, but they also determine the characteristics of future generations as they are passed on from parents to offspring.

Nucleoli are generally spherical in shape and may vary in number from one to many, depending on the type of cell. Like the chromosomes, nucleoli are composed of nucleic acid and protein. They are found in association with specific regions of certain chromosomes and represent extra copies of genes from those regions. These extra copies are apparently necessary for the proper functioning of cells at certain times. As discussed in a later chapter, nucleoli disappear during ordinary cell division and are re-formed after the process is completed. Their disappearance indicates that nucleoli are not indispensable for the cell at all times. However, cells lacking their normal number of nucleoli do not function properly and do not survive long.

Although most of the hereditary material of the cell is found in its chromosomes, many cytoplasmic structures have also been shown to contain genes that govern their

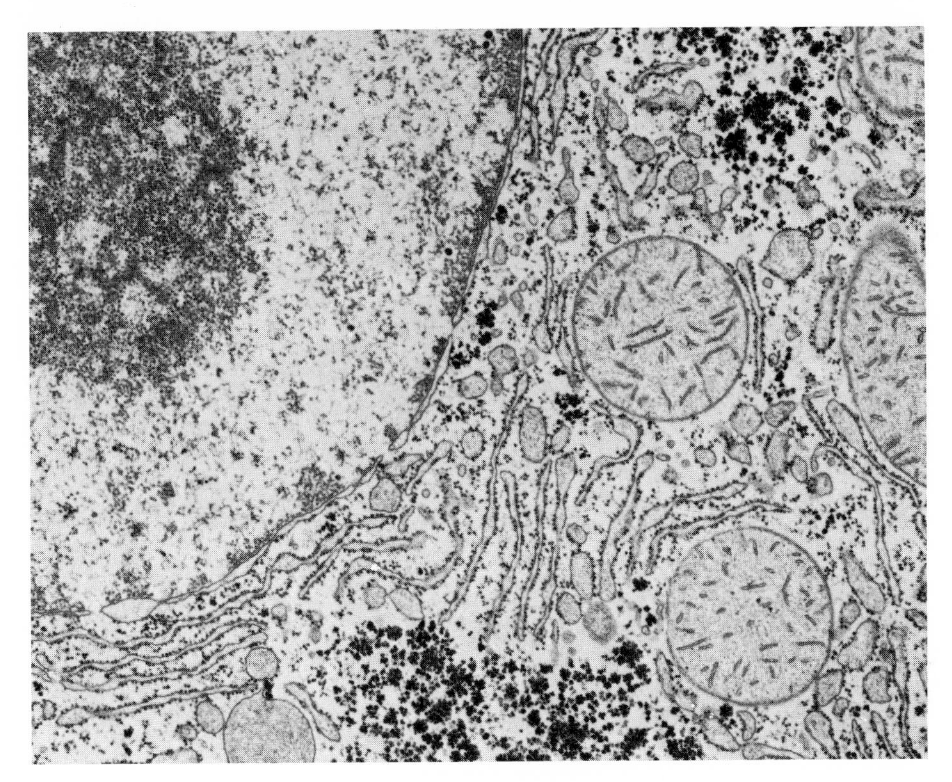

Fig. 7-8
Electron micrograph of part of liver cell showing portion of nucleus (upper left) and surrounding cytoplasm. Endoplasmic reticulum (ribbonlike structures) and mitochondria (oval-shaped bodies) are visible in cytoplasm. Pores of nuclear membrane are also visible. (×14,000.) (Courtesy G. E. Palade, The Rockefeller University, New York; from Hickman, C. P., Hickman, F., and Hickman, C. P., Jr. 1974. Integrated principles of zoology, ed. 5. The C. V. Mosby Co., St. Louis.)

activities. Some of these will be discussed later. The existence of genes in cytoplasmic structures, however, should not diminish our appreciation of the nucleus as an essential cell component that determines most of the cell's activities and is responsible for the transmission of most of the hereditary characteristics to the next generation.

■ MITOCHONDRIA

Another component of cells was discovered toward the end of the nineteenth century; this is the **mitochondrion** (Figs. 7-1 and 7-9). It is roughly the size of a bacterium, as indicated in Fig. 7-2. There may be one mitochondrion in a cell (as in certain single-celled plants), or there may be over a thousand (as in human liver cells). Electron microscope studies have revealed that a mitochondrion, like the nucleus, is bounded by two membranes, each composed of the same alternating layers of protein and lipid that characterize the other protoplasmic membranes. The outer mitochondrial membrane is smooth, but the inner one is extensively folded. The folds are called **cristae.** The interior of a mitochondrion is filled with fluid. The

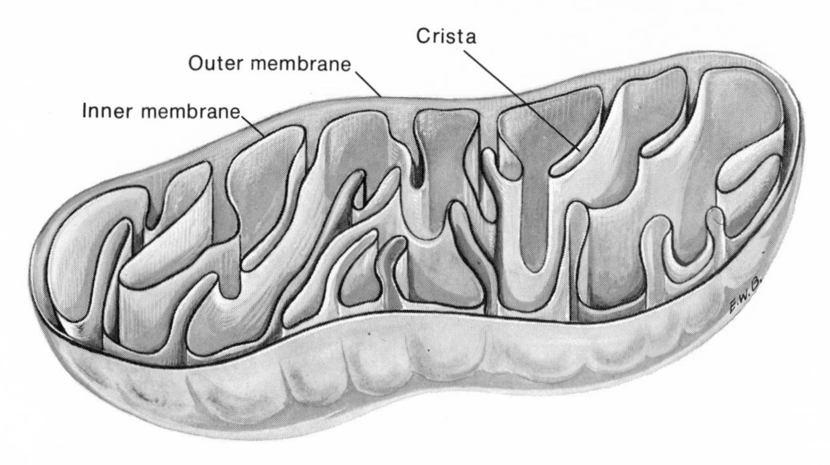

Fig. 7-9
Diagram of mitochondrion, showing inner and outer membranes. (From Schottelius, B. A., and Schottelius, D. D. 1973. Textbook of physiology, ed. 17. The C. V. Mosby Co., St. Louis.)

membranes are selectively permeable and, as a result, are able to regulate the flow of materials into and out of this cytoplasmic structure (Fig. 7-10).

It is possible, experimentally, to break open a cell and separate out its mitochondria from the rest of the cytoplasm. By this means, their functions can be determined outside of the cell in a test tube (**in vitro**). Other experiments have confirmed that they function much the same way in the cell (**in vivo**). Some 90% of the chemical energy obtained by the cell from its food molecules results from chemical reactions that take place in the mitochondria. The complexities of some of these chemical reactions are described in a later chapter. For the present, note that most of the energy-producing reactions take place on the surface of the cristae and further that mitochondria tend to aggregate in those regions of the cell that are most actively engaged in energy-requiring activities.

Mitochondria have also been found to contain nucleic acid, the hereditary material. Therefore in addition to the nucleus, mitochondria function in determining some aspects of cell structure and function. Mitochondria have proved to be most important in the overall functioning of all cells.

■ CILIA AND FLAGELLA

Many types of cells have special hairlike structures whose motions either move the cells or produce a current of liquid across their surfaces. These structures are called **cilia** when they are many in number and relatively short in proportion to the size of the cell. When there are only a few of them and they are relatively long, they are called **flagella** (Fig. 7-1).

Electron microscope studies reveal that cilia and flagella have the same internal structure regardless of the type of cell involved (Fig. 7-11). Each cilium or flagellum consists of the stalk, the basal body, and the rootlet. The **stalk** is an outward extension of the cell membrane that forms a slender cylinder. The **basal body** is located at the base of the stalk, within the main portion of the cell. The basal body gives rise to the stalk and controls its functioning. The **rootlet** projects into the cytoplasm of the cell and serves to anchor the cilium or flagellum.

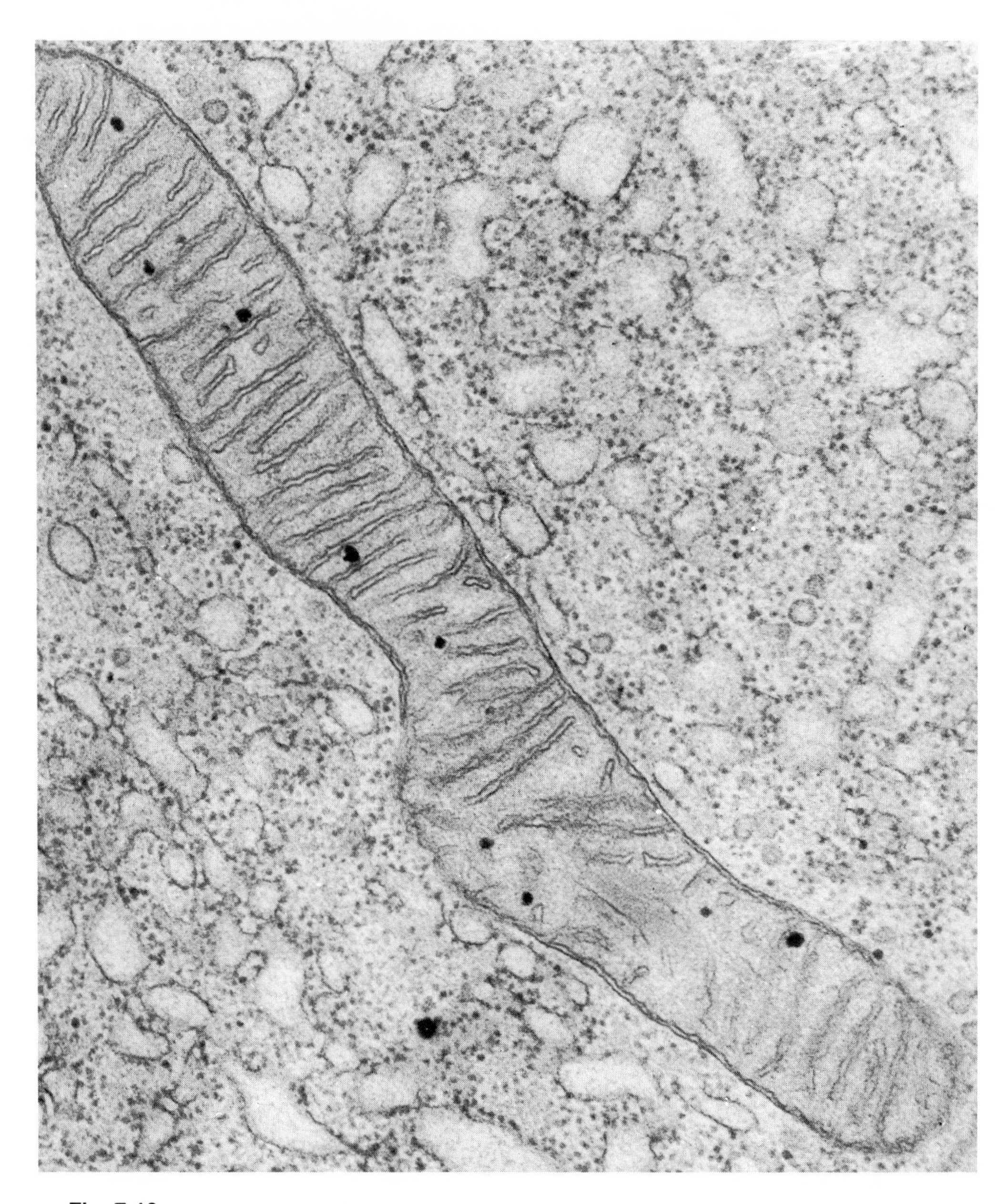

Fig. 7-10
Electron micrograph of elongated mitochondrion in pancreatic cell. Partition-like structures within mitochondrion are cristae formed by inner membrane. (×50,000.) (Courtesy G. E. Palade, The Rockefeller University, New York; from Hickman, C. P., Hickman, F., and Hickman, C. P., Jr. 1974. Integrated principles of zoology, ed. 5. The C. V. Mosby Co., St. Louis.)

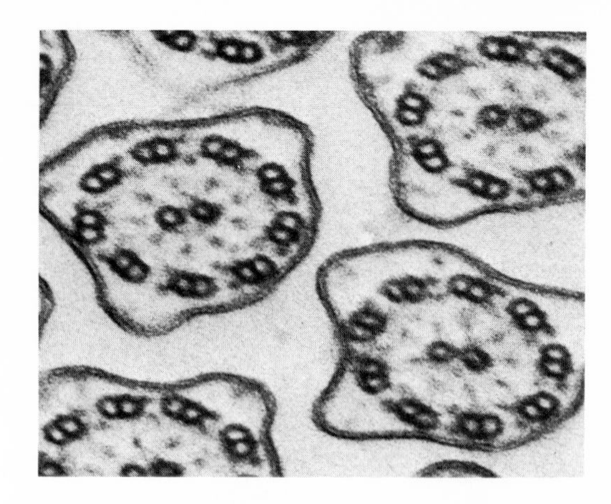

Fig. 7-11
Cross section of flagella. Each flagellum has nine peripheral and two central fibrils, each made up of two microfibrils enclosed in sheath. Cilia have same construction. (Courtesy I. R. Gibbons, Harvard University, Cambridge, Mass.; from Hickman, C. P., Hickman, F., and Hickman, C. P., Jr. 1974. Integrated principles of zoology, ed. 5. The C. V. Mosby Co., St. Louis.)

Cilia and flagella play important roles in the human body both in the production of currents and in locomotion. In the human female, for example, the egg cell is moved from the ovary to the uterus by the coordinated beating of cilia of the cells that line the upper end of the female reproductive tract. In contrast to this, the sperm cell of the human male propels itself through the female's reproductive tract to the egg by the forceful beating of its single flagellum.

CENTRIOLES

In the cytoplasm of almost all cells, there are small cylinder-shaped structures called **centrioles** (Fig. 7-1). They are 300 to 400 nanometers (nm) long and about 150 nm in diameter. Centrioles are often found in pairs, with each member of the pair lying at right angles to its partner. They are usually located near the cell nucleus.

Centrioles can function in two different cell activities. They are important in cell division, as will be discussed in a later chapter. They also form the basal body from which a cilium or flagellum originates. In human males, for example, the flagellum of a sperm arises from one of the centrioles of each cell that becomes a sperm.

As in the case of mitochondria, centrioles are found to contain DNA. This indicates that centrioles play some role in the control of cell structure and function.

LYSOSOMES

Well over a hundred years ago, it was observed that the death of a cell was immediately followed by a chemical breakdown of its protoplasm. This dissolution of the cell was caused by the chemical reactions that occurred within the dead cell; the phenomenon was named **autolysis.** The causative agents of autolysis were later shown to be cellular enzymes, and it was quickly realized that during the life of a cell these enzymes must be somehow prevented from destroying the cell that produced

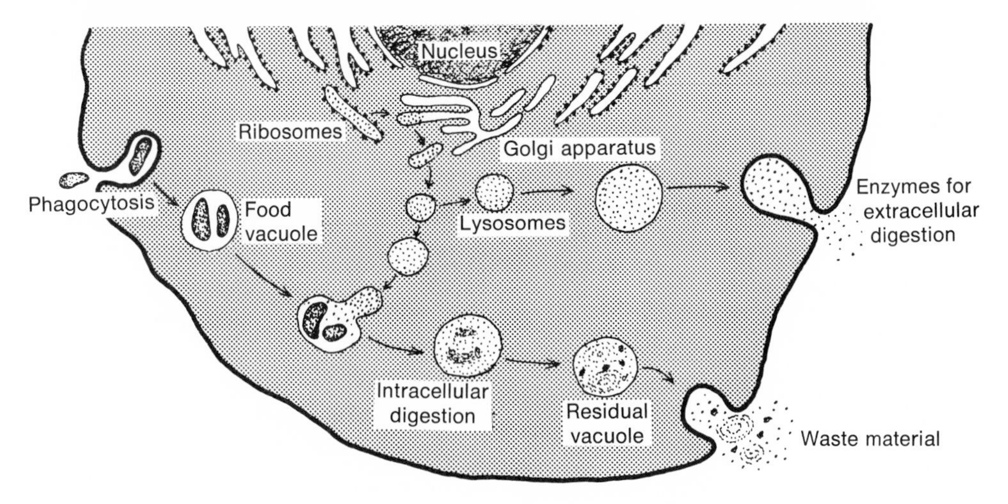

Fig. 7-12
Diagram illustrating lysosome activity.

them. The cell could be saved from self-dissolution if, for example, these enzymes were produced and kept in a chemically inactive form until used and if, in addition, they were stored in special structures that were physically separated from the rest of the cell's protoplasm.

In 1955 Christian de Duve identified special cytoplasmic vesicles that contained a variety of inactive enzymes that, when activated, were capable of chemically breaking down all the major compounds of protoplasm: carbohydrates, lipids, proteins, and nucleic acids. These vesicles were called **lysosomes** (Fig. 7-1). They are usually spherical and about a fourth of a micrometer in diameter. Lysosomal enzymes, like other proteins, are synthesized on the ribosomes of the endoplasmic reticulum. These enzymes then pass through the channels of the endoplasmic reticulum to the Golgi apparatus, where they are packaged in the type of vesicle now known as a lysosome.

As illustrated in Fig. 7-12, lysosomes may either discharge their enzymes (lysozymes) to the outside of the cell for extracellular digestion of materials, or they may fuse with a "food vacuole" and use their enzymes for intracellular digestion. In the latter situation, digested particles pass through the selectively permeable vesicular membrane into the protoplasm of the cell, whereas the undigested residue is discharged to the outside when the vesicle fuses with the cell's membrane.

Human white blood cells, **leukocytes,** contain large numbers of lysosomes. These cells are important in the bacteria-fighting function of leukocytes. Bacteria that enter the body are engulfed by leukocytes through the process of phagocytosis. The vesicles containing the engulfed bacteria represent food vacuoles whose contents are subjected to intracellular digestion, as described previously. The importance of this process can be seen in those individuals whose lysosomes are defective. Such people are said to suffer from the **Chediak-Higashi syndrome,** an hereditary disease that occurs in individuals who carry two genes for this trait. The lysosomes of these affected people do not destroy invading bacteria. This deficiency makes these people extremely susceptible to infection, which usually results in death during childhood.

TISSUES OF THE BODY

The millions of cells of our bodies are specialized into a number of distinct types to perform various functions. Skin cells protect; bone cells support; muscle cells contract; nerve cells generate and relay electrical impulses; etc. The individual cells of the body do not act alone in carrying out their tasks, but rather are associated with many others of the same type. A collection of similarly specialized cells united in the performance of a specific function is called a **tissue.**

A tissue has three components: (1) the cells that are characteristic of the particular tissue, (2) the intercellular space between the cells, and (3) various structures and products that are secreted by the cells into the intercellular space.

The three components of a tissue seldom exist in equal proportions. One component is frequently dominant at the expense of the others. However, all cells exist in an intercellular space that contains at least a small amount of **tissue fluid.** This fluid bathes the cell and permits the exchange of substances between the cells and the capillaries of the circulatory system.

There are four basic types of tissues: epithelium, connective tissue, muscle, and nerve tissue.

Epithelium

Epithelial tissues consist of cells that are tightly fitted together. In epithelium, the cell itself is the dominant component, with little, if any, cell products or intercellular space. This type of tissue covers the surface of our bodies and lines our internal cavities (intestinal tract, lungs, etc.).

Epithelial tissues perform a number of important functions. Those that cover the body act to **protect** the underlying cells from mechanical injury and invasion by harmful bacteria. The epithelial cells that make up the inner lining of the digestive tract function in the **absorption** of food and water, whereas those of the salivary glands are involved in the **secretion** of digestive enzymes, as are many of the digestive tract cells as well.

There are a number of types of epithelial tissues, illustrated in Fig. 7-13, depending on the shape and arrangement of their cells: (1) **squamous epithelium,** composed of flattened cells, is found on the skin's surface and the lining of the mouth; (2)

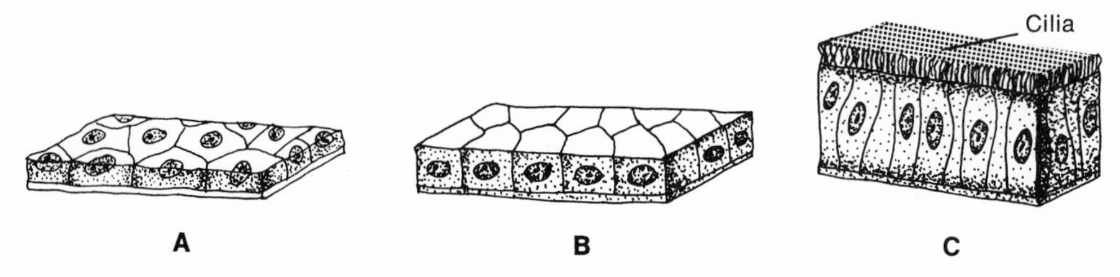

Fig. 7-13
Simple epithelium. **A,** Squamous. **B,** Cuboidal. **C,** Columnar (ciliated).

cuboidal epithelium, composed of cube-shaped cells resembling ice cubes, lines the ducts of the kidney; and (3) **columnar epithelium,** composed of elongated (tall) cells, is found in the lining of the stomach and intestines.

Epithelial tissues may consist of either a single layer of cells, in which case they are referred to as **simple** (squamous, cuboidal, or columnar), or many layers, called **stratified** (squamous, cuboidal, or columnar). Our skin is an example of stratified squamous epithelium, whereas the ducts of our kidneys are composed of simple cuboidal epithelium.

In many cases, columnar cells possess **cilia** on their free surfaces. These cilia function to move materials along the cell surfaces, as discussed earlier.

Epithelial cells, and all other types of cells, may be affected by hereditary diseases that afflict the individual. For example, one of the most devastating disorders among children is **cystic fibrosis.** This disease occurs in individuals who carry two genes for this trait. Among the 3.1 million children born in the United States each year, 1 in every 1000 suffers from this disease. However, the different human racial groups are not equally affected. Virtually all infants with cystic fibrosis are Caucasians; for some presently unknown reason, Negroes and Orientals are rarely afflicted.

Cystic fibrosis affects the mucus-secreting epithelial cells. Instead of these cells forming a thin, freely flowing secretion, they secrete a thick, slow-moving fluid. This thick form of mucus accumulates in and blocks small passageways and ducts of the body. The effect of cystic fibrosis is most noticeable in the ducts leading from the pancreas and in the small passageways (bronchioles) leading to the lungs.

The accumulation of mucus in the ducts of the **pancreas** prevents the pancreatic enzymes from reaching the duodenum. As a result, the digestion and absorption of nutrients, particularly lipids and proteins, is drastically reduced, and the non-digested food is expelled as excessively large, frothy, and foul-smelling stools. Medical research has developed a compensation for this effect of cystic fibrosis; afflicted children can swallow prescribed amounts of pancreatic enzymes each day.

In the **bronchioles,** the thick mucus obstructs the flow of air to and from the lungs. Affected children have a severe hacking cough and labored breathing, and they often suffer a temporary collapse of one or both lungs. The accumulated mucus is an excellent place for bacterial growth, which contributes to frequent and severe upper respiratory tract infections. Before the discovery of antibiotics, most cystic fibrosis victims died in infancy from bacterial infection. Today, they are able to inhale antibiotic-containing vapors that serve to soften the thick mucus of the bronchioles, prevent the occurrence of bacterial infections, and permit the survival of an increasing number of these unfortunate people.

□ Connective tissue

Connective tissues are the most variable and widespread of all the tissues of the body. The cells of this type of tissue are separated from one another, and the spaces between them are filled with materials that are secreted by the cells. These intercellular substances are the dominant components of this type of tissue. The intercellular material gives each type of connective tissue its characteristic color, shape, elasticity, strength, etc. The various types of connective tissue (Fig. 7-14)

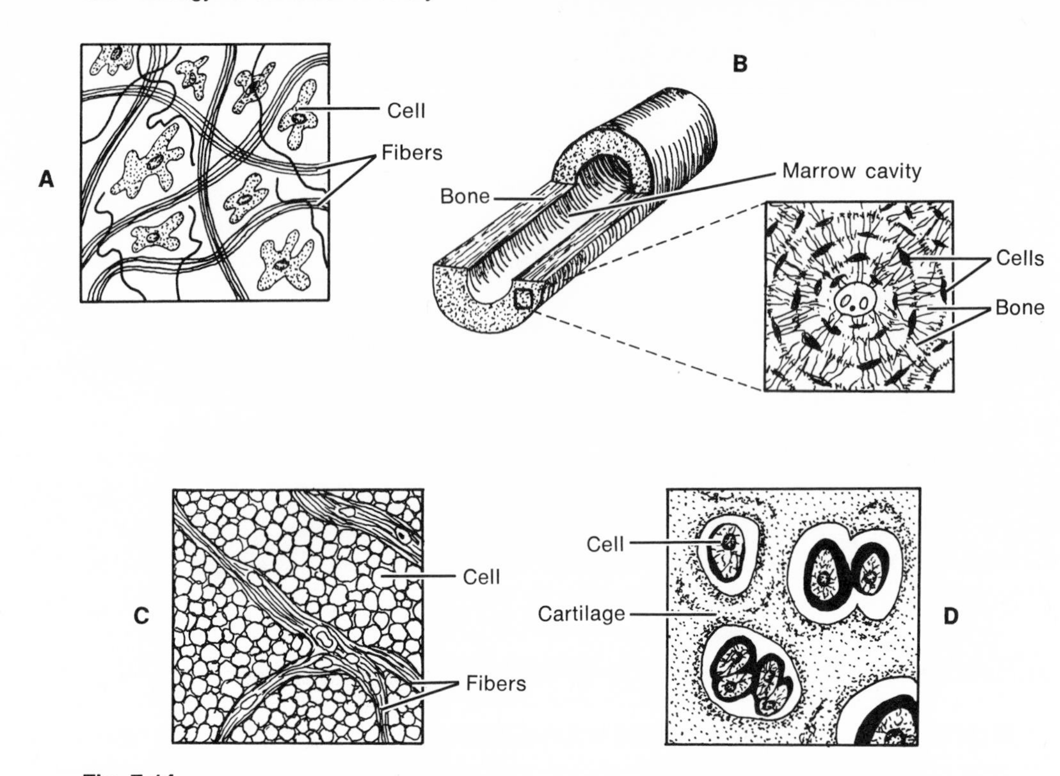

Fig. 7-14
Various types of connective tissues. **A,** Fibrous connective tissue. **B,** Bone. **C,** Fat. **D,** Cartilage.

include fibrous connective tissue, tendons and ligaments, fat (adipose), cartilage, bone, and blood.

The intercellular spaces of **fibrous connective tissue** are filled with a thick, crisscrossing network of microscopic fibers. This tissue occurs throughout the body and functions to bind many structures together. It attaches the skin to the underlying muscles and also forms the sheaths that surround muscles, blood vessels, and most organs of the body. The fibrous connective tissue underneath the skin is especially thick. In animals, such as cows, pigs, etc., this layer of connective tissue can be removed and treated chemically to form **leather.**

In **tendons** and **ligaments,** the intercellular fibers are all oriented in one direction, providing both flexibility and strength. Tendons bind muscle to bone; ligaments bind bone to bone.

Fat (adipose) tissue is a highly modified form of connective tissue in which there is a buildup of fat globules inside the cells. Fat cells are located within the fibrous connective tissue under the skin and around the intestines. Adipose tissue stores energy in the form of reserve food and, in addition, insulates the body against excessive loss of heat.

Cartilage is a tough, somewhat elastic material found at the ends of bones, in the tip of the nose, and in the ear flaps. Its intercellular material is a hard, rubbery substance. In the development of a human fetus, a good deal of the skeleton is first formed of cartilage and is later converted into bone.

Bone is a connective tissue in which the intercellular spaces are filled with calcium salts and organic compounds composed of various protein-polysaccharide combinations. Bone is the supporting, or skeletal, tissue of the body. It is very hard and rigid. Most bones have a large central cavity, called the **marrow cavity.** It is in the bone marrow that red blood cells and certain of the white blood cells are manufactured.

Blood is a connective tissue in which the intercellular spaces are filled with liquid (blood plasma), and the erythrocytes and leukocytes are the cellular elements.

In the chapter on the circulatory system, we discussed a number of diseases of blood. These included two environmentally caused diseases of red blood cells, namely, pernicious anemia and lead poisoning anemia. In addition, two genetic diseases of erythrocytes were analyzed: thalassemia and sickle cell anemia.

There are many other diseases that affect connective tissues. One of these is called **Hurler's syndrome,** after Gertrud Hurler, who described the diagnostic features of the disease in 1919. This disease is genetically determined and, as with many of the genetic diseases, occurs in individuals who carry two genes for this trait. Its frequency in the population is about 1 in 40,000 people.

The most striking features of children suffering from Hurler's syndrome are short stature, grotesque facial appearance (resembling that of gargoyles), protruding abdomen, extremely stiff and partially bent fingers (giving the hands a spadelike or clawlike appearance), and mental retardation. Most individuals also suffer from some form of heart damage involving the various valves of the heart.

At birth, the affected child appears normal. The infant develops normally for a few months, but then progressively deteriorates, physically and mentally. The joints become stiff, growth is extremely slow, and the liver and spleen become enlarged. Death, which in most cases occurs before 10 years of age, usually results from infection or heart failure.

The fundamental defect in those suffering from Hurler's syndrome is an excessive production of **mucopolysaccharides** (compounds composed of polysaccharides and proteins) by the fibrous connective tissue cells. These mucopolysaccharides are stored not only in the connective tissue cells themselves, but also in the cells of the brain, heart, liver, and spleen. The manner in which excess mucopolysaccharides interfere with normal body functions is not known, and, unfortunately, no satisfactory treatment for this disease has yet been developed.

□ Muscle

Muscle is a type of tissue that functions primarily in contraction. In the cytoplasm of muscle cells, there are long thin threads, called **myofibrils,** that are the contractile elements. The intercellular space is drastically reduced in muscular tissues, and there are no true intercellular products. Human beings possess three kinds of muscle tissues (Figs. 7-15 and 7-16): smooth, striated, and cardiac muscles.

Smooth, also called **involuntary, muscle** is associated with various internal organs (stomach, intestines, etc.), blood vessels, and ducts of the body. The individual muscle cells are long and spindle shaped and are completely separable from one another. Smooth muscle cells are usually found in double sheets of parallel cells, with the long axes of the two groups at right angles to each other.

Smooth

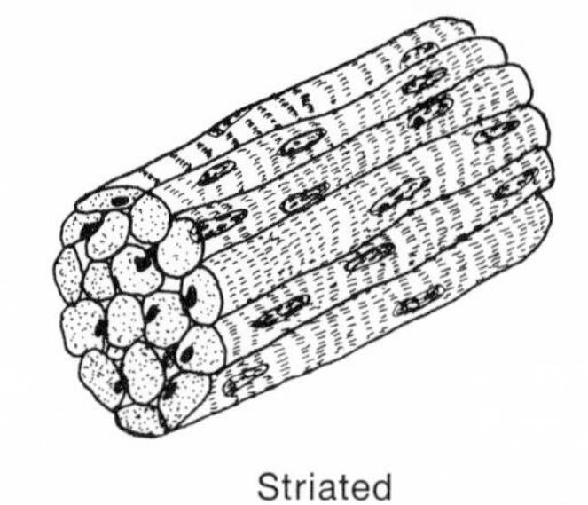

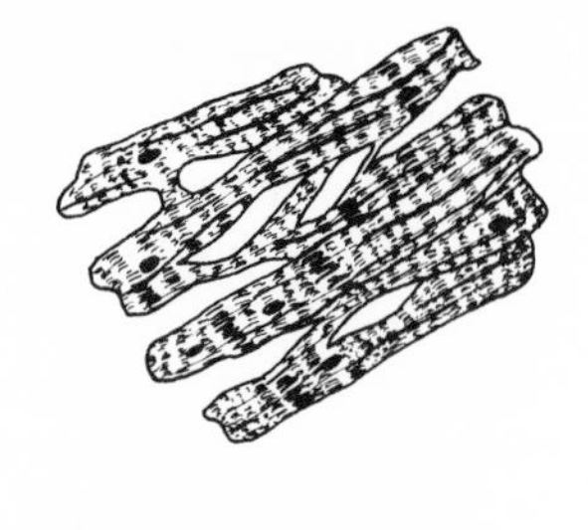

Fig. 7-15
Types of muscle.

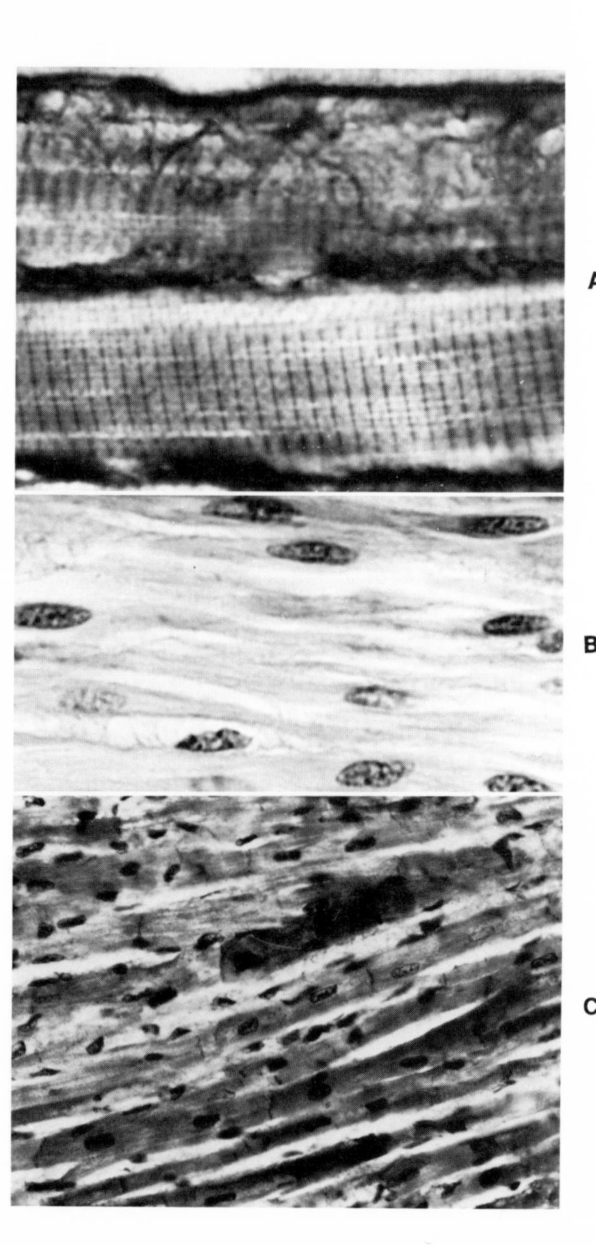

Fig. 7-16
Microscopic appearance of three types of muscle tissue. **A**, Striated. **B**, Smooth, **C**, Cardiac. (×640.) (**A** and **B** from Bevelander, G., and Ramaley, J. A. 1974. Essentials of histology, ed. 7. The C. V. Mosby Co., St. Louis. **C** courtesy Paul Gardner, University of Nebraska Medical Center, Omaha; from Lane, T., ed. 1976. Life the individual the species. The C. V. Mosby Co., St. Louis.)

The contractions of smooth muscles are relatively slow and gentle and generally are **not** under the control of the individual. The waves of muscular contractions of the intestines (peristalsis and segmentation) are rather typical of smooth muscle activities.

Striated, also called **voluntary** or **skeletal, muscle** constitutes the bulk of the body's tissues and is found associated with the skeleton. About two thirds of our body weight is striated muscle.

The individual units of striated muscles are called **fibers.** They differ from typical cells in having many nuclei. The muscle fibers, when viewed under the microscope, show very conspicuous **cross striations** (hence the name). The fibers are arranged in bundles, which are grouped together as muscles. Striated muscle contractions are under the control of the individual and in general tend to be powerful and energetic, as seen in those contractions that produce the movements of the body.

Cardiac muscle is the muscle of the heart. Cardiac muscle cells have properties of both smooth and striated muscles. Like smooth muscle cells, they have only one nucleus per cell. However, their myofibrils form cross striations like those of striated muscle fibers.

Cardiac muscle differs structurally from both smooth and striated muscle in that each of its cells is joined to adjacent cells by cytoplasmic connections. In addition, cardiac muscle differs functionally from both involuntary and voluntary muscles in that its contractions are both automatic and rhythmical.

Many diseases can affect the muscles of the body. Some involve the muscle fibers themselves, whereas others involve the nerves that control the affected muscles. **Muscular dystrophy** is a general name given to a number of closely related diseases that appear to involve the muscle fibers themselves.

All the muscular dystrophic diseases are characterized by a progressive weakening of the striated (skeletal) muscles, with increasing disability and deformity of the body. Muscular dystrophy is the largest and most important single group of muscle diseases of childhood. It is estimated that about 250,000 people in the United States suffer from some form of this disease.

The basic defect in the various types of muscular dystrophy is unknown, and there is no cure. Physical therapy is used as a supportive measure. It tends to delay and minimize the inevitable deformity of a victim's body and assists the affected individual in meeting the demands of daily living.

Of the various forms of muscular dystrophy, the most commonly seen in childhood, and also the most severe, is the **Duchenne (pseudohypertrophic)** type. It generally takes effect before the child is 5 years old and progresses relentlessly until death, which is usually caused by heart failure or respiratory tract infections and occurs somewhere between 20 and 40 years of age. During this period of time, the individual goes from experiencing difficulty in movement to requiring leg braces to the absolute need for a wheelchair.

The Duchenne type of muscular dystrophy affects **males** almost exclusively. As we shall discuss in Chapter 9, hereditary traits that affect males almost exclusively are caused by genes that are located on the X chromosome (one of the sex chromosomes). This type of trait is called **sex linked,** because it occurs in one sex (males) far more frequently than in the other.

Nerve tissue

Nerve tissue consists of cells that possess the properties of **irritability** and **conductivity.** These characteristics enable nerve cells to initiate and transmit the electrical impulses that provide the body with information about its environment and that regulate many of the reactions of the body to environmental changes.

An individual nerve cell, called a **neuron,** is depicted in Figs. 7-17 and 7-18. It consists of the following parts: (1) incoming fibers, called **dendrites,** which carry impulses from either the environment or other nerve cells to the nucleus-containing portion of the particular neuron; (2) the **cell body,** which contains the nucleus of the

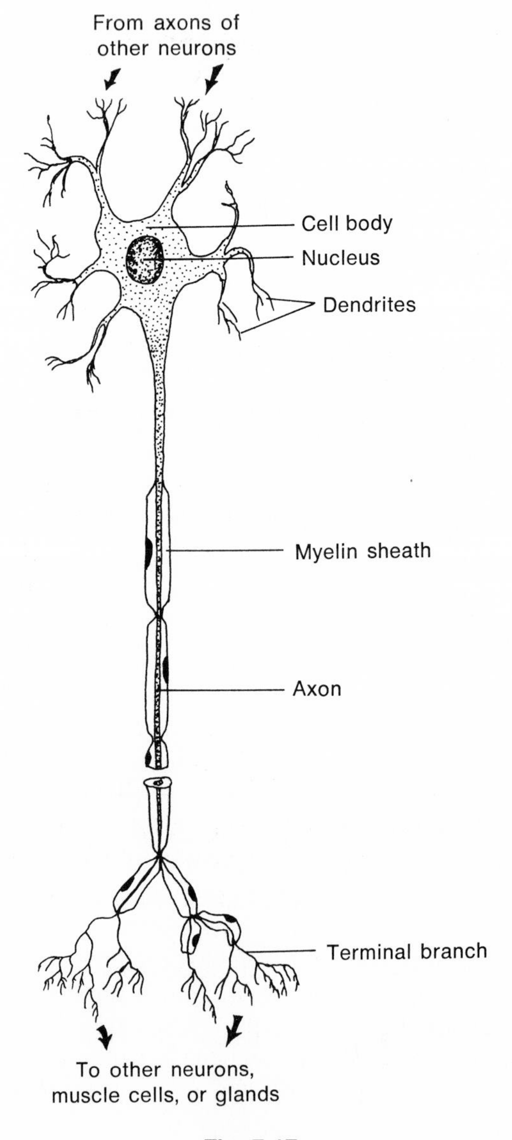

Fig. 7-17
Neuron.

neuron; and (3) an outgoing fiber, called an **axon,** along which impulses are carried either to other nerve cells or to muscles or glands.

What are commonly called **nerves** are actually bundles of nerve processes, namely, dendrites and axons. Nerve processes may be naked (**nonmyelinated**), or they may be surrounded by an insulating sheath (**myelinated**) made of a lipid substance, called **myelin.** In certain regions of the body, we find large numbers of nerve cell bodies. A concentration of nerve cell bodies is called a **ganglion.**

The nervous system consists of a tremendous number of neurons that form a great number of circuits throughout the body. Each circuit consists of two or more neurons. In every circuit, a nerve impulse travels from neuron to neuron by passing from the axon of the first nerve cell to the dendrite of the second. The junction between the axon of one neuron and the dendrite of the next is called a **synapse.** However, the nerve processes do not actually touch at the synapse. Rather, the nerve impulse is passed across the synapse by means of a chemical substance (a neurotransmitter substance).

Many diseases affect nerve cells. Some diseases are infections caused by microorganisms, for example, poliomyelitis is caused by a virus and neurosyphilis by a bacterium; others are known to be genetic in origin, for example, Huntington's chorea and infantile spinal muscular atrophy. In addition, for a considerable number, there is either no known cause, for example, multiple sclerosis, or the disease is known to have many possible origins, for example, epilepsy.

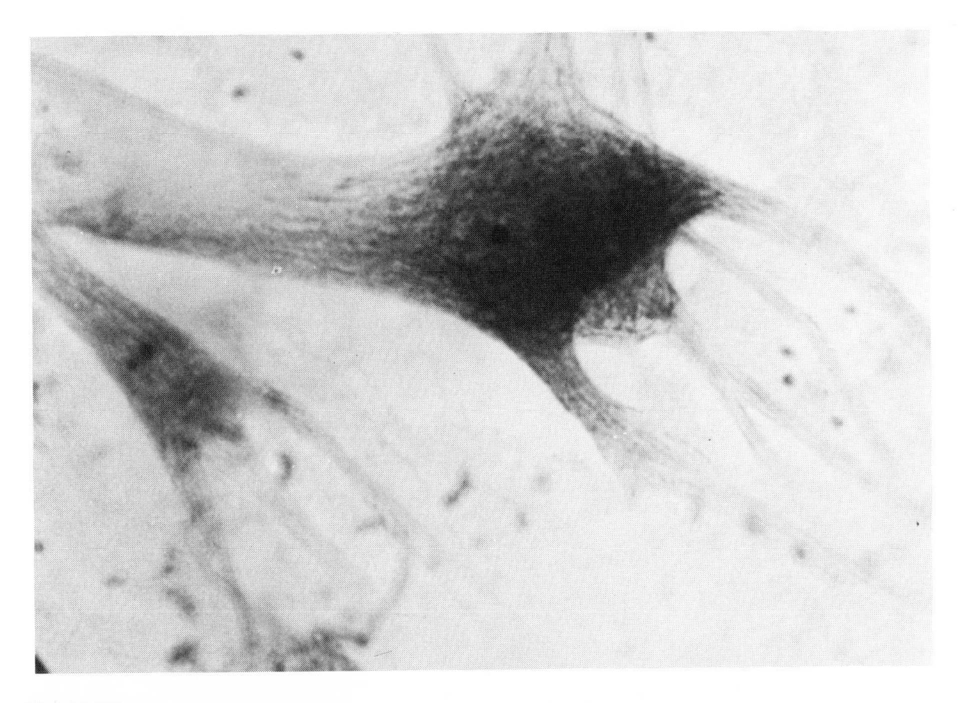

Fig. 7-18
Photomicrograph of spinal cord neuron showing cell body, large axon, and several small dendrites. (From Phillips, J. B. 1975. Development of vertebrate anatomy. The C. V. Mosby Co., St. Louis.)

In most nerve cell diseases, regardless of causative agent, the striated (skeletal) muscles of the body are affected. Muscle involvement in nerve cell disease occurs because the regulation of muscle contractions is a function of many neurons.

Nerve cell diseases may cause the death of the neurons involved, which results in the termination of normal stimulation of the part of the body served by these neurons. When a particular muscle is served by such affected neurons, the muscle will no longer contract (paralysis) and will waste away (atrophy). This is the situation in poliomyelitis (infantile paralysis). It was once a dread disease, but today it is largely controlled by the administration of vaccines.

A number of nerve cell diseases injure, but do not kill, the cell. In many of these cases, the affected neurons initiate and conduct impulses in a random and uncoordinated fashion at unpredictable times. When muscles are the recipients of these impulses, there occurs the uncontrolled twitching and jerking movements so characteristic of Huntington's chorea and epilepsy. There is no known cure for Huntington's chorea, and a large proportion of the affected individuals eventually die of progressive brain damage. In the case of epilepsy, medical research has developed medicines that can at least prevent the occurrence of the erratic muscle spasms and seizures.

■ SUMMARY

In this chapter, we reviewed the various components of cells and their functions. These components include the plasma membrane, endoplasmic reticulum, and Golgi apparatus, which together form an internal transportation and storage system for the cell. We also considered the various factors (diffusion, osmosis, active transport, and membrane selectivity) that determine the passage of materials into and out of the protoplasm of the cell.

Other cellular components studied were nucleus (main hereditary unit), mitochondria (main energy-obtaining unit), cilia and flagella (locomotor and current-producing structures), centrioles (organizational units of cell division), and lysosomes (storage units for inactive forms of digestive enzymes).

Cells of the body are associated into four types of tissues. These include (1) epithelium, the tissues that cover our bodies, line our internal cavities, and secrete many of our body fluids; (2) connective tissue, the tissues that bind together, support, and protect many of our body organs; (3) muscles, the tissues specialized for contraction; and (4) nerve tissue, the tissue specialized for irritability and conductivity.

We shall next turn our attention to the mechanism by which the cell manufactures the chemical compounds necessary for its activities and growth. Chapter 8 will concentrate on the production of proteins, both structural and enzymatic, because these are crucial for the continued survival, functioning, and well-being of every cell and therefore of every human being.

SUGGESTED READINGS

Allison, A. 1967. Lysosomes and disease. Sci. Am. **217**:62-72. Historical account of the discovery of lysosomes and their role in autolysis, foreign particle invasion, and cancer.

Capaldi, R. A. 1974. A dynamic model of cell membranes. Sci. Am. **230**:27-33. Clear, concise discussion of how the structure of the cell membrane determines its functions.

Everhart, T. E., and Hayes, T. L. 1972. The scanning electron microscope. Sci. Am. **226**:54-69. Detailed explanation of how to obtain three-dimensional pictures of objects using the electron microscope.

Lemons, R. A., and Quate, C. F. 1975. Acoustic microscopy: biomedical applications. Science **188**:905-911. Clear account of the use of sound waves instead of light waves for the microscopic study of cells.

Sleigh, M. A. 1971. Cilia. Endeavour **30**:11-17. Well-written account of the internal structure of cilia and how these structures accomplish their primary function of propelling fluids over cell surfaces.

Swanson, C. P. 1969. The cell, ed. 3. Prentice-Hall, Inc., Englewood Cliffs, N.J. Concise, well-illustrated book on the cell.

Whaley, W. G., Dauwalder, M., and Kephart, J. E. 1972. Golgi apparatus: influence on cell surfaces. Science **175**:596-599. Authoritative description of the role of the Golgi apparatus in the assembly of molecules produced by the cell and their transportation to the cell surface.

Wischnitzer, S. 1974. The nuclear envelope: its ultrastructure and functional significance. Endeavour **33**:137-142. In-depth discussion of the structure of the membrane that surrounds the nucleus and how this membrane functions to regulate the exchange of materials between nucleus and cytoplasm.

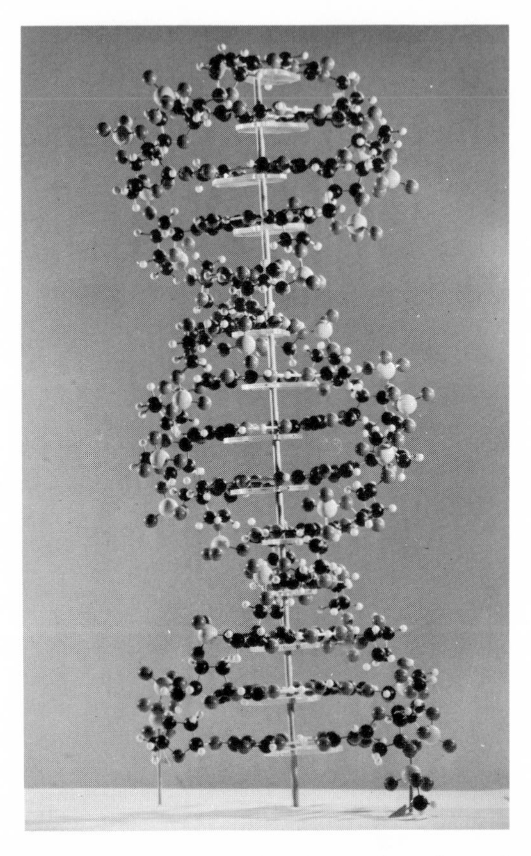

Model of DNA molecule from the laboratory of Professor Alexander Rich of the Massachusetts Institute of Technology. (From Turner, C. E. 1971. Personal and community health, ed. 14. The C. V. Mosby Co., St. Louis.)

CHAPTER 8 How our cells make more of themselves

LEARNING OBJECTIVES

- What is the hereditary material, and of what chemical components is it made?
- What is the physical arrangement of the chemical components of the hereditary material?
- What are the two basic functions of the hereditary material?
- How does the hereditary material determine the production of protein?
- How does the hereditary material make more of itself?
- What is a gene?
- What is the genetic code?
- What changes may occur in the genetic code, and what term is applied to such changes?
- What is an enzyme, and on what principle does it function?
- In what ways does the cell use the energy contained in food?
- What are the characteristics of the chemical compound used by the cell to store energy?
- What steps are involved in obtaining energy from carbohydrates?
- What steps are involved in obtaining energy from lipids?
- What steps are involved in obtaining energy from proteins?

The characteristic of human beings that we call **growth** involves a large number of complex processes by means of which the cells of our bodies manufacture more protoplasm. This ability to make more of themselves is contained within the cells and is also **transmissible** from each dividing cell to both resulting daughter cells.

Because of the tremendous number of structures, enzymes, and chemical compounds that make up a cell, an understanding of the growth process may seem impossible. Fortunately, the problem is greatly simplified by the fact that all cell structure and function is controlled by proteins.

Some of these proteins are **structural proteins,** which, in association with lipids and other compounds, form the component parts of the cell. However, most of the proteins are **enzymes,** which regulate the many different chemical reactions of the cell. These chemical reactions include those that supply energy to the cell, as well as those that result in the synthesis of lipids, glycogen, and various other compounds.

■ THE HEREDITARY MATERIAL

The fact that the ability to make more protoplasm is transmissible during cell division means that the cell's genetic material is the controlling factor, because this is the material that is passed on from cell to cell. It has been known since the early 1900's that the cell's genetic material is contained within the nucleus of the cell and specifically within the **chromosomes** of the nucleus. Chemical analysis showed that

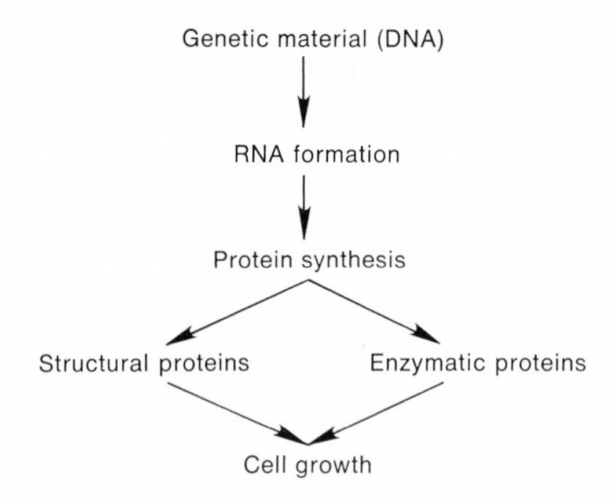

Fig. 8-1
Pattern by which the cell's genetic material controls growth.

the chromosomes consist of proteins and another compound called **deoxyribonucleic acid** (DNA).

In 1944 O. T. Avery, C. M. MacLeod, and M. McCarty convincingly demonstrated that DNA is the genetic material. Further research showed that DNA does not control the production of proteins **directly**, but rather through the use of another compound called **ribonucleic acid** (RNA). The overall process is outlined in Fig. 8-1. In this chapter, we shall be concerned with the structures and functions of the nucleic acids DNA and RNA, especially as related to protein synthesis and DNA replication.

□ DNA

In an earlier chapter, it was pointed out that proteins consist of chains of amino acids. Chemical compounds that are formed by the combination of smaller building blocks are called **polymers.** It is of great importance to stress that DNA is also a polymer.

The building blocks of DNA are called **nucleotides.** Each nucleotide consists of three subunits (Figs. 8-2 to 8-4). One is a nitrogen-containing compound in the form of either a single ring (pyrimidine) or a double ring (purine). In either case, the compound is called a nucleotide **base.** The second subunit of each nucleotide is a 5-carbon **sugar** (pentose), and the third component is **phosphoric acid.**

Each nucleotide of DNA contains **one** of the following four bases: **adenine** (A), **guanine** (G), **thymine** (T), or **cytosine** (C). Of these, adenine and guanine are purines (double rings), and thymine and cytosine are pyrimidines (single rings). The sugar in DNA is deoxyribose.

As stated earlier, DNA is a polymer. Its nucleotides are linked together, as shown in Fig. 8-5, to form a **polynucleotide chain,** in which the sugar of one nucleotide is bound to the phosphate group of another nucleotide, and so on.

Model of DNA structure. Any model of the physical structure of the hereditary material must help explain how DNA carries on its essential functions. One function is the **production of proteins.** The other is to provide for the transmission of its

O
‖
H—O—P—O—H
|
O
|
H

Fig. 8-2
Chemical structure of phosphoric acid.

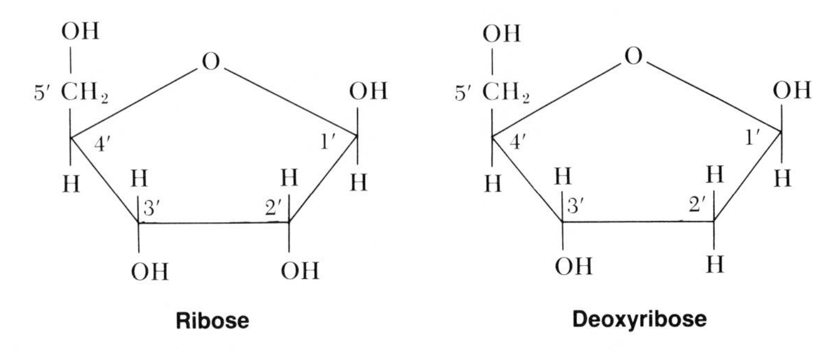

Fig. 8-3
Chemical structures of pentoses found in nucleic acids. (From Levine, L. 1973. Biology of the gene, ed. 2. The C. V. Mosby Co., St. Louis.)

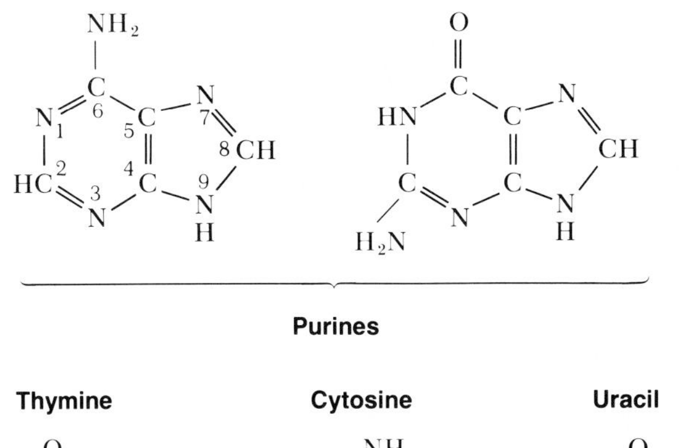

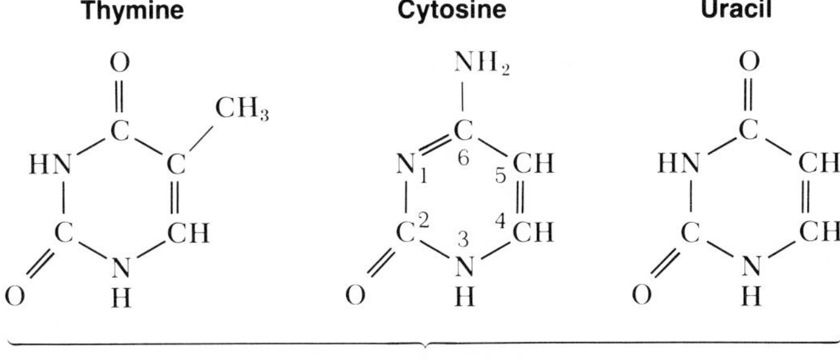

Fig. 8-4
Chemical structures of bases of DNA and RNA. (From Levine, L. 1973. Biology of the gene, ed. 2. The C. V. Mosby Co., St. Louis.)

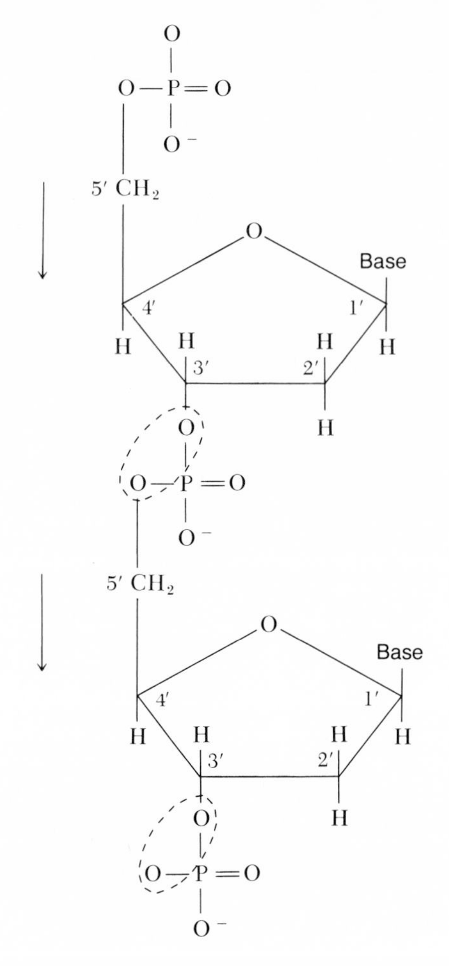

Fig. 8-5
Linkage between nucleotides. "Base" is either a purine or pyrimidine. (From Levine, L. 1973. Biology of the gene, ed. 2. The C. V. Mosby Co., St. Louis.)

protein-synthesizing ability to all cells that are formed from cell division. This second function requires that DNA **replicate itself.**

Chemical analysis of DNA shows that for each nucleotide containing adenine, there is one of thymine, and for each nucleotide containing guanine, there is one of cytosine. A study of the physical characteristics of DNA shows that it has a spiral or helical shape. This information led J. D. Watson and F. H. C. Crick, in 1953, to propose the model for DNA structure shown in Fig. 8-6. They hypothesized that DNA consists of two helically coiled polynucleotide chains (**double helix**). The ribbonlike backbone of each chain is made up of phosphates and sugars, and the nucleotide bases project between them as bars and hold the two strands of the double helix together through chemical bonds.

Based on the fact that adenine and thymine are found in DNA in equal amounts, as are guanine and cytosine, Watson and Crick reasoned that adenine-containing nucleotides must be paired with thymine-containing ones, and guanine must be

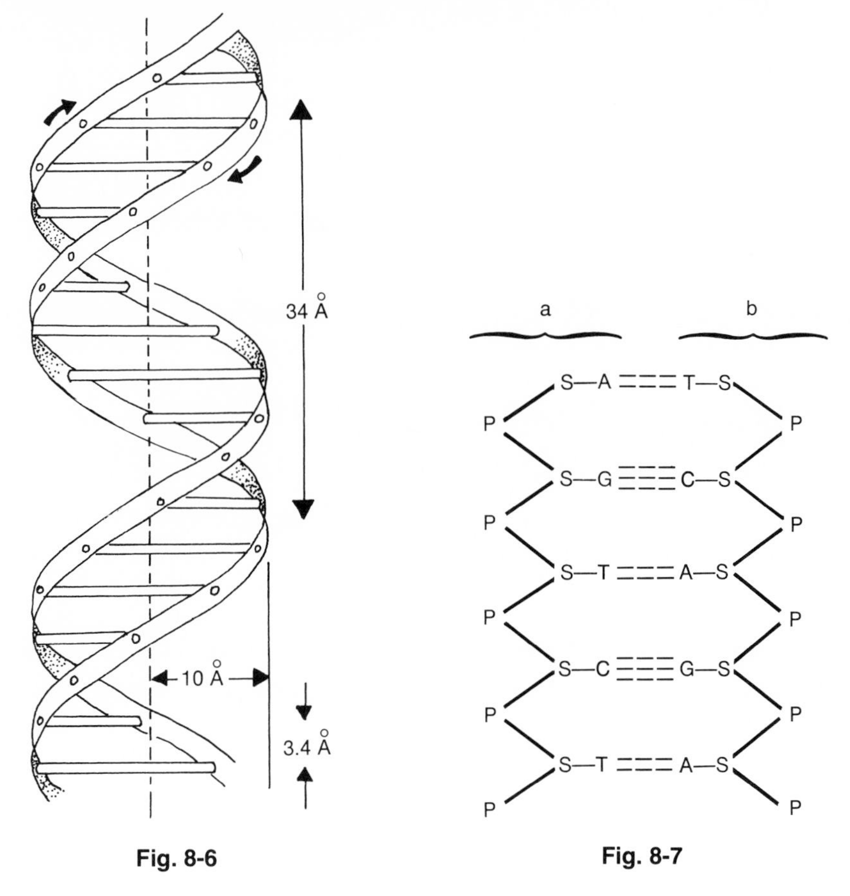

Fig. 8-6

Fig. 8-7

Fig. 8-6
Watson-Crick double-stranded helix configuration of DNA. (From Levine, L. 1973. Biology of the gene, ed. 2. The C. V. Mosby Co., St. Louis.)

Fig. 8-7
Diagram of section of double helix. *P*, Phosphate; *S*, sugar; *A*, adenine; *T*, thymine; *G*, guanine; *C*, cytosine; *a*, one polynucleotide strand; and *b*, the other strand. Double and triple broken lines in center represent hydrogen bonds holding the two strands together. (From Levine, L. 1973. Biology of the gene, ed. 2. The C. V. Mosby Co., St. Louis.)

paired with cytosine. The chemical bonds that hold the nucleotide bases together are **hydrogen** (H) **bonds,** which, as discussed in Chapter 2, are weak and easily broken. A section of a double helix is diagrammed in Fig. 8-7.

The specificity of pairing between adenine and thymine on the one hand and guanine and cytosine on the other is ensured by the number of H bonds involved in each case. A nucleotide of adenine forms two H bonds with thymine, and a nucleotide of guanine forms three H bonds with cytosine.

Complementarity. The arrangement whereby double-ringed purines (adenine and guanine) are always linked to single-ringled pyrimidines (thymine and cytosine) results in a double helix with a uniform diameter of 20 Å. From the Watson-Crick model of DNA, it follows that the base sequences of the two strands are **complementary,** that is, the sequence of either strand may be converted to that of its partner by

—G—T—T—A—C—A—G—G—C—
• • • • • • • • •
—C—A—A—T—G—T—C—C—G—

Fig. 8-8
Segment of hypothetical DNA molecule. (Dots do **not** represent any specific number of hydrogen bonds.)

replacing adenine by thymine, and vice versa, and guanine by cytosine, and vice versa. Schematically, the sequence of bases in a segment of a hypothetical DNA molecule is shown in Fig. 8-8.

□ The genetic code

One of the important functions of the hereditary material is to provide for protein synthesis. The mechanism by which this is accomplished is present in the hereditary material itself. There is a pattern in the structure of the hereditary material that is **translated** into the structure of a protein. The information as to how to form all the cell's proteins is called the **genetic code.**

The discovery that both proteins and DNA are linear arrays of their respective building blocks (amino acids and nucleotides) led to the idea that the linear sequence of the amino acids in a protein is specified by the linear sequence of nucleotides in DNA. This **colinearity** of protein structure and genetic structure constitutes the basis of the genetic code.

Deciphering the code. There is, however, a problem in deciphering the genetic code. How is information contained within a four-letter language (four nucleotides of DNA) translated into a 20-letter language (the 20 amino acids of proteins)? The simplest possible code is a singlet code, in which one nucleotide codes for one amino acid. Such a code is clearly inadequate, because only four different amino acids could be specified. A doublet code, in which each possible combination of two nucleotides codes for an amino acid could code for only 16 (4×4) amino acids and so is also inadequate. The simplest code that can specify all 20 amino acids is the **triplet code,** in which each possible combination of three nucleotides yields a separate code word or **codon.** The triplet code results in 64 ($4 \times 4 \times 4$) codons. Experimental evidence has conclusively demonstrated that the triplet code is, indeed, the operational basis of DNA's ability to produce proteins.

The triplet code provides more codons (64) than are necessary to specify all the amino acids (20). It has been found that some amino acids have more than one codon. It has also been discovered that some codons do not specify any amino acids, but are used to indicate the stopping point of the amino acid sequence (the **polypeptide chain**) that forms all or part of a given protein.

One polynucleotide chain or two? In Fig. 8-8, reading the sequence of nucleotides from left to right, we find that the upper polynucleotide chain calls for three amino acids whose DNA code letters are GTT, ACA, and GGC. These correspond to the amino acids glutamine, cysteine, and proline. The lower polynucleotide chain has the code letter sequences CAA, TGT, and CCG, which correspond to valine, threonine, and glycine. This analysis of codons in a segment of DNA raises a most important question. Does every segment of a double helix provide for the production of two proteins, or is only one polynucleotide chain functional in protein synthesis? Experimental evidence has shown that **only one** polynucleotide chain of any given segment of the DNA double helix provides the information necessary for protein production. The other polynucleotide chain is nonfunctional in protein synthesis, but, as we shall see later, is absolutely necessary in **replication** of the hereditary material of the cell.

What is a gene? Up to this point, we have talked about the genetic material of a cell and some of the hereditary characteristics of human beings, but have not specified what a **gene** is. Now we are in a position to do so, because the definition of a gene is a reflection of its ability to dictate the details of protein synthesis. A gene is that sequence of nucleotides that specifies the amino acid sequence of a **polypeptide chain** (a group of linked amino acids). Why the term polypeptide chain? Why not protein? The reason for specifying polypeptide chain is that many proteins consist of a number of chains, and the genetic code for each may be located in different chromosomes and hence in altogether different segments of DNA. Because more than one gene may be involved in the formation of a given protein, we restrict the definition of a gene to polypeptide chain determination.

□ RNA

Deciphering the genetic code does not complete the picture of how the cell makes more of itself. The hereditary material of the cell is located mainly in the nucleus, whereas protein synthesis occurs in the cytoplasm and specifically on the ribosomes that line the endoplasmic reticulum. This means that there must be some mechanism for transferring the information for protein production from the genes in the nucleus to the ribosomes in the cytoplasm. This is accomplished through the use of another type of nucleic acid, called **ribonucleic acid** (RNA). *The formation of RNA is the basic function of DNA in protein synthesis.* After the RNA is formed, it is transported to the cytoplasm, where it provides for the actual linkage of amino acids into polypeptide chains to produce specific proteins.

Structure. Like DNA, RNA is a polymer whose building blocks are nucleotides. RNA nucleotides also consist of phosphoric acid, sugar, and either a purine or a pyrimidine base. The phosphoric acid is the same in both types of nucleic acid. However, DNA and RNA differ from one another in the other two building blocks.

One difference lies in the sugar portion of their nucleotides (Fig. 8-3). In the pentose of RNA, a hydroxyl (OH) group is present at the 2′ position. In the pentose of DNA, the 2′ oxygen is absent (hence the prefix "deoxy"), and instead of the hydroxyl group, there is a hydrogen atom. Another difference between DNA and RNA lies in the kinds of bases in each (Fig. 8-4). DNA contains the four bases adenine, guanine, thymine, and cytosine. In RNA, the bases are the same except that instead of thymine, we find **uracil.**

The differences in their sugars and bases serve to separate the chemical reactions involving DNA and RNA. As a result, each type of nucleotide carries out its respective functions without interference by the other.

Synthesis of RNA. Previously, it was stressed that the basis for the formation of the DNA double helix is the complementarity exhibited between the nucleotides that contain adenine and thymine, as well as between those that contain guanine and cytosine. Complementarity is also the attractive force that operates in RNA synthesis. However, here complementarity will involve bases from both DNA and RNA nucleotides.

In the nucleus, there are a large number of free nucleotides of the RNA variety. There is also an enzyme called **RNA polymerase** that controls the formation of RNA molecules. The first stage in the formation of an RNA molecule is the separation of the two strands of the existing DNA molecule from each other. Then the free RNA nucleotides become attached temporarily by complementarity to the nucleotides of the particular DNA strand that is the gene.

As a result of the action of complementarity, the combination of a ribose nucleotide base with a deoxyribose base always occurs, as shown in Table 8-1. Once the ribose nucleotides are lined up, the enzyme RNA polymerase causes the formation of chemical bonds between the ribose and phosphoric acid of successive free RNA nucleotides. As this occurs, the newly formed RNA strand separates from the DNA strand and becomes a free molecule of RNA. It is then transported out of the nucleus and into the cytoplasm for use in protein synthesis. One further point: the RNA molecule is derived from only one strand of the double DNA helix. This means that the RNA molecule is a **single** strand of nucleotides.

Types of RNA. An analysis of the RNA in cells shows that there are three distinct types: ribosomal, transfer, and messenger. **Ribosomal RNA** (rRNA) is found in association with proteins; rRNA constitutes some 85% to 90% of the total cellular RNA. The combination of RNA and proteins forms the ribosomes, which are the sites of protein synthesis.

The role of ribosomes in protein synthesis has been discovered through the use of **radioactive** amino acids in the food medium of cells grown in culture. Shortly after these amino acids enter a cell, they become associated with the ribosomes, which then become radioactive. The radioactivity of the ribosomes gradually disappears, and the newly formed protein, containing the "labeled" amino acids, exhibits radioactivity.

Transfer RNA (tRNA) is the term applied to a group of small RNA molecules, of which each one has a specific attraction for one of the 20 different amino acids. tRNA molecules consist of about 80 nucleotides and are in the form of a cloverleaf. They comprise about 5% of the cell's total RNA. The use of radioactively labeled amino acids has revealed that the amino acids arrive at the ribosomes individually, with each amino acid attached to its own specific tRNA.

Messenger RNA (mRNA) is the actual intermediary between the genes located in

Table 8-1
Complementary base pairing between DNA and RNA

DNA base	RNA base
Adenine	Uracil
Thymine	Adenine
Guanine	Cytosine
Cytosine	Guanine

the nucleus and the ribosomes located in the cytoplasm. As its name implies, mRNA carries the genetic code for polypeptide chains that is contained in the sequence of the DNA's bases. The aspect of RNA synthesis that involves transfer of the genetic code from DNA to mRNA is called **transcription.**

The sizes of the different mRNA's are variable, being dependent on the length of the polypeptide chain to be formed. For example, if a polypeptide chain contains 150 amino acids, its corresponding mRNA must contain at least 450 nucleotides in order to provide the necessary number of triplets. In addition, there will be a certain number of **noninformational** nucleotides that are needed for the initial attachment of mRNA to the ribosomes and some triplets that indicate where to begin and where to end protein synthesis. Some polypeptide chains are known to contain as few as 25 amino acids, whereas others are known to have more than 1000. mRNA constitutes about 5% of the total cellular RNA.

In specifying the triplet code letters corresponding to the 20 different amino acids, biologists have generally agreed to use as **codons** the RNA base letters found in mRNA. From the RNA codons we can easily derive the DNA triplets, since RNA is simply the complement of the corresponding DNA. Only one precaution must be observed: DNA always contains thymine in place of the uracil of RNA.

Table 8-2 lists the 20 amino acids found in proteins and their abbreviations. Table 8-3 is the genetic code given in terms of RNA codons. There is one codon, AUG, that acts as the initiator codon of all messages. It also specifies the amino acid methionine. All polypeptide chains are produced with methionine as the first amino acid. In many cases, the methionine is removed by other chemical reactions before the particular polypeptide becomes functional. There are three triplets, UAA, UAG, and UGA, that act as terminator codons for messages; they do **not** specify any amino acid.

■ FUNCTIONS OF HEREDITARY MATERIAL

□ Protein synthesis

The pattern for protein synthesis that emerges from our knowledge of DNA, RNA, and proteins is illustrated in Fig. 8-9. The DNA double helix partially unwinds, permitting the formation of a strand of mRNA complementary to one of the

Table 8-2
The 20 amino acids found in proteins*

Amino acid	Abbreviation	Amino acid	Abbreviation
Alanine	Ala	Leucine	Leu
Arginine	Arg	Lysine	Lys
Asparagine	AspN	Methionine	Met
Aspartic acid	Asp	Phenylalanine	Phe
Cysteine	Cys	Proline	Pro
Glutamic acid	Glu	Serine	Ser
Glutamine	GluN	Threonine	Thr
Glycine	Gly	Tryptophan	Tryp
Histidine	His	Tyrosine	Tyr
Isoleucine	Ileu	Valine	Val

*From Levine, L. 1973. Biology of the gene, ed. 2. The C. V. Mosby Co., St. Louis.

Table 8-3
The genetic code, consisting of 64 triplet combinations and their corresponding amino acids*

Second letter

First letter	U	C	A	G	Third letter
U	UUU Phe	UCU Ser	UAU Tyr	UGU Cys	U
	UUC Phe	UCC Ser	UAC Tyr	UGC Cys	C
	UUA Leu	UCA Ser	UAA Ochre (terminator)	UGA (terminator)	A
	UUG Leu	UCG Ser	UAG Amber (terminator)	UGG Tryp	G
C	CUU Leu	CCU Pro	CAU His	CGU Arg	U
	CUC Leu	CCC Pro	CAC His	CGC Arg	C
	CUA Leu	CCA Pro	CAA GluN	CGA Arg	A
	CUG Leu	CCG Pro	CAG GluN	CGG Arg	G
A	AUU Ileu	ACU Thr	AAU AspN	AGU Ser	U
	AUC Ileu	ACC Thr	AAC AspN	AGC Ser	C
	AUA Ileu	ACA Thr	AAA Lys	AGA Arg	A
	AUG Met (initiator)	ACG Thr	AAG Lys	AGG Arg	G
G	GUU Val	GCU Ala	GAU Asp	GGU Gly	U
	GUC Val	GCC Ala	GAC Asp	GGC Gly	C
	GUA Val	GCA Ala	GAA Glu	GGA Gly	A
	GUG Val	GCG Ala	GAG Glu	GGG Gly	G

*From Levine, L. 1973. Biology of the gene, ed. 2. The C. V. Mosby Co., St. Louis.

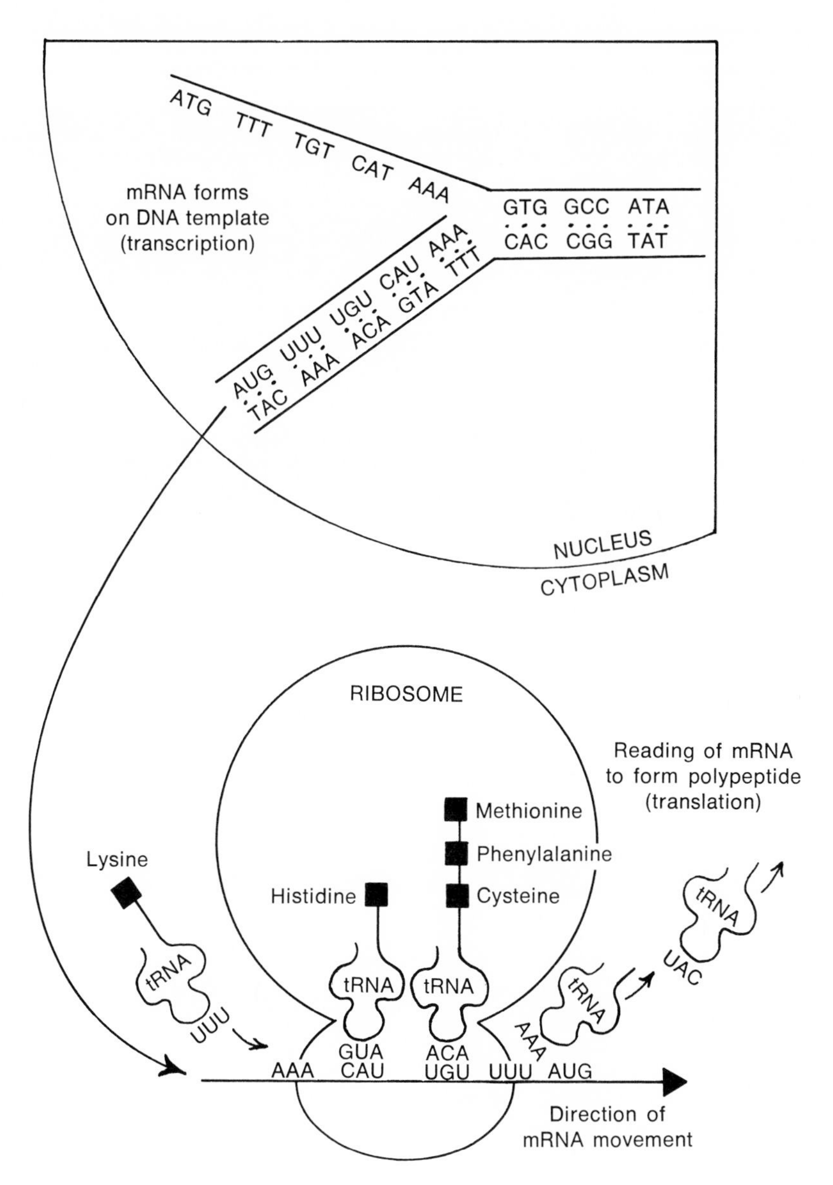

Fig. 8-9
Transcription of DNA and translation of mRNA.

two strands of the DNA. After its formation, mRNA passes out of the nucleus and into the cytoplasm. In the cytoplasm, mRNA becomes attached to a ribosome.

Elsewhere in the cytoplasm, and apparently independent of mRNA formation, amino acids become attached to their specific tRNA's. The amino acids and their tRNA's then move to the mRNA, which is attached to a ribosome. If a particular amino acid corresponds to the one indicated at the beginning of the "message," the amino acid and its tRNA become attached temporarily to the ribosome. After this occurs, the ribosome moves along on the strand of mRNA to the next amino acid–specifying message. At this next message point of the mRNA, another tRNA with its appropriate amino acid attaches temporarily to the ribosome. The two amino acids are then joined by a peptide bond, and the first tRNA is released. This process continues until the entire message is read, and a polypeptide chain is produced. The production of a polypeptide chain according to the specification of an mRNA is called **translation.**

Codon-anticodon pairing. Our discussion of protein synthesis did not indicate the nature of the temporary attraction between the codons of an mRNA and the various tRNA's in the translation process. The basis of this attraction also rests on **complementarity.** As was true of DNA-to-RNA transcription, there is a complementary pairing of purine with pyrimidine in the translation process involving adenine with uracil and guanine with cytosine.

The group of three nucleotides of a tRNA molecule that allows it to recognize a specific mRNA codon is called an **anticodon.** During protein synthesis, the anticodon bases combine loosely by hydrogen bonding with the corresponding (complementary) codon bases of the mRNA. After adjacent amino acids are linked to one another, the hydrogen bonds are broken, and the tRNA is released from the ribosome. It then becomes available for use in the synthesis of some other protein.

Control of cell metabolism. Each cell produces perhaps a thousand or more protein enzymes in the manner outlined in this chapter. These enzymes control all the other chemical reactions that take place in the cell, including the synthesis of lipids, glycogen, purines, pyrimidines, and hundreds of other substances. Thus the genetic material of a cell, through the production of proteins, provides for the cell's own growth and well-being.

Enzymes. Enzymes play an important role in cell metabolism, and their absence or improper function can have serious effects on an individual. Many, if not most, hereditary diseases can be traced to the absence or malfunction of specific enzymes. Enzymes are proteins that function as biological catalysts. They increase the rate at which a chemical reaction takes place. Enzymes provide a mechanism for bringing molecules together, thus promoting chemical change. However, enzymes are not consumed in the chemical reactions they affect.

It is generally believed that enzymes function according to a **lock-and-key principle** (Fig. 8-10). Enzymes are large globular proteins, with many projections and depressions formed by the folding of their protein chains. These projections and depressions are considered to be analogous to the inside surface of a lock. The chemical reactants have corresponding depressions and projections that act as the ridges of a key. Once the key is in the lock, the reactants are together, and the chemical reaction can take place.

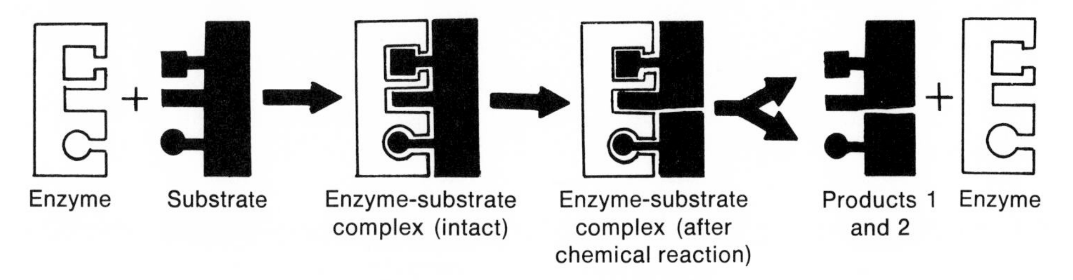

Fig. 8-10
Lock-and-key principle of enzyme activity.

The chemical substances involved in an enzymatically controlled reaction are called the **substrates** of the enzyme. During such a reaction, an enzyme-substrate complex forms, which breaks down to release both the product molecules and free enzyme, as follows:

S	+	E	→	ES	→	P	+	E
Substrate		Enzyme		Enzyme-substrate complex		Product molecules		Enzyme

At the end of the reaction, the enzyme molecule is free to undergo further reactions with other substrate molecules.

Although enzymes are not destroyed during the chemical reactions they catalyze, they do have a limited lifetime, which may be weeks or months. Eventually, every enzyme molecule is broken down and must be replaced with a newly synthesized enzyme.

Enzymes are generally named by adding the suffix "**-ase**" either to the name of the substrate or to the type of reaction catalyzed by the enzyme (**sucrase** is the enzyme that splits sucrose, table sugar, into glucose and fructose; **DNA polymerase** is the enzyme that links together the nucleotides of DNA).

Coenzymes. Many enzymes are inactive in the absence of additional substances known as cofactors, or **coenzymes.** In some cases, the coenzymes are metal ions, such as copper, zinc, iron, cobalt, or magnesium, that help bind the substrate to the enzyme. In other cases, the coenzymes are organic molecules that participate in the chemical reaction catalyzed by the enzyme. An example of an organic molecule coenzyme is **nicotinamide adenine dinucleotide** (NAD). It acts as a carrier of hydrogen atoms in the energy-transferring biochemical pathways that will be discussed later in this chapter.

Most, but not all, vitamins function in the body as coenzymes. For example, the basic unit of the coenzyme NAD is the vitamin niacin (see Table 4-1).

☐ DNA replication

Each cell of our bodies can not only make structural and enzymatic proteins using the instructions contained in its genetic material, but each cell can also make more of its own genetic material. The formation of genetic material is most important in conjunction with cell division. During this process, it provides both daughter cells with all the genetic information they may need for protein synthesis.

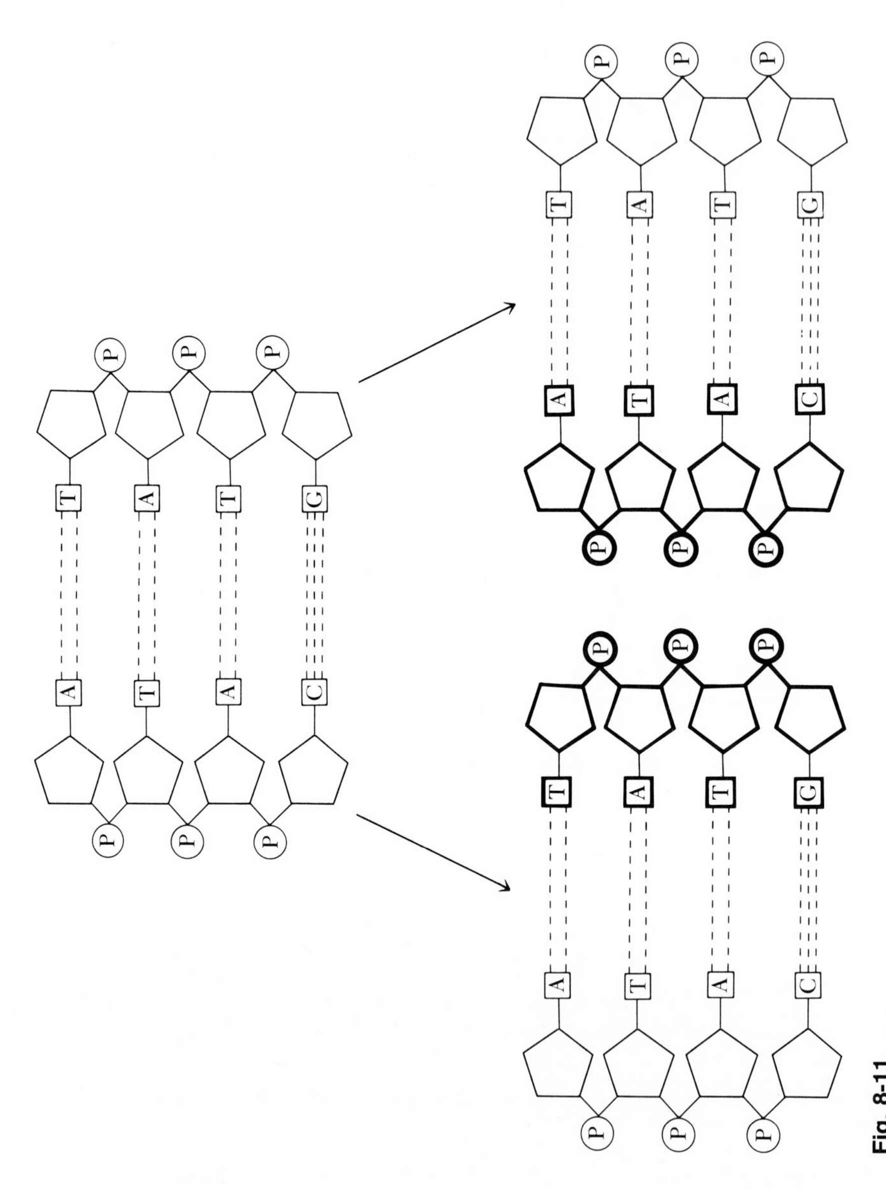

Fig. 8-11
DNA replication, that is, synthesis of new strands of DNA on old strands. (From Levine, L. 1973. Biology of the gene, ed. 2. The C. V. Mosby Co., St. Louis.)

A better appreciation of how DNA replicates itself can be gained by first examining Figs. 8-6 and 8-7. The DNA double helix consists of two strands that are bound to one another through the attraction of their complementary purine and pyrimidine bases. In DNA replication, **complementarity** is again the attractive force that brings the proper nucleotides together.

In the nucleus, there are a large number of free nucleotides of the DNA variety. Also present is an enzyme called **DNA polymerase,** which controls the formation of DNA molecules (in contrast to RNA polymerase, which, as discussed earlier, controls the formation of RNA molecules). The first stage in DNA replication consists of the separation of the two strands of the existing DNA molecule. In the second stage, the free DNA nucleotides become attached to the appropriate complementary bases of both original DNA strands. Then, once the nucleotides are lined up, the enzyme DNA polymerase causes chemical bonds to form between the ribose and phosphoric acid of successive nucleotides. The new strand of DNA remains hydrogen bonded to the old strand, and together they form a new double-stranded DNA helix.

In DNA replication, each polynucleotide chain acts as a **template** for the formation onto itself of a new complementary companion chain, so that eventually there are two pairs of chains where there was previously only one. In this process, the sequence of the base pairs has been duplicated exactly, and each newly formed double helix of DNA contains one "old" and one "new" polynucleotide strand. The results of this process are diagrammed in Fig. 8-11.

The occurrence of DNA replication within each cell sometime before its division and the subsequent distribution of a complete set of genes to both resultant cells mean that all cells of the body possess the same genetic material. If this is the case, how does a muscle cell differ from a nerve cell or a fat cell? As noted in the previous chapter, the differences among the various types of cells lie in their cytoplasms and in their intercellular materials. These differences are caused by variations in gene activity, whereby each type of cell produces certain proteins that are unique to itself. But what causes certain genes in one type of cell to be active in protein synthesis, while other genes are not? The regulation of gene activity is a function of the environment in the particular part of the body where the cell is located.

Mutation. DNA replication normally results in the production of molecules with the exact same nucleotide sequence as the double helix from which they were derived. However, DNA molecules are sometimes formed with altered sequences of nucleotides. This alteration can occur spontaneously or as a result of exposure to some environmental factor, such as x rays, ultraviolet rays, or certain chemicals. A change in DNA nucleotide sequence will be reflected in both the mRNA transcribed from the DNA and the protein subsequently produced. Any alteration of the nucleotide sequence of a DNA molecule is called a **mutation,** and an individual who shows the effect of a mutation is called a **mutant.**

Three types of changes can occur in a DNA nucleotide sequence: a nucleotide can be **added** anywhere in the sequence, a nucleotide can be **deleted** from anywhere in the sequence, or a nucleotide can be **replaced** by one containing a different purine or pyrimidine base. The first two types of nucleotide change will not only alter the particular code word at the point of nucleotide addition or deletion, but they will also cause radical changes in the rest of the mRNA and the subsequently produced protein, because every code word after the affected triplet will be modified. However, the replacement of one kind of nucleotide of a triplet by another will affect only that particular code word. Such an altered code word may designate a different amino acid, which will result in the production of a protein with a single amino acid substitution.

The effects of a mutation on an individual will depend on the particular protein involved and on the amount of change in protein function caused by the altered amino acid sequence. It is quite possible that a relatively small change in a nucleotide sequence can produce a drastic alteration in protein function. Consider, for example, the mutation that produces sickle cell anemia, a disease discussed in Chapter 6. The amino acid sequences of normal and sickle cell hemoglobins differ from one another at only one point, where glutamic acid is found in normal hemoglobin versus valine in sickle cell hemoglobin. Table 8-3 indicates that the mRNA codons for glutamic acid are GAA or GAG, and those for valine are GUU, GUC, GUA, or GUG. If we assume that sickle cell hemoglobin is a mutant form derived from normal hemoglobin, then clearly the substitution of a "U" for the middle "A" in GAA or GAG would result in the mutation that causes the tremendous difference in the oxygen-carrying abilities of these two hemoglobins.

Most mutations are deleterious to their carriers, which is what we would expect of most alterations in the intricate and finely adjusted biochemistry of a living organism. Only very rarely is a mutation beneficial to its carrier. However, it is these rare beneficial mutations that provide the new proteins necessary for evolutionary changes.

■ OBTAINING ENERGY FROM OUR FOOD

In Chapter 3, food was discussed as the supplier of both the building materials and the energy needed for sustaining life and providing for our well-being. Thus far in this chapter, we have reviewed the processes by which the cell utilizes building materials to make more of itself in the form of proteins and DNA molecules. However, in order to perform these and other activities, the cell requires energy. We shall now consider how the cell extracts this energy from the digested material brought to it by the circulatory system.

□ Potential and kinetic energy

The energy we obtain from the food we eat is located in the chemical bonds that hold the atoms of the food together. Each chemical bond has an energy value that is expressed in **calories.** The sum total of these calories represents the **potential energy** present in a particular molecule. As an example, 1 mole of glucose has a potential energy of 686,000 calories (686 kilocalories [kcal]).

Each time two atoms are separated from one another, the energy that held them together becomes available for (1) use in some work-producing activity, (2) transformation into a different chemical bond, or (3) production of heat. Energy that is put to use is called **kinetic energy.** Since energy can neither be created nor destroyed, it is clear that the terms "potential" and "kinetic" refer to the same energy in two different situations.

Potential energy is transformed into kinetic energy as a result of a chemical

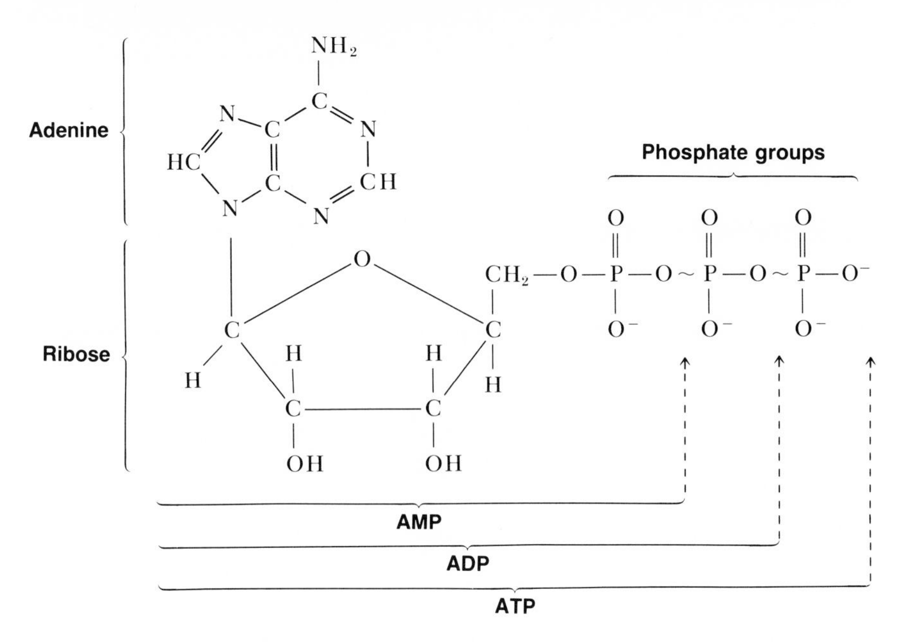

Fig. 8-12
Chemical structure of adenosine triphosphate (ATP).

reaction. This can involve the splitting of a large molecule into smaller ones, or it can involve a rearrangement of the atoms of a compound from one configuration that has a high potential energy to a configuration that requires less energy to hold it together.

Activation energy. Most of the chemical compounds that make up our food are quite stable at body temperature and do not become involved spontaneously in chemical reactions. As a necessary first step in obtaining energy from our food, the stability of a compound's chemical bonds must be disrupted. Only after this has occurred can a new arrangement of chemical bonds be achieved. To upset the balance of forces in a molecule, a quantity of energy known as **activation energy** must be added to it. In our bodies, the transfer and storage of energy involves a particular and most important compound, **adenosine triphosphate.**

Adenosine triphosphate. Adenosine triphosphate (ATP), diagrammed in Fig. 8-12, consists of (1) the purine adenine, (2) a pentose sugar of the ribose type, and (3) three phosphate groups. The bonds by which the second and third phosphate groups are attached to the rest of the molecule are **not** the usual kind of chemical bond, but are **energy-rich bonds** (indicated by the symbol ~).

When an energy-rich bond is broken, far more calories of energy are released (8 kcal/mole) than if any of the other bonds in the ATP molecule are broken. ATP is present everywhere in the cell, and all the cell's activities are dependent on this molecule. In turn, the food in the cell is broken down chemically, and the energy released is used to form ATP. For this reason, ATP has been called the **energy currency** of the body.

In most chemical reactions involving ATP, only the terminal high-energy phosphate bond is broken and its phosphate group removed. The energy contained in the broken bond is transferred to some other molecule or cell constituent for use in a cell activity. The removal of the third phosphate group changes ATP into **adenosine diphosphate** (ADP) (Fig. 8-12). ADP consists of adenine, ribose, and two phosphate groups, of which only the second is attached to the rest of the molecule by an energy-rich phosphate bond.

New ATP can be formed from ADP if a phosphate group is present and if enough energy is available to attach the third phosphate group to the ADP molecule with a high-energy phosphate bond. The phosphate bond energy of ATP supplies the power necessary for protein synthesis, DNA replication, active transport across plasma membranes, muscular contraction, nerve conduction, cilia and flagella movements, etc.

ATP is not only important in biological energy transfers, but, without its second and third phosphate groups, it becomes a ribonucleotide that can be used in the synthesis of an RNA molecule. In fact, all nucleotides, including both the ribose and the deoxyribose types, must first be **activated** by the addition of high-energy phosphate groups before they can be used in their respective types of nucleic acid synthesis. Just prior to the incorporation of the ribose and deoxyribose nucleotides into RNA and DNA, respectively, the second and third phosphate groups of the nucleotides are removed, and the monophosphate-containing nucleotides that remain are joined together by their corresponding polymerase enzyme.

□ Carbohydrate breakdown

A large proportion (about 40%) of the energy we obtain from our food comes from carbohydrates. The complete chemical breakdown of a monosaccharide occurs in three stages: (1) glycolysis, (2) Krebs' citric acid cycle, and (3) cytochrome (electron transport) system. Glycolysis occurs in the cell's cytoplasm, whereas Krebs' citric acid cycle and the cytochrome system are found in the mitochondria. As we shall see, the complete breakdown of each six-carbon monosaccharide molecule requires the use of two high-energy phosphate groups for **activation energy,** with the ultimate production of 40 energy-rich phosphate bonds. The resulting **net gain** is 38 high-energy phosphate bonds.

Glycolysis. The end products of carbohydrate digestion are glucose, fructose, and galactose. These monosaccharides are absorbed from the small intestine into the bloodstream and carried by the circulatory system to the cells of the body, where they are carried across the plasma membrane by **active transport.** In the cell, these hexoses are subjected to complete chemical degradation, starting with a series of reactions called **glycolysis.** Glycolysis involves the splitting of each six-carbon sugar molecule into two three-carbon molecules of a compound called pyruvic acid, with an accompanying transfer of energy to form energy-rich phosphate bonds.

Glycolysis can take place anywhere in the cytoplasm of a cell. On entering a cell, each monosaccharide molecule undergoes **phosphorylation,** that is, a high-energy phosphate group from a molecule of ATP is attached to one of its ends to produce a sugar phosphate. Any galactose phosphate and fructose phosphate that has been formed in this initial phosphorylation reaction is then converted to glucose phosphate. There follows a chemical reaction that converts the glucose phosphate molecule into a related six-carbon compound. The next step of glycolysis involves a second phosphorylation reaction, in which another high-energy terminal phosphate group is attached to the other end of the six-carbon chain, thereby forming a hexose diphosphate. These two energy-rich phosphate groups provide the **activation energy** needed to prepare the hexose for subsequent splitting.

Next, the six-carbon compound (hexose diphosphate) is split in the middle, forming two three-carbon triose phosphate molecules. Several chemical reactions then follow, which result in the removal of two hydrogen atoms from each three-carbon molecule. The removal of the hydrogen atoms releases energy that transforms the low-energy bonds of two phosphate groups into high-energy bonds. Each of these newly formed energy-rich phosphate groups is added to a different ADP molecule to produce two ATP molecules. The newly formed ATP molecules replace the two used for activation energy at the beginning of glycolysis.

Glycolysis is brought to completion by several more reactions that serve to reenergize the phosphate groups on the three-carbon triose phosphate molecules and, ultimately, transfer them to ADP to form two more molecules of ATP. The total energy picture thus far is the **net production** of two energy-rich phosphate bonds that are used in the synthesis of two molecules of ATP.

At the end of glycolysis, for each molecule of glucose, there remains two molecules of **pyruvic acid** (three-carbon compound) and four atoms of **hydrogen.** If oxygen is available to the cell, the pyruvic acid and the hydrogen atoms pass into the **mitochondria** of the cell; there pyruvic acid is further broken down in the Krebs' citric acid cycle, and the hydrogens pass through the cytochrome system, ultimately to combine with oxygen to form water. If oxygen is not available, the hydrogens combine with the pyruvic acid to form **lactic acid,** some of which is stored in the cells and the remainder in the liver, until oxygen becomes available. When oxygen is available again, the lactic acid is converted back to pyruvic acid and hydrogen atoms, which then pass into the mitochondria for further energy production.

One aspect of glycolysis, at least as it occurs in **yeast cells,** has important economic and social consequences for our modern society. These one-celled organisms obtain their energy from the breakdown of carbohydrates, in much the same way as do human beings. But at the end of glycolysis, if oxygen is not available, a chemical reaction occurs in which the hydrogen atoms and pyruvic acid are combined to produce a two-carbon compound, **ethyl alcohol,** and carbon dioxide. The process whereby glycolysis results in the production of alcohol is called alcohol **fermentation.** Fermentation by yeast cells is the biological basis for the beer, wine, and whiskey industries.

Krebs' citric acid cycle. On completion of glycolysis, pyruvic acid and hydrogen pass into the mitochondria, where the three-carbon pyruvic acid is broken down to carbon dioxide and acetic acid (two-carbon compound; vinegar). The acetic acid is then joined to a coenzyme called coenzyme A (CoA), the new compound being called **acetyl-CoA.** In the acetic acid–producing reaction, more hydrogen atoms are released to the cytochrome system of the mitochondria.

The formation of acetyl-CoA begins the **Krebs' citric acid cycle.** The cycle starts with each two-carbon acetyl-CoA molecule combining with a four-carbon compound (oxaloacetic acid) already present in the mitochondria to form a six-carbon compound called **citric acid.**

A series of chemical reactions then removes four pairs of hydrogen atoms and two molecules of carbon dioxide from the citric acid. This process results in the re-formation of the four-carbon compound oxaloacetic acid, which can then pick up more acetyl-CoA to form new citric acid and begin the cycle again.

The overall effect of glycolysis and the Krebs' citric acid cycle is to break up glucose into molecules of carbon dioxide and atoms of hydrogen. The carbon dioxide is released for eventual disposal through our lungs, and the hydrogen atoms are transferred to the cytochrome system. Enough energy is released during each turn of the Krebs' citric acid cycle to produce one more high-energy bond. Since a molecule of glucose yields two molecules of pyruvic acid, the Krebs' citric acid cycle yields two energy-rich phosphate groups that are used in forming two additional molecules of ATP.

Cytochrome system. Even though glycolysis and the Krebs' citric acid cycle result in the complete breakdown of a glucose molecule, it is important to note that none of the chemical reactions discussed thus far has involved oxygen, that is, they have been **anaerobic.** The need for oxygen occurs only at the end of the **cytochrome system** reactions, when the oxygen is combined with hydrogen to form water.

The cytochrome system is a mitochondrial complex of enzymes and coenzymes physically and chemically arranged to facilitate the transfer of hydrogen through a sequence of decreasing energy levels. At each step, energy is released, and at several steps, this energy release is coupled with the formation of high-energy phosphate bonds. Enough energy is obtained in this process to form 34 energy-rich phosphate bonds that are used in the production of ATP molecules. At the end of this energy transfer sequence, hydrogen reaches its lowest energy level and is combined with oxygen to form water.

We may now evaluate the efficiency of the total energy transfer process of carbohydrate breakdown. The complete pathway is summarized in Fig. 8-13. Glycolysis resulted in the net formation of two ATP molecules; the Krebs' citric acid cycle provided for the production of two more; but the greatest transfer of energy came from the cytochrome system, which yielded 34 ATP molecules. Formation of a high-energy phosphate bond requires about 8 kcal/mole. A total of 304 kcal are stored in the 38 moles of ATP formed from the breakdown of 1 mole of glucose. The complete breakdown of 1 mole of glucose releases about 686 kcal of energy, so the overall efficiency of energy transfer is about 44%. The remaining 56% of the energy released is lost as heat and cannot be used by the cells to perform any energy-requiring functions.

However, humans (and other mammals and birds) make use of the heat produced during metabolic reactions to maintain body temperature at a constant level. Organisms that can maintain a constant body temperature are called **warm-blooded** or **homoiothermic** creatures. A steady body temperature permits an individual to maintain a constant level of activity regardless of outside temperature and permits the members of a species to live in varied environments.

□ Lipid breakdown

Lipids provide approximately 40% of the calories in the average American diet (about the same number of calories that are usually obtained from carbohydrates). In

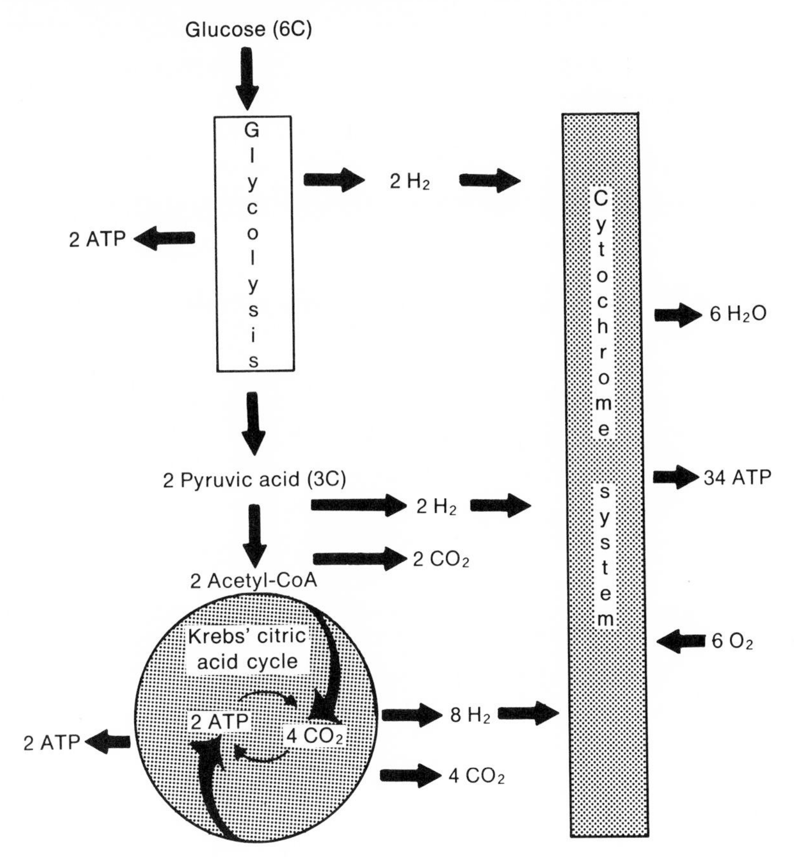

Fig. 8-13
Process by which energy is obtained from glucose through the following formula: $C_6H_{12}O_6 + 6\ O_2 \rightarrow 6\ CO_2 + 6\ H_2O + 38\ ATP$.

addition, about 30% of the carbohydrates consumed by an individual are converted into lipids and stored in fatty connective tissue to serve later energy needs. After periods of even moderate fasting, as much as two thirds of all the energy obtained by the body cells may come from lipids. Thus both directly and indirectly, lipids play a considerable role in meeting the body's needs for energy.

In the digestive process, **lipids** are broken down to their component compounds glycerol and fatty acids. **Glycerol** is a three-carbon compound and, through a number of cellular chemical reactions, is converted to **pyruvic acid,** which then enters the Krebs' citric acid cycle. **Fatty acids** vary as to the number of carbon atoms in each type of chain: they range from 4 to 20 carbon atoms. Fatty acids are broken down into many two-carbon molecules of **acetyl-CoA,** which enters the Krebs' citric acid cycle.

Lipids are a complex group of compounds, and it is not possible to obtain an estimate of the number of ATP molecules formed as a result of the chemical breakdown of an average lipid. But it is known that 1 gram (gm) of lipid molecules contains about 2¼ times as much energy in its chemical bonds as 1 gm of carbohydrate. The complete breakdown of a molecule of the moderately sized fatty acid **stearic acid** ($C_{17}H_{35}COOH$) yields enough energy to produce 156 molecules of ATP.

Protein breakdown

About three fourths of the solid material of our bodies is protein, including structural proteins, enzymes, proteins that transport oxygen, and contractile proteins of our muscles. In the digestive process, **proteins** are broken down to their subunits, **amino acids,** which are then transported by the circulatory system to the individual cells.

Once the immediate needs for amino acids have been met, any additional amino acids in the body fluids are degraded and used for energy or stored as fat. The degradation begins with the removal of the amino group from the amino acid (**deamination**).

After deamination, the remainder of the amino acid can enter any of a number of different biochemical pathways. Some are converted to pyruvic acid and thus enter the Krebs' citric acid cycle. Others are converted to acetyl-CoA and also enter the Krebs' cycle. Still others are changed directly into one of several different compounds of the Krebs' citric acid cycle, and their chemical breakdown is completed from that point.

In general the amount of ATP formed from the energy released by the chemical breakdown of a gram of protein is approximately the same as from a gram of glucose.

SUMMARY

In order for a cell to grow, it must be able to produce proteins and obtain energy from its food. The ability to perform these activities is contained in the genetic material of the cell and is transmissible from one cell generation to the next. Transmissibility of genetic information indicates that the genetic material is capable of replicating itself sometime before each cell division.

DNA is the genetic material and contains the code for the production of proteins in the sequence of its nucleotide bases. The genetic code is in the form of triplets (of bases) and is translated into amino acid sequences through the use of various types of RNA molecules.

Protein synthesis involves the production of both structural and enzymatic proteins. Enzymes regulate the chemical reactions of the body, which include those that supply energy to the cell, as well as those that result in the synthesis of proteins, lipids, glycogen, and other compounds.

DNA has a double-helix structure that is maintained through the mutual attraction of its complementary nucleotides. In the replication of DNA, the two polynucleotide strands separate, and each one forms a complementary chain of nucleotides on itself. Thus in its formation, every DNA double helix consists of one old and one new strand.

Energy for cellular activities is obtained through the transfer of energy from the chemical bonds of food molecules to special high-energy bonds of adenosine triphosphate. This transfer of energy involves a large number of chemical reactions that make up three distinct sequences: glycolysis, the Krebs' citric acid cycle, and the cytochrome system. The various types of energy-releasing food compounds (carbohydrates, lipids, and proteins) enter the energy-transfer process at different points.

The study of cells and how they make more of themselves leads us next to a consideration of cell division, the transfer of genes from one generation to the next, and the various patterns of inheritance of traits.

SUGGESTED READINGS

Barker, G. R. 1968. Understanding the chemistry of the cell, Institute of Biology's Studies in biology No. 13. St. Martin's Press, Inc., New York. Concise account of cell chemistry, energy, and metabolism.

Clark, B. F. C., and Marcker, K. A. 1968. How proteins start. Sci. Am. **218**:36-42. Clear description of DNA transcription and mRNA translation, leading to the formation of proteins.

Frieden, E. 1959. The enzyme-substrate complex. Sci. Am. **201**:119-125. In-depth discussion of the lock-and-key principle of enzyme activity.

Kornberg, A. 1968. The synthesis of DNA. Sci. Am. **219**:64-78. Authoritative account of DNA replication, explaining the role of enzymes in this process.

Lehninger, A. L. 1960. Energy transformation in the cell. Sci. Am. **202**:102-114. Detailed description of how cells obtain energy from food.

Miller, O. L., Jr. 1973. The visualization of genes in action. Sci. Am. **228**:34-42. Description of an exciting electron microscopy study that has obtained photographs showing DNA transcription and mRNA translation.

Sager, R. 1965. Genes outside the chromosomes. Sci. Am. **212**:71-79. Excellently presented discussion of the genetic material that is found in the cytoplasm of all cells.

Watson, J. D. 1968. The double helix. Atheneum Publishers, New York. Account of how the molecular model of DNA was worked out as told by one of the scientists involved.

Identical twins illustrate the importance of heredity. (From Turner, C. E. 1971. Personal and community health, ed. 14. The C. V. Mosby Co., St. Louis.)

CHAPTER 9 Cell division and genetics

LEARNING OBJECTIVES

- Why is it not possible for a human being or other large organism to be composed of a single cell?
- What are the stages of a normal cell cycle?
- How many types of cell division are there, and what are the characteristics of each?
- What are the genetic similarities and/or differences of the different body cells?
- What are the genetic similarities and/or differences of the different sperm and egg cells?
- What are the two types of human chromosomes, and in what way do they differ from each other?
- How can it be ascertained whether a trait is environmentally or genetically determined?
- What are the various types of allelic interaction?
- What are the various patterns by which traits are inherited?
- What types of genetic traits are expressed differently in males and females, and what are the characteristics of these traits?
- How is sex determined in human beings?
- Which genetic diseases are known to be caused by abnormal chromosome numbers?
- In what ways may the genetic material of a cell be rearranged, and what are the consequences for the individual involved?
- Which environmental factors cause chromosomes to break?

Cell growth, as we discussed in the previous chapter, requires the production of structural and enzymatic proteins whose activities will, in turn, result in the manufacture of more protoplasm. However, a cell cannot continue to grow in size indefinitely, because of the relative relationship of the cell's surface to its volume.

All materials enter and leave a cell through its surface (plasma membrane). If we picture a cell in the form of an ice cube that is 1 inch wide, 1 inch high, and 1 inch deep, the total surface area of the cell will be 6 square inches, and its volume will be 1 cubic inch. If a cell were to grow until all dimensions were 2 inches each, its total surface area would be 24 square inches, and its volume would be 8 cubic inches. As a result of doubling each dimension of the cell, its area has increased by a factor of four ($24 \div 6$), whereas its volume has increased by a factor of eight ($8 \div 1$). In other words, the surface-volume ratio decreases as a cell enlarges.

If a cell continues to enlarge, it will eventually reach a point beyond which the surface area of the plasma membrane will not be great enough to bring in adequate nutrients and rid the cell of its waste materials fast enough to meet the needs of the increased volume of protoplasm. The fact that when a cell grows its ratio of surface area to volume decreases means that neither a human being nor any other large organism could consist of one gigantic cell.

The relative relationship of a cell's surface area to its volume requires that the growth of a cell involve not only an increase in protoplasm, but also its subsequent division into two cells. If both resultant cells are to retain the ability to grow and divide, they must receive a full complement of the original cell's genetic material. This requires DNA replication sometime before cell division and the transfer of a complete set of genes to each cell. We shall now examine the life cycle of a cell and the stages of its division process to see how the cell meets its need to divide periodically.

THE CELL CYCLE

The cell cycle includes two distinct stages. One is a relatively long period of time, called **interphase**, during which the cell is actively growing. The second period, called **mitosis,** is much shorter and is devoted to cell division.

Interphase begins immediately after mitosis and is divided into three phases: G_1, S, and G_2. In the **G_1 phase,** the cell is engaged in protein synthesis and is making more protoplasm. During the **S phase,** the production of proteins continues, and, in addition, the cell's DNA is replicated. Also in the S phase, the protein to which DNA is bound in the chromosome is synthesized. In the **G_2 phase,** protein synthesis continues until the cell reaches the appropriate size for cell division. The length of the various phases and of mitosis varies greatly with the particular type of cell. Table 9-1 shows the times involved for human cells that are grown in tissue culture.

More than 90% of a cell's life cycle is devoted to growing and carrying out its specific functions, and only a relatively short period is spent in cell division. However, cell division is absolutely essential for the formation of any large-sized organism, which, of course, includes human beings.

CHROMOSOMES AND TYPES OF CELL DIVISION

In Chapter 8, we discussed the fact that our genetic material consists of a number of double chains of DNA. Each double chain of DNA is bound to protein, making up a **chromosome.** Every species of organism has a typical number of chromosomes; humans have 46.

Each human being is formed through sexual reproduction. In this process, 23 chromosomes are contributed by the mother (by way of the egg) and 23 by the father

Table 9-1
Life cycle of human cells grown in tissue culture

Phase	Time (hours)	% of life cycle
G_1	8.2	42
S	6.2	32
G_2	4.6	23
Mitosis	0.6	3
Total	19.6	100

(from the sperm). In reality, each of us does not contain 46 unrelated chromosomes, but rather 23 pairs of chromosomes. For every chromosome contributed by the mother, there is a corresponding, or **homologous,** chromosome contributed by the father.

To say that the chromosomes are homologous means that if a particular chromosome carries a gene that affects eye color, the corresponding, or homologous, chromosome (from the other parent) will also contain a gene that affects eye color. The homologous chromosomes may or may not carry the gene for the same eye color, but they will both carry genes that affect the trait.

Furthermore, such pairs of genes are located at corresponding points on the homologous chromosomes. If a gene for eye color is located at the tip of a particular chromosome, the corresponding gene in the homologous chromosome will also be located at the tip of its chromosome. This phenomenon of **pairs of genes** applies to genes for all traits, not only eye color.

To understand how cell division provides each resultant cell with the proper number and types of genes, we must study the ways in which chromosomes are transmitted from cell to cell. There are two types of cell division: the more general type, called **mitosis,** that is carried out by all living cells and a special type of cell division, called **meiosis,** that occurs only in the reproductive organs (ovaries and testes) and results in the formation of gametes (eggs and sperm).

□ Mitosis

The general type of cell division, **mitosis,** results in two cells that contain the same genes. In unicellular organisms (*Ameba, Paramecium,* etc.), mitosis is followed by a separation of the two cells, and the unicellular organism is said to have reproduced itself. In human beings, however, the two resultant cells usually remain attached to one another after cell division, and mitosis becomes part of the growth process of the individual.

Occasionally in human beings, after a fertilized egg has divided, the two resultant cells separate from one another. Since both cells have a full complement of genetic material, each cell proceeds to grow and divide, resulting in the formation of **identical twins.** If cells separate after each of a number of successive divisions, **identical triplets,** or **quadruplets,** or **quintuplets** develop.

Mitosis involves the breakdown of the nuclear membrane followed by a splitting of each chromosome. As the membrane breaks, the nucleus temporarily disappears as a discrete cellular structure. A sequence of chromosome movements then results in the distribution of a complete set of chromosomes to each of the opposite ends of the cell. A new nuclear membrane then forms around each set of chromosomes, and for a short period of time, there are two nuclei in the cell. The completion of mitosis involves a splitting of the cytoplasm into two halves to produce two cells. The mitosis process is diagrammed in Fig. 9-1. For the sake of clarity, an animal cell with four chromosomes is used in this diagram.

Stages of mitosis. The brief overview of mitosis just outlined did not discuss the sequence of chromosome movements so important in ensuring the transfer of a full complement of genes to each resultant cell. We shall now consider this aspect of cell division.

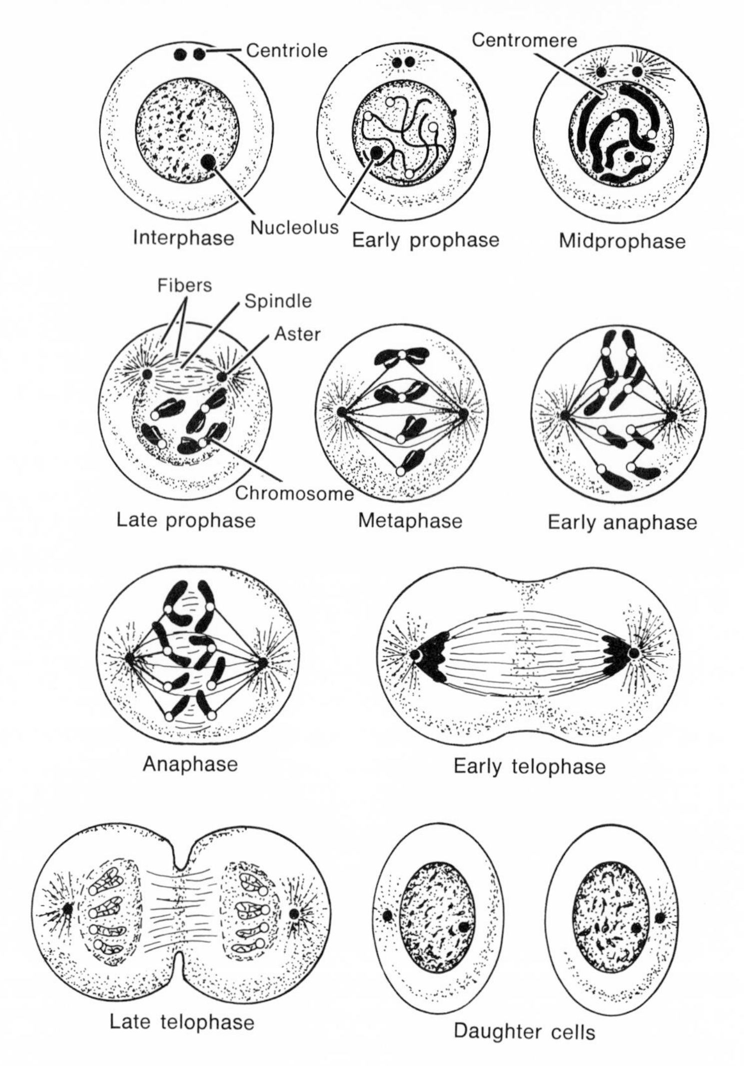

Fig. 9-1
Mitosis in an animal cell with four chromosomes. (Modified from Levine, L. 1973. Biology of the gene, ed. 2. The C. V. Mosby Co., St. Louis.)

The process of mitosis can be viewed as taking place in four stages, although it should be borne in mind that they follow one another in continuous fashion. The four stages are prophase, metaphase, anaphase, and telophase. The period of time between the end of one mitosis (telophase) and the beginning of another (prophase) is called interphase, and for comparison purposes, interphase is the best place to begin a description of mitosis.

Interphase. Cells in interphase have a distinct nucleus, complete with nucleolus, but their chromosomes are only barely visible.

Prophase. Cells are considered to be in prophase when the **centrioles** begin to move away from each other and migrate toward opposite ends of the cell. During this movement, the centrioles are surrounded by raylike fibers, called **asters.** As the centrioles move away from one another, additional, fine, threadlike fibers appear between them and form the **spindle.**

During prophase, the **nucleolus** within the nucleus disappears, and at about the same time, the chromosomes become visibly distinct. The change in visibility of the chromosomes is a result of their coiling, which makes them short and thick.

As soon as the chromosomes can be clearly seen under the microscope, it becomes apparent that each one is actually composed of two chromosomes, which at this point are called **chromatids.** The chromatids were produced by the doubling of the DNA and the synthesis of its bound protein during the previous interphase. The members of each pair of chromatids are attached to one another at a specific point called the **centromere.**

The end of prophase is marked by the disintegration of the nuclear membrane. After this occurs, the chromosomes, together with all other nuclear structures, are actually located in the cytoplasm of the cell.

Metaphase. The metaphase stage of mitosis begins with the completed formation of the spindle, consisting of a series of parallel delicate fibers stretched across the cell from pole to pole.

The spindle functions to bring the chromosomes to the midplane of the cell. On reaching the midplane, the centromere of each chromosome (pair of chromatids) becomes attached to a separate spindle fiber. Thus it can be seen in Fig. 9-1 that only four spindle fibers have chromosomes attached to them.

Anaphase. Metaphase passes into anaphase when the centromeres divide and the chromatids separate from each other (each is now called a chromosome). The newly separated chromosomes, that is, the sister chromatids of prophase and metaphase, move toward opposite poles of the cell. Through this process, the chromosomes separate to produce two groups that are genetically identical. During anaphase, the number of individual chromosomes in a human cell is actually 92.

Telophase. In telophase, the final phase of mitosis, the chromosomes regroup into a nuclear structure. The nuclear membrane is reconstituted, and the nucleoli reappear within the nucleus.

The chromosomes uncoil and become increasingly indistinct. At this time, the spindle and asters disappear. During telophase, there is a division of the cytoplasm resulting in two separate cells, each containing 46 chromosomes. These cells immediately enter the interphase stage of the next cell cycle. Sometime during interphase, the centrioles will divide to form two per cell.

Significance of mitosis. Mitosis results in an exact division and distribution of chromosome material. The longitudinal division of each chromosome into chromatids, followed by a meticulous distribution of a complete set of chromatids to each daughter cell, guarantees that the daughter cells will have, both quantitatively and qualitatively, the same genetic constitution as the original cell from which they arose.

□ Meiosis

In human beings, reproduction involves a fusion of sperm and egg. When egg and sperm fuse, so do their nuclei, bringing their chromosomes together in the same nucleus. If sperm and egg were produced through mitosis, they would each contain 46 chromosomes, and fertilization of an egg would result in a doubling of the chromosome number, generation after generation. But the chromosome number does not double in each generation, which means that some mechanism must keep the chromosome number constant from generation to generation.

The chromosome number for man and all sexually reproducing species remains constant because the number of chromosomes contained in both sperm and egg is reduced in half. This reduction is accomplished by the process called **meiosis.** Thus in human beings, the sperm and egg cells carry 23 chromosomes each, whereas every other body cell has 46 chromosomes.

One further aspect of the reduction of chromosome number is extremely important. Each chromosome of the resultant 23 must be a member of a **different** homologous pair of the original 46. This guarantees that each gamete (sperm or egg) carries a complete set of all the genes of the body.

A cell whose nucleus contains the typical number of pairs of homolgous chromosomes for the species is referred to as a **diploid,** or **2n** cell. After meiosis, the resultant cells are called **haploid,** or **n** cells. Diploid refers to cells with pairs of homologous chromosomes, whereas haploid refers to cells with only single members of each pair. Every sperm and egg contains a haploid set of chromosomes; a fertilized egg is diploid.

Stages of meiosis. Meiosis consists of two cellular divisions accompanied by only one duplication of chromosomes (Fig. 9-2). Each cellular division consists of a prophase, metaphase, anaphase, and telophase. However, the movements of the chromosomes during meiosis differ from those of mitosis; these meiotic chromosome movements will now be examined, beginning with the interphase stage before meiosis.

Interphase I. A cell that is about to undergo meiosis does **not** differ chromosomally from one that will undergo mitosis. Its DNA has been replicated, and the protein to which the DNA is bound has been synthesized. Each chromosome consists of two chromatids that are attached to one another at the centromere.

Prophase I. As in mitosis, the centrioles separate, a spindle is formed, the chromosomes become visibly distinct, and the nucleolus and nuclear membrane disappear. However, the resemblance between the two processes ends here. In meiosis, the members of each homologous chromosome pair come together and remain for a period of time in very tight contact with one another. This close association is called **synapsis.**

During synapsis, the chromatids (four in number) may twist around each other, causing breaks in one or more chromatids. These breaks are healed by specific enzymes of the cell. When breaks have occurred in two or more of the chromatids, the healing process may join together sections of chromatids from the different members of a homologous pair of chromosomes. This joining of "wrong" broken chromatid ends is called **crossing-over** and results in gametes that contain chromosomes derived from both parents of the individual that is producing the egg or sperm (Fig. 9-3).

Toward the end of prophase I, the closely associated pairs of homologous chromosomes move toward the midplane of the cell.

Metaphase I. In meiosis, the members of each pair of homologous chromosomes come to the equatorial plane of the cell in close association with one another. At that point, a spindle fiber becomes attached to the centromeres of each pair of homologous chromosomes and extends from the centromeres to the opposite ends of the cell. In Fig. 9-2, it can be seen that only two spindle fibers have chromosomes attached to them.

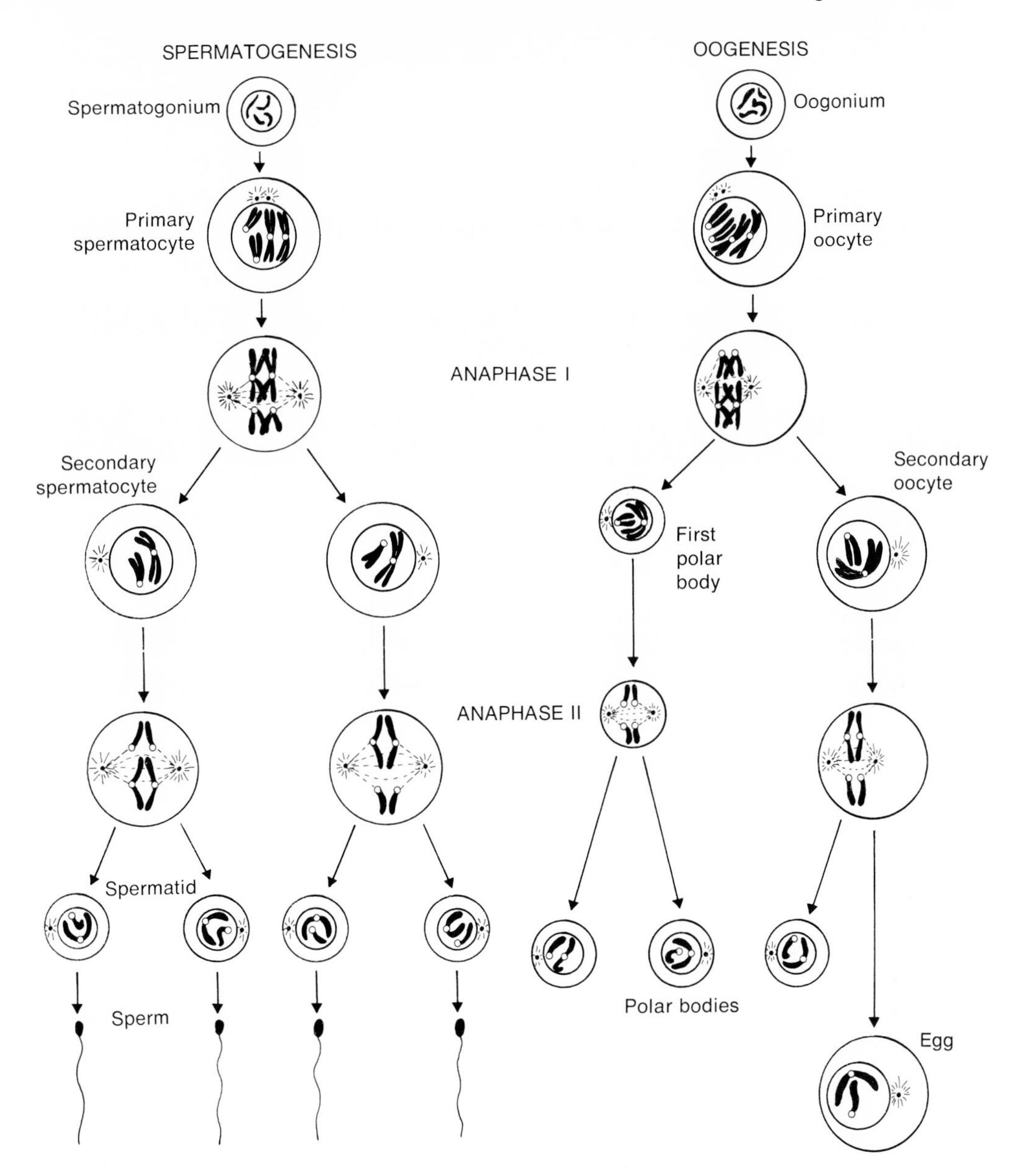

Fig. 9-2
Meiosis in male and female animals. Left, spermatogenesis, resulting in the formation of four sperm. Right, oogenesis, resulting in the formation of one egg and three polar bodies. (From Levine, L. 1973. Biology of the gene, ed. 2. The C. V. Mosby Co., St. Louis.)

Anaphase I. In meiosis, the members of each pair of homologous chromosomes move to opposite poles of the cell. Each chromosome still consists of two chromatids joined together at the centromere. As a result of the meiotic process thus far, two groups of chromosomes are formed; each group consists of one complete (haploid) set of chromosomes.

Telophase I. When the chromosomes reach their respective poles, a nuclear membrane forms around each group, the spindle and asters disappear, and the

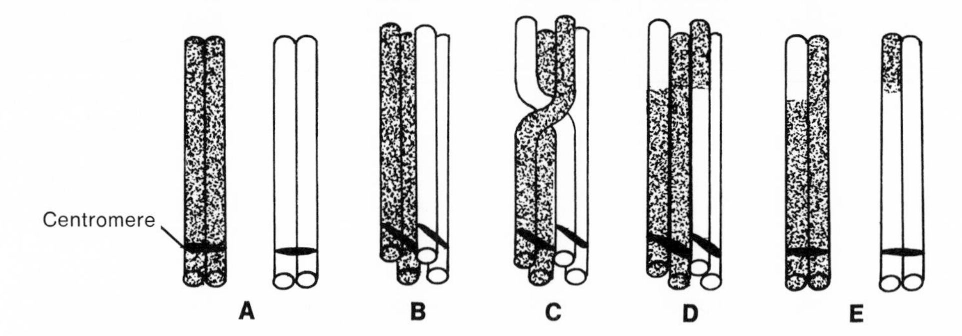

Fig. 9-3
Synapsis and crossing-over during prophase I of meiosis. **A**, Pair of homologous chromosomes (consisting of a total of four chromatids). **B**, Synapsis of homologous chromosomes. **C**, Crossing-over. **D**, Exchange of segments of chromatids. **E**, Separation of homologous chromosomes.

cytoplasm divides. At this time, the total number of chromosomes in each resultant cell is half the number in the original cell. However, since the centromeres did not divide and separate in anaphase I, each chromosome still consists of two chromatids, and meiosis is only half completed.

Interphase II. Interphase II is a short phase between the first and second meiotic divisions. The chromosomes uncoil, but **no** DNA replication occurs.

Prophase II. The nuclear membrane breaks down, the spindle begins to form, and the chromosomes become coiled.

Metaphase II. The chromosomes, still consisting of two chromatids each, align themselves in the center of a newly formed spindle. Each chromosome becomes attached at its centromere to a separate spindle fiber. In Fig. 9-2, it can be seen that only two spindle fibers have chromosomes attached to them.

Anaphase II. In the anaphase II stage, the centromeres divide, and the chromatids (now individual chromosomes) separate and go to opposite poles.

Telophase II. A nuclear membrane forms around each group of chromosomes, the chromosomes uncoil, and the cytoplasm divides, forming two cells. Each daughter cell contains a complete haploid set of chromosomes—half the number found in the original cell.

• • •

We have reviewed the movements of the chromosomes during meiosis and have seen how a diploid cell divides twice to form four haploid cells. The chromosomal pattern is the same in the production of both sperm and eggs. However, the division of the cytoplasm varies depending on which type of gamete is formed during gametogenesis.

Gametogenesis. The formation of gametes, gametogenesis, may be subdivided into **spermatogenesis**, the formation of sperm, and **oogenesis**, the formation of eggs. In spermatogenesis, diagrammed in Fig. 9-2, each cell undergoing meiosis produces **four** functional haploid sperm cells. All four sperm have equal amounts of cytoplasm, and each can function equally well in the fertilization of an egg cell. However, in oogenesis, also diagrammed in Fig. 9-2, each cell undergoing meiosis produces only

one functional haploid egg cell because of the uneven division of the cytoplasm in telophases I and II.

The first meiotic division produces one large cell and one small cell (the first **polar body**). The second meiotic division of the large cell results in a small second polar body and a large cell that becomes the functional egg cell. The first polar body may also go through a second meiotic division, for a total of four cells. But polar bodies are not fertilized, and they disintegrate soon after their formation.

As a result of the unequal cytoplasmic divisions during oogenesis, large amounts of cytoplasm and food are supplied to the egg for use by the developing embryo. In oogenesis, 75% of the potential eggs are sacrificed to ensure the survival of the remaining 25%.

In spermatogenesis, all of the potential sperm cells are actually produced. However, sperm cells contain very little cytoplasm and function solely as contributors of genetic material; they carry essentially no food for the developing embryo.

Significance of meiosis. As just described, meiosis results in a reduction of the chromosome number to one half that of the original cell. Meiosis also requires that one of the chromosomes from each homologous pair be present in each resulting gamete, although, because of the phenomenon of crossing-over, some chromosomes may be derived from segments of both homologous chromosomes.

Equally important for genetic consideration is the fact that all chromosomes derived from a given parent are not necessarily passed along as a group into a given gamete. Returning to our organism with the diploid number of four, let us label the two homologous sets of chromosomes as AA^1 and BB^1. We shall assume that chromosomes A and B were derived from one parent and that chromosomes A^1 and B^1 were derived from the other. If large enough numbers of gametes are considered, this organism should produce equal numbers of gametes with the following chromosome combinations: AB, AB^1, and A^1B, and A^1B^1.

The number of different gametes possible is calculated as $(2)^n$, where n is the number of pairs of homologous chromosomes. Meiosis therefore results in a completely random distribution of chromosomes in the gametes. In the illustration used, with two pairs of homologous chromosomes, the number of different possible gametes is $(2)^2$, or 4. For humans, the corresponding number is $(2)^{23}$, which is more than 8 million different possible gametes. The variability of gametes is further increased by crossing-over.

Meiosis is the process that produces the new combinations of hereditary materials that occur in every generation.

FUNDAMENTALS OF GENETICS

Our study of mitosis and meiosis showed us that chromosomes are present as homologous pairs. It follows from this that genes must also be present in duplicate. Genes that are located at the same places on homologous chromosomes are called **homologous genes.** When homologous genes transcribe different messages, they are called **alleles** (different forms of a given gene).

Alleles may interact in various ways in determining the characteristics of an individual. The effect of one allele may show complete dominance over that of another. On the other hand, for a different trait, the effect of one allele may only partially dominate that of another. We may also find a situation in which the effects of both alleles are expressed completely, a situation called codominance.

An analysis of allelic interactions is complicated by one special chromosome pair. Of the 23 pairs of chromosomes in every cell of our bodies, 22 pairs, called **autosomes,** consists of homologous chromosomes that are equal in their total genic content. However, the twenty-third chromosome pair, called the **sex chromosomes,** consists of chromosomes that are not equal in their genic content.

Sex chromosomes are of two types: X and Y. In human beings, a person with two X chromosomes is a female, whereas an individual with one X and one Y chromosome is a male. In addition to being important in deciding the sex of the person, the X chromosome carries genes that affect other traits of the body (red-green color blindness, hemophilia, etc.). The problem of allelic interaction in sex chromosomes arises from the fact that the Y chromosome appears to be devoid of genes. This apparent absence of genes in the Y chromosome means that any allele, dominant or recessive, in the X chromosome of a male will be fully expressed. As we shall see, the pattern of inheritance of genes on the X chromosome follows a special configuration that has been called X-linked inheritance.

□ Pedigree studies

How do we know whether a trait in human beings is inherited, and, if so, whether the characteristic is autosomal or X linked, dominant or recessive? When a trait appears for the first time and it is important to analyze it, we begin by studying the family of the person showing the trait.

A chart, called a **pedigree chart,** is drawn up, showing all members of the family and whether or not they exhibit the trait. Fig. 9-4 shows some of the symbols used in constructing a pedigree chart.

In deciding whether a trait is inherited or not, we see if the occurrence of the characteristic fits a pattern that is explainable on an environmental or a genetic basis.

Environmental factors tend to affect all people that live or work together, regardless of whether they belong to the same family or not. In contrast to this, we find that

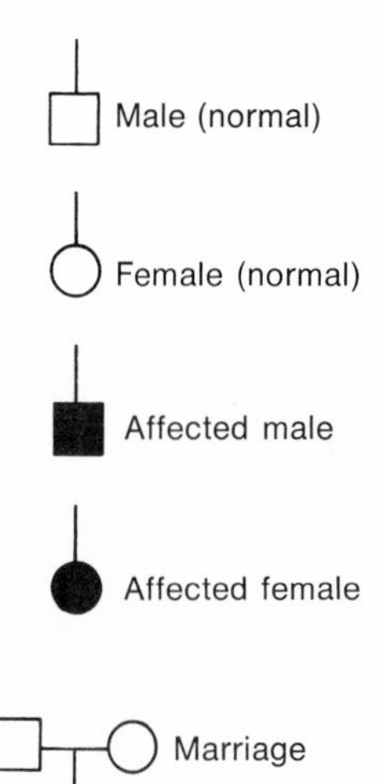

Fig. 9-4
Symbols used in constructing a pedigree chart.

genetic traits run in families and follow a pattern that is in agreement with what we have learned about chromosome movements in meiosis.

Autosomal dominant gene. Fig. 9-5 shows the occurrence of a tooth disease, dentinogenesis imperfecta, in a family. This is a relatively rare disease with an incidence of 1 in 8000 people. The teeth of these people have a brown color, and the crowns of the teeth wear down rapidly.

The frequent occurrence of a relatively rare disease among the members of a single family indicates that we are dealing with a genetic trait. The fact that the trait appears in every generation implies that we are dealing with a dominant gene. The occurrence of the disease with equal frequency in both males and females indicates that the chromosome carrying the gene is an autosome.

The male in generation I must have carried the dominant allele in only one of his chromosomes, whereas the homologous chromosome of that particular pair carried the allele for normal teeth, which in this instance is recessive. We can make this statement because only half of his offspring show the trait. Based on what we know about the separation of homologous chromosomes during meiosis, only 50% of the children in generation II should have received the dominant gene and, as a consequence, exhibited the trait. The same line of reasoning applies to female II-2.

It is quite apparent that this pedigree chart fits the model of an autosomal dominant gene.

Autosomal recessive gene. Fig. 9-6 shows the occurrence of cystic fibrosis, an epithelial tissue disease discussed in Chapter 7, in two families, between whom a marriage occurred in generation III. The sporadic occurrence of the trait and the absence of affected individuals in generations I and III argues against any general environmental factor producing the characteristic.

The occurrence of the trait in the children of unaffected parents and the absence of the trait among the children of an affected parent strongly support the idea of a recessive gene. The autosomal nature of the chromosome carrying the gene is indicated by the fact that both males and females are affected in equal frequency.

The individuals of generation I must all have been carriers of the recessive gene, as must also be true of the members of generation III. The affected people in generation II and IV carry the recessive gene on both homologous chromosomes of the particular pair. The pedigree chart fits the model of an autosomal recessive gene.

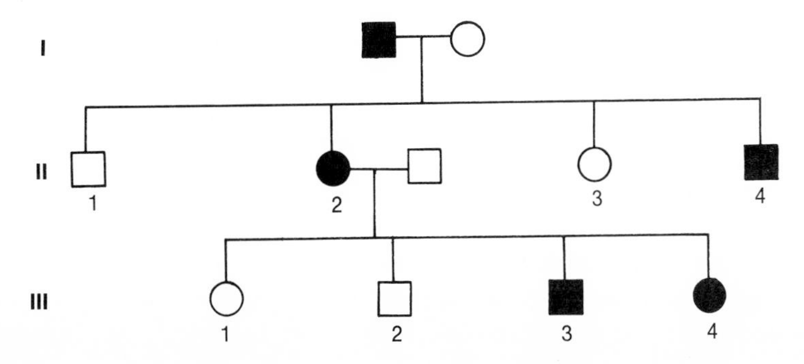

Fig. 9-5
A tooth disease, dentinogenesis imperfecta, in a family pedigree. Roman numerals represent generations; Arabic numbers represent birth order within the particular generation.

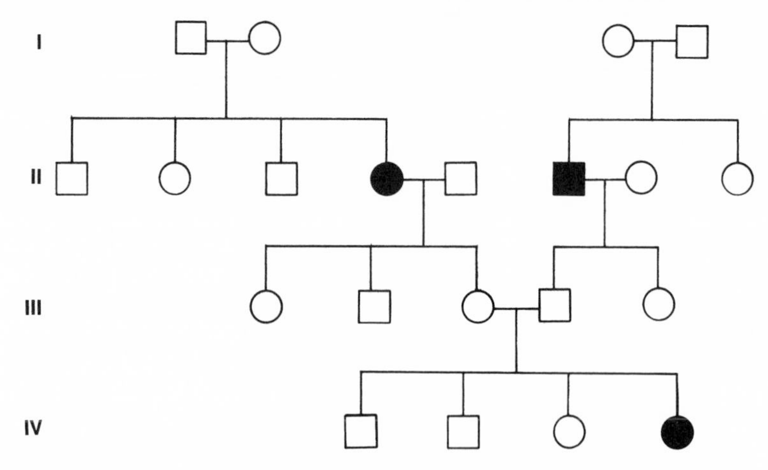

Fig. 9-6
Cystic fibrosis in a family pedigree.

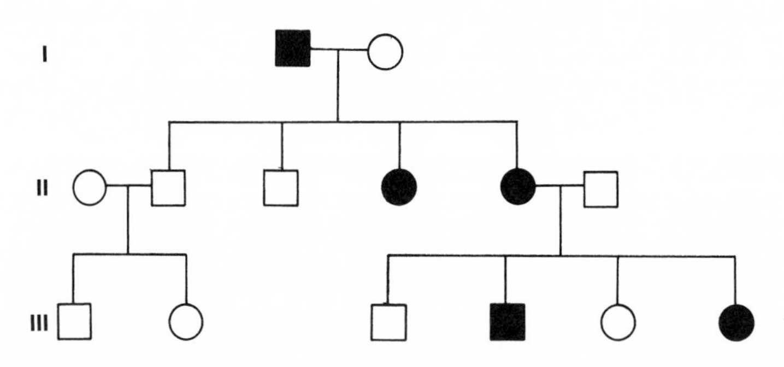

Fig. 9-7
Hypophosphatemia, vitamin D–resistant rickets, in a family pedigree.

X-linked dominant gene. Fig. 9-7 shows the occurrence of hypophosphatemia, also called vitamin D–resistant rickets, in a family. Rickets (see Table 4-1) is a nutritional disease characterized by defective bone formation and caused by an insufficiency of vitamin D in the diets of children. Although most cases diagnosed as rickets respond to vitamin D therapy, there are some that are resistant to this type of treatment. These nonresponsive individuals occur in specific family groups, indicating that the disease is inherited. The appearance of the trait in every generation of the pedigree chart implies that the gene causing the trait is dominant.

The fact that the male in generation I transmitted the genes to some but not all of his offspring means that he carried the dominant allele on only one of his chromosomes. However, the fact that he transmitted the dominant allele only to his daughters places the gene on his X chromosome, for the following reason: Because males are XY and females are XX, all daughters receive one of their X chromosomes from their father, whereas the other of their X chromosomes is one of their mother's two X chromosomes. Sons, on the other hand, receive their Y chromosomes from their fathers, whereas their X chromosome is one of their mother's two X chromosomes. Therefore the absence of father-to-son transmission of a trait places the gene for vitamin D–resistant rickets on the father's X chromosome.

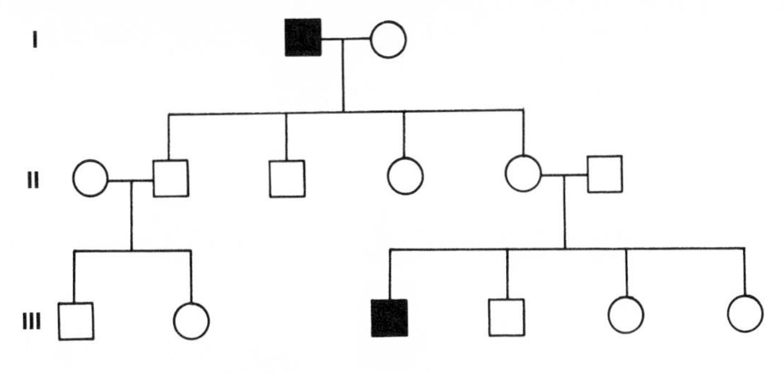

Fig. 9-8
Hemophilia in a family pedigree.

The transmission of the dominant allele from the married female in generation II to both a son and a daughter is consistent with the hypothesis that the gene is on the X chromosome. In Fig. 9-7, we see a pedigree that reflects the presence of an X-linked dominant gene.

X-linked recessive gene. Fig. 9-8 shows the incidence of hemophilia, bleeder's disease, in a family. Hemophilia, as discussed in Chapter 6, results from the absence of a clotting factor in the blood plasma of affected individuals. The incidence of this disease is about 1 in 10,000 male births, hemophiliac females being extremely rare.

The multiple occurrence of a rare disease in a family group immediately suggests that the disease is inherited. Its occurrence in children of unaffected parents and the absence of the trait among the children of an affected parent indicate that the gene involved is recessive. A pattern of transmission from a father **through** his daughters to **half** of their sons fits the model of an X-linked recessive gene. It should be noted that only in X-linked inheritance do males show traits as a result of possessing only a single recessive allele for the particular characteristic.

• • •

In similar fashion, we can analyze pedigree charts for any type of allele interaction.

□ Patterns of inheritance

In our study of pedigrees, we have reviewed the criteria that are used in deciding whether a given trait is environmentally produced or inherited. We have also learned to distinguish autosomal from X-linked traits and dominant from recessive alleles.

For those traits that are genetic in origin, it becomes important to learn what types of offspring will be produced from different sets of parents. This is especially useful in our modern society where we are acquiring an ever increasing amount of information on the genetic determination of many of our traits and diseases.

Simple (complete) dominance. We shall illustrate **simple (complete) dominance** by the gene that causes skin pigmentation in man. For this trait, the gene that produces normal pigmentation of the skin is dominant over the gene that causes albinism (absence of pigment in the skin, the hair, and the iris of the eye).

We shall designate the dominant gene as **C,** while the recessive gene will be designated as **c.** With these two alleles, we can obtain three different genetic combinations: **CC, Cc,** and **cc.** If an individual is **CC** or **Cc,** his appearance (**phenotype**) will be pigmented. If he has the **cc** combination, his appearance will be that of an albino. The genetic constitution (**genotype**) of a **CC** or **cc** individual is said to be **homozygous,** while that of the **Cc** individual is said to be **heterozygous.**

In simple dominance, one gene for the particular characteristic has as much effect as do two genes. The example given in Fig. 9-9 illustrates the cross between a normally pigmented person (Caucasoid, Negroid, or Mongoloid) and an albino (Caucasoid, Negroid, or Mongoloid). The square, called a Punnett square, is used in Fig. 9-9 to ascertain the types and ratios of offspring. It is designed to yield all the possible combinations of gametes that can occur.

As was stressed in the discussion of meiosis, homologous chromosomes must separate and enter different gametes. In the illustrative case, all the gametes from the pigmented parent will contain **C,** while all the gametes from the albino parent will contain **c.** This separation of homologous genes in the formation of gametes was first enunciated by Mendel in 1865 as the **law of segregation** (also known as the law of purity of gametes).

The offspring of the parental cross are called the F_1 (the first filial) generation. In this cross, the offspring are all pigmented, indicating that pigmentation shows simple dominance over albinism. This phenomenon was also described by Mendel in his **"law" of dominance.** This is not a true biological law, since most contrasting traits do

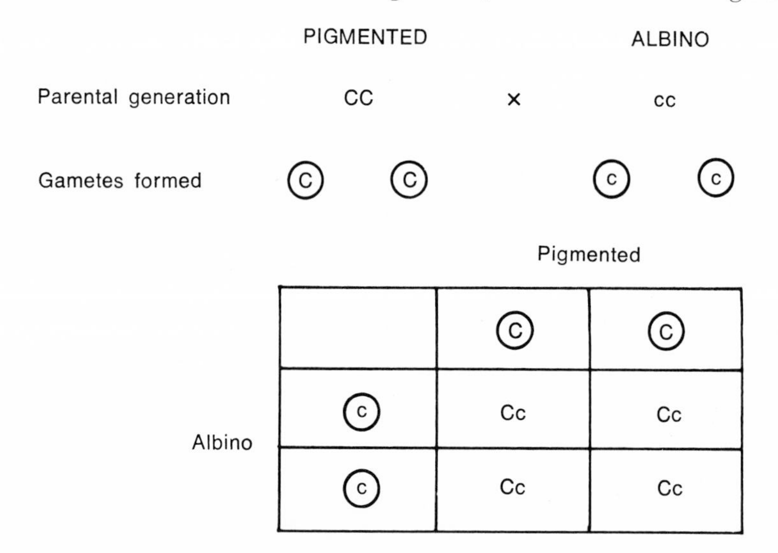

Summary of F_1

Phenotypes	Phenotypic ratio	Genotypes	Genotypic ratio
Pigmented	all	Cc	all

Fig. 9-9
Diagram of a cross between a homozygous pigmented person and an albino. (From Levine, L. 1973. Biology of the gene, ed. 2. The C. V. Mosby Co., St. Louis.)

not show simple dominance. However, Mendel was led to this rule by the action of those traits he had studied, most of which showed simple dominance.

Fig. 9-10 shows the results of a cross of two individuals having the genotype of the F_1 of the previous cross. Again, homologous chromosomes must segregate in the formation of gametes, and every gamete must carry only one gene for each trait.

The offspring of this cross are called the F_2 (second filial) generation. We see here the reappearance of albinism, indicating that the gene for this trait was not lost, but was only hidden in the F_1 generation. The ratios of the phenotypes and genotypes are typical for simple dominance.

Fig. 9-11 shows a typical **testcross,** which consists of a mating between an F_1 heterozygote and a person with the recessive parental trait. Half of the offspring show the recessive trait, while the other half of the offspring exhibit the dominant phenotype.

The testcross is extremely useful in evaluating an individual who shows the dominant phenotype, but whose genotype is not known. This type of cross indicates

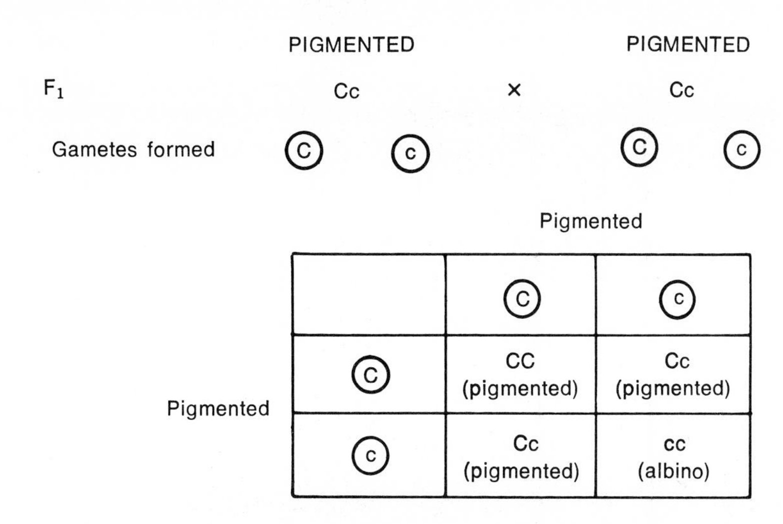

Summary of F_2

Phenotypes	Phenotypic ratio	Genotypes	Genotypic ratio
Pigmented	3	CC	1
		Cc	2
Albino	1	cc	1

Fig. 9-10
Diagram of a cross between two normally pigmented people, both of whom are carriers of the gene for albinism. (From Levine, L. 1973. Biology of the gene, ed. 2. The C. V. Mosby Co., St. Louis.)

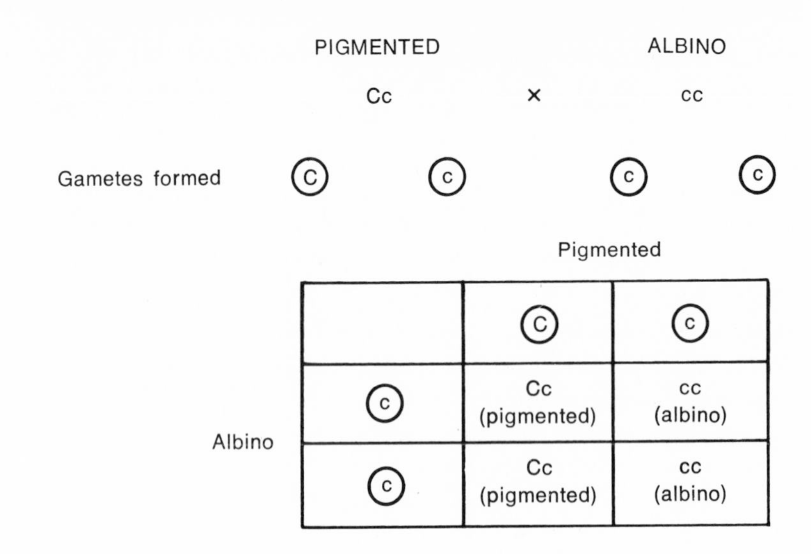

Summary of testcross

Phenotypes	Phenotypic ratio	Genotypes	Genotypic ratio
Pigmented	1	Cc	1
Albino	1	cc	1

Fig. 9-11
Diagram of a cross between a normally pigmented heterozygous person and an albino. (From Levine, L. 1973. Biology of the gene, ed. 2. The C. V. Mosby Co., St. Louis.)

whether the dominant phenotype is caused by a homozygous or a heterozygous genotype. The appearance of a single recessive phenotype immediately signifies that the person in question is a heterozygote. The continued appearance of only pigmented offspring would imply that the individual is a homozygote. Inherent in this type of analysis is the assumption that we can deduce genotypes from observing phenotypes. In reality, the only aspect of an organism that we can see is its phenotype. From its phenotype, we infer its genotype.

Partial (incomplete) dominance. There occurs in humans a gene whose effects result in fingers and toes being abnormally short, a condition known as brachydactyly. Here, too, there are only two alleles: for brachydactyly and for normal-sized fingers.

A gene that yields the normal phenotype is usually designated by the symbol "+." The symbol for each mutant allele, however, is derived from the descriptive name of the mutant trait (**h** for the hemophilia phenotype; **Br** for the brachydactyly phenotype). Recessive alleles are designated by small letters (**h**), and dominant alleles are indicated by capitalizing the first letter of the symbol (**Br**). The gene that produces normal blood clotting is symbolized as $\mathbf{h}^+$ and is dominant to **h**, while the gene that produces normal-sized fingers and toes is symbolized as **Br$^+$** and is recessive to **Br.**

We can also designate the two homologous chromosomes by the use of two slanted lines, "//," or, as is more usually done, by a single slanted line, "/."

Applying these symbols in the case of brachydactyly, we find three possible genotypes: (1) Br/Br, (2) Br/Br$^+$, and (3) Br$^+$/Br$^+$. The genetic combination Br/Br$^+$ (heterozygote) results in the abnormally short fingers and toes mentioned previously, and the genotype Br/Br (homozygote) results in severe crippling with a complete lack of fingers and toes.

For brachydactyly, although there is dominance, it is incomplete, and two genes for this factor have a much greater effect than a single gene. The importance of the partially dominant gene is that if it is deleterious to its possessor, it will produce a handicap even in a single dose, although on a reduced scale.

Codominance. Another type of allele interaction is called **codominance (equal expression);** in this instance, both alleles are expressed in the individual's phenotype. Human blood groups provide an excellent illustration of this type of allelic interaction. We shall consider the ABO series. Here there are three alleles of the same gene. When more than two alleles exist, the series is called **multiple alleles.**

There are six different possible combinations of these three alleles: AA, AO, BB, BO, AB, and OO. Of these six possibilities, four different blood types are recognized: A, B, AB, and O.

Genes A and B exhibit simple dominance when either of these genes is in combination with the O gene. However, when genes A and B are in combination with each other, each one is expressed as if it were present without the other.

This series of alleles therefore exhibits two types of gene interactions: (1) simple dominance between genes A and O or genes B and O and (2) codominance between genes A and B.

The blood groups have played an interesting function in cases of disputed paternity. In most instances, we can show that a particular male could not have fathered a particular child if in fact he was not the father. However, we cannot prove that a particular male did father the child.

As an illustration, let us consider the following case: a male's blood type is O and a female's blood type O. No child of blood type A or B could have this male as a father. However, we cannot be certain, should the child have blood type O, that this particular male was the father. Hence we can disprove paternity, but not prove it.

The courts today need not rely on ABO series alone for evidence of nonpaternity. There are some five or six other blood factors, for example, Rh, MN, Lewis, and Duffy, the combination of which can rule out any improperly accused male.

X-linked inheritance. The patterns of inheritance we have considered thus far have involved allelic interactions of autosomal genes. We now want to examine the special pattern of inheritance that occurs when genes on the X chromosome are studied. The unique aspect of **X-linked inheritance** is that for the XY individual (human male), any gene located in the X chromosome will be fully expressed in the phenotype regardless of whether it acts as a dominant or as a recessive gene in the XX individual (human female).

There are about 80 X-linked traits known in humans. As examples of the effects of

X-linked genes, hypophosphatemia and hemophilia were discussed earlier in this chapter. A pattern of X-linked inheritance, using hemophilia as an example, is shown in Fig. 9-12.

The distinctive crisscross pattern of inheritance, from father through daughter to grandson, of some human traits was known long before the twentieth century. It was apparently well understood by the Jews of the Middle Ages, as illustrated in the talmudic prohibition against circumcision of any male born to a woman whose father, older male child, or even brother was a bleeder. Especially significant was the stipulation that the existence of a bleeder maternal uncle was sufficient grounds for being fearful about the well-being of the child. This stipulation implies the realization that females may act as carriers of X-linked traits for any number of generations,

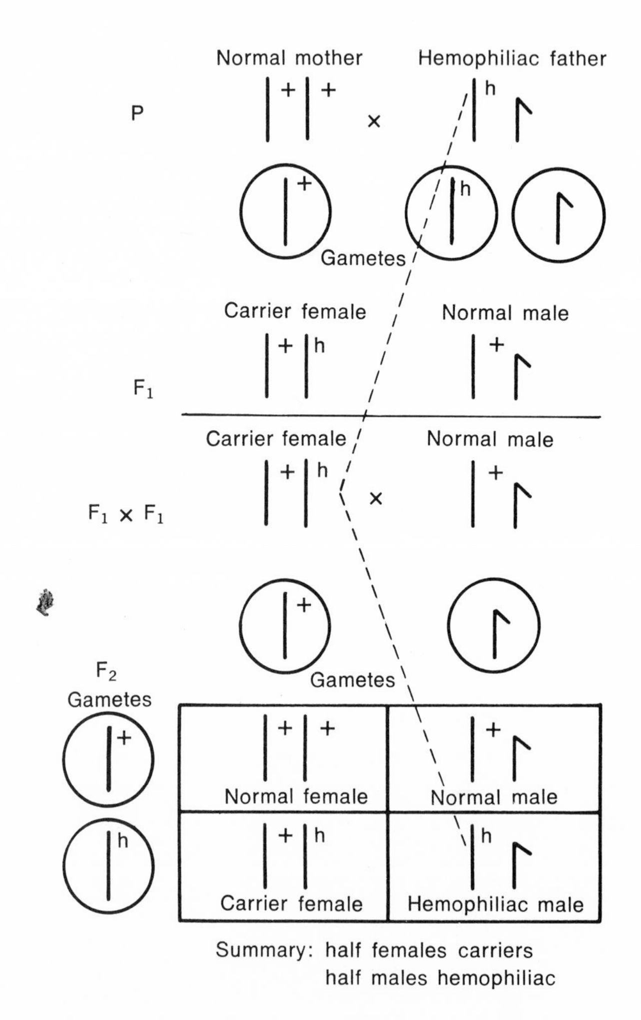

Fig. 9-12
Diagram of genes on the X chromosomes, illustrating a cross between a woman with normal blood clotting and a hemophiliac man. The dotted line through the center illustrates the crisscross pattern of inheritance: from father through daughter to grandson. The symbol "h" represents the X-linked recessive gene for hemophilia. (From Levine, L. 1973. Biology of the gene, ed. 2. The C. V. Mosby Co., St. Louis.)

handing the gene down from mother to daughter without exhibiting any of its effects, yet always capable of transmitting it to a male offspring.

The most celebrated cases of hemophilia occurred in the royal families of England, Russia, and Spain, all of whom were interrelated by marriage. The pedigree begins with Queen Victoria of England, who was a carrier of the gene. Since none of her ancestors or relatives were known to be hemophiliacs, it is assumed that a mutation (change in a gene that transforms it into a different allele) occurred in an X chomosome that was contained in one of the gametes of her parents.

One of her sons, Leopold, Duke of Albany, died of hemophilia at the age of 31. The birth of an affected son rules out Victoria's husband as the source of the gene for hemophilia, since male-to-male inheritance does not occur in X-linked traits. At least two of Victoria's daughters were carriers of the gene for hemophilia, since several of their male descendants were hemophiliacs. The occurrence of hemophilia in the son of the last Czar of Russia and in the princes of Spain, all descendants of Victoria, had considerable political consequences in the history of Europe.

For a female to have hemophilia, or any other X-linked recessive trait, she must receive a gene for the trait from each parent. This can occur in hemophilia, and there are a number of cases that resulted from the marriage of hemophiliac males and carrier females.

As in the case of affected males, hemophiliac females may bleed either continuously or for very long periods of time after only a slight scratch. For the hemophiliac female, there is an added hazard caused by the menstrual flow, and these affected females often require transfusions to replace the great loss of blood accompanying the menses.

Attempts have been made to discover Y-linked traits in man. The most extensive studies conducted thus far have been on the inheritance of ear hair. There are men who have very hairy ears, and the trait appears to be transmitted from father to son. A number of independent studies of different family pedigrees have failed to resolve whether the trait is Y linked or autosomal. To date, we have no unequivocally demonstrated case of a Y-linked gene in man.

Independent assortment. Up to now, we have examined patterns of inheritance that involve alleles. We now want to consider the pattern of inheritance that occurs when genes on nonhomologous chromosomes are transmitted from one generation to the next.

We shall take as our examples both the gene that determines whether an individual can taste the chemical compound phenylthiocarbamide (PTC) and the gene that determines hair color. PTC is a bitter-tasting substance that can be detected by persons carrying the allele **T**, which is completely dominant over its counterpart **t**. Persons who are **T/T** or **T/t** are referred to as **tasters**, whereas **t/t** individuals are **nontasters.** For human hair color, the gene **B** (for **black hair**) is completely dominant over its allele **b** (for **red hair**). Thus persons who are **B/B** or **B/b** are black haired, whereas those that are **b/b** are red haired.

If individuals who are **T/T, B/B** marry persons who are **t/t, b/b,** their children will all be tasters and black haired **(T/t, B/b).** Should double heterozygotes from different families marry, the offspring produced would fit the pattern shown in the Punnett square of Fig. 9-13.

	Taster Black-haired		Nontaster Red-haired
P	TTBB	×	ttbb
Gametes formed	(TB) (TB)		(tb) (tb)
F_1	TtBb	×	TtBb

F_2

Gametes	TB	Tb	tB	tb
TB	T/T B/B	T/T B/b	T/t B/B	T/t B/b
Tb	T/T B/b	T/T b/b	T/t B/b	T/t b/b
tB	T/t B/B	T/t B/b	t/t B/B	t/t B/b
tb	T/t B/b	T/t b/b	t/t B/b	t/t b/b

Summary: 9/16 tasters, black-haired
3/16 tasters, red-haired
3/16 nontasters, black-haired
1/16 nontasters, red-haired

Fig. 9-13
Diagram of crosses involving the genes for PTC tasting and hair color.

The 9:3:3:1 ratio of phenotypes found in the F_2 generation is typical of crosses involving nonhomologous chromosomes. We must remember that although in meiosis each pair of homologous chromosomes undergoes segregation, nonhomologous chromosomes distribute themselves independently of one another. This type of inheritance pattern was also studied by Mendel in 1865; in explaining the results, he formulated the **law of independent assortment.** This genetic rule states that every characteristic is inherited independently of every other characteristic.

Although independent assortment results in a relatively simple phenotypic ratio, the genotypic ratio is much more elaborate. There are actually nine different genotypes, some represented only once, while others are found twice and one four times.

Sex-influenced genes. There are a number of known autosomal genes whose expression will be either dominant if the individual is a male or recessive if the individual is a female. Such genes are said to be **sex influenced.** In humans, baldness appears to be such a sex-influenced autosomal trait. Heterozygous males become bald, while the gene must be in a homozygous state for baldness to occur in women. However, baldness can occur in a heterozygous woman who develops a masculinizing tumor of the ovary.

Sex-limited genes. The **sex-limited genes** are those autosomal genes that can be expressed in only one of the two sexes. Genes that control the characteristics of the male reproductive system have obviously no opportunity to function if the individual is a female, and vice versa.

In birds where the plumage pattern varies with the sex, there are a large number of genes for feather colors that can, in some cases, function only in the male, while in other cases, only the female is affected.

Sex determination

It was pointed out earlier in this chapter that human males were XY in chromosome constitution, whereas human females were XX. A question arises. Is maleness determined by the Y chromosome or by the presence of a single X chromosome? The answer to this question was obtained from a chromosomal study of certain types of individuals who are sterile.

Some human females (1 in 5000) exhibit what is called **Turner's syndrome.** These females are sterile as a result of a complete lack of development of their ovaries. When the chromosomes of their cells were studied, it was found that these females contained 45 chromosomes in their cells. There were 44 autosomes and one X chromosome (designated as 45,XO). This discovery, made in 1959, showed that a single X chromosome does not produce a male and indicated that maleness in humans depended on the presence of a Y chromosome.

Support for this hypothesis of sex determination in humans came from a different study. Some human males (1 in 700) exhibit what is called **Klinefelter's syndrome.** These males are sterile because of a defective development of their testes. When the chromosomes of their cells were studied, it was found that these males usually contained 47 chromosomes in their cells. There were 44 autosomes, two X chromosomes, and a Y chromosome (designated as 47,XXY). Some males with Klinefelter's syndrome were found to be 48,XXXY, and a very few were 49,XXXXY. In all cases, the presence of a Y chromosome made the individual a male. In humans, the Y chromosome is absolutely male determining.

Chromosome numbers

Every species has a characteristic number of chromosomes. In human beings, this number is 46, or, more specifically, 23 pairs. We have already discussed two conditions, Turner's syndrome and Klinefelter's syndrome, that result from alterations in the chromosome number of humans.

Unfortunately, the addition or deletion of a chromosome upsets the normal development of an individual and usually leads to a physically and/or mentally retarded person. Each homologous pair of human chromosomes is fairly distinct in its size, position of centromere, and overall shape. A complete set of human chromosomes is shown in Fig. 9-14.

Because the individual whose chromosomes are shown in Fig. 9-14 is a male, both the X and Y chromosomes are included. If the cell had come from a female, the only differences would be the absence of the Y chromosome and the presence of a second X chromosome.

The importance of chromosome numbers in health and development is apparent when we consider that 40% of all spontaneous abortions during the first 90 days of pregnancy involve embryos with abnormal chromosome numbers.

Down's syndrome (mongolism). A number of abnormal conditions are known to result from the presence of extra autosomes. The first proved association of a human defect with an extra autosome was **Down's syndrome** (mongolism) in 1959.

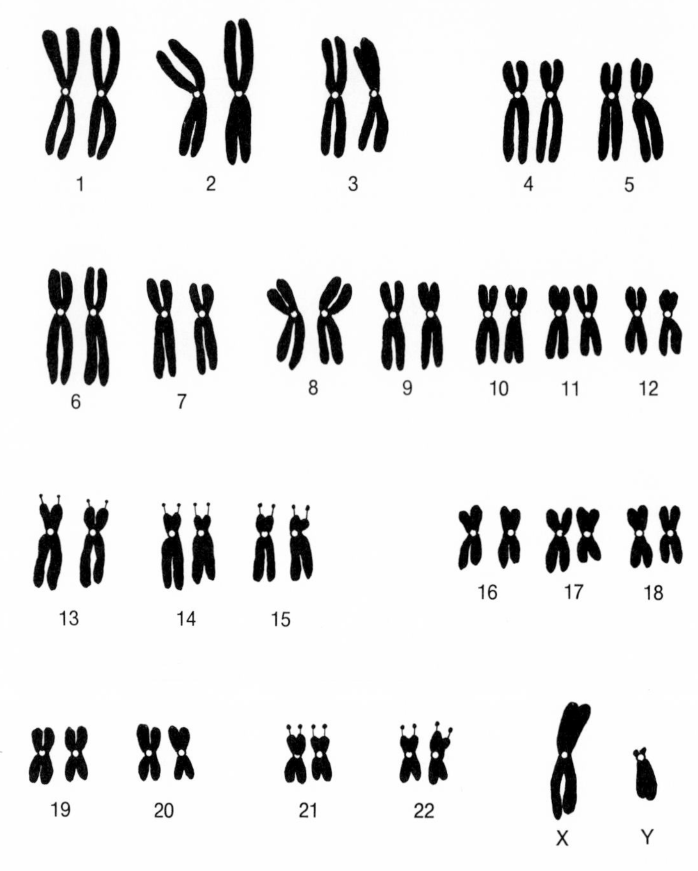

Fig. 9-14
Ideogram of the mitotic metaphase chromosomes of a human male. (From Levine, L. 1973. Biology of the gene, ed. 2. The C. V. Mosby Co., St. Louis.)

An individual with mongolism is characterized by mental retardation, short stature, stubby hands and feet, and often malformation of the heart. Prior to the discovery of antibiotics, these children rarely lived past their tenth year because of their sensitivity to infections. Today, their life span has been greatly extended.

One of every 600 children born in the United States suffers from Down's syndrome. Since slightly more than 3 million children are born in the United States each year, this rate of incidence of Down's syndrome results in about 5000 such afflicted infants being born annually.

A study of the chromosome complement of the cells from Down's syndrome individuals revealed that these people had 47 chromosomes in their cells and that the extra chromosome was number 21 (designated as 47,21+) (Fig. 9-14). It is not known how the extra chromosome 21 causes the various defects of mongolism, nor is there as yet any known cure for this genetic disease.

Investigations of Down's syndrome have revealed a relationship between the incidence of the disease and the age of the mother. There is a marked increase in the frequency of the births of afflicted children with increasing age of the mother (Table 9-2). The age of the father does not influence the incidence of affected children at all.

Table 9-2
Incidence of Down's syndrome children born to variously aged mothers

Age group of mother (years)	Incidence of affected children
Less than 30	1/1500 births
30-34	1/750 births
35-39	1/600 births
40-45	1/300 births
More than 45	1/35 births

Why should the age of the mother affect the frequency of births of children with Down's syndrome? The answer to this question lies in a study of the female reproductive system. All the eggs that will be produced by a woman during her reproductive period are present in her ovary at the time she herself is born. As a result, an egg produced by a 20-year-old woman is, in fact, 20 years old itself. An egg produced by a 40-year-old woman is 40 years old.

For reasons that are not understood, older ovarian cells do not undergo meiosis as perfectly as do younger cells. This results, on occasion, in a lack of separation of sister chromatids to opposite poles of a dividing cell. The egg that is formed at the end of this imperfect meiotic process will contain an extra chromosome.

Should not an impaired meiosis also, at times, produce an egg that is lacking a particular chromsome? Undoubtedly this does occur and presumably as frequently as the production of eggs with extra chromosomes. However, the absence of such individuals in chromosomal studies leads us to conclude that the lack of a chromosome must in most cases prove lethal to the developing embryo.

The existence of females with Turner's syndrome does indicate that an individual with only one X chromosome can survive. However, for autosomes, it is apparently more disruptive to have too few than too many.

Patau's syndrome. Are chromosomes other than 21 ever involved in imperfect meiosis? There are a number of genetic diseases that result from an excess of some chromosome other than 21. One of these conditions, called **Patau's syndrome,** results from the presence of an extra chromosome 13 (designated as 47,13+). It occurs in 1 of 7600 births, and most infants die within a few weeks after being born as a result of multiple developmental defects. Unfortunately, there is as yet no cure for the conditions that result from the presence of excess chromosomes.

Mosaicism. The separation and migration of chromosomes during cell division is called **disjunction.** Any failure of this process is termed **nondisjunction.** Nondisjunction can occur either during meiosis or mitosis.

When nondisjunction occurs during meiosis, the resultant gamete helps form an individual in whom all the cells contain an abnormal chromosome complement. We have already discussed this type of situation in the case of Down's syndrome, and the same pattern of events has been shown to result in Turner's syndrome, Klinefelter's syndrome, and Patau's syndrome.

When nondisjunction occurs in mitosis, there results two or more cell types within the individual. This condition is called **mosaicism** and may involve either autosomes or sex chromosomes.

Nondisjunction during the first mitotic division of a fertilized egg produces two cell types: one with 45 chromosomes and the other with 47. Should the first mitotic division be normal and nondisjunction occur in one of the cells during the second mitotic division, three types of cells will be produced: two cells with 46 chromosomes, one cell with 45, and another with 47. The complexity of mosaicism is virtually unlimited, as it may involve different chromosomes in various parts of the body at different times during the individual's life.

Possibly the most interesting case of nondisjunction was found in a pair of "identical" twins. One twin is a female with Turner's syndrome (45,XO) and the other is a normal male (46,XY). They are presumed to have originated as an XY fertilized egg and to have lost the Y chromosome from one of the cells during the first mitotic division. The resultant cells then separated, each forming an entire individual. The evidence for their being "identical" twins includes their complete identity with regard to blood groups and all other genetic traits.

☐ Chromosomal rearrangements

Earlier in this chapter we learned that during meiosis chromosomes sometimes break, and, subsequently, there may be a fusion of segments from homologous chromosomes (**crossing-over**). In addition to crossing-over, other patterns of fusion of broken chromosome segments can occur.

We shall now examine various patterns of chromosome breakage and fusion and see how they alter the genic contents and structural arrangement of a chromosome. Some of the chromosomal rearrangements that can occur include deficiency, inversion, and translocation.

Deficiency (deletion). When two different breaks occur in the same chromosome, two end fragments and a middle piece are formed. In healing, it is possible for the two end fragments to fuse, leaving the middle section as an **acentric** (without a centromere) piece (Fig. 9-15).

The acentric fragment cannot become attached to the spindle that is formed at the next cell division. Because of its lack of spindle attachment, the acentric fragment remains in the middle of the cell during anaphase and is extruded from the cell when the cytoplasm divides.

The two end fragments that fused with one another will form a chromosome that is **deficient** for the genes in the acentric fragment. The type and degree of damage suffered by a person because of a deficiency will depend on its size and importance and the extent to which those genes present in the undamaged homologous chromosome are able, in a single dose, to perform functions ordinarily carried out by the two genes.

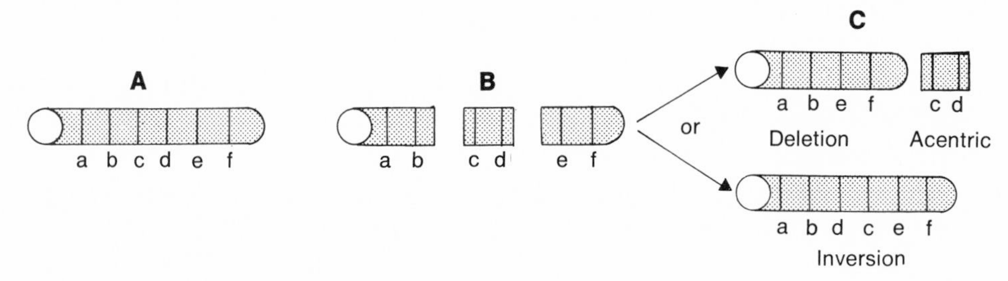

Fig. 9-15
Steps involved in the formation of a deficiency (deletion) or an inversion.

Should a deletion occur in a chromosome during gamete formation or in a fertilized egg, the resultant individual will carry the deficiency in every cell of the body. If, however, a deletion occurs in a chromosome of a somatic (body) cell, only that cell and the cells derived from it will be affected. The type and severity of any deleterious effects of a deficiency will vary with the type and amount of body tissue carrying the altered chromosome.

A number of human genetic diseases have been traced to deficiencies in chromosomes. Probably the best known example is the syndrome called **cri du chat.** As its French name implies, the crying of the affected infant resembles that of a suffering kitten. The abnormal crying sound is caused by a failure of the infant's vocal cords to close properly.

These children are found to have a deletion in one of the chromosomes from pair 5. This deficiency is found in cells from all parts of the body, indicating that the deficiency occurred either during the formation of one of the gametes or in the fertilized egg stage of the child's development. The affected infants exhibit severe mental and physical retardation and, in most cases, have to be institutionalized.

Inversion. As discussed previously, a deficiency may result from the union of broken fragments of a chromosome. Another possible outcome of the reunion of two breaks occurring in the same chromosome is for all three chromosomal sections to join together again, with the middle piece remaining in the middle but turned end-for-end. This situation is called an **inversion** and is also illustrated in Fig. 9-15.

An inversion causes no addition or loss of chromosome material. Only the order of the genes in the middle segment has been changed. The rearrangement of gene order may have important consequences, since adjacent genes are often involved in related steps of the same biochemical pathway. Although inversions are known to be important in the well-being of other organisms, there is no evidence that they affect the normal functioning of those human beings who have a chromosome with this type of rearrangement.

Translocation. Thus far we have discussed chromosome breaks and fusions that have involved either a single chromosome (deletion, inversion) or homologous chromosomes (crossing-over). Breaks may also occur in chromosomes that are not homologous. When the fusion of broken segments of chromosomes results in the transfer of genes from one chromosome to a nonhomologous chromosome, the event is called a **translocation.**

People who carry translocated chromosomes in the cells of their bodies are known to produce a greater number of malformed children than the rest of the population. As an example of this type of unfortunate situation, we shall examine the special circumstances that produce about 5% of the cases of Down's syndrome (mongolism).

In our earlier discussion of Down's syndrome, we learned that afflicted individuals usually have three 21 chromosomes, and, as a result, they contain 47 chromosomes in the cells of their bodies. However, a small number of persons have been found who exhibit all the characteristics of Down's syndrome, but who have 46 chromosomes in their cells. An analysis of their chromosomes reveals that they have two normal chromosomes 21, one normal chromosome 15, and one longer than normal chromosome 15 that was formed from a fusion of a chromosome 15 and a third chromosome 21. This condition is usually referred to as "translocation Down's syndrome 15/21."

A study of the parents of these affected individuals reveals that although the parents exhibit a normal phenotype, one of them does have only 45 chromosomes in the cells of the body. However, one of the 45 chromosomes is the translocation chromosome 15/21, thus giving that parent a complete genetic complement. An individual who is 45,15/21 functions genetically as the equivalent of a carrier of a dominant gene for the production of Down's syndrome children. Half of the children born to such an individual will suffer from Down's syndrome. For female carriers of the translocated chromosome 15/21, the production of afflicted children will **not** be mother's age dependent.

It is quite apparent that the occurrence of a translocation can have serious consequences for the future generations of a family, as can other chromosomal rearrangements. It is important, therefore, to be aware of those environmental factors in our modern society that are known to cause breaks in chromosomes.

Environmental factors causing chromosome breaks. Various environmental factors have been found to cause chromosome breaks. One of these is exposure to x rays. A large number of investigations have demonstrated not only that x rays can break chromosomes, but also that the percentage of breaks produced is **directly proportional** to the dosage of the radiation. The increasing use of x rays for medical and dental diagnostic purposes carries with it the increased hazard of producing chromosome breaks.

Virus diseases have also been shown to produce chromosome breaks. These breaks and the chromosomal rearrangements that result are involved in many of the abnormalities that occur in a developing embryo whose mother has had a viral disease during early pregnancy. Among the diseases that have been found to have this effect are rubella (German measles), measles, chickenpox, mumps, and infectious hepatitis. The birth of malformed children to mothers who have had viral diseases during early pregnancy has been the biological basis for one of the arguments for liberalization of the abortion laws in many states.

Exposure to chemicals, especially drugs, has also been demonstrated to cause chromosome breaks. Our modern society has witnessed an increasingly widespread use of drugs as a mechanism to achieve pleasure or to escape from problems. One of the more widely used drugs is the hallucinogen LSD (lysergic acid diethylamide). As the use of LSD increased, questions were raised about the possible genetic damage that might result from the ingestion of this drug. Studies of individuals who have ingested the drug find that about 25% of the cells contain one or more chromosome breaks. People who do not use LSD are found to have one or more chromosome breaks in less than 2% of their cells. It is clear that two types of genetic damage may result from the use of LSD: (1) malfunctioning of the cells of the user of the drug and (2) production of malformed children.

■ SUMMARY

After studying our cells, their functions (Chapter 7), and how they make more of themselves (Chapter 8), we turned our attention, in the present chapter, to cell division and genetics. We learned that there are two types of cell division: (1) **mitosis,** which results in cells that are genetically identical with each other and with the cell from which they were derived, and (2) **meiosis,** which produces cells that contain half the genetic material of the cell from which they were derived. Mitosis results in cells that take part in the growth process of the individual. Meiosis produces cells (gametes) that function in sexual reproduction.

Our genetic material is organized into pairs of chromosomes. The corresponding genes on homologous chromosomes may be identical, or they may be alleles. The members of each pair of alleles interact with one another in one of the following ways: (1) the effects of one allele may **completely dominate** that of the other, (2) there may be only **partial dominance**, or (3) there may be **codominance**, in which both alleles are completely expressed. These patterns of dominance are modified in males for those genes located on the X chromosome, because the Y chromosome appears to be devoid of genes.

To decide whether traits of human beings are environmentally or genetically determined, we must conduct **pedigree studies.** Environmental factors affect all people that live or work together regardless of their family affiliations, whereas genetic traits run in families.

After discussing patterns of inheritance, we examined the genetics of **sex determination** and found that in human beings, it is the Y chromosome that is the male-determining factor.

We next considered the effects of changes in chromosome number on an individual. We found that any addition of an autosome or an X chromosome leads to a malformed person. A deletion of an autosome is apparently lethal, as no such individual has yet been found. A viable but sterile female develops from a fertilized egg that has 44 autosomes, but only one X chromosome.

The last subject we reviewed was the effects of chromosomal rearrangements. Deficiencies and translocations result in gross malformations of the individuals involved, whereas inversions have thus far been found to be benign.

Our study of cells in these past three chapters leads us to a further consideration of the various organ systems of the body and how they function. In Chapter 10, we shall consider those organ systems that provide us with protection, support, and movement: skin, skeleton, and muscle.

SUGGESTED READINGS

Baserga, R., and Kisieleski, W. E. 1963. Autobiographies of cells. Sci. Am. **209:**103-110. Description of how radioactive atoms can be used to trace the life cycle of cells.

Cohen, S. N. 1975. The manipulation of genes. Sci. Am. **233:**25-33. Detailed description of how to transfer specific genes from one organism to another.

Davis, B. D. 1970. Prospects for genetic intervention in man. Science **170:**1279-1283. Provocative article on the extremely controversial topic of changing man's heredity.

Jacobs, P. A. 1969. Structural abnormalities of the sex chromosomes. Br. Med. Bull. **25:**94-98. Review of the various types of chromosomal rearrangements found in human sex chromosomes.

Mazia, D. 1974. The cell cycle. Sci. Am. **230:**54-68. Review of the experiments that revealed the stages of mitosis and meiosis.

Nagle, J. J. 1974. Heredity and human affairs. The C. V. Mosby Co., St. Louis. Clearly written book that covers reproduction and evolution, as well as genetics.

Peters, J. A., ed. 1959. Classical papers in genetics. Prentice-Hall, Inc., Englewood Cliffs, N.J. Compilation of the papers that mark the principal achievements in the development of the science of genetics.

Ruddle, F. H., and Kucherlapati, R. S. 1974. Hybrid cells and human genes. Sci. Am. **231:**36-44. Very fine explanation of how the location of genes on human chromosomes is determined.

Scheinfeld, A. 1950. The new you and heredity, ed. 2. J. B. Lippincott Co., Philadelphia. Accurate account of human heredity written in pleasant, everyday language.

Tjio, J. H., and Levan, A. 1956. The chromosome number of man. Hereditas **42:**1-6. Original paper giving definitive evidence that the diploid chromosome number in man is 46.

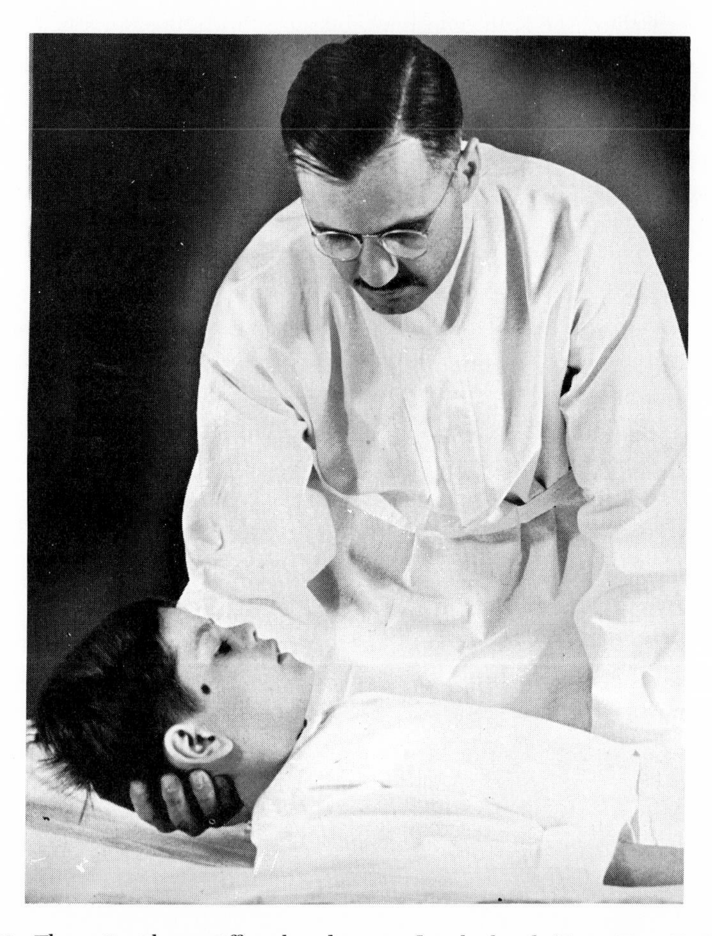

Poliomyelitis. The patient has a stiff neck and cannot flex the head. (From Top, F. H., Sr., and Wehrle, P. F., eds. 1976. Communicable and infectious diseases, ed. 8. The C. V. Mosby Co., St. Louis.)

CHAPTER 10 Body protection, support, and movement

LEARNING OBJECTIVES

- What are the functions of our skin?
- How is our skin structured to perform its various functions?
- What are the common skin diseases, and what are the causes of each one?
- What are the functions of our skeletal system?
- How is our skeletal system organized to perform its various functions?
- What are the various types of joints found in our skeletal system, and what are the characteristics of each type?
- What are the common diseases that affect our skeletal system, and what are the characteristics of each disease?
- How do our muscles produce body movements?
- What is the mechanism of muscle contraction?
- How does the body provide the energy needed in those physical activities that require sustained muscle activity?
- What are the common diseases that affect the muscles of the body, and what are the characteristics of each disease?

Protection, support, and movement of the human body are largely functions of three of its organ systems: skin, skeleton, and muscle. The skin covers the body, protecting it from invasion by disease-producing organisms and giving it pigmentation that varies with geographical population. Our skeletons provide protection and support for our vital organs and attachment sites for our muscles. Our muscles, especially the skeletal muscles, form the bulk of our bodies and function in locomotion and other bodily functions. We shall now consider each of these organ systems.

■ SKIN

Human skin has a number of different functions. These include, among others, (1) protection against injury and disease-producing organisms, (2) production of skin pigmentation, (3) formation of hair, nails, and glands, and (4) regulation of body temperature. As is true of all organ systems, a study of the arrangement of the parts of the skin will provide an explanation of how it performs its various functions.

□ Parts of the skin

The skin is composed of two main parts: a comparatively thin, upper layer called the epidermis and a thicker, lower layer called the dermis (Fig. 10-1). The **epidermis** is composed of stratified squamous epithelial cells, which are divided into a number of distinctive layers. The basal layer of the epidermis consists of somewhat rounded cells that undergo constant division to form new cells. After these cells are produced,

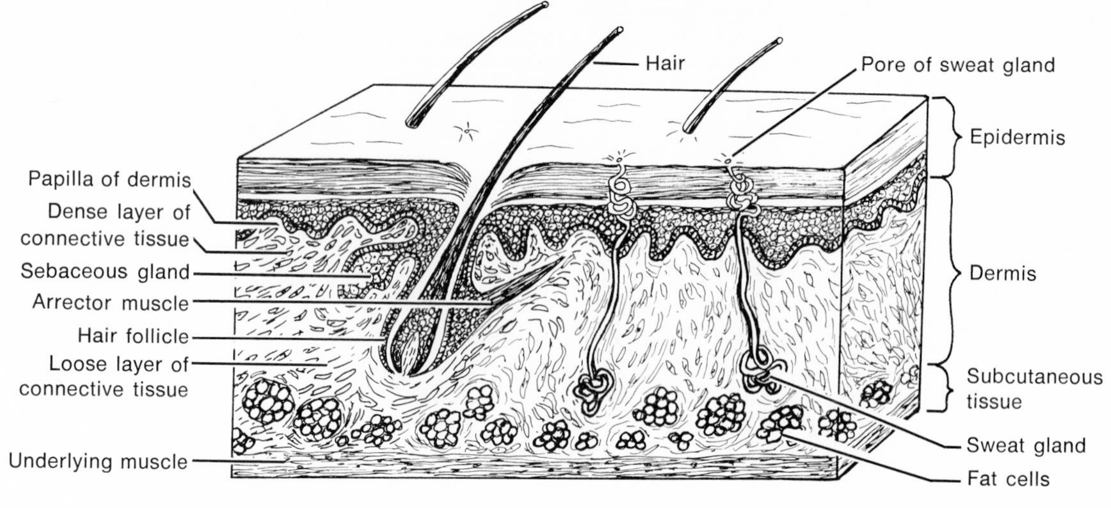

Fig. 10-1
Section through human skin, illustrating various layers and structures.

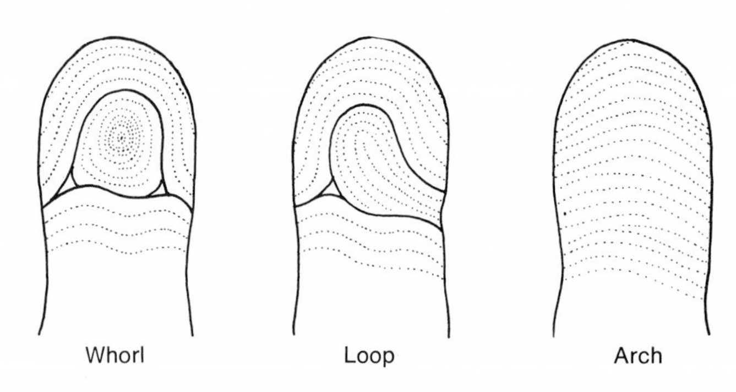

Fig. 10-2
Three basic types of fingerprints: whorl, loop, and arch. (From Levine, L. 1973. Biology of the gene, ed. 2, The C. V. Mosby Co., St. Louis.)

they are gradually pushed upward by younger cells formed below them. As they move upward, they flatten, and their nuclei degenerate. The cells eventually die, and their contents are converted into the protein **keratin,** which also forms the main component of hair, nails, and other epidermal structures. The keratinized cells are worn away or shed continuously, but are replaced by the keratinization of the newer cells that move up from the basal layer.

The **dermis** of the skin is thick and consists mainly of fibrous connective tissue. In the dermis, we find the glands of the skin, hair follicles, blood vessels, lymphatic vessels, and nerve processes (dendrites). The dermis is divided into an upper and a lower layer. The upper level presses against the epidermis and indents it with fingerlike projections called **papillae.** Papillae serve to bring capillaries and nerve fibers into close contact with the epidermis. The epidermis has no blood vessels, and nutrients and oxygen can reach its cells only by diffusion through the tissue fluid of the papillae.

The papillae of the dermis are arranged in circular patterns that form ridges.

Table 10-1
Percentage of fingerprint patterns among races of man

Race	Arches (%)	Loops (%)	Whorls (%)
Caucasoid	4	65	31
Negroid	7	63	30
Mongoloid	3	48	49

These ridges, although covered by the epidermis, produce raised patterns of arches, loops, and whorls on the skin surfaces of the undersides of the hands, fingers, and feet (Fig. 10-2). Ridge patterns aid the individual in grasping and holding objects. The patterns of arches, loops, and whorls are unique in each individual and are used in identification as **fingerprints,** palm prints, or sole prints.

Fingerprint patterns are present at birth and do not change throughout the remainder of life. Identical twins, having the same genetic constitution, have almost identical fingerprints. The hereditary nature of dermal ridge patterns can also be seen in the distribution of arches, loops, and whorls in the various races of man. The data shown in Table 10-1 indicate that the loop pattern predominates in Caucasoid and Negroid populations, whereas Mongoloids have slightly more whorls than loops.

□ Functions of the skin

Protection. Protection is afforded against a variety of external agents. The upper layer of the epidermis is composed of dead cells that are resistant to abrasion and shield the underlying cells from mechanical injuries caused by pressure, friction, or blows. The epidermis is virtually germ proof as long as it remains unbroken, and it protects the body against invasion by most disease-producing organisms.

Skin pigmentation. Skin pigmentation is the result of the blending of three colors found in the skin: (1) melanin, a brown pigment also found in the eyes and hair; (2) carotene, a yellowish pigment; and (3) hemoglobin, the red pigment found in blood. The shades of yellow-brown, brown, brown-black, and black that are found in most of the world's population are a result of the production of different amounts of **melanin** by the skin of the inhabitants from different regions of the world. **Carotene** gives the skin a yellow appearance that is modified by the red color of the blood's **hemoglobin** in the capillaries of the dermis to produce a **flesh pink** color.

The skin color of Caucasoids varies from the pale white of the Scandinavians through the Mediterranean tan of the Greeks and Italians to the brown of the Arabs and, finally, to the brown-black of the people from India. Mongoloid skin color ranges from nearly white in northern Chinese through yellow in southern Chinese, Koreans, and Japanese, to brown in the various American Indian groups of North, South, and Central America. Negroid skin color varies from brown in Ethiopia, Mozambique, and Angola to black in the Congo, Sudan, Kenya, and Tanzania.

There is a tremendous overlap of skin color among the various geographical populations of the world, and racial classifications are not based solely on this one characteristic. In general, it is found that the lighter pigmented groups of a race tend to live in more northerly areas, whereas the darker pigmented groups tend to live in more southerly regions.

Skin pigmentation is genetically determined. The complete absence of pigment production by an individual results in an **albino.** The trait occurs in those people who are homozygous for the recessive gene for albinism. Albinos are found among Caucasoids, Mongoloids, and Negroids.

Formation of hair, nails, and glands. Another basic function of the epidermis, although the dermis does become involved as well, is the formation of hair, nails, and glands. **Hair** is formed when epidermal cells extend downward into the dermis to form a small tubelike sac called a **hair follicle.** The cells at the base of a follicle produce a hair, which is a fusion of dead, keratinized epidermal cells (Fig. 10-3).

Hair has two basic characteristics that have interest for an individual. One is the type of hair, and the other is its color. An examination of **hair type** shows that, with variations, we can distinguish three main categories: straight, wavy or curly, and kinky or wooly. Each strand of hair consists of a tube with dead epidermal cells on the outside and a partially filled space in the center. A strand of straight hair is round in shape, wavy hair and curly hair are oval, and kinky hair and wooly hair have a flattened oval shape.

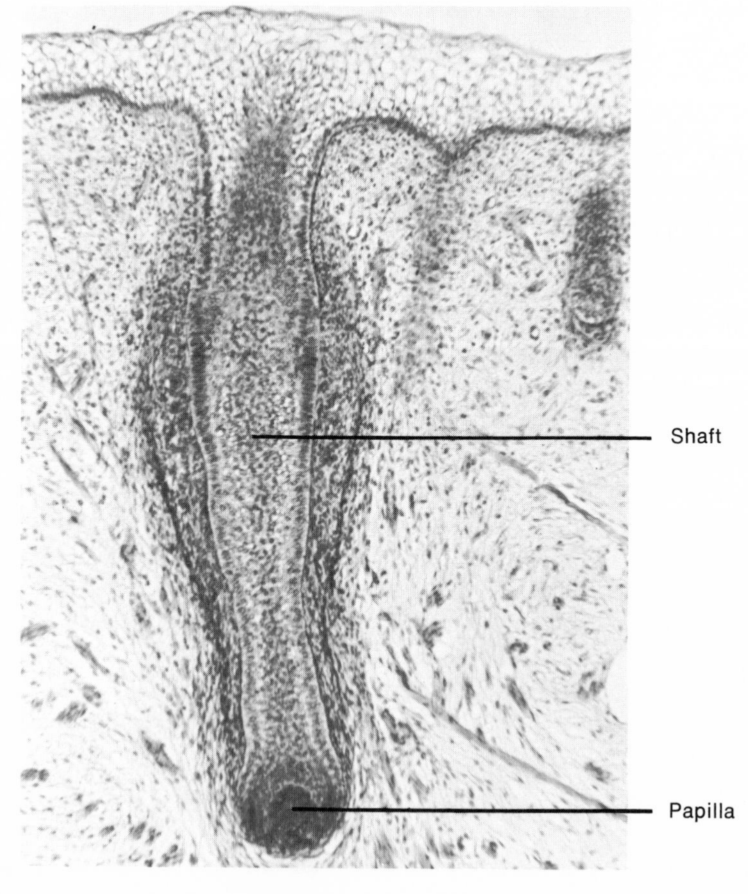

Fig. 10-3
Longitudinal section of developing hair. (×60.) (From Bevelander, G., and Ramaley, J. A. 1974. Essentials of histology, ed. 7. The C. V. Mosby Co., St. Louis.)

Hair type is genetically determined and tends to vary with different human populations. Mongoloids tend to have **straight, coarse** hair, Negroids have chiefly **kinky** or **wooly** hair, and Caucasoids have **wavy, curly,** or **straight, fine** hair. The types of straight hair found in Mongoloids and Caucasoids are quite different. Mongoloid hair is thicker, heavier, and coarser than that of Caucasoids. There is as yet no known advantage for a particular racial group having one or the other type of hair.

As mentioned previously, hair type is genetically determined. When a different hair-type gene is received from each parent, wooly and kinky dominate curly, which dominates wavy, which dominates straight. An exception to this sequence is found in the case of Mongoloid straight hair, which dominates all other hair types, including wooly and kinky. The dominance of Mongoloid straight hair is because it is thicker and stiffer than the hair of other racial types.

An understanding of **hair color** requires that we again consider the structure of a strand of hair. In the center of the tube, we find deposits of pigment granules and tiny air bubbles distributed in random fashion. The cells of the outside of the tube also contain pigment granules. The color we judge the hair to be is actually an interaction of the pigment granules, the air bubbles, and the type of lights shining on the hair.

There are two types of pigment that can be produced by an individual, providing that person has the proper genes. One is a red pigment that, when present alone, will produce a **red-haired** individual. The other pigment is melanin, which, as we have discussed, is important in skin pigmentation. In the **absence** of red pigment and where the amount of melanin produced is very small, the person's hair will be **blond** in color. If the amount of melanin produced is large, the hair will be **brown,** and in those cases where there is an intense deposit of melanin, the hair is **black.** In instances where melanin production is heavy (brown or black hair color), the presence of a gene for red pigment will have no noticeable effect.

Extremely **light** hair color can be found among people with dark skin pigmentation, although light-colored hair is in most cases associated with light-colored skin and eyes. This is especially true of the Scandinavians.

As we discussed in the section on skin color, the presence of two genes for albinism will result in an individual without pigmentation. An albino will have **white** hair.

The various hair-color genes can be found in all human populations, although not in the same frequencies. Red-haired individuals are found occasionally among Negroids and are quite frequent among the usually black-haired people of Latin America. The highest proportion of red-haired people, 11%, is found among the Scotch Highlanders. In similar fashion, some blond-haired individuals are found in a large number of predominantly black-haired populations.

Nails, both fingernails and toenails, also develop from epidermal cells that extend inward into the dermis. The growth of nails is similar to that of hair. Nails are composed of densely packed dead epidermal cells that are translucent, allowing the underlying capillaries to show through and give the nails their normal pink color.

Glands of the skin include two types that are found throughout the entire body: sebaceous and sweat glands (Fig. 10-1). **Sebaceous** (oil) **glands** are formed by the downward extension of epidermal cells into the dermis. These cells are usually associated with hair follicles. The oil produced by sebaceous glands passes into the hair follicle and up to the skin surface. The oil keeps the hair and skin from becoming dry and also prevents excessive loss of water from the skin.

The other type of widely distributed skin gland is the **sweat** (perspiration) **gland.** It, too, forms from an invagination of epidermal cells, and it, too, is located in the dermis of the skin. A long, straight duct connects the gland to the surface of the skin.

Sweat glands are located in all parts of the skin, but are most numerous under the arms, on the forehead, on the soles of the feet, and on the palms of the hands. There are over 2 million sweat glands in the skin, and, under normal conditions, they eliminate more than a quart of water each day. Perspiration, the secretion of the sweat glands, functions to reduce body temperature as a result of its evaporation from the surface of the skin.

A more specialized type of skin gland, with a very restricted distribution, is found in the canal of the outer ear. This canal leads to the eardrum, is about 1 inch long, and is lined with glands that secrete a waxlike substance, called **cerumen,** which protects the eardrum from dust and other foreign substances.

There are two types of cerumen, or earwax: one form is crumbly and dry, whereas the other is moist and sticky. Whether a person's ear canal cells manufacture dry or moist cerumen is genetically controlled, and it takes the presence of two genes for moist earwax to produce that trait.

A study of human populations shows that the different races have different frequencies of the two types of earwax. The data collected are shown in Table 10-2, which illustrates a striking difference when Mongoloid populations are contrasted with Caucasoid and Negroid populations. There is as yet no known advantage for a particular racial group having one or the other type of earwax.

Another specialized skin gland is the **mammary gland,** which produces milk during the nursing (lactation) period following childbirth. The glandular tissue is embedded in a mound of fat cells and connective tissue fibers derived from the chest (pectoral) region. The **nipple** of the breast is a modification of the skin at the point at which the ducts of the mammary gland open onto the surface. The nipple is surrounded by an area of pigmented skin, the **areola,** which usually darkens in color following pregnancy.

The mammary gland of a male is a rudimentary organ that consists of a nipple surrounded by a small areola, but without internal glandular development.

Regulation of body temperature. The regulation of body temperature is another function of the skin. This includes those mechanisms that aid in releasing body heat

Table 10-2
Percentage of population with moist earwax among races of man

Race	% of population
Negroid	93
Caucasoid	84
Mongoloid	8

on hot days and conserving body heat on cold days. The evaporation of body perspiration, as we have discussed, is a mechanism for cooling the body. Another cooling device involves the rich supply of blood vessels in the skin. Dilation of the skin's blood vessels increases the blood flow to the extensive capillary network of the dermis. This serves to speed the release of body heat to the environment and has the effect of cooling the body.

In the cold, a contraction of the skin's blood vessels decreases the amount of blood coming to the surface of the body and results in heat conservation. Another response to low temperatures is a contraction of the **arrector muscles** that connect to the hair follicles. The contraction of these muscles causes a tightening of the skin that reduces the surface area of the body and prevents the loss of a considerable amount of body heat. Arrector muscle contraction gives the individual the sensation known as "goose pimples" (Fig. 10-4).

□ Diseases of the skin

The skin, like every other part of the body, can be involved in various diseases. These can result either from infections by other organisms or from developmental abnormalities.

Among the **infectious** diseases are those caused directly by **bacteria,** for example, sties, boils, and abscesses, or **viruses,** for example, cold sores, measles, and chickenpox, and those bacterial or viral infections that are transmitted by **skin parasites,** for example, mites, ticks, lice, and fleas (Fig. 10-5).

There are any number of developmental diseases of the skin. The most dangerous of these is **skin cancer.** Of the approximately 650,000 cases of cancer that will be discovered in the United States this year, about 120,000 will involve skin cancers,

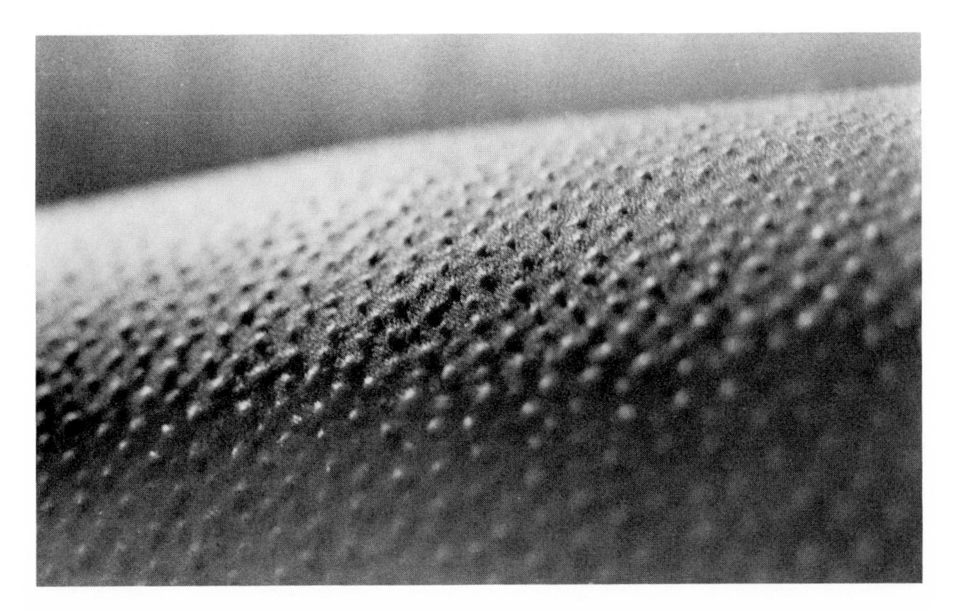

Fig. 10-4
Goose pimples, a response of the skin to low temperatures. (Copyright © 1976 by Theodore R. Lane.)

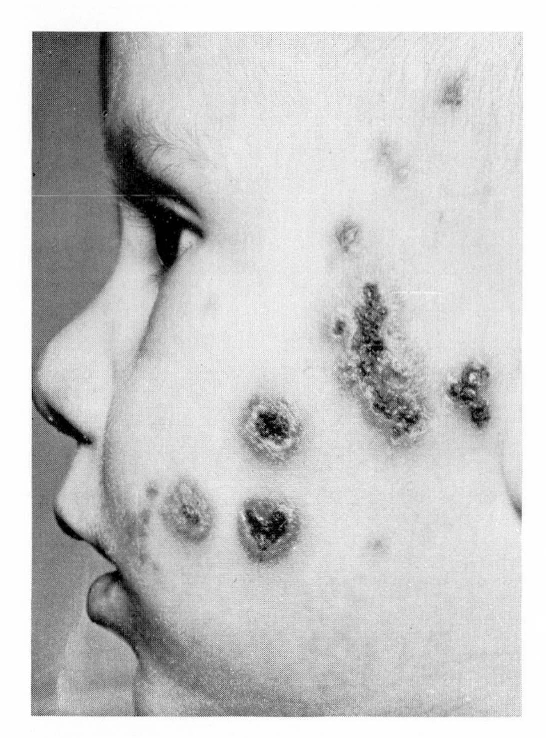

Fig. 10-5
Impetigo, a skin disease caused by bacterial (staphylococci) infection. Older skin lesions are dark and encrusted. (From Top, F. H. 1947. Handbook of communicable diseases. The C. V. Mosby Co., St. Louis.)

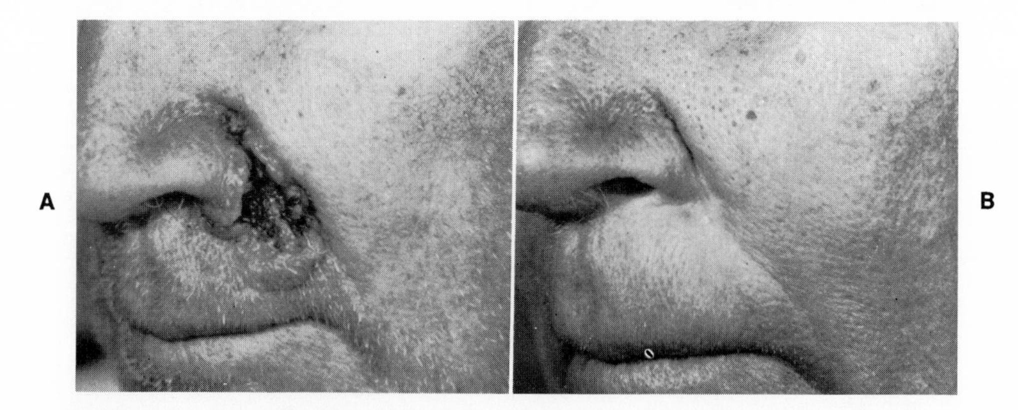

Fig. 10-6
A, Basal cell carcinoma. **B,** Following x-ray therapy. (From Ackerman, L. V., and del Regato, J. A. 1970. Cancer—diagnosis, treatment, and prognosis, ed. 4. The C. V. Mosby Co., St. Louis.)

and about another 70,000 will be breast cancers. In the case of skin cancers, three times as many cases will occur in the southern United States as in the northern states, implicating the ultraviolet rays of the sun as a causative agent. Skin cancer may involve the lowermost cells of the epidermis (producing a **basal cell carcinoma**), the uppermost cells of the epidermis (producing a **squamous cell carcinoma**), or the melanin-producing cells located at the juncture between the dermis and epidermis (producing a **melanocarcinoma**).

Basal cell carcinoma is the most frequently occurring skin cancer (about 75%). It tends to be relatively benign because it has a very slow rate of spreading, which permits its early detection and treatment (Fig. 10-6). Squamous cell carcinoma is fast spreading and hence dangerous. It is frequently found in the lower lip of pipe smokers, indicating a possible cause-and-effect relationship between the carcinoma and either tobacco or the constant physical irritation of the lower lip.

Melanocarcinoma has a moderate growth rate, and its fatal effects depend on whether the cells have spread to a lymph node before the cancer is detected and treated. With lymph node involvement, 90% of affected people die because the cancer can spread rapidly throughout the body via this system. Without lymph node involvement, 50% of those with the disease can hope to survive.

■ SKELETON

All the bones of the human body, when organized in their normal relationships to each other, compose the skeletal system. Our skeletons provide rigidity and support for the body, protection for its vital organs, attachment sites for its striated muscles, and production centers for certain of its blood cells.

In human beings, as is true of fish, amphibians, reptiles, birds, and other mammals, the skeleton is located within the body and is called an **endoskeleton.** It is made of living cells that grow and divide as the individual develops. The main constituent of the skeletal system, bone, is one of the types of connective tissue discussed in Chapter 7.

□ Parts of the skeleton

The human skeleton (Fig. 10-7) is composed of two divisions: (1) an axial skeleton, which, as its name implies, determines the axis or symmetry of the body, and (2) an appendicular skeleton, which is attached to the axial skeleton and functions in locomotion and related activities.

Axial skeleton. The **axial skeleton** is constructed of (1) skull (cranium), (2) vertebral column, (3) ribs, and (4) breastbone (sternum). The skull consists of 22 separate bones that are fused tightly together. A separate bone, called the **mandible,** forms the lower jaw.

The **vertebral column** (Fig. 10-8) is composed of 33 to 34 bones called vertebrae. They are separated from each other by discs of cartilage (**intervertebral discs**) and are bound closely together by **ligaments.** These bones are arranged in the following sequence:

1. **Cervical vertebrae,** seven in number, located in the neck region
2. **Thoracic vertebrae,** 12 in number, located in the chest region, have a pair of ribs attached to each one

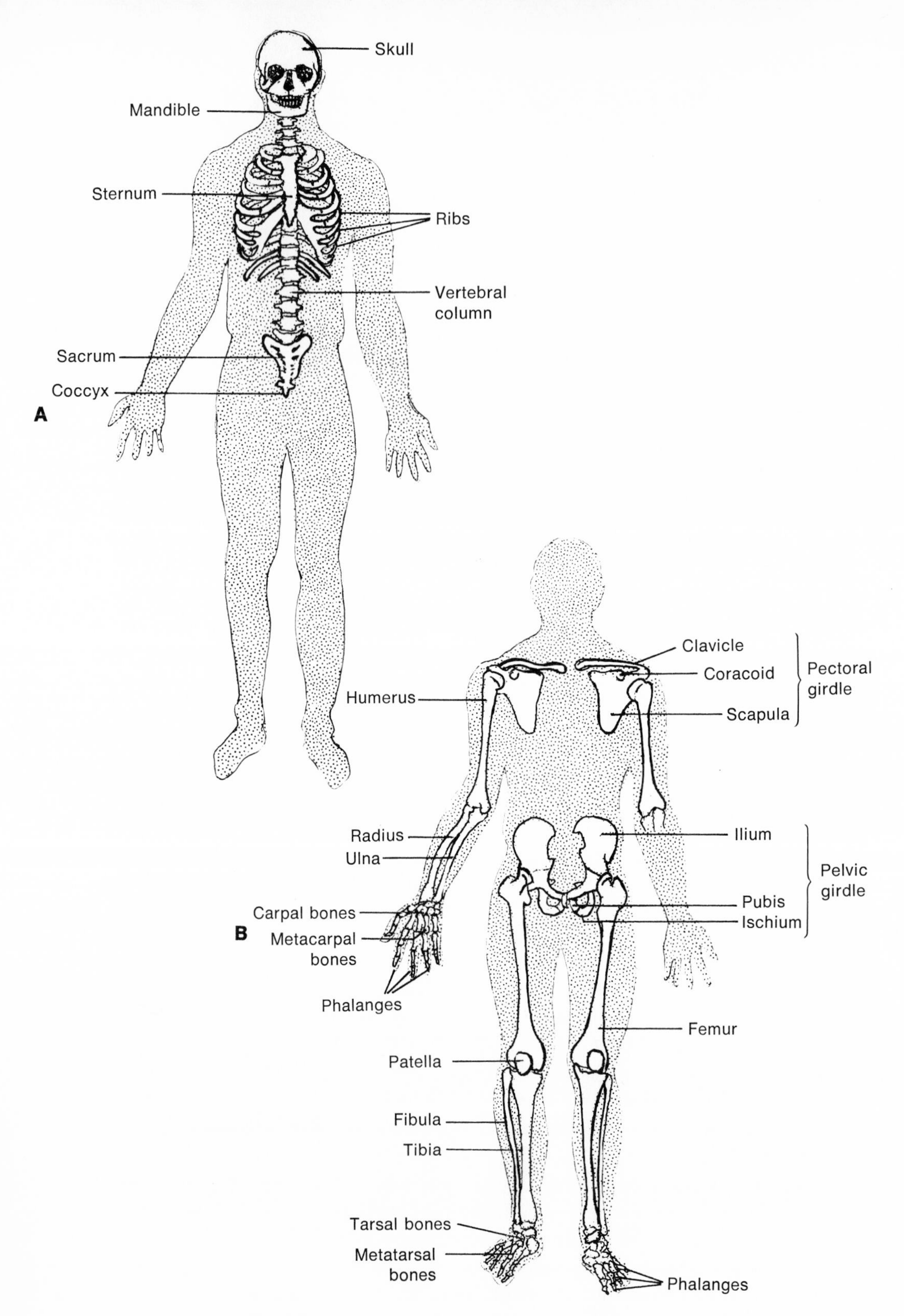

Fig. 10-7
Human skeleton. **A**, Axial skeleton. **B**, Appendicular skeleton.

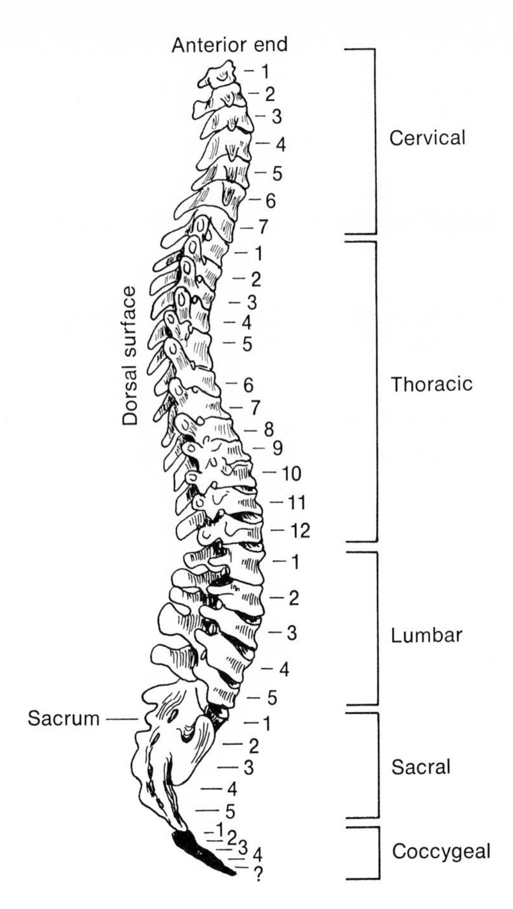

Fig. 10-8
Human vertebral column (side view).

3. **Lumbar vertebrae,** five in number, located in the lower back region
4. **Sacral vertebrae,** five in number, fused together for increased support, found in the hip region
5. **Coccygeal vertebrae,** four or five in number, largely rudimentary, found in what is comparable to the tail region of other animals

Ribs consist of a series of long slender bones that support the chest wall and keep it from collapsing. A pair of ribs is attached to each thoracic vertebra. Each of us has 12 pairs of ribs.

The **breastbone (sternum)** also serves to keep the chest wall from collapsing. In man, the first seven pairs of ribs, called **true ribs,** are attached directly to the breastbone. The next three pairs of ribs, called **false ribs,** are attached to the seventh pair of ribs and thus indirectly to the sternum. The last two pairs of ribs, called **floating ribs,** are so named because they are very short and are attached only at one end to the vertebrae, having no attachment whatsoever to the breastbone.

Appendicular skeleton. The **appendicular skeleton** is composed of two groups of bones called girdles (pectoral and pelvic) and their respective appendages (arms and legs). The girdles encircle the axial skeleton both at its front (**anterior**) and at the hind (**posterior**) end.

The **pectoral girdle** is located anteriorly and includes the following components:

1. **Shoulder blades (scapulae),** broad flat bones, two in number, located in the back (**dorsally**) of the body
2. **Collar bones (clavicles),** long slender bones, two in number, located in the front region (**ventrally**) of the body
3. **Coracoids (vestigial bones),** two in number, attached to the ventral and lateral (side) surface of the scapulae

The pevic girdle is located posteriorly and consists of the following:

1. **Ilia,** two in number, attached dorsally to the fused sacral vertebrae (**sacrum**)
2. **Ischia,** two in number, located posterior to the ilia
3. **Pubes,** two in number, located ventral to the ischia

On each side of the body, the respective ilium, ischium, and pubis are fused together to form a **hip bone.** The pelvic girdle consists of the two hip bones, which meet each other ventrally and are fused to the sacrum dorsally. The term "**pelvis**" refers to the combination of the two hip bones and the sacrum.

The arm forms a joint, that is, it articulates, with the pectoral girdle at a socket called the **glenoid fossa,** which is formed by the scapula (the shoulder blade). Each arm consists of the following:

1. **Humerus,** a single upper arm bone
2. **Radius** and **ulna,** two forearm bones
3. **Carpals,** eight wrist bones
4. **Metacarpals,** five hand bones
5. **Phalanges,** 14 finger bones (two in the thumb and three in each of the other fingers) (Fig. 10-9)

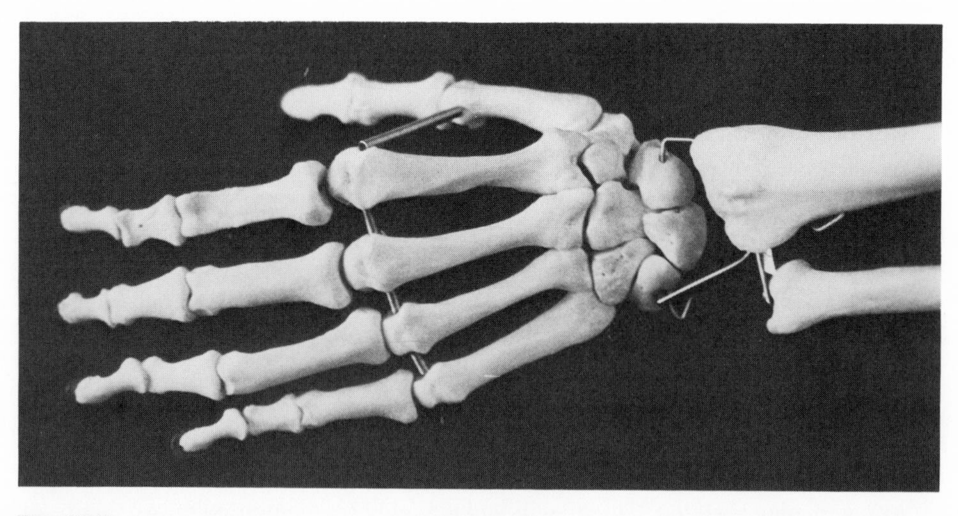

Fig. 10-9
Bones of the hand and wrist. The ends of the radius (larger bone) and ulna (smaller bone) of the forearm are also visible on the right. In the body, adjacent bones are fastened to one another by ligaments. (Copyright © 1976 by Theodore R. Lane.)

The leg articulates with the pelvic girdle at a socket called the **acetabulum,** which is formed at the juncture of the ilium, ischium, and pubis of the hip bone. Each leg consists of the following:

1. **Femur,** a single thigh bone
2. **Tibia** and **fibula,** two shank bones
3. **Tarsals,** seven ankle bones
4. **Metatarsals,** five foot bones
5. **Phalanges,** 14 toe bones

The **knee cap,** also called the **patella,** is a separate bone of the leg. It has no counterpart in the arm.

□ Joints

The point of contact between two bones is called a **joint** (Fig. 10-10). There are various types of joints, including the following:

1. **Sutures** are joints between the bones of the skull. They are immovable because of the intricate dovetailing of the edges of the bones.
2. A **ball-and-socket joint** occurs at each shoulder between the humerus and the scapula (glenoid foss) and also at each side of the hip between the femur and the hip bone (acetabulum). These joints permit free movement in several directions.
3. **Hinge joints** occur at the knee and elbow. This type of joint restricts movement to one plane only, that is, it does not permit rotation.
4. **Gliding joints** occur at the wrists and ankles. This type of joint is intermediate in freedom of movement between the hinge and the ball-and-socket joints.
5. A **pivot joint** occurs between the radius and the lower end of the humerus (the inside of the elbow). This type of joint permits movement in the long axis of the radius, resulting in rotation of the hand.

The two bones of a moveable joint do not touch directly. The end of each bone is covered with a layer of smooth cartilage. Into the space between the cartilages a lubricating fluid, the **synovial fluid** (similar to lymph), is secreted by cells that form a membrane around the joint. The joint is further enclosed by a capsule formed by **ligaments** that bind the two bones to one another (Fig. 10-11).

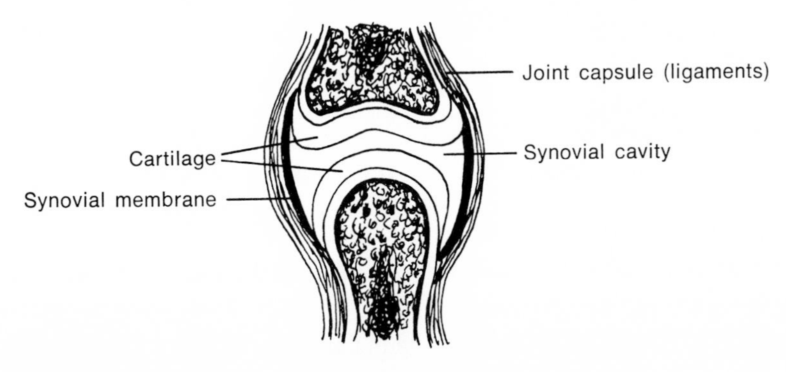

Fig. 10-10
Human hinge joint.

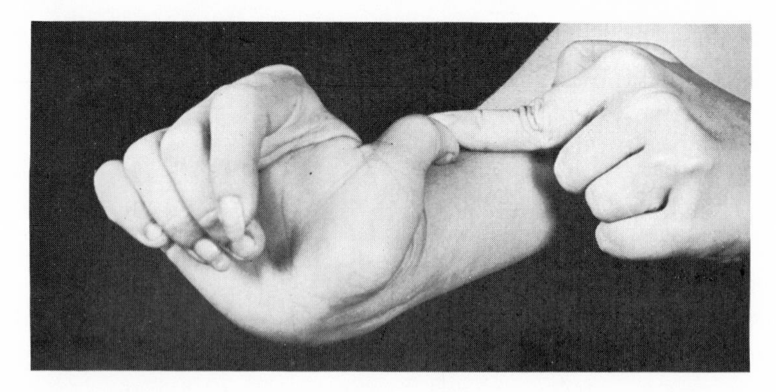

Fig. 10-11
Loose-jointedness, a manifestation of the Ehlers-Danlos syndrome, which is caused by an autosomal dominant gene. Note hyperextensibility of the thumb of the right hand, and, in the left hand, the hyperextension of the index finger and the abnormal separation of the knuckles. (From McKusick, V. A. 1972. Heritable disorders of connective tissue, ed. 4. The C. V. Mosby Co., St. Louis.)

□ Functions of the skeleton

Rigidity and support. One of the many important functions of the skeletal system is to provide **rigidity** for the body and **support** for the organs. Most of the structures of the body are soft and would clump together in a shapeless mass if a rigid framework were not present. The skeletal system provides a supporting structure from which other systems of the body may be suspended or derive support.

The vertebral column plays a major role in providing rigidity and support. It combines strength with flexibility. In addition, it provides a base for the support and movement of the head and gives the pectoral and pelvic girdles a firm anchor for limb articulation and action. In doing this, the vertebral column causes the body to have equal right and left halves, thus establishing its bilateral symmetry.

Protection. Another function of the skeletal system is **protection.** Virtually all the vital organs of the body are encased by the skeleton. The **skull** protects the brain and its associated sense organs for vision, hearing, taste, and smell. Openings of the skull admit light rays to the eyes, sound waves to the ears, food to the digestive system, and air to the lungs.

The **vertebrae** encase the spinal cord, which extends posteriorly from the brain, whereas the heart and lungs are protected by the vertebrae, ribs, and sternum that make up the **thoracic case.**

Muscle attachment sites. Virtually every bone of the skeleton acts as an **attachment site** for one or more muscles. In fact, the contraction of muscles results in work only insofar as the muscles either bring together or separate different parts of the skeletal system.

Blood cell production. Still another function of the skeletal system is the **production of certain blood cells:** red blood cells, neutrophils, eosinophils, and basophils. During early embryonic development, the manufacture of blood cells takes place in the liver. However, in late fetal life and during childhood, this task is taken over by the cells in the marrow cavities of the bones of the body. As the individual develops

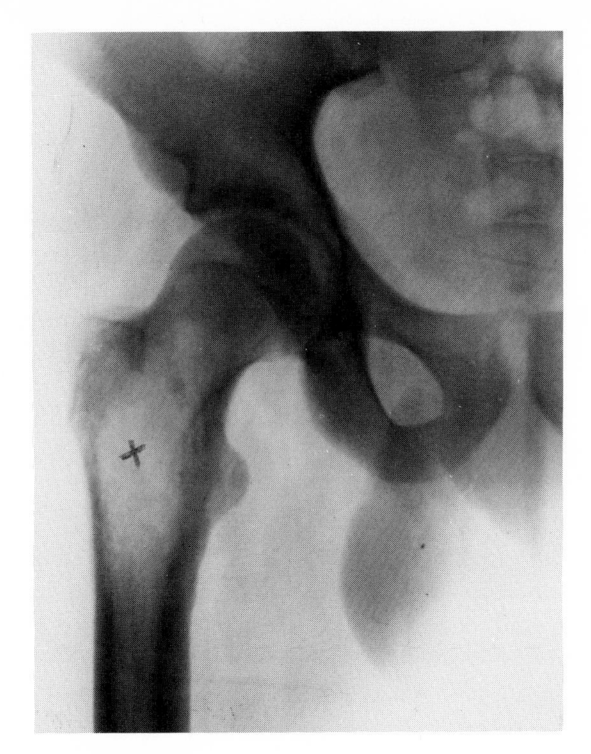

Fig. 10-12
Osteosarcoma (indicated by X) of the upper femur. The light region results from bone destruction. (Courtesy Charles F. Geschickter, M.D., Georgetown University Hospital, Georgetown University Medical Center, Washington, D.C.; from Brooks, S. M. 1975. Basic science and the human body: anatomy and physiology. The C. V. Mosby Co., St. Louis.)

into adulthood, the long bones of the body cease to manufacture blood cells, and this vital function is continued in the marrow cavities of the hip bone, vertebrae, sternum, and ribs.

□ Diseases of the skeleton and joints

A vast number of skeletal defects can occur. Some are genetic in origin, others are caused by infections produced by microorganisms, still others are developmental diseases, and a goodly number are of as yet unknown origin.

One genetic disease that affects the skeletal system is **achondroplasia,** which occurs with a frequency of approximately 2/100,000 births. It is caused by a dominant mutation that retards the growth of the long bones of the arms and legs, but not those of the rest of the skeleton. The individual involved, therefore, has a normal-sized body, but shortened arms and legs. This type of person is called a **dwarf** and should not be confused with a **midget,** who, although extremely short in overall stature, is normally proportioned in body, arms, and legs. The small size of the midget is caused by a pituitary gland hormone insufficiency.

Bones may be infected by microorganisms either indirectly through the bloodstream or directly following a compound fracture in which the bone is exposed to the air. The bacterium *Staphylococcus aureus* is frequently found in such infections, and the disease caused is called **osteomyelitis.** Before the discovery of antibiotics, such bone infections frequently proved fatal. Today, early treatment usually results in a complete cure of the disease.

Among the most fearful of the developmental diseases of the skeletal system are the two types of cancer that can attack its tissues: **chondrosarcoma,** a fast-growing cancer of the cartilage, and **osteosarcoma,** a fast-growing cancer of the bone (Fig. 10-12). Although cancer, particularly leukemia, is the leading cause of death from disease in children, bone cancers are relatively rare. Only about 150 cases of bone cancer are reported annually in the United States. In all cases of bone cancer, amputation of the affected limb is absolutely required. This is followed by chemical therapy designed to destroy any stray cancer cells that might have reached other areas of the body.

A very frequent skeletal system defect is **scoliosis,** whose cause is unknown. This disease is characterized by an abnormal lateral curvature of the vertebral column (spine) and, in varying degrees of severity, occurs in about 2% of the U.S. population. Scoliosis strikes in childhood, affecting girls five times more frequently than boys. It is a progressive disease, and, without treatment, the lateral curvature becomes worse with time, disfiguring the affected youngster. Eventually, the increasing spinal curvature can distort the vital organs within the chest cavity, contributing to the premature death of the individual. Most cases of scoliosis are treatable either by braces or, in severe and advanced cases, by surgery.

The most prevalent disease associated with the skeletal system affects not the bones themselves, but the joints at which bones articulate. The general term used to describe the many different defects at the joints is **arthritis.** About 13 million Americans (more than 5% of the total population) suffer from this disease. Although arthritis is not a leading cause of death, it is probably the leading cause of pain resulting from illness. The three most frequent forms of the disease are osteoarthritis, gout, and rheumatoid arthritis.

Osteoarthritis occurs chiefly in older people and is characterized by a degeneration of the cartilage coverings of the articulating bones, an enlargement of the bones themselves, and changes in the synovial membrane surrounding the joint. It very often appears following an injury to a joint and, for unknown reasons, affects more men than women. The persons affected suffer from pain and stiffness at the joints, especially after prolonged activity.

Gout appears to be an hereditary disease, caused by the interaction of a number of different mutant genes in the same individual, that is, it is of multifactorial or polygenic inheritance. It affects chiefly men and usually involves the big toe, although it can occur in other joints of the feet and hands. The onset of an attack of gout is likely to be sudden, exceedingly painful, and without apparent reason. These attacks may last days or weeks and may then be followed by long periods of time without any discomfort. Gout victims have high levels of uric acid in their blood, and an attack of gout results from the deposition of uric acid crystals in and around the joints. Medicines are available to alleviate the pain and reduce the severity of these attacks.

Rheumatoid arthritis is a chronic, progressive disease of the joints. Its cause is unknown. It usually appears before age 30 years and occurs three times more frequently in women than in men. In 10% to 15% of the cases, the person becomes

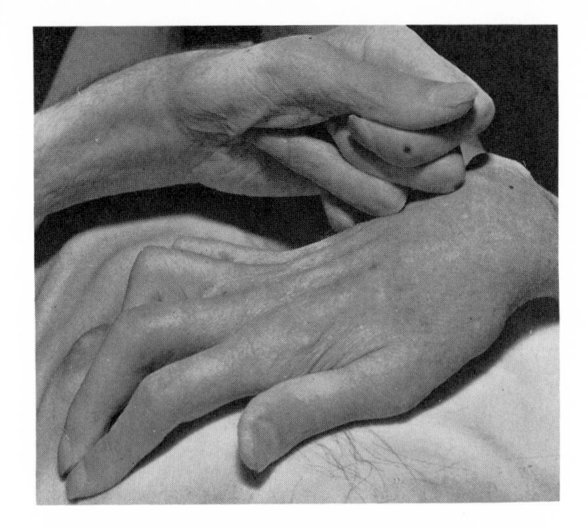

Fig. 10-13
Deformed and stiffened hands in chronic rheumatoid arthritis. Note evidence of muscular atrophy and smooth glossy skin. (From Anderson, W. A. D., ed. 1971. Pathology, ed. 6. The C. V. Mosby Co., St. Louis.)

totally disabled. As yet there is no drug that will effect a cure, but there are drugs that will provide relief from the pain (Fig. 10-13).

■ MUSCLE

The ability of man to move about, grasp objects, pass food through his digestive tract and blood through his circulatory system, blink his eyes, and even speak depends on the action of muscles. As was discussed in Chapter 7, muscles are of three types: (1) **smooth,** making up the wall of the digestive tract and certain other internal organs; (2) **cardiac,** making up the wall of the heart; and (3) **striated,** attached to and moving the bones of the skeleton. In this chapter, we shall deal specifically with striated muscles and their associations with the skeletal system.

Striated, also called **voluntary** or **skeletal,** muscle makes up the bulk of our body tissues and accounts for about two thirds of our body weight. The overall shape of the human body is to a large degree a reflection of the development of its skeletal musculature, allowing for the important role played by the individual's skeletal system itself.

□ Structure of muscle

Skeletal muscles are attached to bones, at both ends, by **tendons.** The end of a muscle that remains relatively stationary when the muscle contracts is called the **origin** of the muscle. The end of a muscle that moves when the muscle contracts is named the **insertion** of the muscle. The thick portion of the muscle between the origin and the insertion is called the **belly.**

□ Muscle antagonism

Muscles can perform work only when contracting. In relaxing, they cannot exert any effective force. Thus muscles can **pull,** but cannot push. To overcome this limitation, muscles are arranged in pairs or paired groups called **antagonists.** Each member

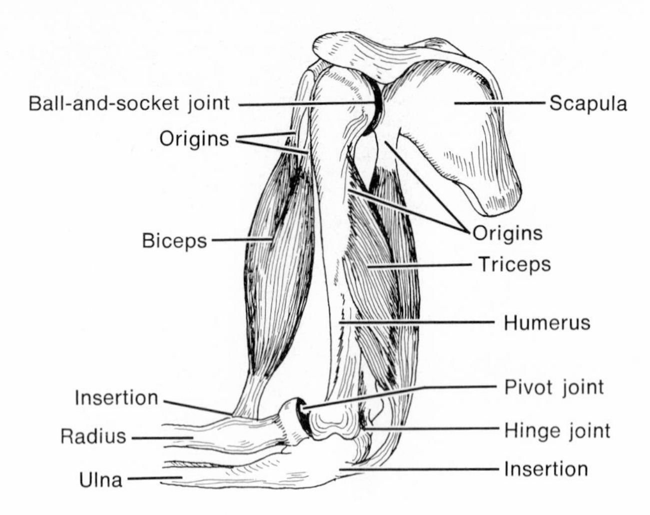

Fig. 10-14
Antagonistic muscles of the forearm, involved in flexion and extension.

of an antagonistic pair produces movement in a direction opposite to that produced by the other muscle. As an example, we can consider the two antagonistic upper arm muscles, the biceps and the triceps (Fig. 10-14).

The **biceps** has its origin on the scapula of the pectoral girdle, and its insertion is on the radius of the forearm. When it contracts, the biceps pulls the forearm toward the upper arm, thereby decreasing the angle at the elbow joint. A decrease of an angle at a joint is called **flexion,** and the muscle involved is termed a **flexor.** When the biceps contracts, its antagonist, the triceps, relaxes.

The **triceps** has two origins, one on the scapula and the other on the upper end of the humerus, and its insertion is on the ulna. When it contracts, the triceps pulls the forearm away from the upper arm, thereby increasing the angle at the elbow joint. This type of action is called **extension,** and the muscle involved is called an **extensor.** When the triceps contracts, its antagonist, the biceps, relaxes.

A different type of muscle action can be seen in two antagonistic shoulder muscles, the deltoid and the pectoralis major (Fig. 10-15). The **deltoid** has its origins on the scapula and the lateral, that is, toward the side of the body, one third of the clavicle, and its insertion is on the lateral surface of the upper end of the humerus. When it contracts, the deltoid moves the arm away from the body. The movement of a limb (arm or leg) away from the body is called **abduction,** and the muscle involved is called an **abductor.** When the deltoid contracts, its antagonist, the pectoralis major, relaxes.

The **pectoralis major** has its origins on the sternum and adjacent ribs and on the medial, that is, toward the midline of the body, two thirds of the clavicle, and its insertion is on the medial surface of the upper end of the humerus. When it contracts, the pectoralis major moves the arm toward the body. The movement of a limb (arm or leg) toward the body is called **adduction,** and the muscle involved is called an **adductor.** When the pectoralis major contracts, its antagonist, the deltoid, relaxes.

Although there are considerable periods of time when muscles are not contracting to effect a movement, they are not completely relaxed either. All so-called resting muscles are partially contracted, a phenomenon called **tonus.** Tonus is very impor-

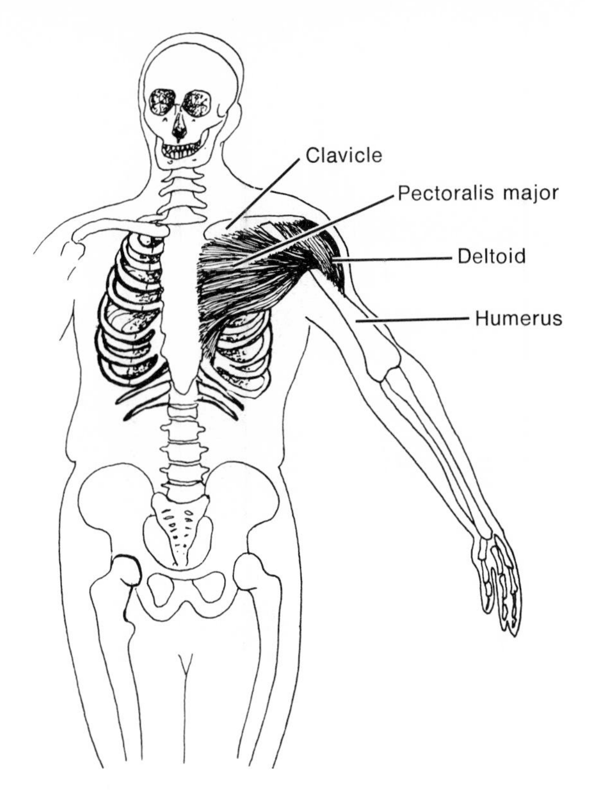

Fig. 10-15
Antagonistic muscles of the chest and upper arm, involved in abduction and adduction.

tant in a number of situations. For example, posture is maintained by the partial contraction of the muscles of the back, neck, and legs. Especially important when in a standing position is the simultaneous contraction of the flexors and extensors of the shank that lock the knee in place and hold the leg rigid to support the body. Also necessary is the simultaneous contraction of the flexors and extensors of the thigh that prevents the body from swaying backward or forward on the legs.

□ Anatomy of a muscle

A typical skeletal muscle (Fig. 10-16), such as the biceps or triceps, is composed of millions of individual **muscle fibers.** Each muscle fiber, in turn, is composed of thousands of smaller units called **myofibrils.** Myofibrils are the actual working units of a muscle.

Each myofibril consists of a long ribbonlike series of alternating light bands (I bands) and dark bands (A bands). In the middle of each light I band, there is a dark line (Z line), while in the middle of each dark A band, there is a light zone (H zone). The myofibrils, in turn, are subdivided into a large number of contractile subunits, composed of protein filaments, that extend from each Z line to both its adjacent ones, as shown in Fig. 10-16.

There are two main proteins in muscle: myosin and actin. **Myosin** forms thick filaments, whereas **actin** forms thin ones. Note in Fig. 10-16 that the light I bands contain only actin filaments, while the dark A bands contain the thick myosin filaments and the ends of the thin actin filaments.

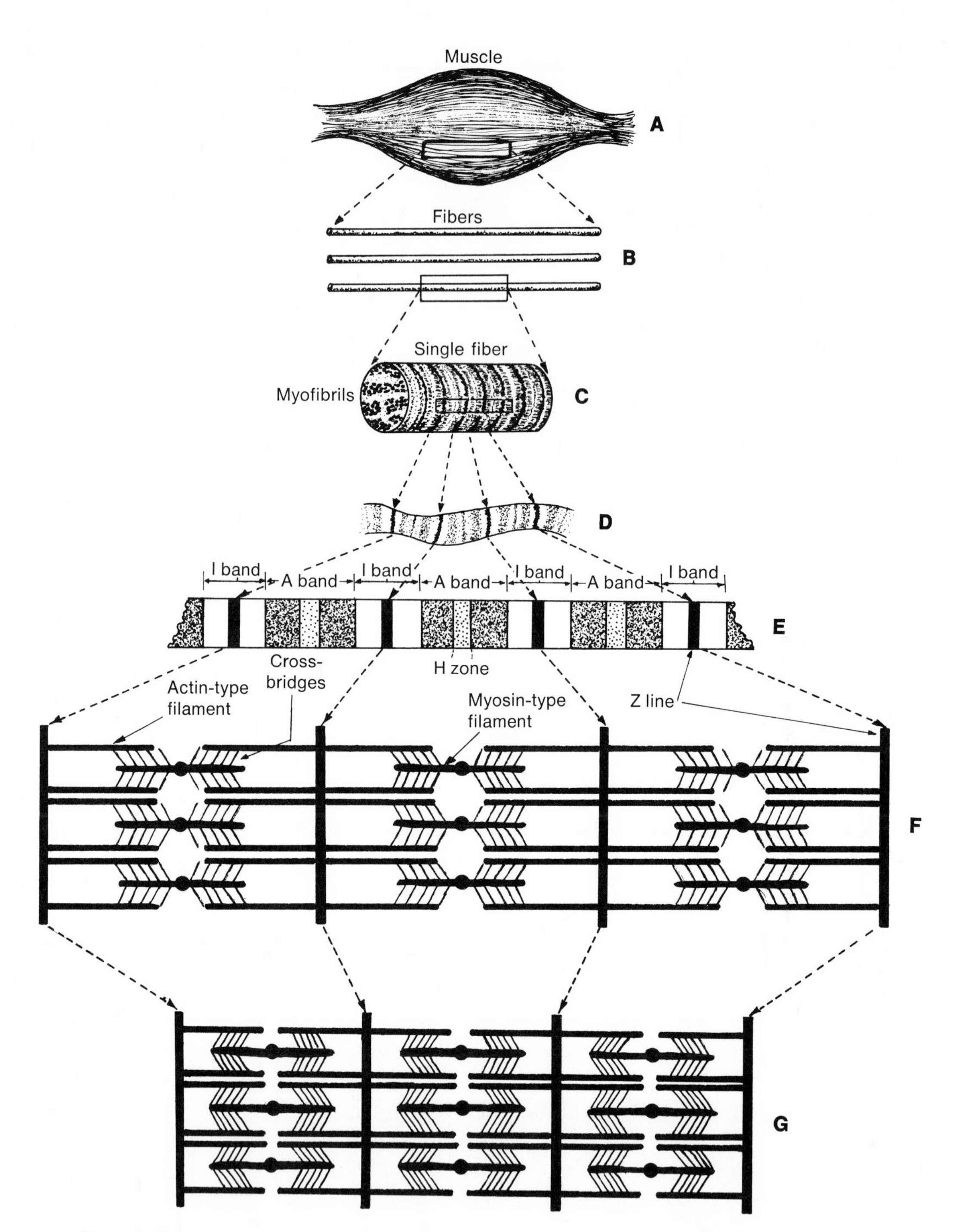

Fig. 10-16
Striated muscle anatomy and contraction. **A**, Striated muscle. **B**, Muscle fibers. **C**, Single muscle fiber. **D**, Segment of a myofibril. **E**, Detailed structure of a myofibril, showing three contractile subunits. **F**, Contractile subunits in a relaxed state. **G**, Contractile subunits in a contracted state.

☐ Muscle contraction

Although the myofibril is the working unit of a muscle, all the myofibrils in a given muscle fiber contract simultaneously when the muscle fiber is properly stimulated by a nerve impulse. This means that the muscle fiber itself is the unit of response to nerve stimulation.

Each muscle fiber requires a certain minimal level (**threshold**) of stimulation for contraction. However, the minimal stimulus that causes contraction causes complete contraction; a stronger stimulus will case **no** greater contraction. This type of response to a stimulus is referred to as the **all-or-none law.**

In contrast to its muscle fibers, the whole muscle does **not** follow the all-or-none law. When a nerve impulse stimulates a muscle, a variable number of muscle fibers will contract, depending on the strength of a nerve impulse. The strength of a nerve impulse is usually a reflection of the task to be accomplished; it takes less muscle activity to lift the paperback version of a book than its hardcover edition.

As shown in Fig. 10-16, the actual contraction of a myofibril consists of a sliding inward of the actin filaments among the myosin filaments. In the resting, elongated state of the myofibrils, the ends of the actin filaments are separated from each other. However, during contraction, these ends approach each other, bringing the two Z lines close to the ends of the myosin filaments.

ATP and muscle contraction. Since every activity of a cell requires energy, we should ask, "What process provides the energy for myofibril contraction?" Here, too, the **energy currency molecule, adenosine triphosphate** (ATP), plays the major role.

During the resting, elongated state of a myofibril, molecules of ATP bind to the myosin filaments. When the electrical stimulation of a nerve impulse reaches a muscle fiber, the impulse passes internally, causing the release of a large number of calcium ions that are normally bound to the endoplasmic reticulum of the myofibrils. These calcium ions combine with the ATP molecules, which subsequently split off their terminal energy-rich phosphate group, thereby providing energy for the movement of the actin filaments. In this process, ATP becomes adenosine diphosphate (ADP).

After the electrical energy of the nerve impulse has been dissipated, the calcium ions are again bound to the endoplasmic reticulum, the ADP is restored to ATP through an energy-transfer process, the actin filaments are returned to their previous position, and the myofibril is once more in a resting, elongated state.

☐ Energy and muscle activity

Muscular activity, such as lifting and running, requires a great deal of energy, and active muscles quickly exhaust their supply of ATP. If the demands on the muscles are not great, the energy needed to reconvert ADP to ATP can come from the breakdown of carbohydrates to carbon dioxide and water in the myofibrils through glycolysis, the Krebs' citric acid cycle, and the cytochrome system, as discussed in Chapter 8. During continued muscular activity, however, the energy demands of the muscles exceed the energy-providing capacity of their own cells. The limiting factor is **oxygen,** which is required for the continued functioning of the cytochrome system and whose availability is limited by the oxygen-providing capacity of a person's circulatory and respiratory systems.

If no alternative biochemical pathway were available, an insufficiency of oxygen in the muscle cells would soon lead to a blocking of the entire energy-providing metabolic process, and the individual could no longer continue his activities. However, an alternative biochemical pathway is available for continued muscular activity under conditions of insufficient oxygen.

In the absence of a sufficient amount of oxygen, pyruvic acid, the end product of glycolysis, is converted to lactic acid, which diffuses from the muscle cells into the adjacent blood capillaries and is carried to the liver. Muscles are thus freed of their pyruvic acid and can operate even in the absence of oxygen, that is, **anaerobically,** for some time. However, the accumulated lactic acid must eventually be reconverted to pyruvic acid, which is then completely broken down to yield carbon dioxide, water, and energy. This will, of course, require oxygen.

The accumulated lactic acid in the blood and liver represents an **oxygen debt** that must be repaid through heavy breathing at the end of any sustained physical activity. The oxygen thus obtained is supplied to the liver, where about one fifth of the lactic acid is first converted to pyruvic acid, which is then completely broken down. The energy obtained from the breakdown of the one fifth of the lactic acid is used to convert the other four fifths to glucose, which is brought to the muscle cells by the circulatory system and stored as glycogen.

In sprinting, a runner uses energy at a rate that requires that his muscle cells receive 22 liters of oxygen/minute. However, the maximum rate of oxygen delivery to muscles is about 4 liters/minute. Thus in 1 minute of sprinting, a runner would accumulate an oxygen debt of 18 liters. This is just about the maximum oxygen debt possible, and such a runner would be completely "out of breath" in 1 minute. For long races, a runner must "pace himself" so that he does not accumulate an oxygen debt of 18 liters. It is important to note that one of the severely limiting factors in competing in athletic events, for example, in running, swimming, or rowing, is the oxygen-providing capacity of the body (Fig. 10-17).

How does the body "know" to increase its rate of breathing and heartbeat when a person is engaged in heavy muscular activity? Experimental research has shown that as muscle cells pour more and more lactic acid into the blood, the acidity of the blood increases slightly. This increased acidity stimulates the brain centers that control breathing and circulatory rates, causing an increase in these body activities and thus bringing more oxygen to all the body cells. As a result of this increased oxygen supply, the oxygen debt is eliminated. It is important to realize that in the workings of the body, the presence of an undesirable compound (lactic acid) serves as the stimulus for increasing the body's efforts (heavy breathing and increased heartbeat) to obtain the necessary means (oxygen) to eliminate the undesired compound.

□ Diseases of muscle

Muscular diseases are as varied in their origins as the diseases of any other part of the body. In Chapter 7, **muscular dystrophy,** a muscle disease of genetic origin, was discussed.

Acute alcoholic myopathy is a muscle disease of dietary origin. It is characterized by muscle pain, tenderness, and edema (accumulation of fluid in tissues). A microscopic examination of the muscle tissues reveals some necrosis (death of individual cells), intracellular edema, and destruction of mitochondria. The disease occurs in chronic alcoholics, who usually consume about 42% of their total calories in the form

Fig. 10-17
Oxygen hunger. (Courtesy Wide World Photos; from Turner, C. E. Personal and community health, ed. 14. The C. V. Mosby Co., St. Louis.)

of alcohol. Many of the effects of the disease tend to disappear if the person is able to avoid alcohol consumption over a prolonged period of time.

Another muscular disorder, **tetany,** marked by muscle twitchings, cramps, and convulsions, can also be dietary in origin, although other causes are known. Any situation that produces a drop in the blood calcium level can cause tetany. It is found that a vitamin D–deficient diet will result in a drop of the blood calcium level and will lead to this muscle disorder.

A debilitating muscle disease of unknown origin is **myasthenia gravis.** It is characterized by progressive muscular paralysis. Although it may affect any muscle of the body, it usually involves those of the face, lips, tongue, throat, and neck. This accounts for the expressionless face and drooping eyelids of a person who has this disorder. No cure is available for the disease, but some drugs prove helpful in temporarily improving the patient's condition.

Many disorders of muscles are not caused by diseases of the muscles themselves, but rather by diseases of the nerves that innervate the affected muscles. The best example of this is **poliomyelitis**, or **infantile paralysis.** It is caused by a viral infection of the nervous system that leads to the death of the nerve cells that normally control skeletal muscle contractions. In the absence of nervous stimulation, muscles begin to **atrophy**, that is, lose myofibrils and suffer a reduction in diameter of the remaining myofibrils. After about 4 months without nervous stimulation, a replacement of muscle fibers by fat and fibrous connective tissue occurs. After that, there is no possible return of function in the muscle.

Since the poliomyelitis virus can invade and kill any nerve cell, all muscles are at risk of losing their nervous system innervation. When the muscles of the legs are involved, paralysis results. When the muscles involved in breathing, that is, rib muscles and diaphragm, become nonfunctional, death results.

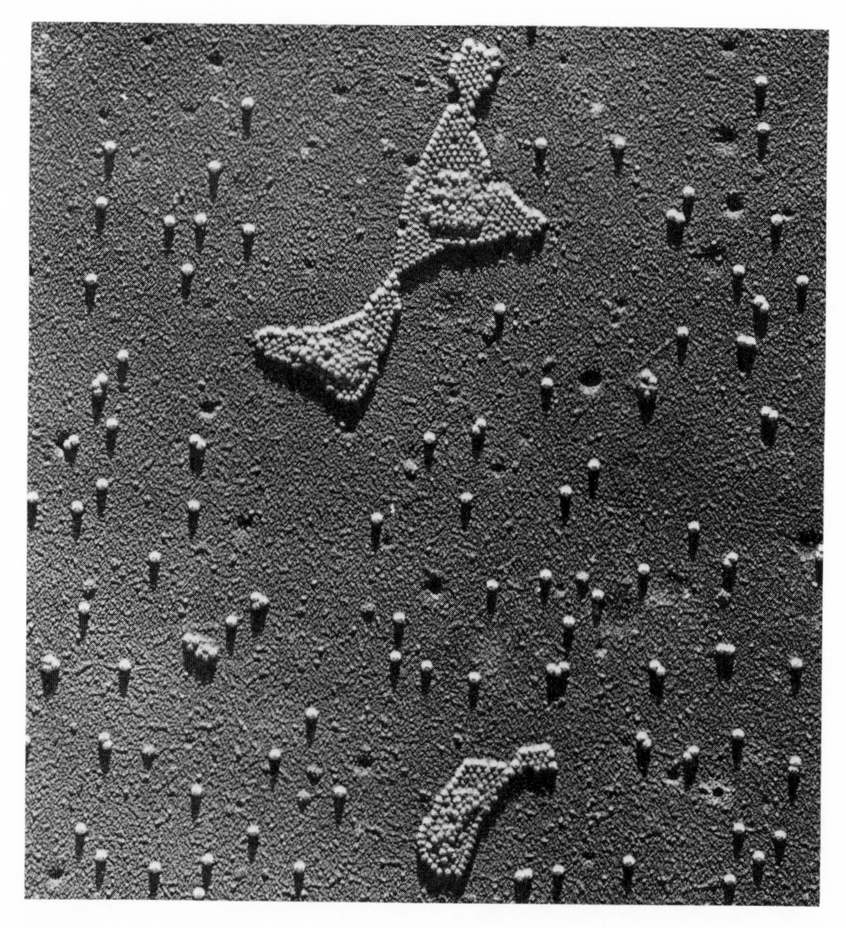

Fig. 10-18
Electron micrograph of poliomyelitis virus crystal (type 1 strain purified and crystallized). Individual viral particles are 28 mμ in diameter. In the absence of any immunity, one crystal of poliovirus contains enough particles to infect the population of the world. ($\times$64,000.) (Courtesy Parke, Davis & Co., Detroit; from Smith, A. L. 1976. Microbiology and pathology, ed. 11. The C. V. Mosby Co., St. Louis.)

We can appreciate the magnitude of the poliomyelitis problem by looking at some data. In 1952 there were 21,300 cases of paralytic polio reported in the United States and 1400 deaths. In 1955 the Salk vaccine against polio was licensed for use. In 1969 there were 19 cases of paralytic polio reported and no deaths. Over a period of 40 years, for which there are adequate records, this disease has attacked more than 500,000 children and young adults, killing 57,000 and permanently crippling 300,000 (Fig. 10-18).

■ SUMMARY

In this chapter, we have examined three systems of the body that play major roles in the protection, support, and movements of our bodies. These organ systems are skin, skeleton, and muscle.

The skin serves to protect us from mechanical injury and invasion by disease-producing organisms. It also produces the hair and glands that function in temperature regulation and the nails that function in grasping and tearing objects. Skin also provides the body with its external pigmentation, which varies with geographical population.

Our skeletons provide protection and support for our vital organs (brain, heart, lungs, etc), attachment sites for our skeletal muscles, and production centers for various types of blood cells.

Although the skeleton provides rigidity and support, it is muscle contraction that produces all the movements of the body. Muscles work in antagonistic pairs and depend on ATP for the movements of the protein filaments (myosin and actin) that are the ultimate contractile subunits.

The ability of an individual to compete in athletic events or engage in sustained muscular activities is to a large extent determined by the amount of oxygen available to his muscle cells for completion of the energy-transfer biochemical pathways. Another factor important not only in muscular activities but in all living processes is the ability of the organism to rid itself of waste materials. It is to these two vital body functions that we shall now address ourselves as we study how we obtain oxygen for our cells and how we eliminate metabolic wastes.

SUGGESTED READINGS

Bendall, J. R. 1969. Muscles, molecules and movement. American Elsevier Publishing Co., Inc., New York. Well-organized and detailed book on muscle structure and physiology.

Chapman, C. B., and Mitchell, J. H. 1965. The physiology of exercise. Sci. Am. **212**:88-96. Discussion of the involvement of nervous, respiratory, and circulatory systems in muscular activity.

Daniels, F., van der Leun, J. C., and Johnson, B. E. 1968. Sunburn. Sci. Am. **219**:38-46. Discussion of the effects of exposure to ultraviolet rays on the various layers of our skin.

McLean, F. C. 1955. Bone. Sci. Am. **192**:84-91. In-depth consideration of the structure and functions of bone.

Merton, P. A. 1972. How we control the contraction of our muscles. Sci. Am. **226**:30-37. Discussion of the feedback mechanism through which we control the contractions of our skeletal muscles.

Montagna, W. 1956. The structure and function of the skin. Academic Press, Inc., New York. Authoritative review of what our skin is and what it does.

Morton, D. J. 1952. Human locomotion and body form. The Williams & Wilkins Co., Baltimore, Description of the mechanics of movement with relation to body form.

Murray, J. M., and Weber, A. 1974. The cooperative action of muscle proteins. Sci. Am. **230**:59-71. Detailed discussion of muscle proteins and the role of calcium ions in controlling muscle contraction.

Photograph of a sneeze taken with an exposure of $^{1}/_{30,000}$ second. (Courtesy M. W. Jennison; from Turner, C. E. 1971. Personal and community health, ed. 14. The C. V. Mosby Co., St. Louis.)

CHAPTER 11 Obtaining oxygen and eliminating metabolic wastes

LEARNING OBJECTIVES

- What three meanings does the term "respiration" have?
- What is the sequence of structures of the respiratory system, going from nostrils to lungs?
- What are the mechanical steps involved in breathing?
- How is the rate of breathing controlled by the body?
- What are the common diseases of the respiratory system, and what are the characteristics of each?
- What is excretion?
- What waste products are formed as a result of the chemical breakdown of carbohydrates, lipids, and proteins?
- What are the excretory organs of the body, and what does each organ excrete?
- How do the kidneys function?
- What three processes are involved in the formation of urine by the kidney, and what are the characteristics of each process?
- How is the rate of urine production regulated?
- What are the common diseases of the kidneys, and what are the characteristics of each?

Our discussion of body movement, in Chapter 10, emphasized the crucial need for oxygen in muscle contraction. The requirement for oxygen applies equally to all other body activities as well. What we learned about muscle metabolism complements our study of the circulatory system (Chapter 6), in which we found that the overall circulatory pathway (see Fig. 6-10) ensured the delivery of a maximum amount of oxygen to all parts of the body.

As we have already noted, all biochemical pathways produce some compounds that cannot be used by the body and whose accumulation would endanger the proper functioning of the organism. These are true metabolic waste materials, and it is necessary that the body rid itself of them. We shall, in this chapter, study both how the body obtains the essential element oxygen and how it eliminates its metabolic wastes.

RESPIRATORY SYSTEM

Before describing the human respiratory system, we must point out the various usages of the term "**respiration.**" This word has had and continues to have three different meanings. Originally, it was synonymous with breathing, and even today the term "artificial respiration" refers to the process of forcing an individual who has stopped breathing to once more inhale and exhale. The term is also used to describe

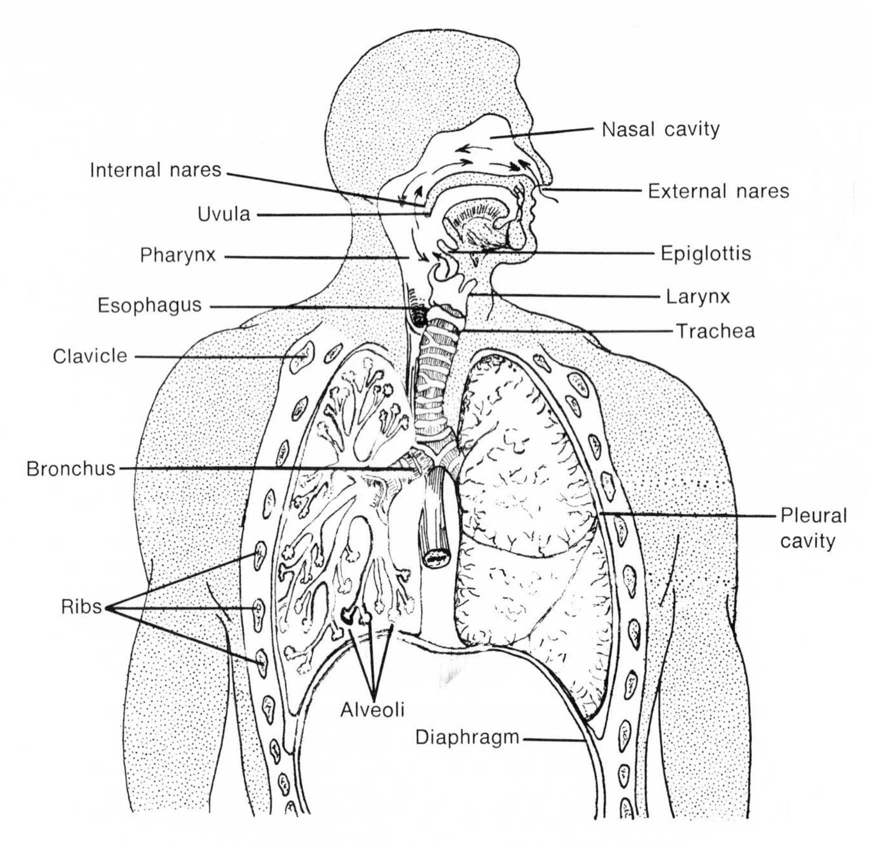

Fig. 11-1
Human respiratory system.

the exchange of gases that occurs in both the lung cavity (external respiration) and at the cellular level (internal respiration). The third and most recent use of the word has been to denote the involvement of oxygen in the energy-transfer system of the cell. Thus the cytochromes are referred to as "respiratory enzymes."

The term **"respiratory system"** refers only to those structures that are involved in the exchange of gases in the lung cavity (external respiration). Thus the respiratory system (Fig. 11-1) comprises the lungs, the series of passageways leading to the lungs, and the chest structures responsible for the movement of air into and out of the lung cavity.

☐ Structures involved in respiration

Nose. Although air can enter the respiratory passageways through the mouth, the nose is the normal route. The nostrils lead into nasal cavities, which open into the pharynx. Each nasal chamber is lined with a layer of highly vascular connective tissue, on top of which is a ciliated epithelial membrane.

The nasal membranes contain many mucus-secreting cells that produce a film of mucus that coats the entire surface of the membranes. The combination of moist mucus and the warmth of the nasal chambers, because of their rich blood supply, serves to (1) warm and humidify the incoming air and (2) trap fine particles contained in the air. The cilia of the epithelial membrane move the mucus toward the nostrils from which it and its contained foreign particles are periodically expelled through the act of "blowing the nose."

In addition to warming, moistening, and filtering the incoming air, the nose performs other functions. There are special olfactory cells in the nasal cavities, making the nose the organ of smell. The nose also aids in speaking by resonating many of the sounds produced by our vocal cords. This last function is very noticeable in the loss of quality of the voice when a person is suffering from a "stuffy nose" caused by a cold or an allergy.

Sinuses. During the development of the nasal membranes in the embryo, certain groups of cells extend outward from the margins of the nasal cavities. The bones of the skull are forming during this same embryonic period, and, in their development, they encase these patches of cells, which produce sacs, called **sinuses,** within the bones. The sinuses maintain their connections with the nasal cavities by short ducts or openings and have the same type of glandular epithelium and vascular supply as the nasal membranes from which they originate.

The sinuses function (1) to produce additional mucus to coat the inner lining of the nasal chambers and (2) to act as resonators in speech production. The sinuses can be extremely troublesome during infections. Since the nasal cavities are connected to the sinuses, infections of the nose may spread into these bone-locked chambers. Inflamed sinuses produce vast amounts of mucus that may not be able to drain into the nasal cavities if the membranes of the sinus ducts are also swollen. The pressure of the accumulated mucus in the sinus chambers produces the feeling of misery associated with sinus infections. Studies indicate that about 17 million Americans suffer from chronic sinus infections.

Pharynx. As was discussed in Chapter 5, the pharynx is the cavity behind the mouth. Through it, materials pass into both the digestive and respiratory systems. Food enters the pharynx ventrally from the mouth and leaves the pharynx dorsally through the esophagus. Air, on the other hand, enters the pharynx dorsally from the internal openings (**nares**) of the nasal cavities and leaves the pharynx ventrally through the glottis. A person cannot swallow and inhale at the same time (as will be explained later).

A good deal of the pharynx is also lined with a ciliated epithelial membrane that humidifies and frees the incoming air of its particles. The walls of the pharynx are well supplied with skeletal muscles, which function to change the shape and size of the pharyngeal cavity. These changes alter the resonating effect of the pharynx in speech production.

The dorsal wall of the pharynx contains two types of lymph glands, the **tonsils** and the **adenoids.** As was discussed in Chapter 6, lymph glands are sites both for the manufacture of lymphocytes and for the filtering out of bacteria and other foreign agents that have entered the lymphatic system. The tonsils and adenoids also function in capturing and destroying bacteria that enter the nose and mouth. Tonsils and adenoids are relatively large in young children, but tend to decrease in size with age. If they are repeatedly or chronically infected, or if they become enlarged to the point where they obstruct breathing, their surgical removal may be indicated. However, with the advent of antibiotics, many of the problems formerly experienced with these lymph glands have tended to disappear.

Larynx. On leaving the pharynx and passing through the **glottis,** incoming air enters the larynx. As a protective device for preventing food and drink from entering the larynx and going down the "wrong way," there is a flap of tissue called the **epiglottis,** that covers the entrance to the glottis whenever food is swallowed. The mucous membrane from the pharynx continues into and through the larynx.

The **larynx,** also called the **voice box,** is the organ of sound production. It contains two fibrous bands of tissue, the **vocal cords,** that are strung horizontally across its passageway. When air rushes through the larynx, the vocal cords begin to vibrate, and a sound is produced. The greater the frequency of vibration of the vocal cords, the higher is the **pitch** of the sound produced. The pitch of the human voice is controlled by the amount of tension placed on the vocal cords by the muscles of the larynx. Tense cords yield high-pitched notes, whereas relaxed cords produce low-pitched notes.

An important consequence of the relation of pitch to vocal cord tension is seen in males. At sexual maturity (**puberty**), the vocal cords of the male approximately double their original length, with the result that the average adult male voice "changes" and becomes about an octave lower in pitch than the average adult female voice. This degree of difference in pitch does not exist between very young boys and girls.

Trachea. From the larynx, incoming air passes into the **trachea,** or windpipe. The trachea is a cylindrical tube about 5 inches long and 1 inch in diameter. Cartilaginous C-shaped bands are spaced horizontally at regular intervals in the tracheal wall. These keep the tracheal tube open at all times. The cartilaginous bands do not form complete rings, but are open on their dorsally located ends. These open dorsal ends are joined by strips of smooth muscle.

Because of the open dorsal ends of its cartilaginous bands, the windpipe is flexible enough so that when food is swallowed from the pharynx, the soft, collapsed esophagus can expand and receive the food without being compressed by the trachea that is immediately ventral to it.

Like the other respiratory passageways, the trachea is lined with a ciliated mucus-secreting epithelial membrane. In the trachea, as in all other parts of the respiratory system posterior to, that is, below, the pharynx, the beat of the cilia moves the mucus toward the pharynx. On reaching the pharynx, the mucus and its contained particles are unconsciously swallowed into the esophagus and passed through the digestive tract. When large amounts of mucus are being produced, as during an infection, the mucus is brought to the pharynx periodically by "coughing" and is then expelled from the body through the mouth. The importance of the presence and movement of mucus in the respiratory system is not only to keep the lungs clean, but also to prevent bacterial infection, since many bacteria enter the body on inhaled dust particles. A major cause of lung infection is thought to be the paralysis of the epithelial cilia of the respiratory tract by noxious agents, including substances in cigarette smoke. This results in a buildup of mucus in the air passages, as more of it is constantly being produced by the epithelial cells. The bacteria are then able to lodge and reproduce in the stationary mucus layer and subsequently invade the respiratory system tissues.

An additional problem brought on by smoking is the smoker's "early morning cough." This characteristic of chronic smokers is the body's attempt to clear the air passages of obstructive stationary mucus.

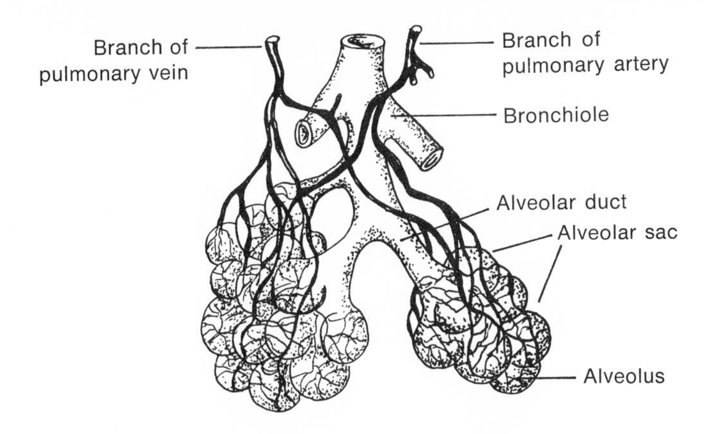

Fig. 11-2
Bronchioles, alveoli, and associated blood supply.

Bronchi. At about the level of the fifth thoracic vertebra, the trachea divides into the **bronchi.** The right bronchus is somewhat larger than the left, and any air particles that pass through the trachea are more likely to enter and lodge in the right bronchial system than in the left. At their upper ends, both bronchi have the same structure as the trachea. However, as they divide and subdivide into smaller and smaller tubes, the bronchi gradually lose their cartilaginous supporting bands. Finally, they become a network of microscopic tubes, called **bronchioles,** composed only of smooth muscle and fibrous connective tissue.

Alveoli. Every bronchiole terminates in a series of ducts, called **alveolar ducts** (Fig. 11-2). Each duct leads into an alveolar sac, a structure that somewhat resembles a bunch of hollow grapes. Each "grape" of the sac is called an **alveolus.** It has been estimated that there are 250 million alveoli in the human lungs. The cellular wall of each alveolus consists of a single layer of squamous cells, which is surrounded by a thick network of capillaries. The alveolus is the working unit of the lung, the place in which the exchange of gases (external respiration) occurs.

□ Mechanics of breathing

We have traced the path of incoming air from the nostrils, its point of entry, to the alveoli, the area of gas exchange. We shall now consider the mechanism by which the air is brought into and expelled from the lungs in the breathing process.

Chest cavity. The lungs, two in number and consisting of the smaller bronchi, the bronchioles, and the alveoli, are located in a section of the body, called the **chest cavity,** that is completely enclosed and has no communication with the outside atmosphere or with any other body cavity. The chest cavity is closed off on the top and sides by the chest wall, which contains the ribs, and on the bottom by a strong, dome-shaped sheet of skeletal muscle, called the **diaphragm.**

The outside of each lung is covered by a thin sheet of smooth epithelium, called a **pleuron,** that separates each lung from the other and from the inner wall of the chest cavity. The inner wall of the chest cavity is also covered by a pleuron. The two pleura around each lung are separated from one another, and the space between them, which forms a **pleural cavity** around each lung, is filled with air that is at a pressure less than that of the outside atmosphere. This difference in air pressure between the pleural cavity and the outside atmosphere tends to cause air to enter the lung and keep it inflated.

When the pleural linings around a lung become infected and inflamed, they secrete a fluid that accumulates in the pleural cavity, producing a condition known as **pleurisy.** Pleurisy often accompanies **pneumonia,** a bacterial infection of the bronchioles and alveoli. Luckily, these infections respond well to antibiotic treatment.

Another bacterial disease of the lung, although it may also attack the kidneys and bones, is **tuberculosis.** In severe cases of pulmonary tuberculosis, it is sometimes necessary to collapse one lung and give the infected tissues a chance to rest and overcome the infection. This is done by puncturing the chest wall and injecting sterile air into the pleural cavity. The injected air increases the pressure in the pleural cavity to the point where it is greater than that of the surrounding atmosphere, and, as a consequence, the lung collapses. Over a period of time, the excess air in the pleural cavity is removed by the bloodstream, and the lung gradually returns to its normal level of activity (Fig. 11-3).

Inhaling and exhaling. The periodic expansion and contraction of the chest cavity is controlled mainly by muscles of the diaphragm and secondarily by muscles of the rib cage and shoulders (Fig. 11-4). When the **diaphragm** is **relaxed,** it arches upward against the lungs. As the **diaphragm contracts,** it stretches out horizontally, thereby increasing the size of the chest cavity. Simultaneously, certain muscles of the shoulder and rib cage contract, drawing the ventral (front) ends of the ribs upward and outward. The movements of the ribs complement the action of the diaphragm in enlarging the chest cavity.

Because the chest cavity is completely closed off, an increase in its volume results in a lowering of the air pressure inside the lungs. When the pressure in the lungs falls below atmospheric pressure, air from the outside rushes through the respiratory system into the alveoli. As air fills the alveoli, the lungs are distended.

When the **diaphragm relaxes,** it returns to its previous convex (dome) shape, thereby decreasing the volume of the chest cavity. Simultaneously, there is a relaxation of the shoulder and rib cage muscles that had previously contracted, and a simultaneous contraction of their antagonists, causing the ribs to move downward and inward. Here, too, the movements of the ribs complement the action of the diaphragm, but in this case, it is to decrease the size of the chest cavity.

A decrease in volume of the chest cavity results in an increase of the air pressure within the lungs. When the lung pressure becomes greater than the atmospheric pressure, the air is pushed from the alveoli through the respiratory system to the outside.

The amount of change in lung pressure during the inhaling and exhaling of air is relatively small. Air pressure is measured in terms of millimeters of mercury (mm Hg), with 760 mm Hg taken as atmospheric pressure. Between breaths, lung pressure is the same as atmospheric pressure. As the diaphragm contracts, lung pressure decreases to about 3 mm Hg below atmospheric pressure, causing inhalation; when the diaphragm relaxes, lung pressure increases to approximately 3 mm Hg above atmospheric pressure, causing exhalation.

Control of rate of breathing. During exercise or heavy manual labor, the body

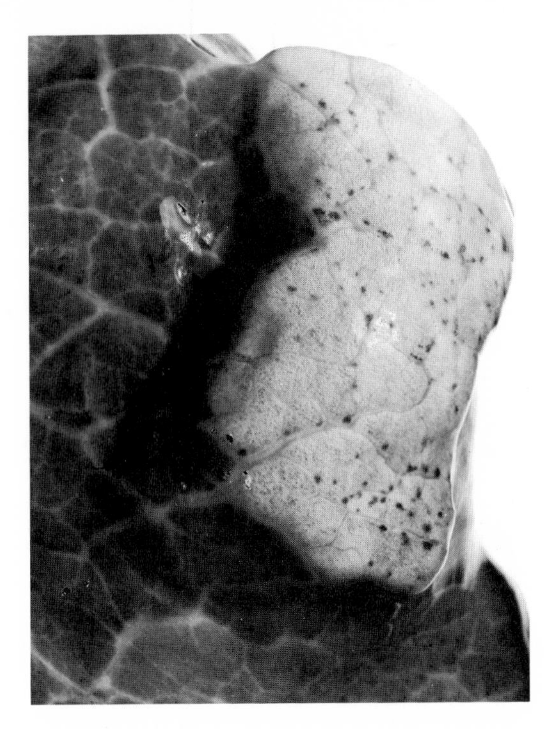

Fig. 11-3
Partially collapsed lung caused by obstruction of a number of bronchioles in patient with cystic fibrosis (Chapter 7). There is a clear contrast between pale, still aerated portion of lung and the dark collapsed section beside it. (From Anderson, W. A. D., and Kissane, J. M., eds. 1977. Pathology, ed. 7. The C. V. Mosby Co., St. Louis.)

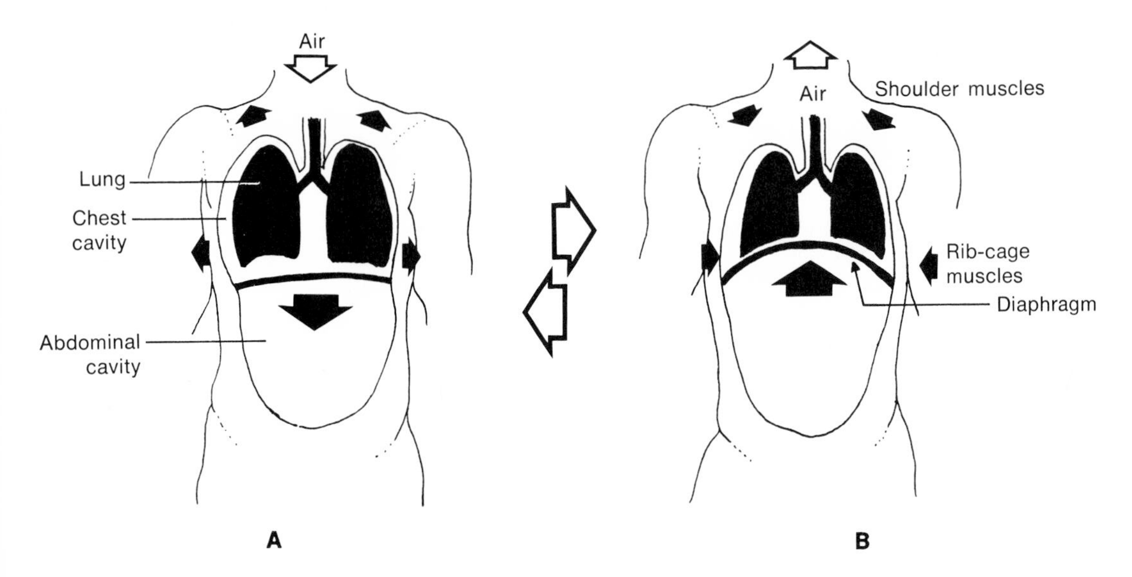

Fig. 11-4
Respiratory movements in man. **A**, Inhaling. **B**, Exhaling. Arrows indicate the effects on the chest cavity of contraction and relaxation of the diaphragm and various sets of shoulder and rib-cage muscles.

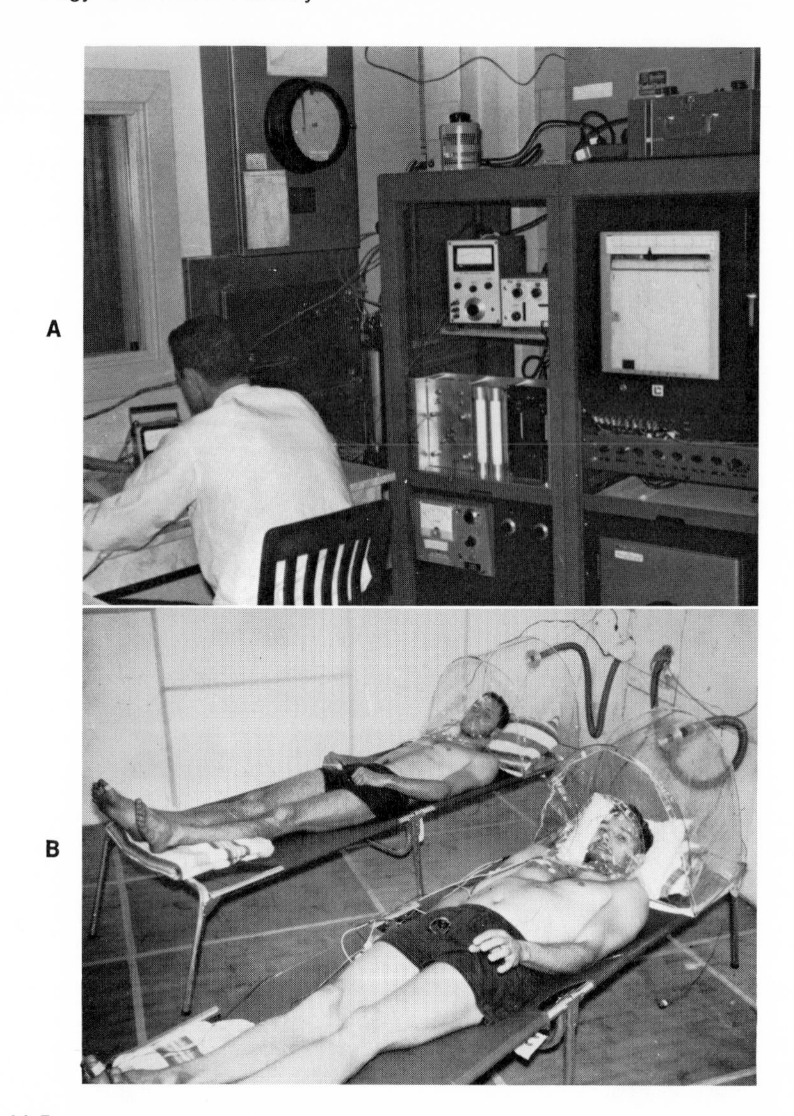

Fig. 11-5
A, Technician at control panel outside metabolic chamber at the Human Performance Laboratory at The Pennsylvania State University monitoring the oxygen consumption, carbon dioxide excretion, and changes in surface temperature of subject in chamber behind glass. **B,** Subjects in metabolic chamber in which the effects of environmental temperature changes on oxygen consumption and skin temperature are being determined. (Courtesy Public Relations Department, The Pennsylvania State University; from Guthrie, H. 1975. Introductory nutrition, ed. 3. The C. V. Mosby Co., St. Louis.)

must supply increased quantities of oxygen to its tissues, especially its muscles, and must remove increased amounts of carbon dioxide from them. This is accomplished in part by an increase in the rate of the heartbeat and in part by an increase in the rate of breathing.

A person's rate of breathing is controlled by the nervous system, mainly by the respiratory center of the brain. In Chapter 10, we learned that an increase in the level of lactic acid in the blood during sustained muscle activity results in an increase in the rate of breathing. The same effect is produced by an increased blood level of carbon dioxide. As the amount of carbon dioxide in the blood rises above 5.3%, the respiratory center of the brain is stimulated to increase the individual's rate of inhaling and exhaling. Here, as with lactic acid, we find that a waste product of cell metabolism serves as the stimulus for increased body activity to remove the waste product (Fig. 11-5).

□ External respiration

The exchange of gases between the respiratory and circulatory systems is called **external respiration.** The gases involved are oxygen and carbon dioxide, and the body organs involved are the alveoli of the lungs and the capillaries of the pulmonary circulatory circuit.

As we discussed in Chapter 7, gases pass through cell membranes solely under the influence of **diffusion,** namely, movement from areas of higher concentration to areas of lower concentration. The concentration (**partial pressure**) of a gas either in air or in a liquid (blood or tissue fluid or protoplasm) can be calculated and is usually stated in terms of millimeters of mercury. Table 11-1 lists the partial pressures of the respiratory gases in alveolar air and in the pulmonary capillaries.

Under the conditions shown in Table 11-1, oxygen diffuses from the alveoli into the blood, whereas carbon dioxide does the reverse. In the blood, most of the oxygen becomes bound to the hemoglobin of the red blood cells and is carried around the body as oxyhemoglobin. A small amount (2%) of the oxygen is carried in the blood plasma in simple solution. Carbon dioxide, on the other hand, is carried mainly (80%) in the blood plasma as sodium bicarbonate, whereas the remainder is carried in the red blood cells.

□ Internal respiration

When the blood reaches the various tissues of the body, an exchange of respiratory gases also occurs. However, in **internal respiration,** the flow of oxygen will be

Table 11-1
Partial pressures of respiratory gases in alveolar air and pulmonary capillaries

	Partial pressure (mm Hg)	
Respiratory gases	*In alveolar air*	*In pulmonary capillaries*
Oxygen	104	40
Carbon dioxide	40	45

Table 11-2
Partial pressures of respiratory gases in tissue capillaries and tissue fluid

	Partial pressure (mm Hg)	
Respiratory gases	*In tissue capillaries*	*In tissue fluid*
Oxygen	95	40
Carbon dioxide	40	45

out of the capillaries, while the flow of carbon dioxide will be into the capillaries. Table 11-2 lists the partial pressures of the respiratory gases in the tissue capillaries and in tissue fluid.

After internal respiration takes place, there is an exchange of respiratory gases between the tissue fluid and the cells themselves. The tissue fluid transfers oxygen to the cells and receives carbon dioxide from them.

□ Diseases of the respiratory system

Respiratory system diseases include some that are caused by viruses, others that are caused by bacteria, and still others that are developmental in nature. The causative agents of most of the developmental diseases are thus far unknown.

Influenza. Among the viral diseases, the most deadly has been **influenza (flu).** In the worldwide epidemic of 1918, the disease killed 30 million persons. Since that time, there have been epidemics of lesser magnitude in 1957 ("Asian flu"), 1968 ("Hong Kong flu"), and 1973 ("London flu"). In each case, it was a different strain of influenza virus that produced the epidemic. The particular term applied to each epidemic refers to the region or city from which the epidemic is believed to have originated. The 1968 epidemic affected more than 30 million people in the United States alone and resulted in more than 200 deaths. Undoubtedly, more deaths would have occurred but for the availability of antibiotics to combat secondary bacterial infections such as pneumonia.

Influenza itself starts out as an infection of the nasal epithelium and pharynx. In severe cases, the bronchi and alveoli become involved. Unfortunately, the various

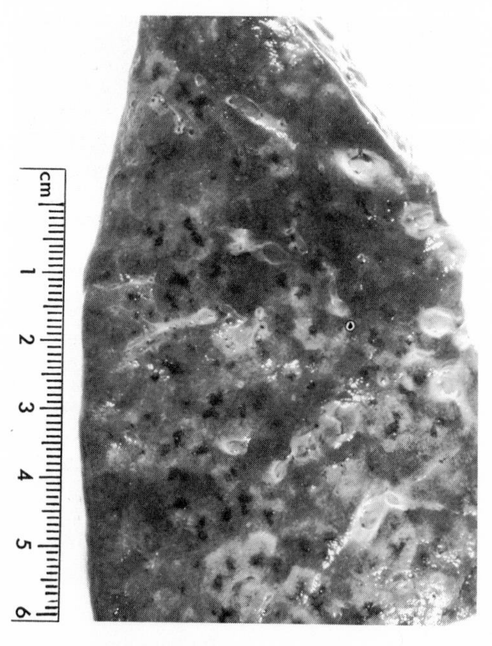

Fig. 11-6
Portion of lung of pneumonia victim. Lighter areas are groups of alveoli filled with pus. (From Anderson, W. A. D., and Kissane, J. M., eds. 1977. Pathology, ed. 7. The C. V. Mosby Co., St. Louis.)

strains of the virus mutate quite frequently, producing new strains that are unaffected by antibodies formed against the older strains.

In 1900 influenza and pneumonia were the most frequent causes of death in the United States (200/100,000 people) (Fig. 11-6). By 1970 these respiratory system diseases had dropped to sixth place (31/100,000 people), heart disease and cancer being first and second.

Tuberculosis. One of the bacterial respiratory diseases is tuberculosis. Although modern medicines have drastically reduced the death rate from this disease, it still remains a threat to man's welfare, particularly in underprivileged areas of the world where crowding and poverty exist in large measure.

In 1900 the death rate from tuberculosis in the United States was 194/100,000 people, and it was the second most frequent cause of death in the country, exceeded only by influenza and pneumonia. By 1967 the death rate from tuberculosis was reduced to 3/100,000 people. However, 40,000 new cases are discovered each year, and half of these are in the "inner cities" of our large metropolitan areas.

Any organ of the body may be affected by tuberculosis, but the lung is the major seat of the disease and the usual point of origin from which the infection reaches other organs. In the lungs, the tuberculosis-causing bacteria destroy the alveoli, causing pulmonary bleeding and loss of respiratory function.

Emphysema. Of the many developmental diseases of the respiratory system, we shall consider two: emphysema and cancer. **Emphysema** is a disease in which the alveoli become enlarged, and their walls become torn. It occurs four to five times more frequently among men than among women, especially those over 40 years of age. It causes over 16,000 deaths annually in the United States.

The disease usually starts with difficulty in breathing, which soon becomes a major effort. Chronic coughing is a prominent feature. The heart must pump harder to supply the cells of the body with oxygen, and death from heart failure is not unusual.

Although the cause of emphysema is not entirely clear, all forms of air pollution are clearly implicated. This is especially true of cigarette smoking. The death rate from emphysema is six times greater for smokers than nonsmokers.

Lung cancer. Lung cancer was a relatively rare disease in the United States 50 years ago. Today, new cases occur at the rate of 75,000/year, and the death rate from this form of cancer is 70,000/year. Of these 70,000, about 55,000 will be men, making lung cancer the most common cause of death from cancer in men.

As with emphysema, all forms of air pollution are implicated as causative agents, especially cigarette smoking. Approximately 90% of all lung cancer cases occur in people who smoke cigarettes, and the death rate from lung cancer is eight times greater for smokers than nonsmokers (Fig. 11-7).

Laboratory studies have isolated from tobacco smoke a number of chemical compounds that can cause cancer (**carcinogens**). In addition, there are other compounds in tobacco smoke called **cocarcinogens,** that cannot produce cancer when alone, but do form carcinogenic combinations with other molecules.

The health hazards involved in cigarette smoking depend to a large extent on the effect of cigarette smoke on the cilia of the mucus-secreting epithelial membranes of the respiratory system. Cigarette smoke paralyzes the cilia and prevents the respira-

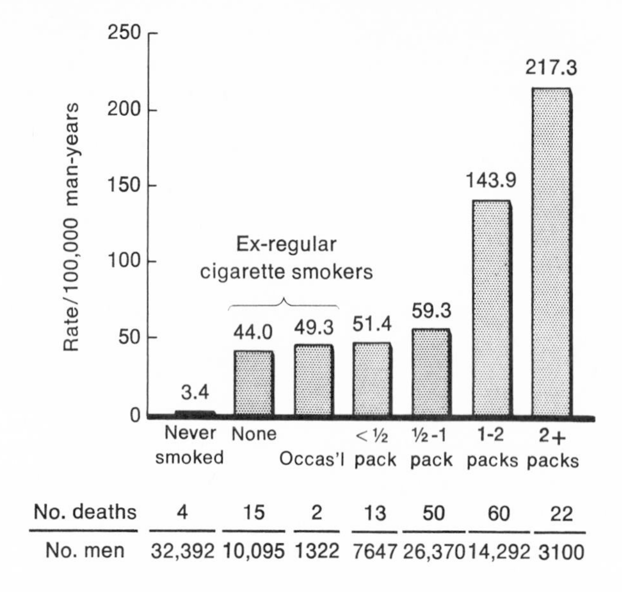

Fig. 11-7
Age-standardized death rates for lung cancer related to current amount of cigarette smoking. (Redrawn from Hammond and Horn, J.A.M.A. **168**:1294, 1958; from Turner, C. E. 1971. Personal and community health, ed. 14. The C. V. Mosby Co., St. Louis.)

tory system from cleaning itself of all foreign particles through the movement of mucus to the outside. Thus cancer-causing and cancer-promoting chemical compounds can accumulate on the epithelial membranes, especially in the bronchial tubes where most cases of human lung cancer originate.

■ EXCRETION

All the metabolic reactions that occur in the body produce various chemical by-products. Some of these can be used in other physiological processes. However, a large number of the by-products are useless for any other biochemical activities, and some are actually harmful if retained by the body for any considerable period of time. The removal of surplus and dangerous metabolic by-products occurs through the process of **excretion.** It should be noted that excretion eliminates the waste products of cellular metabolism, not the residues of food from the digestive tract. The remains of the digestive process have never been used by the body and are eliminated as feces.

The waste products of cellular metabolism will vary, depending on the type of chemical compound that is broken down. Thus the breakdown of carbohydrates and lipids results in the formation of carbon dioxide and water. Although these compounds may be of use to the body (carbon dioxide as a stimulant for the brain's respiratory center and water as a component of protoplasm and as a medium for chemical reactions), both are in reality waste products and must be eliminated. The breakdown of amino acids also gives rise to carbon dioxide and water and, in addition, compounds containing nitrogen, sulfur, and the other elements found in amino acids. These compounds can be toxic to the body and must be excreted as quickly as possible.

In Chapter 3, the uses of amino acids in the body were discussed. Their principal function is in the synthesis of proteins for growth and repair of cells. Any amino acid that is not used for either of these purposes is **deaminated** (the amino group is removed), and the remainder of the molecule, containing carbon, hydrogen, and oxygen, is either converted to sugar for immediate use as an energy source or for storage (as glycogen) or is converted to lipid for storage.

Deamination results in the formation of ammonia (NH_3). In high concentration, ammonia is toxic to the body, and it is quickly converted to **urea** ($CO[NH_2]_2$), a relatively nontoxic substance, as follows:

$$2NH_3 + CO_2 \rightarrow H_2N{-}\underset{\underset{O}{\|}}{C}{-}NH_2 + H_2O$$

The liver is the organ primarily involved in both deamination and urea formation. Urea is the chief nitrogen-containing waste product found in our bodies.

□ Excretory organs

Several parts of the body are involved in excretion. These include the lungs, the skin, the digestive tract indirectly, and, most importantly, the kidneys.

Lungs. The lungs function as excretory organs in ridding the body of all the excess carbon dioxide and about 15% of the excess water. The role of the lungs in eliminating waste carbon dioxide has already been discussed. The elimination of water occurs through breathing. When air is inhaled, it is saturated with moisture before reaching the alveoli. When the air is expelled, it carries the moisture out with it. Our bodies lose an average of 2200 milliliters (ml) of water each day. (See Table 4-5.) Of this amount, 300 ml is lost from the lungs.

Skin. The skin eliminates water, salts, and some body wastes. Some water is lost directly through the skin surface, since the epidermis is not completely watertight. This is called **insensible perspiration.** In addition, the long, coiled sweat glands eliminate water as a component of sweat. About 600 ml of water is lost daily through the skin.

Although the primary function of sweat is to cool the body, some salts and other body wastes are eliminated in the process. To a very limited extent, the sweat glands function in the removal of urea. The composition of perspiration is very much like that of dilute urine. However, even with maximum perspiration, a person would lose no more than a tenth of the urea that is eliminated daily in the urine.

Digestive tract. Some excretory substances are eliminated through the digestive tract. An example of these is the pigments found in bile. Bile is manufactured by the liver and passed into the small intestines. The **bile pigments** are derived from the hemoglobin of worn-out red blood cells. Other examples of excretory substances that are eliminated through the digestive tract are excess **calcium** and **iron,** which are excreted into the large intestine by the cells of its internal lining.

Kidneys. The kidneys are the primary excretory organs of the body. In addition to ridding the body of urea, the kidneys function to maintain the water content of the blood at a constant level and to regulate the chemical composition of body tissues. The kidneys are organs in which structure and function are very closely correlated, so that many deductions about the kidneys' functions can be made from a study of their structure.

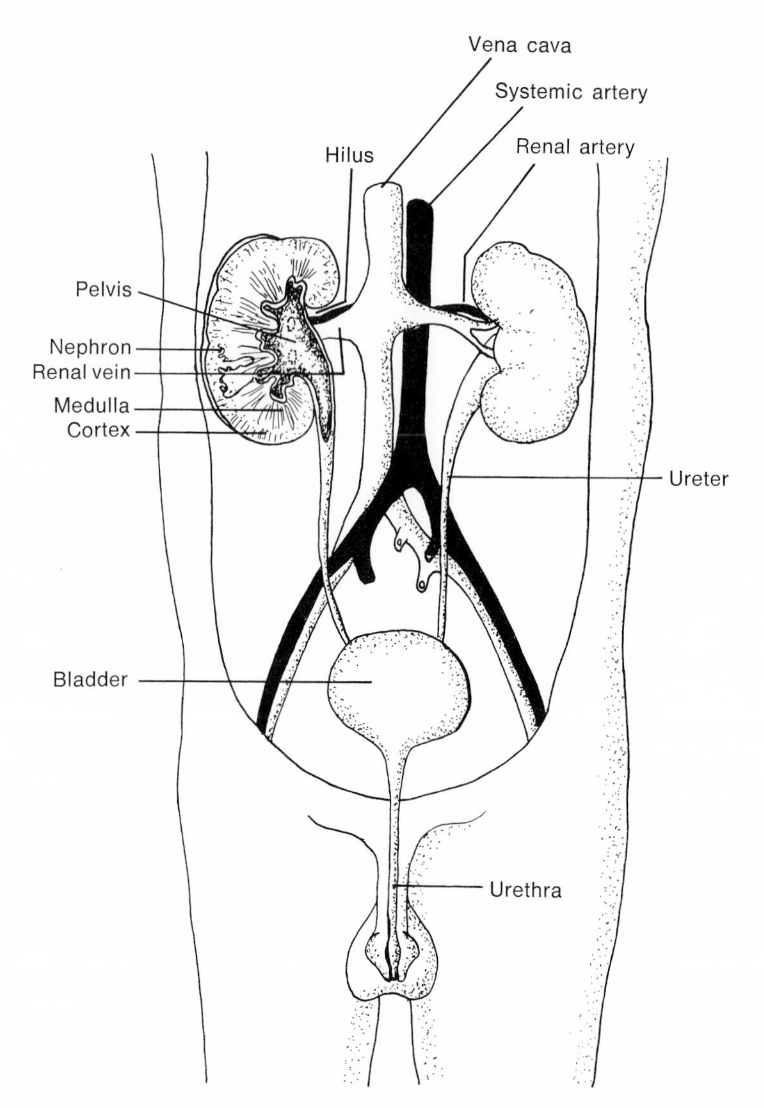

Fig. 11-8
Human excretory system, including associated blood vessels.

Man's kidneys are dark red, bean-shaped organs about 10 centimeters (cm) long (Fig. 11-8). They are located on the inner surface of the dorsal body wall just below the stomach, one on either side of the middorsal line. Leading from the indentation in each kidney is a tube, called the **ureter,** that carries the urine formed in the kidney to the **urinary bladder.** Another tube, the **urethra,** passes the urine from the urinary bladder to the outside. In the human female, this duct is rather short, leading directly to the external surface of the body just ventral to the opening of the vagina. In the male, however, the urethra is much longer, leading to the outside through the penis.

When a kidney is cut lengthwise, we find that within the kidney, the ureter is expanded to form a large cavity, called the **pelvis.** The rest of the kidney is divided

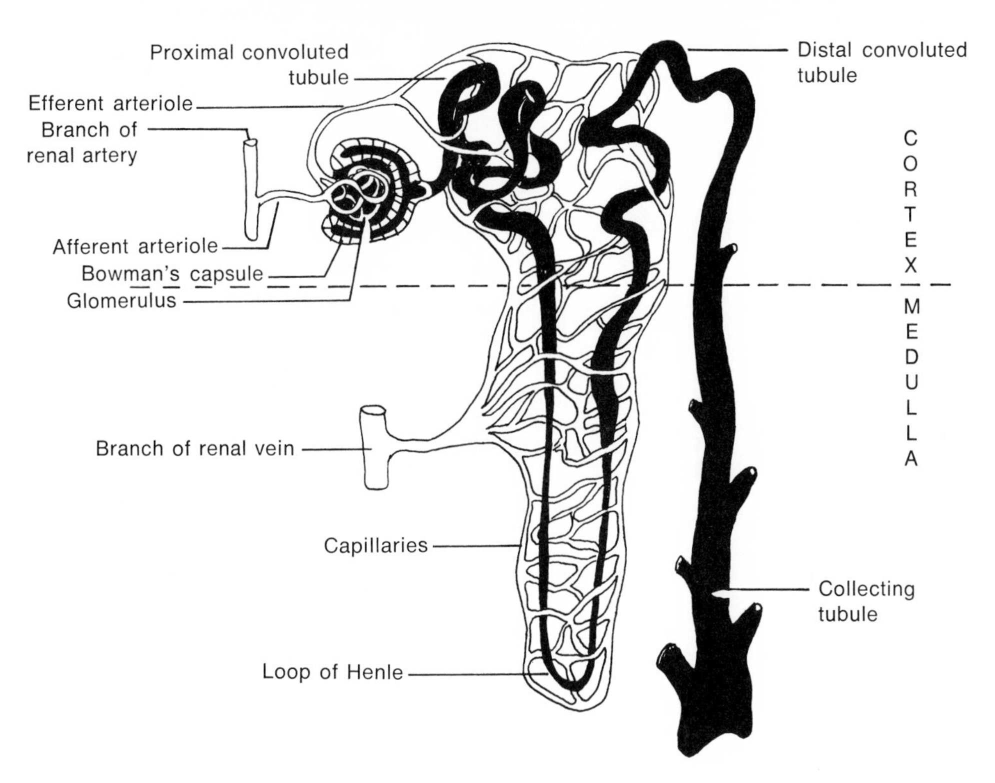

Fig. 11-9
Renal corpuscle and associated blood vessels.

into an outer layer, the **cortex**, and an inner layer, the **medulla.** The kidneys are supplied with rather large renal arteries and renal veins.

Renal corpuscle. The **renal corpuscle** is the structural and functional unit of the kidney (Fig. 11-9). Each renal corpuscle consists of two subunits: a glomerulus, derived from the circulatory system, and a collecting tubule (nephron), derived from the kidney itself. A kidney contains about 1 million renal corpuscles.

Each renal artery divides many times within the kidney to form a large number of **afferent arterioles.** Each afferent arteriole is further subdivided to form a network of about 50 parallel capillaries, called a **glomerulus.** The capillaries of the glomerulus are clustered into a spherical tuft. On leaving the glomerular tuft, the capillaries fuse to form an **efferent arteriole,** which in turn divides to form capillaries that surround the kidney tubules. Subsequently, these capillaries fuse to form the renal vein. As we shall see, the formation of the double set of capillaries between the renal artery and the renal vein permits the kidney to perform its function in a most efficient manner.

At its upper end, each **collecting tubule** makes contact with a glomerulus. The cells of the tubule invaginate, forming a cup, called **Bowman's capsule,** around the glomerular capillaries. The close association of Bowman's capsule and the glomerulus permits the diffusion of substances from the capillaries through the cells of Bowman's capsule into its cavity. The cavity of Bowman's capsule leads directly into the **proximal convoluted tubule,** which initially is very twisted, but later becomes short and

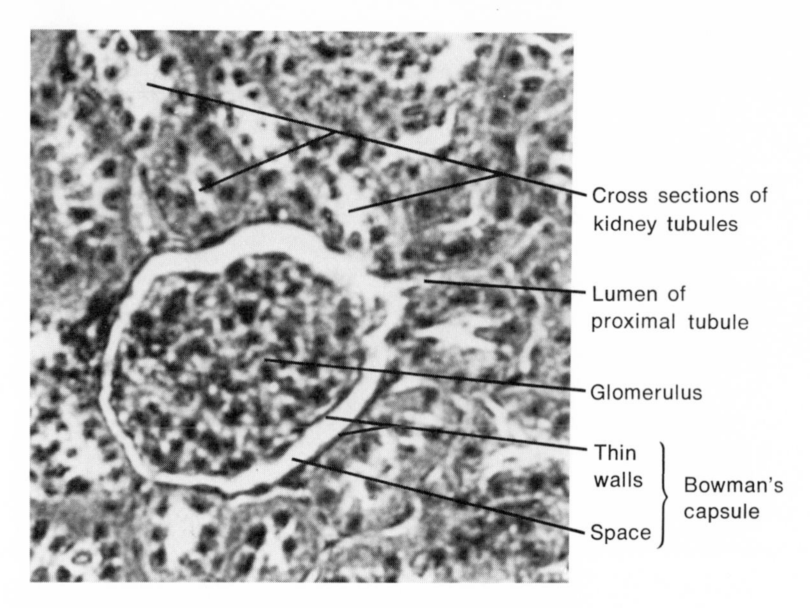

Fig. 11-10
Longitudinal section through cortex of mammalian kidney. Glomerulus is prominent in center. Space of Bowman's capsule surrounds glomerulus and can be seen to be continuous with lumen of proximal tubule. Cross sections of many convoluted kidney tubules also visible. (×760.) (Photo by Victor B. Eichler, Wichita State University, Wichita, Kansas; from Lane, T., ed. 1976. Life the individual the species, The C. V. Mosby Co., St. Louis.)

straight and leads to a looped section, called the **loop of Henle.** Henle's loop turns upward to a second twisted portion, called the **distal convoluted tubule.** This joins a large **collecting tubule** that brings the urine to the pelvis of the kidney, which then continues as the ureter to the urinary bladder (Fig. 11-10).

□ Formation of urine

The excretory function of the kidney involves removing the wastes, but conserving the useful components of the blood. This is achieved by first forcing nearly all the small-sized molecules of the blood (including water) into the kidney tubules and then reabsorbing all the useful materials back into the blood. A combination of three processes enables the kidneys to perform its excretory function. These processes are filtration, reabsorption, and tubular secretion. The first two processes, filtration and reabsorption, depend, respectively, on hydrostatic pressure and osmotic pressure, two phenomena discussed in Chapters 6 and 7. Tubular secretion is a specialized process of greater importance in other animals than it is in man.

Filtration. Kidney filtration is the passage, under pressure, of small-sized molecules from the glomerulus into the cavity of Bowman's capsule. The strength of the force causing filtration is equal to the blood pressure in the glomerulus **minus the sum** of the osmotic pressure in the glomerulus and the reverse hydrostatic pressure of the fluid in the cavity of Bowman's capsule. Blood entering the glomerular capillaries is under a hydrostatic pressure of 70 millimeters of mercury (mm Hg). The osmotic pressure within the capillaries in 32 mm Hg, while the reverse hydrostatic pressure of the fluid in the cavity of Bowman's capsule is 14 mm Hg. The net filtration pressure is therefore 24 mm Hg.

Under a net filtration pressure of 24 mm Hg, the blood passing through the glomerulus loses most of its water, salts, amino acids, sugar, urea, and other small-sized molecules into the cavity of Bowman's capsule. Remaining in the blood capillaries are the blood cells and such large molecules as the plasma proteins. Therefore glomerular filtrate is the same as blood plasma except that it has no proteins.

The amount of glomerular filtrate produced each day is extremely large. Our two kidneys receive about one fourth of the total output of the heart, amounting to 1300 ml of blood/minute. This blood contains 700 ml of plasma, the remainder being cells. Of this amount of plasma, 125 ml/minute is filtered off through the glomeruli into Bowman's capsules. This adds up to a total daily filtrate of approximately 180 liters. Under normal conditions, only about 1.2 liters of urine is produced each day. Therefore 178.8 liters of glomerular filtrate (99%) must be reabsorbed back into the bloodstream by the kidney.

Reabsorption. The reabsorption of 99% of the water, most of the salts, and, normally, all of the sugar and amino acids from the glomerular filtrate occurs while the material is passing through the convoluted tubules and the loop of Henle. As stated earlier, the capillaries of the glomerulus fuse to form an efferent arteriole that subsequently divides to form capillaries around the kidney tubules. As the blood flows through the efferent arteriole into the capillaries, its hydrostatic pressure falls from 70 mm Hg to 13 mm Hg. However, because its protein contents remain constant, the osmotic pressure within the capillaries remains high at 32 mm Hg, and this results in the reabsorption into the capillary blood of most of the useful substances in the tubules. The percent of water in the glomerular filtrate that is reabsorbed from each segment of the nephron tubule is shown in Table 11-3. Most of the water, as is also true of the other useful substances, is reabsorbed from the proximal convoluted tubule.

Although filtration of blood plasma through the glomerular capillaries is limited solely by the size of the plasma components, the reabsorption of substances from the tubules is highly selective. Most substances have what is called a **kidney threshold level.** If the concentration of a substance in the blood is less than its kidney threshold level, the substance will be absorbed from the tubules back into the surrounding capillaries.

Table 11-3
Percent of water in the glomerular filtrate reabsorbed from each segment of the nephron tubule

Segment of nephron tubule	Water reabsorbed (%)
Proximal convoluted tubule	80
Loop of Henle	6
Distal convoluted tubule	9
Collecting tubule	4
Total	99

Glucose is an example of a compound that has a kidney threshold level. Normally, all glucose in the glomerular filtrate is reabsorbed, because the kidney threshold value (150 milligrams [mg]/100 ml of blood) is higher than the blood glucose level (100 mg/100 ml). However, if the blood sugar level exceeds 150 mg/100 ml, as occurs in diabetes, sugar will appear in the urine. In diabetes, the kidney is functioning to reduce blood sugar levels as much as it can. Unfortunately, the elimination of some of a diabetic's excess blood sugar through the kidney does nothing to cure the basic disease.

In a similar fashion, the kidneys help regulate the concentrations of such inorganic ions as sodium, potassium, and chloride in the blood plasma. Whenever the concentration of one of these ions in the blood exceeds its kidney threshold value, the excess of the ion is not reabsorbed, but appears in the urine. This results in a relatively constant blood plasma composition and a considerable variation in urine composition.

Inasmuch as our blood and tissue fluid interact continuously, the function of the kidney in maintaining the constancy of the blood components in effect ensures the constancy of our tissue fluid and the proper functioning of our bodies. The tendency in an organism to maintain a constant internal environment is called **homeostasis,** and the kidneys are the main homeostatic organs in our bodies.

Tubular secretion. Tubular secretion is the direct addition of waste materials to the urine by the cells of the kidney tubules. These cells obtain the waste materials directly from the surrounding capillaries and add these substances to the fluid passing through the tubules.

The processes of reabsorption and tubular secretion serve to concentrate the waste materials in the urine. A comparison of the components of blood plasma, glomerular filtrate, and urine is presented in Table 11-4. As stated earlier, glomerular filtrate has the same composition as blood plasma except for the lack of proteins. However, urine differs markedly from glomerular filtrate in having about 70 times more urea and 10 times more uric acid. In addition, urine contains virtually no glucose or amino acids.

□ Regulation of kidney function

All homeostatic mechanisms depend on feedback control for their regulation. The kidney is an excellent example of this. If a person drinks a great deal of water, there is a dilution of the individual's blood and therefore a reduction in the blood's osmotic

Table 11-4
Comparison of components of blood plasma, glomerular filtrate, and urine

Component	% of blood plasma	% of glomerular filtrate	% of urine
Urea	0.03	0.03	2.0
Uric acid	0.004	0.004	0.05
Glucose	0.10	0.10	Trace
Amino acids	0.05	0.05	Trace
Inorganic salts	0.90	0.90	1.50
Proteins	7.0	0	0

pressure. This reduced osmotic pressure results in less reabsorption of water from the kidney tubules, which in turn results in the formation of an increased amount of more dilute urine. Through this process, the blood returns to its normal levels of concentration of salts, glucose, proteins, etc.

When water consumption is reduced below normal, the individual's blood increases in osmotic pressure. This leads to an increased reabsorption of water from the kidney tubules into the blood and the formation of small amounts of a more concentrated urine.

Antidiuretic hormone. The homeostatic mechanism that controls the reabsorption of water does not depend solely on osmotic pressure. Also involved is the **pituitary gland,** which is sensitive to changes in the concentration of water in the blood. When the water concentration in the blood drops, the pituitary gland cells are stimulated to produce and release a hormone called **antidiuretic hormone** (ADH). This hormone causes the pores of the epithelial cells of the kidney tubules to increase in size. As a result, there is an increased flow of water molecules out of the kidney tubules and into the surrounding capillaries.

When the concentration of water in the blood increases, that is, when the blood becomes diluted, the pituitary gland cells cease producing ADH. In the absence of ADH, the pores of the cells of the tubules become reduced in size, and there is a reduced reabsorption of water.

Role of solids. The quantity of urine formed by the kidneys depends not only on how much liquid is consumed, but also on the amount of solids to be excreted from the blood. For example, in the case of diabetes, the passage of an increased amount of sugar out of the glomerular capillaries into the cavity of Bowman's capsule results in an increase in the osmotic pressure of the glomerular filtrate. This increased osmotic pressure within the kidney tubules tends to keep water in the tubules (decreases the amount of water reabsorbed from the tubules), resulting in an increase in urine volume. That is why diabetes is characterized not only by the presence of sugar in the urine, but also by the passage of large amounts of urine at frequent intervals and constant thirst.

□ Diseases of the kidney

Infection is the most common disease of the kidney and its associated structures (ureters, urinary bladder, urethra). It is estimated that there are at present more than 3 million undetected cases of kidney infection in the United States and that 25,000 to 75,000 Americans die of kidney disease each year.

An often fatal disease, **nephritis,** occurs when bacteria attack the glomeruli of the kidneys. Because of the bacterial infection, the glomeruli become more permeable than normal, so that proteins and red blood cells pass through into the urine. As this disease progresses, there is a marked reduction in the amount of urine formed, and the waste products normally excreted by the kidneys accumulate in the blood. This condition is called **uremia.** Because many of these accumulated metabolic waste products are toxic to the body, death occurs unless the condition is corrected.

If a person loses one kidney through injury or disease, the remaining kidney can do the work previously done by both kidneys, as long as it is free of disease. Should a person lose both kidneys, that individual is faced with an extremely serious situation.

Previously, such a person was doomed. In recent years, however, medical research has developed two new methods of therapy—kidney transplantation and the artificial kidney—that can maintain the lives of these individuals.

A kidney may be transplanted into a patient who has lost most or all kidney function. As was discussed in Chapter 6, although the surgical techniques for transplanting kidneys have been developed to a high degree of excellence, the body's production of antibodies to the foreign organ limits the success of this method of therapy.

Until kidney transplantation becomes a more certain form of therapy, the **artificial kidney** remains the only hope of survival for the majority of people who have lost both kidneys or who suffer from chronic kidney failure. Although artificial kidneys have been known since 1945 and were used in cases of acute, temporary kidney failure, it was not until 1960 that they could be used repeatedly in cases of permanent kidney failure or loss.

In using an artificial kidney, blood from an artery in the person's arm is channeled through a cellophane tube that is immersed in a bath containing all the normal blood chemicals except urea. As the blood flows in the tube, urea passes through the cellophane membrane into the bath, because of a difference in concentration of urea in the solutions on either side of the membrane. The purified blood is returned to the body through a vein in the arm. This type of treatment, called **dialysis,** is given several times a week for periods of 6 to 14 hours each. Some individuals have been maintained through this treatment for 15 years. Many of these people have been able to carry on full-time employment and others, part-time.

A kidney disease of lesser consequence is the formation of **kidney stones.** Some of the normal constituents of urine—uric acid and calcium phosphate—when present in large amounts may precipitate to form so-called stones. Kidney stones can form anywhere in the kidney or its associated structures and may produce painful obstructions that block the drainage system of the kidneys. These obstructions also increase the probability of infection in the kidney. If these stones become large enough to completely block the passage of urine, they must be removed surgically, because it is impossible to dissolve them.

SUMMARY

The proper functioning of the human body depends on a vast number of physiological processes, each with its own special requirements. However, there are some common needs that apply to all body activities. One of these is the necessity to obtain a continuing supply of oxygen, and another is to rid the body of its metabolic wastes.

Oxygen is obtained through the respiratory system, which consists of a series of passageways and tubes, starting from the nasal cavities and ending at the alveoli. The pumping mechanism for breathing involves the diaphragm and certain muscles of the rib cage and shoulders. Regulation of the rate of breathing is a function of the "respiratory center" of the brain, which is, in turn, controlled by the level of carbon dioxide in the blood.

Gases are exchanged, both in the lung cavity and at the tissue level, in accordance with the principle of diffusion. The pulmonary capillaries obtain oxygen from and release carbon dioxide to the air in the alveoli, while the tissue capillaries give oxygen to and receive carbon dioxide from the tissue fluid.

Although the respiratory system is subjected to microorganismal as well as developmental diseases, a common contributing factor in both types of disease is the paralyzing of the cilia of the mucus-secreting epithelium by air pollutants. Especially implicated in respiratory system diseases is cigarette smoking.

The metabolic waste products produced by a particular biochemical reaction vary with the type of compound that is broken down. Carbohydrates, lipids, and amino acids all yield carbon dioxide and water as by-products of their metabolism. However, amino acid metabolism also produces compounds containing nitrogen, sulfur, etc. The elimination of the various metabolic waste products is accomplished by the lungs, skin, digestive tract, and kidneys. Of these, the most important excretory organ for nitrogenous wastes is the kidney.

The basic unit of structure and function of the kidney is the renal corpuscle. Each corpuscle consists of a group of capillaries and a collecting tubule. Three processes—filtration, reabsorption, and tubular secretion—are involved in removing waste products from the blood while conserving its useful components. Filtration is the passage, under hydrostatic pressure, of permeable substances from the glomerular capillaries into the cavity of Bowman's capsule. Reabsorption is the pulling back, through osmotic pressure, of usable materials from the collecting tubules to their surrounding capillaries. Tubular secretion is the direct addition of waste materials to the urine by the cells of the kidney tubules.

Kidney function is regulated by the amount of water intake, the amount of solids in the blood to be excreted, and ADH. The most common disease of the kidney is nephritis, which may be fatal or lead to the need for an artificial kidney. A less common disease involves the formation of kidney stones.

In the next chapter, we shall study the human body's two systems of internal coordination: the nervous and endocrine systems.

SUGGESTED READINGS

Comroe, J. H., Jr. 1966. The lung. Sci. Am. **214**:56-68. Very clear description of the structure and functioning of the human lung.

Fenn, W. O. 1960. The mechanism of breathing. Sci. Am. **202**:138-148. In-depth discussion of the human respiratory system and the mechanics of breathing.

Hock, R. J. 1970. The physiology of high altitude. Sci. Am. **222**:52-62. Discussion of the mechanism by which people living at high altitudes adapt physiologically to the low levels of oxygen in the air they breathe.

Pitts, R. F. 1968. Physiology of the kidney and body fluids, ed. 2. Year Book Medical Publishers, Inc., Chicago. Authoritative book on how our kidneys function.

Slonim, N. B., and Hamilton, L. H. 1972. Respiratory physiology, ed. 2. The C. V. Mosby Co., St. Louis. Authoritative book on how the body obtains and uses oxygen.

Smith, H. W. 1953. The kidney. Sci. Am. **188**:40-48. Well-written account of the structure, function, and evolution of the kidney.

Snively, W. D., Jr. 1960. Sea within: the story of our body fluid. J. B. Lippincott Co., Philadelphia. Interestingly written book on our body fluid and the role of the kidneys in its regulation.

Solomon, A. K. 1962. Pumps in the living cell. Sci. Am. **207**:100-108. Article dealing with the process of active transport in kidney tubules.

Sports are the best example of the need to synchronize the body's various systems in order to perform a given activity. (Courtesy Underwood and Underwood; from Turner, C. E. 1971. Personal and community health, ed. 14. The C. V. Mosby Co., St. Louis.)

CHAPTER 12 Coordination of body activities

LEARNING OBJECTIVES

- What are the two coordinating systems of the body, and what are the characteristics of each one?
- Into which three interconnecting subdivisions is the nervous system organized, and what aspects of control are exercised by each subdivision?
- What are the specialized parts of the brain, and what does each control?
- What are the different types of nerve cells, and what is the function of each?
- What is the unit of function of the nervous system, and how does it work?
- What is a nerve impulse, and how is it transmitted along a nerve cell?
- How is a nerve impulse transmitted from nerve cell to either nerve cell or muscle cell?
- What are brain waves, and how are they used in modern medicine?
- Which sensations are caused by mechanical stimulation, which are caused by chemical stimulation, and which sense organs are involved in each sensation?
- How does the human eye work, and what defects in vision may occur?
- How does the human ear work, and what defects in hearing may occur?
- Which drugs are involved in drug abuse, and what are the characteristic effects of each drug?
- Which are the endocrine glands of the body, and what do their hormones control?
- How is hormone production regulated by the body?
- Which diseases are caused by endocrine gland malfunction, and what can be done to cure the affected individual?

The various activities of the human body, for example, digestion, circulation, respiration, and excretion, are complex and intricately dependent on one another. In addition, the processes responsible for metabolism, growth, and development must be carefully coordinated if an individual is to function successfully.

Over and above the need for continuous regulation of internal activities, the body is constantly faced with challenges from the outside environment that require the coordinated response of various organs and systems. The necessity, at times, to escape from a dangerous situation may call for maximum effort from the circulatory, respiratory, and muscle systems. It may, at the same time, require a severe reduction in the activities of the digestive and excretory systems.

The coordination of all body activities, both of a stressful and normal nature, is controlled by two systems of the body: nervous and endocrine. In general, but with a number of important exceptions, the nervous system coordinates the swift responses of the body to the outside environment, while the endocrine system coordinates the relatively slow processes involved in metabolism, growth, and development.

Nervous system

The unit of structure of the nervous system is the nerve cell, or neuron (Chapter 7). This type of cell consists of three parts: one or more **dendrites,** a nucleus-containing **cell body,** and an **axon.** (See Fig. 7-17.) The dendrite carries impulses to the cell body, and the axon carries them away from it. A nerve impulse travels from neuron to neruon by passing from the axon of one nerve cell to the dendrite of the next; the junction between the axon of one neuron and the dendrite of the next neuron is called a **synapse.**

Three different kinds of neurons are found in the nervous system: sensory neurons, motor neurons, and association neurons (Fig. 12-1). **Sensory neurons** conduct impulses toward the nervous system from sense organs and sensory receptors elsewhere in the body, such as the skin. The dendrites of sensory neurons are connected to these sense organs or receptors, and their axons synapse with the dendrites of other neurons. **Motor neurons** conduct impulses away from the nervous system. The dendrites of motor neurons are synapsed to the axons of other neurons, and motor neuron axons are connected to some effector, such as a muscle or gland. **Association neurons** conduct impulses within the nervous system. Both their dendrites and axons synapse with other neurons.

Dendrites and axons are often referred to as nerve cell fibers. The dendrite of a sensory neuron is very long, while its axon is relatively short. Both dendrites and axons of association neurons are quite short. On the other hand, motor neurons have short dendrites and long axons. The cordlike structures we call nerves are each composed of many nerve cell fibers, that is, the dendrites and axons of many neurons.

■ IMPULSE TRANSMISSION

□ Reflex arc

The function of the nervous system is to coordinate the actions of the individual in response to changes in the environment. Although the neuron is the unit of structure of the nervous system, it is the sequences of neurons (nerve cell pathways) that are the units of function in controlling body actions.

A relatively simple type of nervous system circuit includes a **receptor** (sense organ or cell) that perceives a change in the environment (stimulus) and initiates the nerve impulse, a **sensory neuron** that carries the impulse to the spinal cord, an **association neuron** that transmits the impulse to other neurons, such as a **motor neuron** that carries the impulse to an **effector,** which is a muscle or gland whose action constitutes the body's reaction (response) to the stimulus. This sequence is called a **reflex arc** and represents the minimal amount of nervous system involvement in coordinating an action. The brain is **not** involved in a reflex arc.

An example of an extremely simple reflex arc can be seen in the **knee jerk reflex,** a familiar test made by physicians to check a person's reflex reactions. In this test, the person sits on a table with the legs dangling down, and the physician, using a small

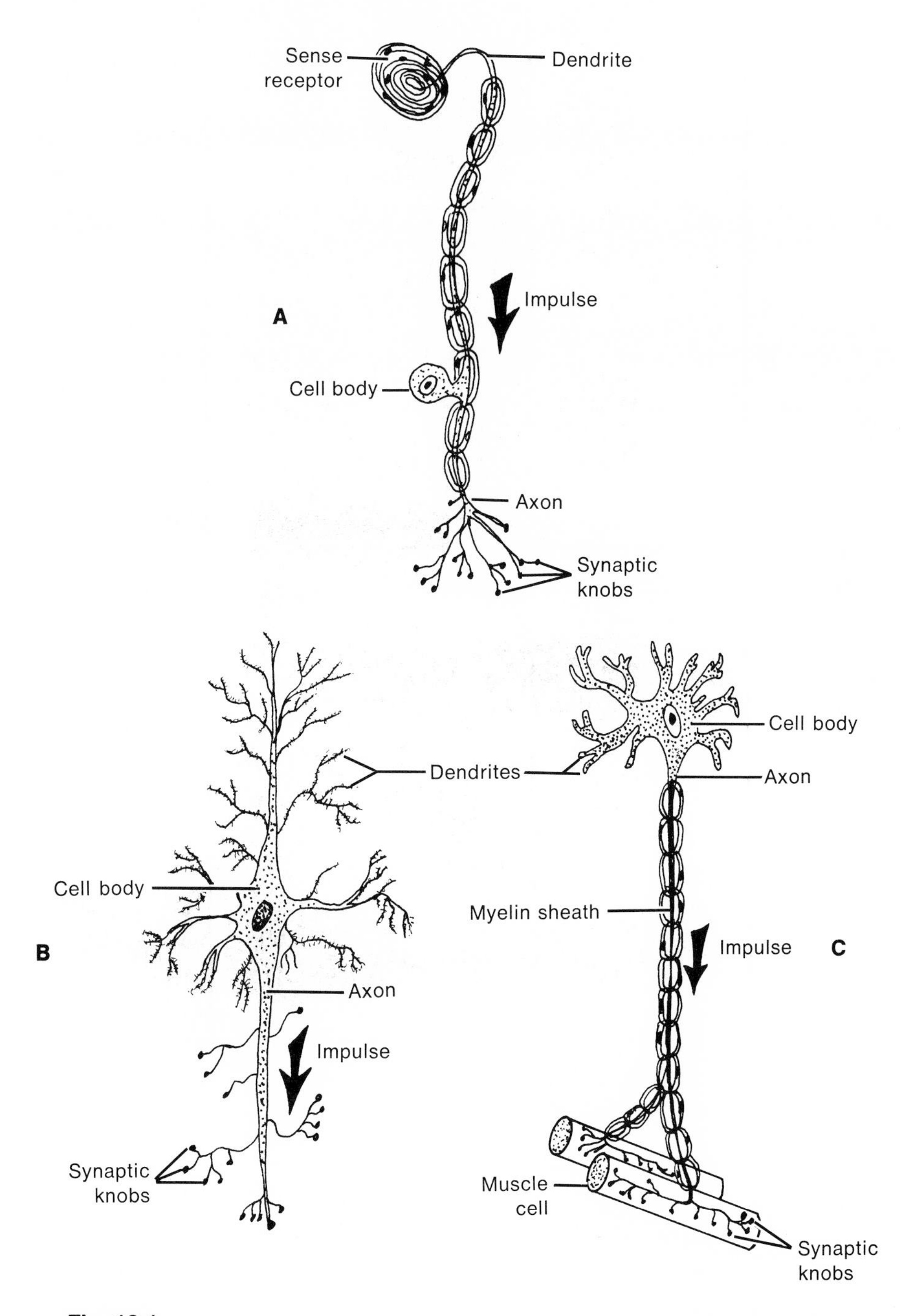

Fig. 12-1
Kinds of neurons. **A**, Sensory neuron. **B**, Association neuron. **C**, Motor neuron.

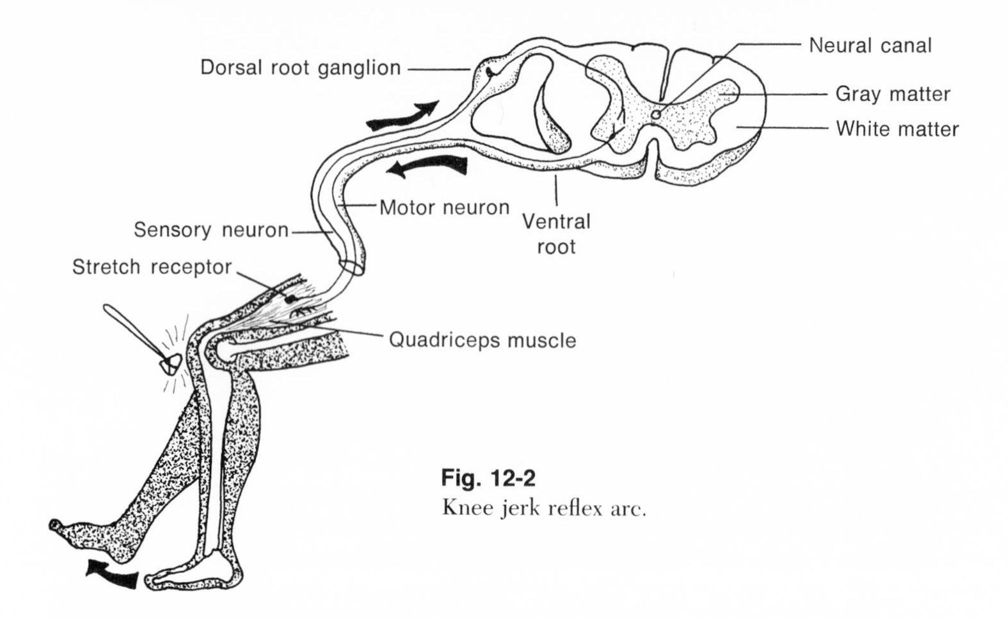

Fig. 12-2
Knee jerk reflex arc.

rubber hammer, taps the tendon just below the knee cap. This gentle striking of the tendon has the effect of stretching the quadriceps muscle (located on the front of the thigh) and results in an impulse being carried by a sensory neuron to the spinal cord. Here a synapse is made directly with a motor neuron, sending an impulse back to the quadriceps muscle, which contracts, raising the lower leg with a "jerking motion." This reflex involves only two neurons, and the receptor and effector are the same body structure, the quadriceps muscle (Fig. 12-2).

Other reflex patterns are more complex and involve one or more association neurons. A few details about the neurons of the reflex arc are worth noting. All the cell bodies of the **sensory neurons** of a given region of the body are located in a special area near the dorsal region of the spinal cord. An aggregation of cell bodies is called a ganglion, and this particular complex is called the **dorsal root ganglion.** It is the location of this ganglion that explains why sensory neurons have long dendrites (which come from sense organs) and short axons (which travel to association neurons). **Association neurons** are completely contained within the spinal cord and, as a result, have dendrites and axons of about equal length. **Motor neurons** have their cell bodies in the spinal cord, but their axons reach out to muscles and glands, which explains why their dendrites are short and their axons long.

□ Nerve impulse

Nerve cells possess the capability of transmitting **impulses.** But what is an impulse, and how is it propagated along a neuron? In order to understand this process, it is necessary to analyze an important characteristic of nerve cells, namely, their ability to selectively admit and exclude various ions so that a difference in electric charge, that is, **electrical potential,** is formed across their membranes.

A neuron has two active transport systems (Chapter 7) that are important in establishing and maintaining its membrane's electrical potential; one is called a **sodium pump** and the other, a **potassium pump.** These "pumps" operate to move

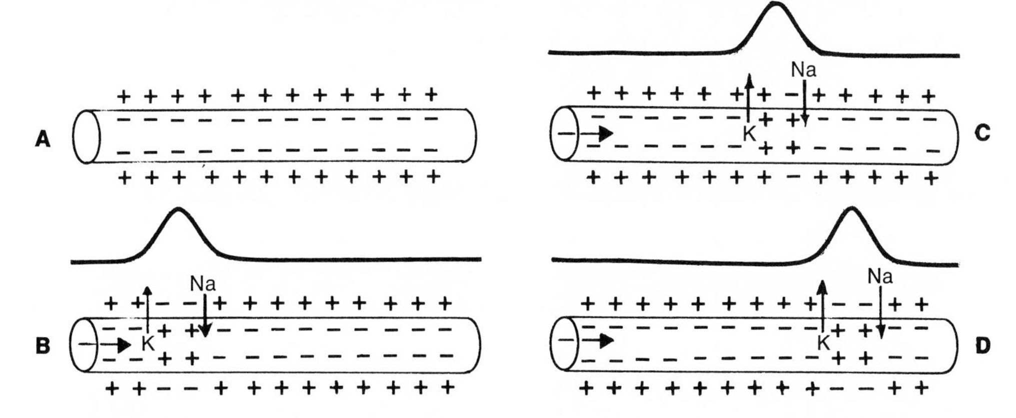

Fig. 12-3
Nerve transmission. **A,** Resting fiber, showing polarization of membrane. **B** to **D,** Passage of impulse along the membrane of the fiber, showing depolarized and repolarized regions. K is the symbol for potassium and Na is the symbol for sodium.

sodium ions to the outside of the cell, while concentrating potassium ions inside the membrane. The action of the sodium pump is aided by the cell membrane, which most of the time is impermeable to sodium, thereby preventing these ions from moving back into the cell. However, the cell membrane is permeable to potassium, some of which diffuses out of the cell and adds its positive charge to that of sodium. A further factor that aids in establishing an electrical potential across the nerve cell membrane is the presence, inside the neuron, of many large negative ions that cannot diffuse through the membrane. As a result of this selective concentration of differently charged ions, the nerve cell membrane at rest is positively (+) charged on its outside and negatively (−) charged on its inside (Fig. 12-3). Such a membrane is said to be **polarized.**

The polarized state of a neuronal membrane is maintained as long as nothing occurs that increases the permeability of any point of the membrane to sodium. Should an increase in the cell's permeability to sodium ions occur, they will enter the cell, causing the disappearance of the normal resting potential at the point of their entry. This phenomenon is called **depolarization** (Fig. 12-3). Among the factors that can increase the permeability of the normal resting nerve cell membrane are mechanical pressure, temperature change, chemicals, and electrical stimulation.

The flow of sodium ions into the cell (depolarization) is quickly followed by the passage of additional potassium ions out of the cell, effectively neutralizing the cell membrane at that point. This change in the membrane's permeability at one point causes the adjoining region to change in a like manner, and this sequence is repeated until the end of the neuron is reached. A **nerve impulse,** therefore, is in reality **a wave of depolarizations** along the length of the nerve cell membrane.

At each point of depolarization, the influx of sodium ions reaches its peak in about 0.001 second (1 millisecond). Then that segment of the membrane loses its permeability to sodium, and the sodium and potassium pumps return the membrane to its previous state. This part of the process is called **repolarization.** Once the entire neuron has been repolarized, it is capable of transmitting another impulse.

The propagation of an impulse along a neuron does not complete our consideration of the nerve impulse. Two points still to be considered are (1) the mechanism by which the impulse is transmitted from neuron to neuron and (2) how it is finally transmitted to an effector (muscle or gland).

□ Nerve impulse transmission across the synapse

The various types of neurons of the nervous system are placed end to end, forming long pathways. The area of contact between the axon of one neuron and the dendrite of another is called the synapse.

At its tip, each axon is swollen to form a **synaptic knob** (Fig. 12-4). Synaptic knobs contain many small sacs, called **synaptic vesicles,** filled with chemical compounds that act as impulse transmitters. When a nerve impulse reaches a knob, the synaptic vesicles empty their contents into the gap (**synaptic cleft**) between the axon and the

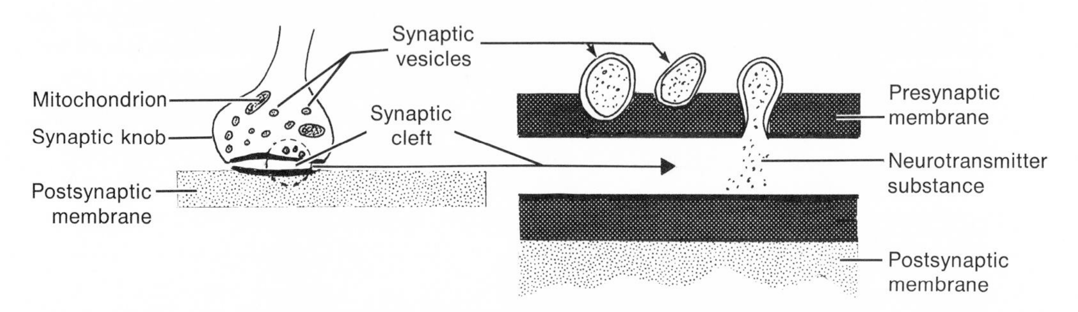

Fig. 12-4
Synapse, showing release of neurotransmitter substance.

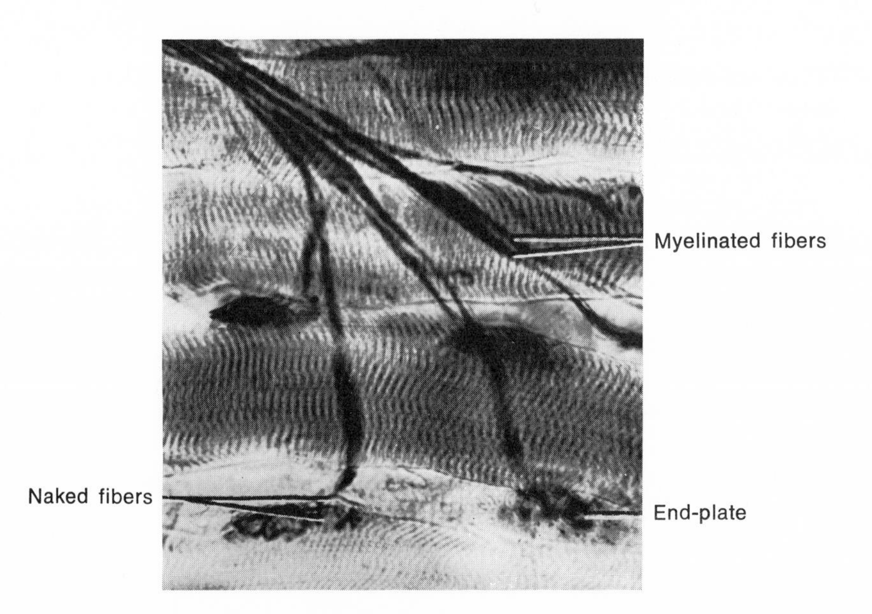

Fig. 12-5
Connection of axon of motor neurons to muscle fibers. (From Bevelander, G., and Ramaley, J. A. 1974. Essentials of histology, ed. 7. The C. V. Mosby Co., St. Louis.)

dendrite of the next neuron in the pathway. The released chemical compound, called a **neurotransmitter,** then depolarizes the dendrite, thereby initiating an impulse in that neuron. Because neurotransmitters are only found in axons, nerve impulses can only travel in one direction through the synapse, causing it to act as a **one-way valve.**

There are a fair number of chemical compounds that have been found to act as neurotransmitters in different parts of the nervous system. The best known include epinephrine (adrenaline), norepinephrine (noradrenaline), dopamine, and serotonin. Shortly after being released into a synaptic cleft, the neurotransmitters are destroyed by enzymes; otherwise, their persistent presence would continually cause the neuron to fire, and muscular movements would be uncontrolled and spasmodic.

A dendrite may have as many as 60,000 synaptic connections with other neurons. It would be chaotic if that many circuits were activated at any one time. A protective arrangement exists to prevent this type of catastrophe. The neurotransmitters of some neurons actually inhibit the depolarization of dendrites. Whether a particular dendrite is depolarized or not depends on the relative amounts of stimulating and inhibiting neurotransmitters acting on it. This has the effect of selectively switching on and off certain circuits of the nervous system as required by a particular situation. Much of the complexity of the brain is the result of the immense variety of circuits that may be formed by switching particular neurons on and off.

□ Neuromuscular junction

Eventually, a nerve impulse reaches a point where the axon (of a motor neuron) connects with an effector (muscle or gland). The connection between an axon and a muscle cell is called a **neuromuscular junction** (Fig. 12-5). When a nerve impulse reaches a neuromuscular junction, the axon involved releases **acetylcholine,** a neurotransmitter that is able to excite the muscle fiber, causing the actin and myosin filaments in the contractile units of the myofibrils to slide past each other (Chapter 10).

■ DIVISIONS OF THE NERVOUS SYSTEM

The nervous system is composed of three parts: the central nervous system, the peripheral nervous system, and the autonomic nervous system. These divisions are connected to one another, both structurally and functionally, but each does have its own distinctive characteristics.

The **central nervous system** consists of the brain and the spinal cord and is primarily the integrating unit of the entire nervous system. The **peripheral nervous system** consists of the nerves that go to and from the brain and spinal cord. Some of these nerves carry impulses to the central nervous system from the sense organs, while others carry impulses from the central nervous system to the skeletal muscles of the body. The **autonomic nervous system** is associated with the glands of the body and the various organs of the digestive system. The autonomic nervous system is involved in many body activities over which we have no conscious control.

There are about 12 billion neurons in the human nervous system. Approximately three fourths of them are in the brain. The remaining 3 billion are found mostly in the spinal cord, with some scattered among various tissues and organs of the body. The almost 12 billion neurons of the central nervous system, however, constitute only 10% of its cells. The other 90% are the **neuroglia cells,** which are very small in size and function in nourishing, supporting, and filling the spaces between the neurons of the brain and spinal cord.

□ Central nervous system

Spinal cord. Our spinal cord is about 46 centimeters (cm) (18 inches) long and extends from the base of the brain down through the vertebral column to the level of the first lumbar vertebra. (See the discussion of the skeleton in Chapter 10.) A cross section of the spinal cord (Fig. 12-2) reveals that it consists of an outer white portion and an inner gray portion. The white section consists of long, myelinated nerve fibers running up and down the cord. (See the discussion of the nerves in Chapter 7.) In contrast, the gray section contains the cell bodies of association and motor neurons.

Running vertically through the gray matter of the spinal cord is the **neural canal,** which connects with the cavity of the brain. The neural canal and the brain cavity are filled with a liquid called **cerebrospinal fluid.**

The spinal cord and the brain are wrapped in three sheets of connective tissues called **meninges.** The meninges can be infected by some bacteria, viruses, or other microorganisms, usually introduced from elsewhere in the body. An acute inflammation of the membrane surrounding the brain or spinal cord or both is called **meningitis.** The most common form of meningitis is caused by the bacterium *Neisseria intracellularis* and can occur in epidemic form. Affected individuals suffer from high fever, severe headache, delirium, and convulsions. A specific diagnosis of the disease is achieved through an examination of a sample of cerebrospinal fluid obtained by a "spinal tap" of the neural canal. A spinal tap also permits the direct introduction of antibiotics into the neural canal to combat the infection (Fig. 12-6).

The spinal cord carries out two main functions in the coordination of body actions. First, it connects the peripheral nervous system to the brain. Thus information

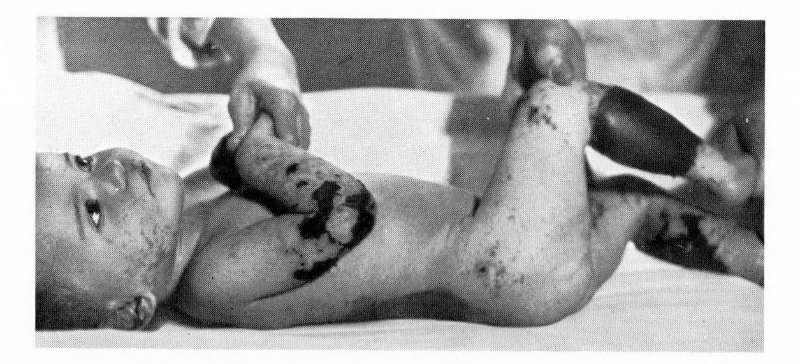

Fig. 12-6
Acute fatal meningococcal infection in a 14-month-old child. Note vacant stare on face of child and large areas of hemorrhage over child's body. One day after the onset of sore throat, small pinpoint hemorrhages (petechiae) appeared in the skin. Three days later, larger areas of skin became hemorrhagic, necrotic, and then gangrenous. Meningitis developed on the twelfth day of illness. (From Top, F. H., Sr., and Wehrle, P. F., eds. 1976. Communicable and infectious diseases, ed. 8. The C. V. Mosby Co., St. Louis.)

reaching the spinal cord through sensory neurons can be transmitted up the cord through association neurons to the brain for interpretation and possible action. Any impulses coming from the brain travel back down the cord through other association neurons and leave the spinal cord to an effector by way of motor neurons. A second function of the spinal cord, as has been discussed already, is to act as the coordinating center for simple reflex actions, such as the knee jerk reflex. The brain does not need to receive or initiate any nerve impulses for this type of action to be carried out successfully.

Brain. The brain is one of the most active metabolic organs of the body. It replaces nearly all of its protein every 3 weeks, it receives 20% of the blood flowing from the heart, and it is the prime user of glucose in the body. (See the discussion of starvation in Chapter 4.)

The brain is the enlarged, anterior end of the spinal cord. Basically, the brain receives nerve impulses from the spinal cord and from nerves, called **cranial nerves,** that lead directly to it from the eyes, ears, nose, etc. These various impulses are interpreted by the brain as conditioned by memory and motivation, and a response by the body is organized and initiated. The brain (Figs. 12-7 and 12-8) is divided into three regions: (1) hindbrain, (2) midbrain, and (3) forebrain.

Hindbrain. The hindbrain consists of two specialized portions: medulla oblongata and cerebellum. The **medulla oblongata** is the most posterior part of the brain, connecting directly to the spinal cord. Although small in size, the medulla is extremely important in controlling breathing, heartbeat, dilation and constriction of blood vessels, swallowing, and vomiting. Destruction of the medulla immediately results in death.

Above the medulla is the **cerebellum,** which consists of two deeply convoluted hemispheres. The cerebellum regulates and coordinates muscle contraction. Injury

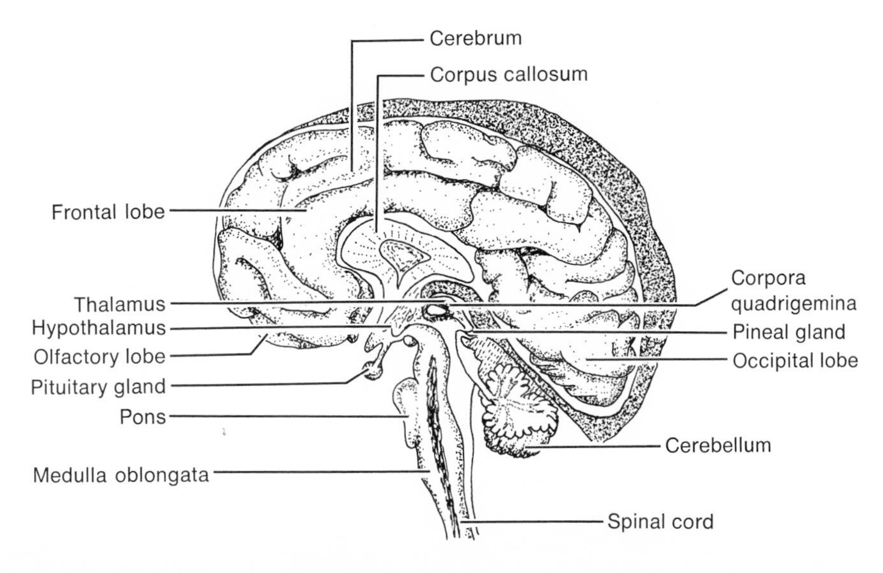

Fig. 12-7
Median section of human brain.

A

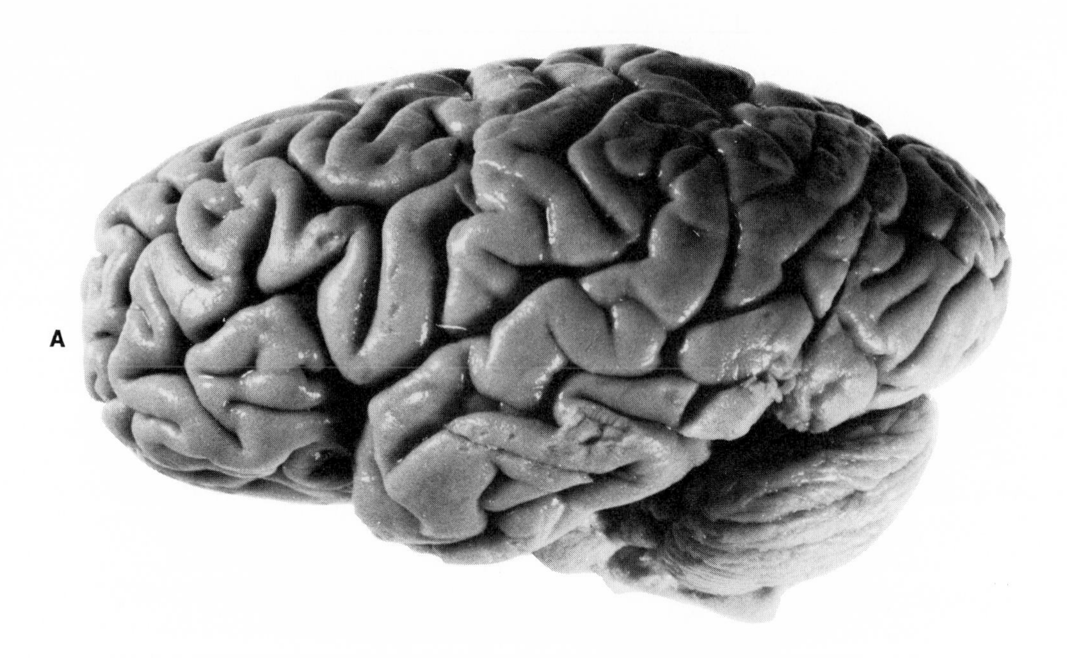

B

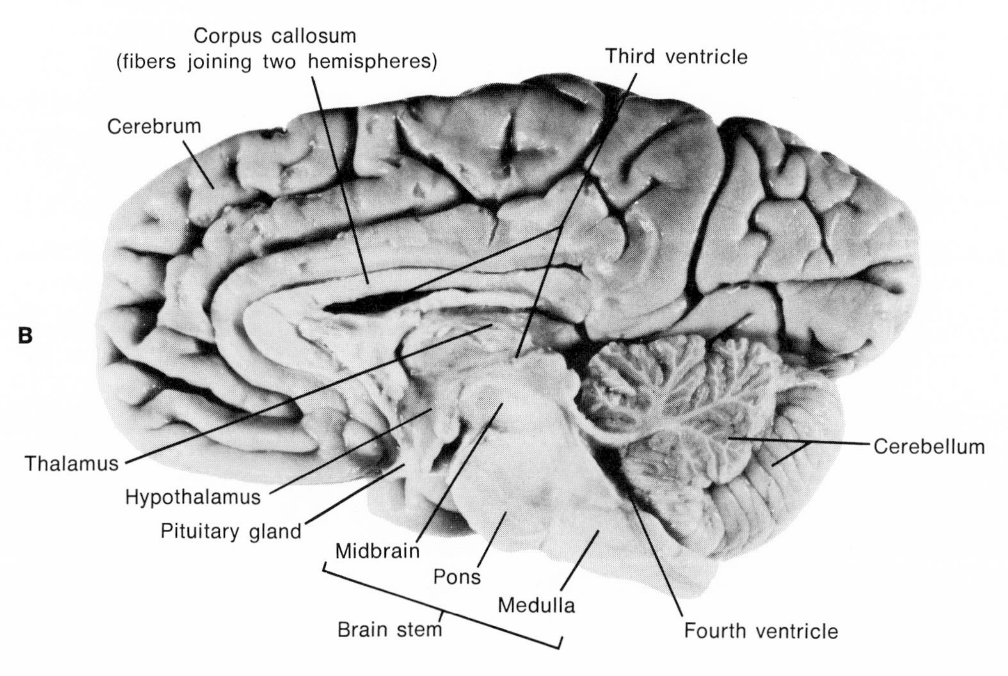

Fig. 12-8
Photograph of human brain. **A,** Whole brain, showing grooved surface of massive neocortex. Length is approximately 17 cm. **B,** Same brain cut along midline to reveal internal structures. (From Lane, T., ed. 1976. Life the individual the species. The C. V. Mosby Co., St. Louis.)

to the cerebellum results in a loss of muscle coordination, and an activity such as threading a needle becomes an impossible task.

On the ventral side of the cerebellum is a thick bundle of fibers called the **pons,** or bridge. These fibers carry impulses from one hemisphere of the cerebellum to the other, thus coordinating muscle movements on both sides of the body.

Midbrain. The midbrain lies in front of the cerebellum. It is quite small and contains four rounded bodies, called the **corpora quadrigemina.** These structures contain the centers for visual and auditory reflexes, such as the reduction in size of the pupil of the eye when light is shined on it and the raising up of a dog's ear in response to a sound.

Forebrain. The forebrain lies in front of the midbrain. Proceeding in an upward and forward direction, we find the thalamus, hypothalamus, cerebral hemispheres, and olfactory lobes. The **thalamus** serves as a relay station for sensory impulses going from "lower" brain areas to the cerebrum. Below the thalamus is the **hypothalamus,** which controls body temperature, sleep, water balance, hunger, thirst, and such emotions as pleasure and pain. Attached to the bottom of the hypothalamus by a stalk is the **pituitary gland,** the "master" endocrine gland, to be discussed later.

Surrounding most of the brain structures discussed thus far is the cerebrum. Most of the dorsal and anterior portion of the brain consists of the two **cerebral hemispheres.** The outer surface, or **cortex,** contains the cell bodies of neurons and nonmyelinated nerve fibers, giving it a gray color. The cerebrum functions in conscious sensations, voluntary movements, learning, memory, and thinking. The various functions have been localized in specific regions of the cerebral cortex, as shown in Fig. 12-9. The destruction of the human cerebrum results in total blindness and extensive paralysis. Although such an individual can for a while carry out vegetative functions such as breathing and swallowing, death follows within a short period of time.

The two cerebral hemispheres are connected by tracts of nerve fibers in the **corpus callosum.** At the extreme anterior end of the brain are the **olfactory lobes,**

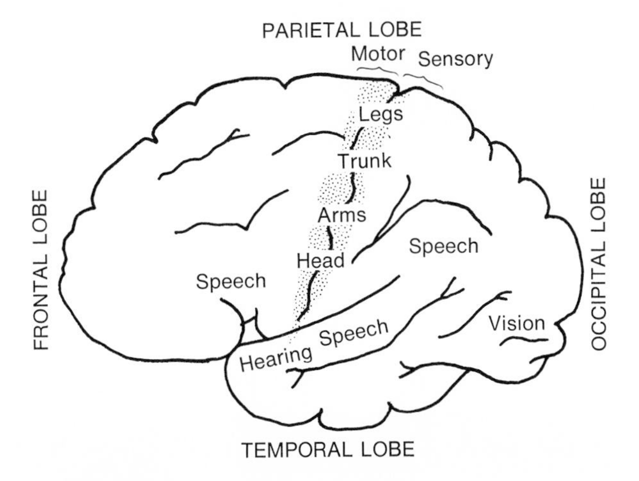

Fig. 12-9
Localization of various functions in the cerebrum.

which control the sense of smell. Extending from these lobes are the olfactory nerves that run forward from the main mass of the brain into smaller chambers of the skull, eventually reaching the nasal passages.

Brain waves. In carrying out their various activities, the brain cells are constantly producing electrical impulses. This electrical activity can be recorded by a machine known as an **electroencephalograph** (EEG). To obtain EEG recordings, electrodes are placed on different parts of the scalp, and the activity of the underlying regions of the cerebral cortex is measured. Examples of brain wave recordings are shown in Fig. 12-10.

When we are at rest, with our eyes closed, the EEG records a regular pattern of activity, called **alpha waves,** which are produced at a synchronized rate of about 10/second. These relatively slow waves also characterize most of our sleeping hours. When we awaken, the impulses in the cerebral cortex become much more rapid and are desynchronized. They are then called **beta waves.**

During that portion of sleep in which dreams occur, the brain waves are of the beta type. In addition, during the dream period, there are characteristic rapid, involuntary movements of the eyeballs. Hence this period of time is known as **rapid eye movement** (REM) sleep.

If the brain is damaged by hemorrhage, tumor, infection, epilepsy, or injury, the EEG will record a distinctive, recognizable wave pattern. The site of injury or tumor formation can be detected by noting the part of the brain showing abnormal waves.

In our modern society, brain waves have become an important indication of whether a person is alive or dead. The absence of detectable brain waves is taken as proof that an individual has died, even if machines are keeping the person breathing, since without brain function, life is impossible. The use of brain waves rather than heartbeat as a criterion for "living versus dead" has great importance when the removal of an organ (kidney, heart, etc.) after the death of an individual and its transplantation to a needy person are contemplated. Such situations arise from time to time, for example, as a result of a fatal accident.

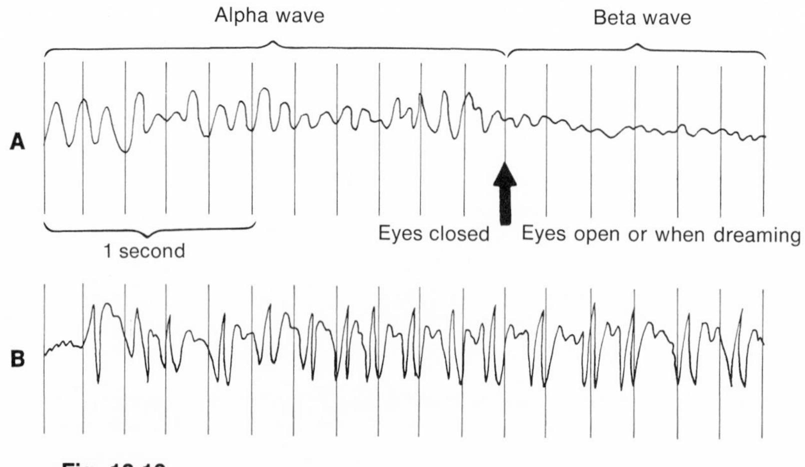

Fig. 12-10
EEG recordings. **A,** Normal rhythms. **B,** Epileptic seizure rhythms.

□ Peripheral nervous system

Every receptor and every effector in the body is connected through a pair of nerves to either the brain or the spinal cord. These paired cranial and spinal nerves make up the **peripheral nervous system.** The nerves are bundles of nerve fibers—axons and dendrites.

There are 12 pairs of **cranial nerves** in man. Three of these nerves are purely sensory in function: (1) the **olfactory,** which originates in the nasal mucosa and conducts the impulses that the brain interprets as a particular odor; (2) the **optic,** which originates in the retina of the eye and is essential for sight; and (3) the **auditory,** which originates in two places, the cochlea of the ear, where it is involved in hearing, and the semicircular canals of the ear, where it is concerned with the senses of movement, balance, and rotation.

All the other cranial nerves are "mixed," that is, they carry both sensory and motor fibers. The mixed cranial nerves are mainly involved in receiving sensations and effecting reactions for the head and shoulder regions of the body. In addition, one of the nerves, the **vagus** nerve, receives impulses from many of the internal organs—lungs, stomach, blood vessels—and in turn sends impulses to the heart, stomach, and small intestine.

There are 31 pairs of **spinal nerves** in man. All the spinal nerves are mixed nerves, having roughly equal amounts of both sensory and motor fibers. Each spinal nerve makes contact with the receptors and effectors of a particular region of the body. Each spinal nerve emerges from the spinal cord as two strands or roots (Fig. 12-11). All the sensory neurons enter the spinal cord through the dorsal root, whereas all the motor neurons leave the cord through the ventral root. The swelling on each dorsal root, the **dorsal root ganglion,** is the location of the cell bodies of the sensory neurons. The cell bodies of the motor neurons are located in the spinal cord itself.

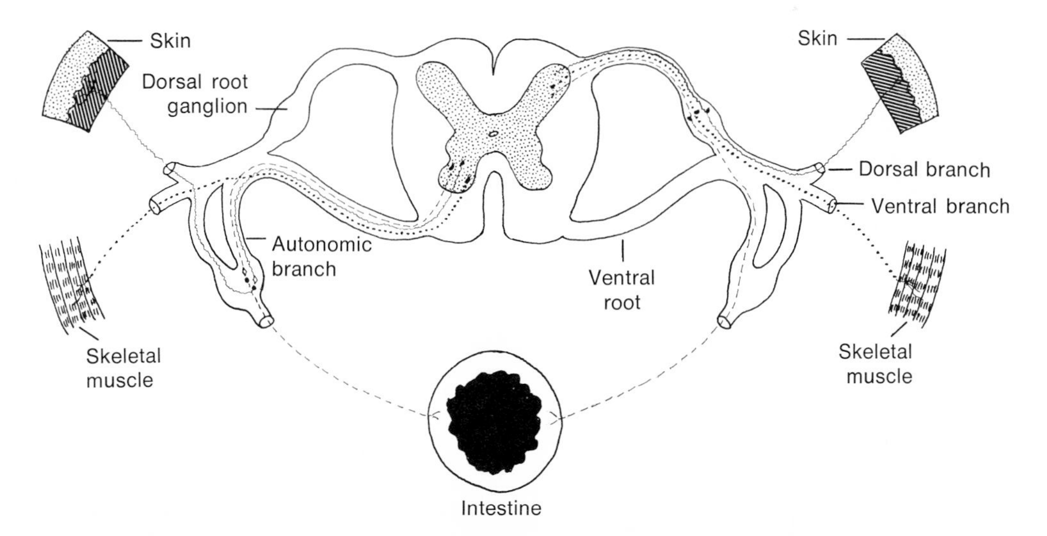

Fig. 12-11
Spinal nerve with associated neurons. Sensory neurons are shown on right and motor neurons on left, although both kinds are present on each side of body.

If the dorsal root of a spinal nerve is severed, the part of the body innervated by that nerve suffers complete loss of sensation, but retains its muscle action. On the other hand, if the ventral root is cut, there is complete loss of muscle action in that part of the body, but the senses of touch, pressure, temperature, and pain are unaffected.

The dorsal and ventral roots of a spinal nerve unite shortly after they leave the cord. Then each spinal nerve divides into three branches: a dorsal branch that serves the skin and muscles of the back of the particular body region; a ventral branch that innervates the skin and muscles of the sides and belly; and an autonomic branch that serves the viscera.

□ Autonomic nervous system

There are a large number of body activities that occur without our being aware of them: heartbeat, breathing movements, peristalsis and segmentation of the alimentary tract, change in size of blood vessels, secretory activities of various glands, changes in size of the pupil of the eye, and many others. These activities are coordinated by reflexes that are controlled by the **autonomic nervous system.** This system is composed of two parts: the sympathetic and parasympathetic divisions.

The autonomic nervous system contains only motor neurons. Each internal organ is serviced by a double set of nerve fibers: one set coming from the sympathetic division and the other coming from the parasympathetic division. Impulses from these two divisions always have antagonistic effects on the organ innervated. If one division speeds up an activity, the other slows it down.

The impulses of the autonomic nervous system reach the effector organs from the brain or spinal cord not through a single motor neuron, as do those to all other parts of the body, but through a sequence of two or more neurons. The cell body of the first motor neuron in the sequence is located in the brain or spinal cord; the cell body of the second neuron is located in a ganglion somewhere outside the central nervous system. The ganglia of the sympathetic division are close to the spinal cord; those of the parasympathetic division are close to or actually within the organs they innervate.

Sympathetic division. In the sympathetic division, the cell bodies of the first motor neuron in each sequence are located in the lateral portions of the gray matter in the thoracic and lumbar regions of the spinal cord. Their axons become the autonomic branch of a spinal nerve that leads to a sympathetic ganglion. These

Table 12-1
Effects of the autonomic nervous system

Organ	Action of sympathetic division	Action of parasympathetic division
Arteries	Constricts	Dilates
Heart	Speeds beat	Slows beat
Iris of eye	Dilates pupil	Constricts pupil
Stomach	Slows peristalsis	Speeds peristalsis
Sweat glands	Increases secretion	Decreases secretion
Urinary bladder	Dilates	Constricts

ganglia are paired, and there is a chain of 18 of them on each side of the spinal cord from the neck to the abdomen. Some of the effects of the sympathetic division are listed in Table 12-1.

Parasympathetic division. The neurons of the parasympathetic division originate either in the brain or in the lower region of the spinal cord. The axons of the anteriorly located neurons leave the brain as parts of various cranial nerves, and those of the posteriorly located neurons emerge from the spinal cord as parts of various spinal nerves. Some of the effects of the parasympathetic division are listed in Table 12-1.

■ SENSES OF THE BODY

Our awareness of the world around us depends on the kinds and amount of information we gather from the environment and the interpretation we give to this information. All nerve impulses are the same as to their basic nature. Therefore the interpetation we give to a nerve impulse becomes a function of our past experience as affected by our memory and the coordinating areas of the brain. The dependency of interpretation of stimuli on the brain is apparent from studies of the effects of hallucinogenic drugs (LSD, marihuana, etc.). Users of these drugs often "see" music as a particular color, and colors may appear to have a "taste," both effects the result of the action of the drug on the brain.

Our ability to obtain information from our environment depends on the types and distribution of receptors throughout our bodies. Some of these receptors are located in the skin and function to detect changes in the outside environment, whereas others are located deep within our bodies and yield information on our internal organs. The actual excitation of a receptor cell appears to be either mechanical or chemical. Examples of sensations caused by **mechanical stimulation** include touch, pressure, hearing, and balance. Examples of sensations caused by **chemical stimulation** include sight, smell, taste, pain, heat, and cold. We shall now consider some of the senses that make us aware of our environment.

□ Cutaneous senses

Our skin contains receptors that are specialized for the sensations of touch, pressure, pain, cold, and heat. Each of these types of receptors is found in separate specialized sense organs that are widely distributed throughout the skin (Fig. 12-12). In the case of touch and pressure, mechanical distortion of the cell membranes of the respective receptors is thought to initiate the nerve impulse, that is, make the nerve cell membrane permeable to sodium ions. The pain receptors are stimulated by the action of chemical substances released from damaged cells, and the heat and cold receptors are affected by chemical changes induced in them by changes in temperature.

□ Taste and smell

The receptors for taste and smell are located, respectively, in the mouth and nasal chambers. The receptors for both taste and smell are affected by chemical stimulation, and the two sensations often interact. Hot foods are tastier than cold foods, because they vaporize and the vapors enter the nasal passages, stimulating smell

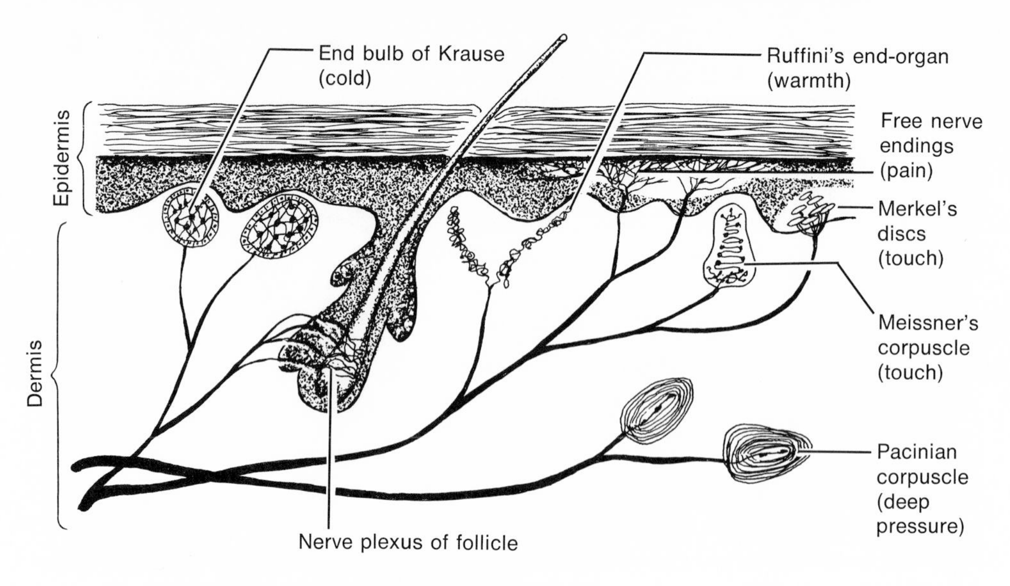

Fig. 12-12
Sense organs of skin.

receptors. If the nasal passages are coated with mucus, as when a person has a cold, it becomes difficult to judge the taste of foods, because the smell receptors are covered and essentially nonfunctional.

Receptor cells for taste are located on the upper surface of the tongue, pharynx, and soft palate. There are four basic taste sensations—sour, salty, bitter, and sweet—and each is caused by the stimulation of a different kind of taste bud. These various types of taste buds are distributed in a characteristic pattern over the upper surface of the tongue, as shown in Fig. 5-3.

Receptor cells for smell are located in two grooves in the upper part of the nasal passages. There has been no universally accepted classification of basic odors. One of the more generally agreed on groupings of primary odors includes camphoraceous, ethereal (like ether), floral, musky, pepperminty, pungent, and putrid. It is believed that all other odors are combinations of these.

□ Vision

Our sense of vision depends on the eye (Fig. 12-13). The adult human eye is globe shaped, measures about 1 inch in diameter, and is composed of three layers. The outermost coat of tissue is called the **sclera.** It is a tough, opaque layer of connective tissue that protects the inner structures and helps maintain the shape of the eyeball. In the front part of the eye, the cells of the sclera become very thin and form the transparent **cornea,** through which light enters.

The middle layer of eye tissue is the **choroid coat.** It is a black-pigmented layer that absorbs those light rays that are internally reflected within the eye. This layer also contains some of the blood vessels that supply the eye with blood. In the front part of the eye, the choroid coat becomes the **iris** and a muscle-containing structure, the **ciliary body.** In the center of the iris is an opening called the **pupil.** The color of the eye is the result of pigments in the iris. The **lens** of the eye is attached to the ciliary body by the **suspensory ligament.**

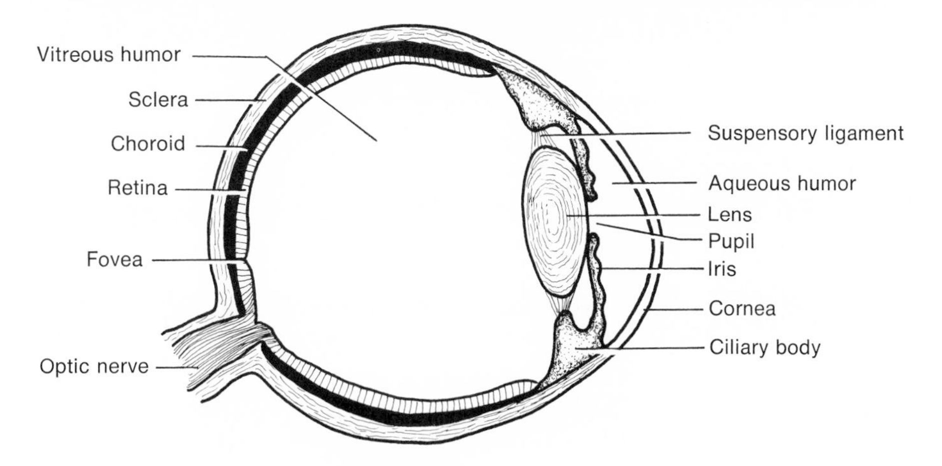

Fig. 12-13
Section through human eye.

The inner layer of the eye is the **retina.** It contains the light receptors, which consist of **rods** and **cones.** Rods function mainly in image detection, whereas cones are involved in both image and color detection. Rods are exceedingly light sensitive and enable us to see in light that is too dim to stimulate cone cells. The retina contains about 125,000,000 rods and 6,500,000 cones. In the center of the retina, directly in line with the center of the cornea and lens, is the region of sharpest vision, the **fovea.** It is a small depressed area containing a heavy concentration of cones.

There are two cavities within the eye: the **anterior cavity,** which is located in front of the lens and is filled with a watery fluid called **aqueous humor**, and the **posterior cavity**, which is located behind the lens and is filled with a jellylike liquid called **vitreous humor.** The liquids filling the cavities of the eye maintain a pressure within the eye, keeping the layers in place and preventing the eye from collapsing.

How does the human eye work? The eye operates very much like a camera, with the exception that changes in the focus of the lens are brought about not by changing the distance between lens and retina, but by changing the **curvature** of the lens. The shape of the lens is controlled by the ciliary body, a muscular ring to which the suspensory ligament is attached. All changes in lens shape are accomplished automatically through the autonomic nervous system whenever a blurred image is focused on the retina. Impulses, which cause the **ciliary body** to contract or expand the lens, are sent more or less at random until a clearly focused image falls on the retina, at which time the adjustments cease.

The iris regulates the amount of light that is allowed to enter the eye and thus acts like a camera diaphragm. There are two sets of muscles in the iris, one set arranged circularly and one set arranged radially. Contraction of the circular muscles constricts the pupil, whereas contraction of the radial muscles dilates the pupil. When a very strong light falls on the retina, an impulse transmitted by the parasympathetic branch of the autonomic nervous system results in the contraction of the circular muscles of the iris. When the light falling on the retina becomes too weak, an impulse transmitted by the sympathetic branch of the autonomic nervous system causes the radial muscles of the iris to contract, enlarging the pupil.

The response of the iris muscles to changes in light intensity is relatively slow, requiring 10 to 30 seconds. As a result of this slow reaction time, when we step from a light to a dark area, some time is needed for our eyes to adapt to the dark. Conversely, when we go from a dark room to a brightly lighted area, our eyes are dazzled by the bright lights until the size of the pupils is reduced.

Defects in vision. Defects in vision are all too frequent. The commonest eye defects are farsightedness, nearsightedness, and astigmatism. Approximately 100 million people in the United States wear glasses to correct for these or other defects in their vision. In the normal eye, the shape of the eyeball is such that the retina is the proper distance from the lens for the light rays to converge on the retina. In a **farsighted eye,** the eyeball is too short and the retina too close to the lens, causing the light rays to reach the retina before they have converged and resulting in a blurred image. This condition can be corrected by the use of **convex lenses** that bend the light rays before they reach the eye and thus enable the eye to converge the light rays on the retina. The amount of correction needed for the particular eye will determine the strength of the lens used in the glasses (Fig. 12-14).

In a **nearsighted eye,** the eyeball is too long and the retina too far from the lens, causing the light rays to converge at a point in front of the retina, also resulting in a blurred image. This condition can be corrected by the use of **concave lenses** that separate the light rays before they reach the eye and thus enable the eye to converge the light rays on the retina. Here, too, the strength of the lens in the glasses will be determined by the amount of correction needed.

Astigmatism is a condition in which the cornea or the lens is curved irregularly, so that light rays in one plane are focused at a different point from those in another

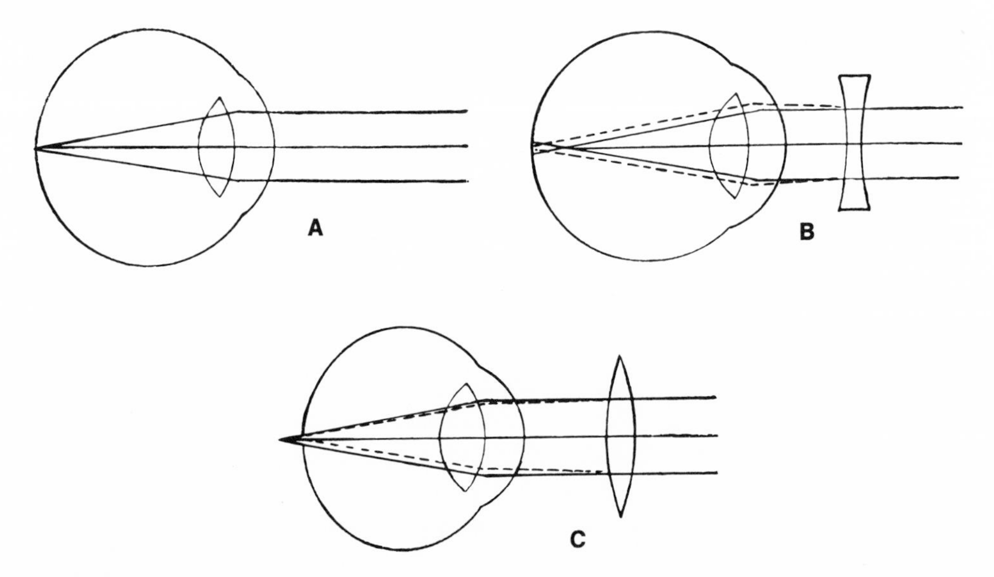

Fig. 12-14
Errors in refraction. **A,** Normal or emmetropic eye is of the proper depth, and the image forms on the retina. **B,** In nearsightedness, or myopia, the eyeball is too long, and a concave lens is needed to make the image fall on the retina. **C,** In farsightedness, or hyperopia, the eyeball is too short, and a convex lens is needed to make the image fall on the retina. (From Bard, P., ed. 1961. Medical physiology, ed. 11. The C. V. Mosby Co., St. Louis.)

plane. To correct for astigmatism, lenses must be ground unequally across their surface to compensate for the unequal curvature of the cornea or lens.

Blindness. Blindness, the complete loss of vision, is the unhappy condition of about 2 million people in the United States, 17% of whom were either born blind or inherited a disease that produces blindness. Most of the others became blind from such disorders as cataracts, glaucoma, retinal degeneration, and corneal disease.

The leading cause of blindness in the United States is the development of **cataracts** in the lens of the eye. About 300,000 people in the United States—most of them over the age of 60 years—develop cataracts each year. Fortunately, most of these cases are treatable and need not lead to blindness.

Cataracts occur when there is a change in the protein fibers of the lens. The lens is composed of cells that form transparent protein fibers. Associated with aging or as a result of infection or injury, the protein fibers become compacted in the center of the lens, which begins to lose its transparency. As the condition progresses, the loss of vision increases, and blindness results.

At the present time, the surgical removal of cataracts is the only method of treatment available. There is no medical treatment that will dissolve the cataract or prevent its development and progression. However, from 90% to 95% of all cataract operations are successful, and normally, with the use of specially prescribed glasses after the operation, full vision is restored.

The second most frequent cause of blindness is **glaucoma.** It is characterized by a loss of vision associated with increased pressure within the eye and damage to the optic nerve. In the normal eye, aqueous humor, which is constantly being produced by the cells of the ciliary body, is continuously removed from the eye through a tiny drainage system and deposited in the bloodstream. When there is an obstruction to this normal outflow, a buildup of pressure within the eye occurs that damages the optic nerve. If left untreated, glaucoma can lead to blindness.

In some cases, surgery can be performed to provide a new outflow channel for the aqueous humor. However, most cases are controlled by the use of specially prescribed eyedrops that reduce the pressure within the eye. Unfortunately, any vision already lost cannot be restored. Glaucoma is also a disease associated with aging; it rarely occurs before the age of 35 years.

The third leading cause of blindness in the United States is **retinal degeneration.** This usually results from an interference with the blood supply to the retina, because of infection, injury, or metabolic disease. The most frequent cause of this type of blindness is a disorder of the blood vessels in the retina stemming from diabetes (diabetic retinopathy). Unfortunately, retinal disorders are the least understood and also the least treatable conditions.

An eye disease that causes about 5% of all cases of blindness is **corneal disease.** It is characterized by the cornea becoming opaque as a result of infections or injuries. In many cases, sight can be restored by replacing the nonfunctional tissue with a healthy cornea. Corneal transplantation is the most successful of all transplant operations, because the cornea normally does not have a blood supply, and there is no opportunity for antibody production and subsequent tissue rejection (Chapter 6). Donor corneas are taken posthumously from people who arrange before death for the donation of their eyes for this purpose.

• • •

The preceding discussions of eye defects, eye diseases, and blindness indicate the need for immediate attention to all eye injuries and infections. It is also very important to have periodic, preferably yearly, eye examinations to discover and treat the early stages of many eye diseases that normally go unnoticed and hence untreated, resulting in unnecessary cases of blindness.

□ Hearing and equilibrium

Hearing and equilibrium (balance) are sensations initiated by mechanical stimulation. The organs of both these senses are located in the ear. The human ear (Fig. 12-15) is divided into three parts: the outer ear, the middle ear, and the inner ear.

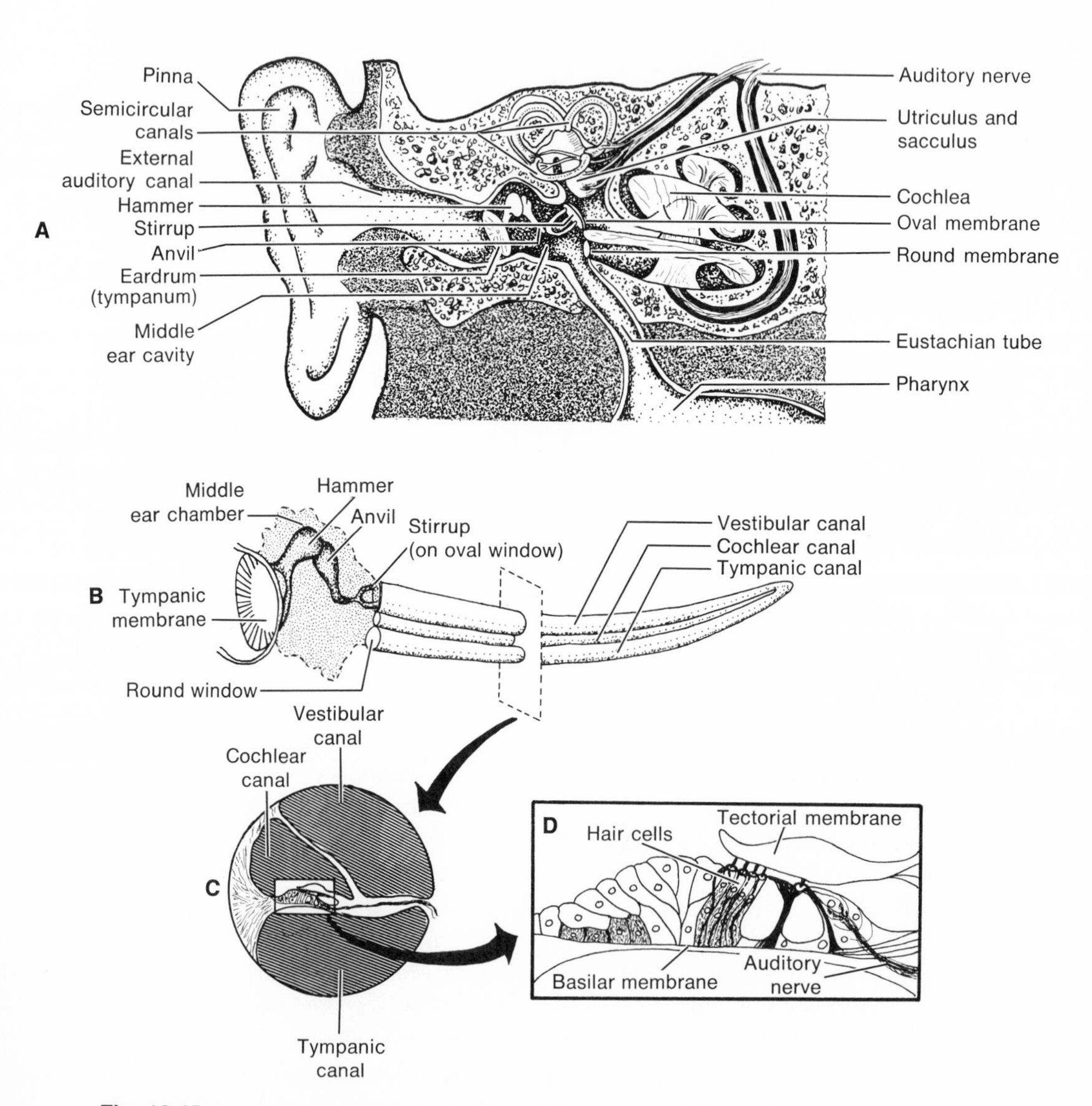

Fig. 12-15
How we hear. **A,** Section through human ear. **B,** Middle ear and cochlea (uncoiled to show its canal system). **C,** Cross section through cochlea. **D,** Organ of Corti.

The **outer ear** consists of the ear flap (pinna) and the auditory canal that funnels sound to the eardrum (tympanic membrane). On the inner side of the eardrum is the chamber of the **middle ear.** This chamber is bounded by the eardrum at one end and by the membranes of both the oval window and the round window at the other. Connecting the eardrum and the membrane of the oval window are three bones: (1) the hammer (malleus), (2) the anvil (incus), and (3) the stirrup (stapes). These bones transmit the vibrations of the eardrum to the inner ear. The chamber of the middle ear is connected to the pharynx by the eustachian tube, which functions as a duct to equalize the air pressure between the outer and the middle ears.

On the inner side of the membranes of the oval and round windows is the **inner ear,** which is a complicated series of chambers and canals. The upper group of chambers (utriculus and sacculus) and canals (semicircular canals) is concerned with the sense of equilibrium. The lower portion of the inner ear consists of a long tube coiled like a snail shell, called the **cochlea.** It contains the hearing organ. Inside the cochlea are three canals: (1) the vestibular canal, which begins at the oval window and connects, at its apex with the tympanic canal; (2) the tympanic canal, which continues from the vestibular canal and ends at the round window; and (3) the cochlear canal, which lies between the other two. All three canals are filled with fluid.

The cochlear canal rests on the tympanic canal and is separated from it by a membrane, called the **basilar membrane.** The cochlear canal contains the actual organ of hearing, the **organ of Corti,** which consists of rows of cells that are located on the basilar membrane. These cells have hairs projecting up from them. Overhanging these hair cells is another membrane, the **tectorial membrane.** Attached to the hair cells of the basilar membrane are fibers of the cranial auditory nerve. All these structures run the entire length of the cochlear canal.

How do we hear? Vibrations in the air (sound waves) pass down the auditory canal of the outer ear and strike the eardrum, causing it to vibrate at the same frequency as the impinging air waves. The chain of small bones in the middle ear (hammer, anvil, stirrup) are so arranged that the vibrations of the eardrum are passed to the membrane of the oval window, causing it, in turn, to vibrate. The vibrations of the oval window membrane produce movements of the fluid in the vestibular and tympanic canals. The movements of the fluid in these canals are at the same frequencies as the sound waves that entered the outer ear and cause the basilar membrane to move up and down, rubbing the hairs of its cells against the tectorial membrane. The mechanical stimulation involved in rubbing the hairs against the tectorial membrane initiates nerve impulses in the dendrites of the sensory neurons lying at the base of each hair cell. These impulses are then carried to the auditory center in the brain for interpretation (Fig. 12-16).

The sense of equilibrium originates in the fluid-filled chambers and canals of the upper portion of the inner ear. There are two types of equilibrium: static and dynamic. **Static equilibrium** refers to the position of the head with respect to gravity. The sense of static equilibrium is a function of two chambers, called the **utriculus** and the **sacculus.** These chambers are lined with sensitive hairs and contain small stones, **otoliths,** made of calcium carbonate. Normally, the pull of gravity causes the otoliths to press against particular hair cells. At the bases of all hair cells are sensory nerve fibers, and the pressure from the otoliths initiates impulses to the brain. When the

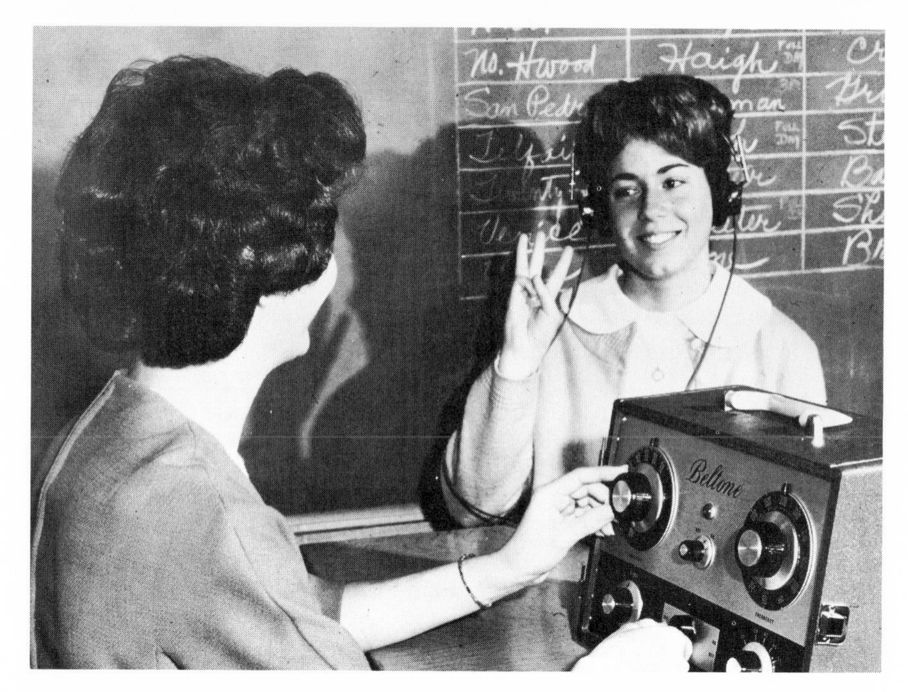

Fig. 12-16
Individual, pure tone audiometry. The audiometrist is presenting a number of "beeps" and asking the pupil how many she hears. (From Turner, C. E. 1971. Personal and community health, ed. 14. The C. V. Mosby Co., St. Louis.)

head is tipped, the otoliths press on the hairs of other cells, initiating impulses that produce the sensation of a "tipped head." **Dynamic equilibrium** refers to the sense of continuing movement and originates in the three **semicircular canals** of the inner ear. These canals are filled with fluid, and each canal is oriented in a different plane in space. At the base of each canal is a small chamber that contains a cluster of sensory hairs, but **no** otoliths. When the head is moved in any direction, either as a result of whole body movement or head rotation, there is a slight lag in the movement of the fluid in the canals as compared to the body itself. This delayed flow of fluids acts as a pressure wave on the hair cells and stimulates them to initiate impulses to the cerebellum of the brain. The cerebellum, by integrating the impulses from each canal, determines the direction and speed of body or head movement.

Noise. Our modern society has introduced a new factor in man's environment: **noise**—loud, disagreeable sound. Unfortunately, noise has come with technology, and many of our very helpful mechanical devices, for example, vacuum cleaners, automobiles, construction machinery, and airplanes, also pollute our environment with unwanted sounds that may actually impair our sense of hearing and result in deafness. It is estimated that more than 16 million people in the United States already suffer from hearing loss caused by noise and that another 40 million are exposed to its potential hazards.

The loudness of sound is measured in units called **decibels,** a unit that compares the power of a given sound with that of a barely audible one. Fig. 12-17 gives the

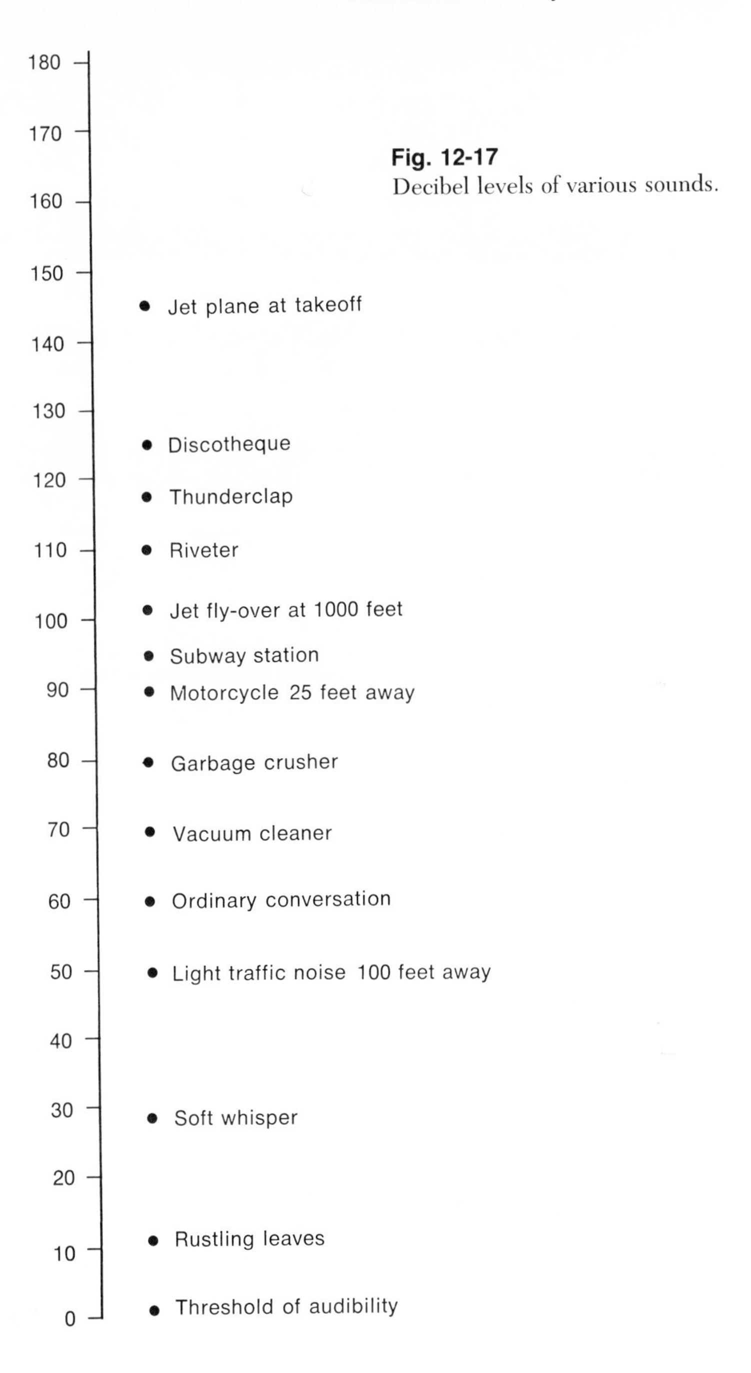

Fig. 12-17
Decibel levels of various sounds.

decibel values for some of the more familiar sounds in our environment. In general those whose decibel levels are between 0 and 50 decibels are considered "quiet" sounds. Decibel levels between 50 and 100 are characteristic of "loud" sounds. Sounds that make a person hearing them feel uncomfortable have levels between 100 and 125 decibels, and decibel levels above 125 characterize sounds that are painful to the listener.

Which decibel level is considered as "noise" varies with the individual. A given sound may be music to one person, but noise to another. A particular individual may feel very uncomfortable when close to a construction site and exposed to the sound of a riveter (110 decibels), but yet feel quite content in a discotheque (125 decibels). Most people consider sounds of greater than 75 decibels to be noise. Continuous exposure to such sounds can result in damaged hearing and even deafness.

Continuous exposure to sounds of high decibel levels almost never damage the outer ear, eardrum, or middle ear, although blasts of great intensity can rupture the eardrum. The primary site of auditory injury from excessive exposure to noise is the organ of Corti in the inner ear. Excessive exposure to noise can result in the destruction of the hair cells of the organ of Corti and also of the sensory neurons attached to these cells.

There are many occupations that expose people to deafening noise: construction work, factory work, airport work, etc. In technologically developed societies, the fact that women have better hearing than men is undoubtedly related to the greater exposure of men to occupational noise. If this deleterious aspect of our modern society is to be corrected, a great deal of attention must be paid to the reduction of noise pollution in our environment.

■ DRUGS

A **drug** is defined as any chemical agent, other than food, that significantly affects body structure or function. Drugs may be used for medicinal purposes (antibiotics, morphine, etc.), or they may be used for social purposes (caffeine in coffee, alcohol in beer, wine, and hard liquor). **Drug abuse** is defined as the use of a chemical agent to the point where it seriously interferes with the health or with the economic or social functioning of the user.

Our modern society has seen a greatly increased use of drugs in recent years. The reasons for their expanded use are many and complex. For some, drug use has a social value in being connected with acceptance into certain groups. For others, it is an attempt to escape from personal problems. For still others, it represents a defiance of authority, both parental and societal. Whatever the individual's reasons for using drugs may be, it is important to understand the effects of the various drugs on the body and the problems that can arise.

All drugs affect the nervous system. Some depress the activities of the system by interfering with its normal functioning, while others stimulate the system by increasing its rate of activity. Every drug can also be characterized as to whether an individual using it becomes dependent on it and as to whether its user builds up a tolerance to it.

Dependency on a drug may take one or two forms: psychological and physical. An individual is **psychologically dependent** on a drug if he has a periodic or continuous desire to be in whatever state the drug induces. The person feels at a loss without the drug. If the drug is permanently withdrawn, the individual experiences emotional trauma in adjusting to life without the drug. However, if the dependency is psychological and not physical, there will be no feeling of sickness following discontinuance of the drug.

An individual is **physically dependent** on a drug if discontinuance of the drug brings on **withdrawal symptoms.** These vary somewhat with the drug and the severity of the dependency, but generally include nervousness, anxiety, sleeplessness, sweating, twitching of muscles, diarrhea, abdominal cramps, and a feeling of desperation that results from an obsessive desire to secure more of the drug ("get another fix"). Withdrawal occurs because some drugs produce physiological changes in the body, making it impossible for the body to function smoothly without the drug. In large part, it is the fear of the withdrawal symptoms that causes an individual to continue using a drug. After going through the withdrawal process, a person will no longer be physically dependent on the particular drug. However, that individual may still be psychologically dependent on it.

Tolerance to a drug means that the body has become accustomed to a certain level of drug dosage and no longer responds to that dose level. The dose of the drug must be increased to achieve the desired effect. Tolerance to a drug can develop regardless of whether the drug induces psychological or physical dependence in its users.

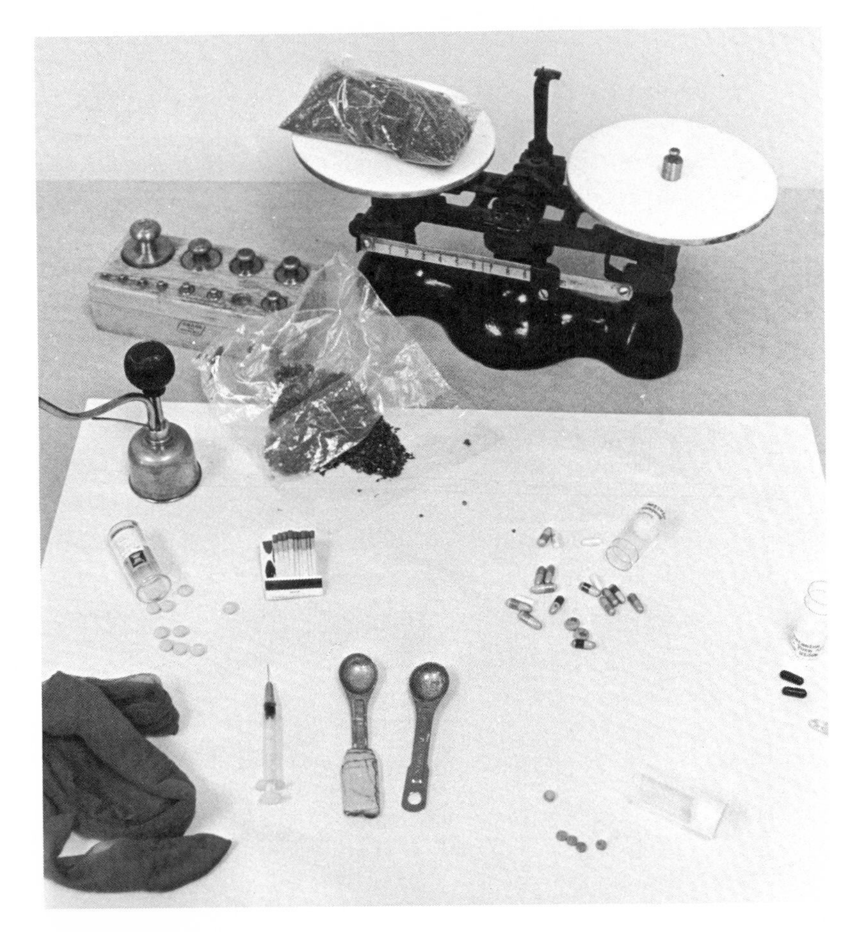

Fig. 12-18
Confiscated drugs. Clockwise, from top, scales for measuring marihuana; drugs available only by prescription; amphetamines; pills containing LSD; heroin paraphenalia, including spoon, hypodermic syringe, and a stocking to be used as a tourniquet; class A narcotics and marihuana pipe. (From Kirk, R., Mayshark, C., and Hornsby, R. P. 1972. Personal health in ecologic perspective. The C. V. Mosby Co., St. Louis.)

Although the effect of a drug is either to depress or stimulate the nervous system, drugs are usually classified into four categories: depressants, narcotics, stimulants, and hallucinogens. As will be discussed, narcotics are a specialized group of depressants, and hallucinogens are a specialized group of stimulants. The classification of drugs is somewhat further complicated by the fact that some drugs produce behavioral effects that are characteristic of more than one of these groups (Fig. 12-18).

□ Depressants

Depressants act to slow down, or anesthetize, the central nervous system. The most commonly used depressants are alcohol and barbiturates.

Alcohol. Without doubt, the most abused drug in the world is alcohol. In the United States, the amount of alcohol consumed averages 2.6 gallons/adult/year. Approximately 95 million Americans drink alcohol in some form, and there are 9 million alcoholics or problem drinkers. The magnitude of this form of drug usage becomes apparent when we compare the number of alcoholics with the number of hard drug addicts in the United States, which is estimated at a quarter of a million.

In half of all murders in the United States, either the killer or the victim, and sometimes both, have been drinking. A fourth of all suicides are found to have significant amounts of alcohol in the bloodstream, and more than half of the automobile accidents involve drivers or pedestrians "under the influence" of alcohol. In most states, a person with a blood concentration of 0.1% or more of alcohol is considered legally intoxicated. In Table 12-2 are listed the blood alcohol levels and behavior for a 155-pound individual who rapidly consumes the indicated quantities of whiskey on an empty stomach. As was discussed in Chapter 5, alcohol is not broken down in the digestive process, but is transported, as alcohol, to all the tissues of the body. In the brain, it depresses the motor centers. When blood alcohol levels are low, their effects are usually mild sedation, relaxation, and tranquility. Slightly higher levels produce behavioral changes that appear to be stimulatory in nature: people become talkative, aggressive, and excessively active. However, these changes actually result from a depression of the brain centers that normally inhibit or restrain such behavior. At still higher levels, an increasing amount of brain function is depressed, producing lack of coordination, confusion, disorientation, unconsciousness,

Table 12-2

Blood alcohol levels and behavior after consumption of various quantities of alcohol

Quantity (ounces of whiskey)*	Blood alcohol level (%)	Resulting behavior
3	0.05	Sedation and tranquility
6	0.10	Talkative and noisy
12	0.20	Obvious intoxication
15	0.30	Unconsciousness
30	0.50+	Death possible

*Three ounces of whiskey equals two "shots"; 1½ ounces of whiskey (one shot) contains the same amount of alcohol as 5½ ounces of wine or two bottles of beer.

and sometimes even death. Alcohol produces both psychological and physical dependency, as well as tolerance, in its users. The withdrawal symptoms are so characteristic that they have been given a special name, **delirium tremens** (also known as "the DT's"). Delirium tremens is a serious and sometimes fatal condition, in which chronic drinkers, often after withdrawal, are confused, trembling, fevered, and sometimes convulsive. They have terrifying hallucinations, usually of cockroaches, rats, or snakes crawling on them. It is estimated that 20% to 25% of the people who suffer delirium tremens when alone and unattended die as a result. Suicide is sometimes the cause of death.

Barbiturates. Barbiturates are the most common of the synthetic drugs classified as **depressants,** or **sedatives.** Sedatives belong to a large group of drugs manufactured for the medical purpose of relaxing the central nervous system. Doctors prescribe sedatives to treat high blood pressure, insomnia, peptic ulcers, and epilepsy.

Taken in normal, medically supervised doses, barbiturates mildly depress the action of nerves, skeletal muscle, and heart muscle. They slow down heart rate and breathing and lower blood pressure. They also produce drowsiness and sleep.

In larger doses, however, instead of producing drowsiness or sleep, barbiturates may produce restlessness, excitement, and even delirium, resembling the excitation of an alcoholic. People intoxicated with barbiturates may be mistaken for "drunks." Their coordination is poor, their speech is slurred, and they become irritable, confused, and unsteady in walking. Judgment and memory are impaired. In extreme cases, aggressive behavior and hallucinations may develop. Those who take an overdose usually go into a coma and may die, because the barbiturate has depressed the brain center that controls breathing.

In slang terminology, barbiturates are often referred to by the color of the tablet: sodium pentobarbital (Nembutal) is called a "yellow jacket"; sodium secobarbital (Seconal) is a "red devil"; and amobarbital (Amytal) is a "blue angel." As a group, the sedatives are referred to as "goofballs," "sleepers," and "downers."

Barbiturates are very dangerous drugs. They produce a strong psychological and physical dependency, as well as a high degree of tolerance, in their users. In the United States, about 3000 people die each year, intentionally (suicide) or unintentionally (accidental poisoning), from barbiturate overdose. A very dangerous situation arises when barbiturates and alcohol are used together. Both alcohol and barbiturates are depressants, and each acts to intensify the effect of the other. The two acting together may cause coma and even death in an individual, despite the fact that the separate levels of each are within the nontoxic range.

Barbiturate withdrawal is very painful and dangerous and should be undertaken only under medical supervision. The withdrawal process takes about 2 days and, in addition to the symptoms discussed earlier, may include hallucinations, delirium, and, finally, convulsions that can be fatal.

□ Narcotics

The term "narcotic" refers to drugs that deaden pain (analgesics) and cause sleepiness. These include opium and those drugs made from opium: morphine, paregoric, and codeine. Opium is the thick dark juice squeezed from the seed pods of the opium poppy, *Papaver somniferum.* Heroin, another narcotic, is a derivative of morphine.

Several synthetic drugs such as meperidine (Demerol) and methadone are also classified as narcotics.

Narcotics are widely used in medicine. Paregoric, for example, is used to counteract protracted cases of diarrhea. Codeine, another opium derivative, is a component of many cough syrups. Narcotic drugs are **depressants.** As such, they relieve tension and are sought for this effect. The narcotic of choice for the drug abuser is heroin.

Heroin. Heroin, referred to by such slang terms as "smack," "scag," "horse," "junk," and "H," is not used in medicine, although its parent drug, morphine, is prescribed to relieve pain. Heroin can be sniffed, but is usually administered by injection either just under the skin ("skin-popping") or more generally directly into a vein ("mainlining").

The heroin "fix" follows a fairly predictable course. Immediately after injection, the individual experiences an intense "high" that lasts about 1 minute. This high is sought after for the sense of tranquility and well-being (euphoria), as well as the feeling of warmth. It is usually followed by several hours of drowsiness, accompanied by an inability to concentrate, lessened physical activity, and lethargy. Most of the time the individual is under the drug's influence, he is "on the nod," in a dreamy state. This general sleepiness and state of unawareness is caused by the fact that narcotics depress various functions of the brain.

Continued use of heroin leads to the development of tolerance, psychological dependency, and physical dependency. The three are interrelated. Psychologically, the addict may initially seek the high, or he may be looking for the escape from his problems that being on the nod provides. However, once tolerance develops, the person will no longer feel these effects with the same dosage. If he cannot obtain enough money to increase his dosage level, he will try desperately to maintain his level of dependency in order to avoid going through withdrawal. The withdrawal symptoms for heroin addiction are quite severe and last for a period of 10 days, the worst part occurring 24 to 48 hours after the last fix.

In addition to the problem of withdrawal symptoms, users of heroin are constantly faced with the possibilities of receiving an overdose of the drug and of infection from unsterile needles and syringes. Death from overdosage results from a reduction in the activities of the portion of the brain that controls breathing. Death from contaminated injections results from liver infection (hepatitis) and blood infections that subsequently settle in the brain or heart valves or spread throughout the body.

Methadone. Methadone is a synthetic narcotic that has been used in the treatment of heroin dependency. A heroin addict can avoid withdrawal symptoms by substituting orally administered methadone for injectable heroin. Methadone can be used in long-term treatment and rehabilitation because, while on full doses of methadone, the addict experiences neither the high nor the lethargy that normally follows the high. As a result, the individual can hope to function normally both at work and in his personal life. However, methadone is still a narcotic and does produce its own level of psychological and physical dependency. Complete freedom from all narcotics still requires that an addict go through a carefully supervised withdrawal process.

☐ Stimulants

As the term itself implies, stimulants are chemicals that increase the level of activity of the nervous system. Some stimulants are quite weak and their effects are relatively mild. These include **nicotine**, found in tobacco, and **caffeine**, found in coffee and tea. It is the caffeine that causes some people to avoid coffee late at night, because it results in their having difficulty falling asleep.

In addition to the just mentioned plant-derived stimulants, there is a group of synthetically produced stimulants, called **amphetamines.** These chemicals are quite strong and have marked effects on the central nervous system.

Amphetamines. Amphetamines increase the level of activity of the nervous system by mimicking the effects of epinephrine and norepinephrine on nerve endings. Amphetamines stimulate those areas of the nervous system that control blood pressure, heartbeat, breathing rate, and metabolic level, all of which are increased. On the other hand, appetite is markedly decreased, while the senses are hyperalert. The body is placed in a general state of stress, as if it were extremely threatened or expecting a violent fight.

Various types of people use amphetamines. Truck drivers take them to stay awake on long trips, athletes take them to excel in contests, and students take them while cramming for examinations. Physicians usually prescribe them for depression and appetite control.

The various amphetamines are referred to by such slang terms as "bennies" (amphetamine sulfate [Benzedrine]), "dexies" or "pep pills" (dextroamphetamine [Dexedrine]), and "speed" (methamphetamine [Methedrine]). These drugs are usually taken in pill or capsule form, except for methamphetamine, which is also found in crystal or liquid form and is often injected.

Administration of a stimulant by injection produces the most immediate and intense effects. The user is hit by a surge of alertness and energy. This stimulating and exciting high is the primary goal sought by those who take stimulants for kicks. Large doses of stimulants result in extreme tension, loud talkativeness, and nervousness. This is known in slang terminology as being "strung out." "Speed freaks" (those who use methamphetamine regularly) often do not sleep or eat for days at a time.

Unlike depressants and narcotics, stimulants do not usually lead to physical dependency or to the development of tolerance to the drug. However, a strong psychological dependency on the drug often occurs. The sense of power, self-confidence, and exhilaration artificially created by amphetamine use is so attractive, and the fatigue and depression that follow discontinuance of the drug is so foreboding, that the user is greatly tempted to continue the habit.

A most unfortunate use of both a stimulant and a depressant has developed in our modern society. It is called "the housewife syndrome" and consists of alternate use of an amphetamine and a barbiturate, in cyclical fashion. In a typical case, a woman takes amphetamines to suppress her appetite in order to lose weight and improve her appearance. The amphetamines overstimulate her and make her jittery at bedtime. She then takes barbiturates to induce sleep. The next morning amphetamines are again swallowed, both to suppress her appetite and to overcome the grogginess caused by the depressant. Such a pattern usually leads to the problems of both amphetamine and barbiturate dependency.

The syndrome just described can develop just as easily in a man who takes amphetamines during the day to help improve his performance at his job and then requires barbiturates to fall asleep at night. However, thus far this type of drug pattern has been seen mostly in women.

□ Hallucinogens

Hallucinogens represent a subcategory of the stimulant type of drug. These drugs are called "mind-affecting drugs," because they produce strong and bizarre mental reactions in people and striking distortions in their physical senses—what and how they see, touch, smell, and hear. The effects of hallucinogens (or psychedelics) range from pleasant images and feelings to terror and dread.

There are a large number of known hallucinogens: lysergic acid diethylamide (LSD), marihuana, peyote, mescaline, psilocybin, and others. LSD and marihuana will be considered in some detail, as their use is quite prevalent in our modern society.

LSD. LSD is synthesized from ergot, a fungus (bacteria-like organism) that grows on rye plants. LSD is a tasteless, odorless, colorless liquid that can be put in sugar cubes, cookies, or crackers. The liquid can also be injected, but is usually taken in tablet or capsule form. LSD is so potent that the amount equivalent to two aspirins will provide about 6500 average doses. Because of the high degree of potency of the drug, a small variation in dosage may produce a large variation in effect.

The physical effects of LSD are quite uniform. It increases pulse and heart rate, causes a rise in blood pressure and body temperature, and produces a dilation of the eye pupils, shaking of the hands and feet, shivering, irregular breathing, and loss of appetite.

The psychological effects of LSD vary widely from person to person, although marked changes in sensation are typical. Flat objects seem to stand out in three dimensions. One sensory impression may be translated or merged into another, for example, music may appear as a color, and colors may seem to have taste. Hallucinations (experiencing nonexistent sights and sounds) can occur, and delusions (false beliefs) are sometimes expressed. Emotional variations are marked, ranging from bliss to horror, sometimes within a single experience.

Users of LSD often refer to "good" and "bad" trips. A good trip involves pleasant sensations and pleasing images, while a bad trip, or "bummer," is the opposite. During a bad trip, the images are terrifying, and the emotional state is one of dread and horror. Unfortunately, there is no way to guarantee a good trip, and a history of having had good trips is no assurance that the next drug experience will not be horrifying.

LSD seems to produce its effects by altering the levels of certain chemicals, like serotonin and norepinephrine, in the brain, thereby producing changes in the brain's electrical activity. It is believed that the brain's normal filtering-out process becomes blocked, causing the brain to become flooded with unselected sights and sounds. Chronic LSD users indicate that they continue to suffer from an overload of stimulation to their senses long after a trip has been completed.

LSD does not produce physical dependence, although it does result in psychological dependency and tolerance. Although there are no physical withdrawal symptoms on discontinuance of the drug, "flashbacks"—sudden recurrences of the drugged state—may persist for months after the drug's use has been discontinued.

Marihuana. Marihuana is a drug found in the leaves and flowering tops of the Indian hemp plant, *Cannabis sativa.* The plant grows wild in many parts of the world, including the United States, and is frequently cultivated commercially for its stem, which contains tough fibers that are used in the manufacture of rope and cloth.

In slang terminology, marihuana is known by such names as "pot," "tea," "grass," "weed," and "Mary Jane." For use as a drug, the leaves and flowering tops of the plant are dried and crushed and then typically rolled into thin homemade cigarettes, called "joints," "reefers," and "sticks." It may also be smoked in small pipes and is occasionally ingested in the form of a drink, cakes, or other food.

When smoked, marihuana quickly enters the bloodstream and acts on the brain and other parts of the nervous system. It affects the user's mood and thinking, but medical science has not yet discovered just how the drug works in the body.

The first effects of marihuana may be a feeling of restlessness accompanied by fear and anxiety. This first period is followed shortly by a feeling of well-being (euphoria). Uncontrolled laughter often occurs, brought on for no apparent reason. The individual's senses of time and space frequently become distorted, making driving extremely dangerous. Minor hallucinations or delusions may occur. After these effects wear off, the user generally falls into a dreamless sleep and awakes without a hangover. Marihuana does not seem to incite people to aggressive or violent behavior.

Although some of the effects of marihuana resemble those of a depressant, it is classed as a mild hallucinogen, because it may cause hallucinations and delusions. Marihuana is the second most popular intoxicant in the world, alcohol being the first. It has been estimated that as many as 8 to 12 million Americans have used marihuana at least once in their lives. Of these, perhaps as many as 1 million are chronic users ("potheads").

Marihuana does not produce physical dependency in the user, nor does the body develop a tolerance to the effects of a particular dosage. Discontinuance of marihuana use does not produce physical sickness. However, as with any drug, a psychological dependency on marihuana can be developed.

□ Who takes what drugs?

At the beginning of this discussion on drugs, it was pointed out that the reasons for taking drugs varied considerably from individual to individual. However, a question remains as to whether different groups of our modern society have a tendency toward the use of certain types of drugs.

Heroin addiction is found primarily in the younger people of the economically and socially deprived areas, such as the Negro and Mexican communities. It would appear that, as a group, these young people are using heroin as a means of temporary escape from their surroundings.

The major drug problem in the youth of higher income families is the use of stimulant drugs, including the hallucinogens. The most widely used of the ordinary stimulants are the **amphetamines.** Among the hallucinogens, the most widely used drugs are **LSD** and **marihuana.** As a group, these individuals find life rather meaningless and boring and are seeking excitement and new adventures.

It is interesting to find that the adults of the different income groups do not use the same drugs as their children. **Alcohol** is the preferred means of escape from reality for adults in economically depressed areas. In higher income groups, **barbiturates,** as well as alcohol, are used to relieve feelings of tension and anxiety. Adults seeking a high, regardless of income level, are usually satisfied with that produced by the moderate consumption of alcohol.

□ Is there a sequence of drug usage?

One of the questions that has arisen concerning the taking of drugs is whether the use of so-called legal drugs (beer, wine, hard liquor) does or does not lead to the use of illicit drugs (marihuana, LSD, amphetamines, heroin). Another question is whether marihuana is or is not a crucial stepping-stone to the use of other illicit drugs.

A 5-year study was made of a group of students, starting at the time they entered high school. For most individuals today, it is during this period of life that drug usage begins and develops to its final form. In this study, it was found that there is a sequence of drug usage and that marihuana is a critical transitional stage for most of those who will later use other illicit drugs.

The sequence of drug usage for most individuals is (1) legal drugs, (2) marihuana, and (3) other illicit drugs. Of those who are using marihuana, 93% have a history of drinking, whereas only 7% have not previously used any legal drugs at all. Of those who are using LSD, amphetamines, or heroin, 85% have a history of involvement with marihuana, 12% have previously used only legal drugs, and 3% have never used any type of drug at all. Although these data show a definite pattern in drug usage, involvement with a particular drug does not invariably lead to the use of others further along in the sequence. Of the original group of high school students, 36% became users of legal drugs. Of those who drank, 27% became users of marihuana, and of this latter group, 26% became involved with other illicit drugs. The continuous reduction in the size of the group of students as we go from one stage of the drug usage sequence to the next means that many individuals stop at a particular stage and remain there. There are also those persons who actually reverse themselves and stop using illicit drugs. However, illicit drug users do not revert directly to nonuse. Initially, they revert to legal drugs, at which stage they may either remain, or they may progress subsequently to complete nonuse of all types of drugs.

Endocrine system

The second coordinating system of the body is the endocrine system. This system utilizes special chemical messengers, called **hormones,** that are carried around the body by the circulatory system and exert their effects at points that may be far removed from the site of hormone production. In general, but with a number of important exceptions, the endocrine system coordinates the relatively slow body processes involved in metabolism, growth, and development.

Although most of the endocrine tissues of the body are found as separate organs, some are part of organs that also have nonendocrine functions. Many of the endocrine glands influence the function of other endocrine glands, resulting in balance and coordination within the endocrine system itself.

THYROID GLAND

The thyroid gland (Fig. 12-19) lies on the trachea, just beneath the larynx. The thyroid gland secretes three hormones: thyroxin, triiodothyronine, and calcitonin. **Thyroxin** and **triiodothyronine** are both derived from the union of two amino acid tyrosine molecules. Both hormones contain iodine, which their precursor molecule, tyrosine, does not; thyroxin has four atoms of iodine, whereas triiodothyronine has three. These hormones regulate general body metabolism. A deficiency of these hormones in children results in **cretinism,** a disease characterized by retarded physical and mental development. In adults, the lack of a proper amount of thyroxin

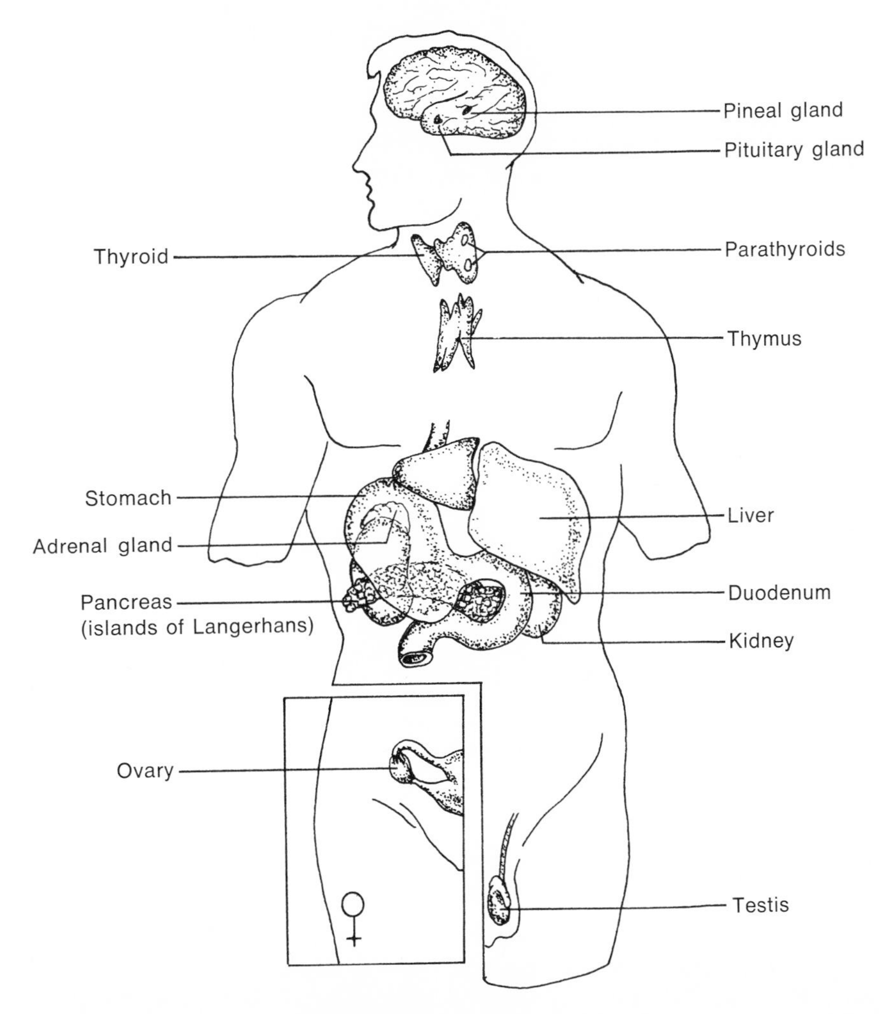

Fig. 12-19
Endocrine glands in man.

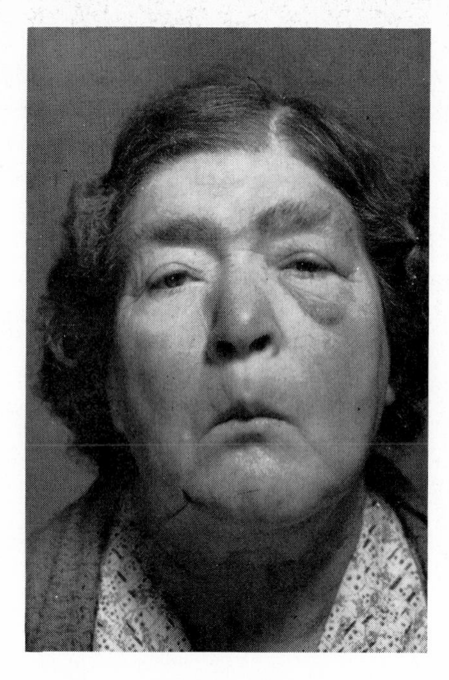

Fig. 12-20
Woman with myxedema. (From Schottelius, B. A., and Schottelius, D. D. 1973. Textbook of physiology, ed. 17. The C. V. Mosby Co., St. Louis.)

and triiodothyronine leads to **myxedema,** which is characterized by a loss of physical and mental vigor, a loss of hair, and a thickening of the skin (Fig. 12-20).

A deficiency of thyroxin and triiodothyronine results from either an insufficient development of the thyroid gland or an insufficient amount of iodine in the diet. In cases where the diet contains an insufficient amount of iodine for the synthesis of these hormones, the gland tends to compensate for the insufficiency by increasing in size. The resulting enlargement is called a **simple goiter.** As discussed in Chapter 4, this type of goiter occurs in areas where the soil lacks iodine, and the situation is best controlled by the introduction of iodized salt into the diets of the people living in such areas. Unfortunately, an increase in dietary iodine has virtually no effect on an already established goiter.

An oversecretion of thyroxin and triiodothyronine results in an increase in body metabolism accompanied by a loss of weight, nervousness, and high body temperature. The most characteristic symptom of an excess of these hormones is the protrusion of the eyeballs, called **exophthalmos,** which gives the individual a wild, staring expression. Overproduction of thyroid hormones results either from the overactivity of a normal-sized gland or from an increase in the size of the gland itself. In the latter case, the swelling of the gland is called an **exophthalmic goiter.** Treatment of hyperactive thyroid conditions can involve the use of drugs that block the synthesis of the thyroid hormones or the surgical removal of a portion of the gland.

Production of thyroxin and triiodothyronine is regulated by the **pituitary gland,** as is true of many other hormones. The pituitary gland secretes a thyroid-stimulating hormone (TSH) that increases the activity of the thyroid gland cells in synthesizing these hormones. This is a **self-regulating system** in that an excess of thyroxin and triiodothyronine in the blood suppresses the production of TSH by the pituitary gland, and, as a result, thyroid gland activity diminishes. When the blood level of these hormones falls, the pituitary gland secretes more TSH, thereby stimulating thyroid gland activity. This is an example of **negative feedback control,** in which an increase in the level of a substance (thyroid hormone) results in a decrease in the synthesis of the hormone (TSH) that regulates the production of the substance.

The third hormone secreted by the thyroid gland, **calcitonin,** is functionally unrelated to the other two. It is one of two hormones in the body (the other, to be discussed, is parathormone, which is produced by the parathyroid glands) that regulate the level of calcium in the blood and tissue fluids. About 99% of the body's calcium is contained in the bones. The remainder is found in the blood and tissue fluids, where it functions in blood clotting, in neuromuscular function, and in fusing cells to one another. The effect of calcitonin activity is to **remove** excess amounts of calcium from the blood and deposit it in the bones of the body. The production of calcitonin by thyroid gland cells is controlled by the level of calcium in the blood. As the blood calcium level increases, so does the production of calcitonin. Here, too, there is a "feedback" mechanism that controls hormone production. However, this is an example of **positive feedback control,** in which an increase in the level of a substance causes an increase in the synthesis of the hormone that controls the level of the substance.

PARATHYROID GLANDS

The parathyroid glands (Fig. 12-19) are four small organs embedded within the thyroid gland (Fig. 12-21). They secrete the hormone **parathormone** that helps regulate calcium metabolism. Parathormone produces an effect opposite to that of calcitonin and results in the removal of calcium from bone and its transfer to the blood and tissue fluids. Since calcium in the bone is bonded to phosphate as calcium phosphate, there is a simultaneous release of both into the blood. This results in the presence of excess phosphate in the blood. Parathormone compensates for this by stimulating the kidneys to excrete the excess phosphate. At the same time, parathormone inhibits the kidney from excreting calcium and thus increases the blood calcium level.

When the blood level of calcium is low because of dietary insufficiency or lack of calciferol, vitamin D (Chapter 4), the parathyroid glands increase their production of parathormone, and, as a result, the calcium level of the blood increases. The reverse effects of parathormone and calcitonin serve to maintain the proper blood calcium level.

It takes very little resorption or deposition of bone calcium to maintain an adequate level of calcium in the body fluids. However, overactivity of the parathyroid glands can cause the removal of too much calcium from the bones, resulting in their becoming fragile and easily fractured. On the other hand, a deficiency of parathormone results in low levels of blood calcium. This causes the nerves and muscles of the body to become very irritable, responding to even minor stimuli and producing muscle spasms. In extreme cases, the individual may die as a result of erratic contractions of the muscles that control breathing.

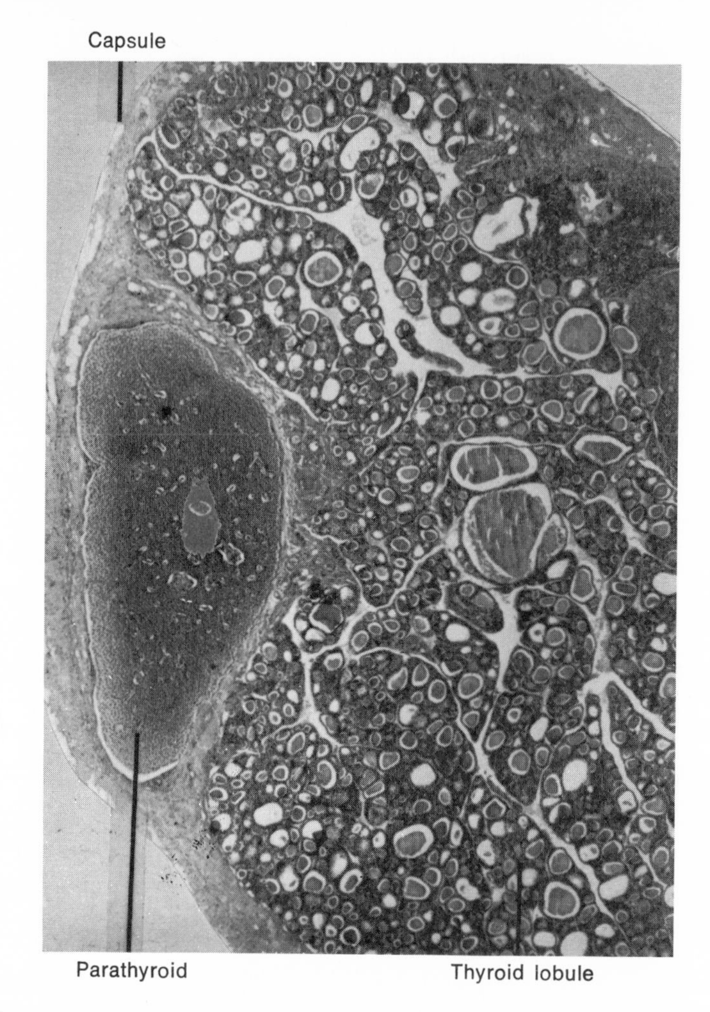

Fig. 12-21
Section of thyroid and parathyroid glands. (×40.) (From Bevelander, G., and Ramaley, J. A. 1974. Essentials of histology, ed. 7. The C. V. Mosby Co., St. Louis.)

■ ADRENAL GLANDS

The adrenal glands (Fig. 12-19) lie just above the kidneys, and each gland consists of an inner **medulla** that is not essential for life and an outer **cortex** that is. The **adrenal medulla** arises from nervous tissue that is, in its origin, part of the sympathetic division of the autonomic nervous system. However, instead of transmitting nerve impulses, the adrenal medulla releases two hormones, **epinephrine** (**adrenaline**) and **norepinephrine** (**noradrenaline**). Both hormones are derived from the amino acid tyrosine, have very similar functions, and produce a great variety of effects on the body.

These hormones cause a rise in blood pressure, an increase in heartbeat, a release of glucose into the blood by the liver, an increase in oxygen consumption, an increase of blood flow in skeletal and heart muscle, a decrease of blood flow in the digestive tract, and a dilation of pupils of the eyes. These effects of the adrenal medulla hormones taken together constitute the reaction of the body to stress caused by physical exertion, pain, fear, anger, or other emotional states. Epinephrine and norepinephrine produce what is called the **emergency response,** or the **fight-or-flight reaction.** These effects are essentially the same as those produced by the sympathetic nerves. The complete removal of the adrenal medulla causes little noticeable change in the individual's behavior.

The **adrenal cortex** is absolutely essential to life. Its destruction leads to death from Addison's disease, which is preceded by a severe fall in the concentration of sodium and chloride in the blood and tissue fluids, a sharp rise in the concentration of potassium in these fluids, and a loss of water from the blood, resulting in diminished blood volume and lowered blood pressure, faulty kidney function, loss of weight, and a general muscular weakness. These symptoms are **not** caused by a deficiency of only a single hormone, but are the combined effects of the many hormones produced by the adrenal cortex.

There are at least 50 hormonelike substances that are produced by this portion of the adrenal gland. All these molecules are similar in chemical structure. They are **steroids,** which means that they are lipids arranged in a pattern of four interlocking rings. The starting molecule for the adrenal hormones is **cholesterol,** a compound implicated in heart diseases (Chapter 3). By a series of chemical reactions involving the addition and deletion of atoms of oxygen, hydrogen, or small molecules, a large variety of steroids is produced. In man and other mammals, only the hormones of the adrenal cortex and those of the reproductive organs are steroids. All the hormones of the other endocrine glands are amino acid derivatives, short polypeptide chains, or proteins.

The adrenal cortex hormones may be grouped into three categories on the basis of their functions: (1) **glucocorticoids,** for example, cortisone and hydrocortisone, which help regulate carbohydrate and protein metabolism, especially the storage of glycogen by the liver and the conversion of amino acids into glucose; (2) **mineralocorticoids,** for example, aldosterone, corticosterone, and deoxycorticosterone, which help control salt and water balance by stimulating the cells of the convoluted tubules of the kidney to resorb sodium, which leads to the simultaneous reabsorption of chloride ions and water; and (3) **sex hormones,** for example, androsterone, testosterone, estrone, and estradiol, which affect primarily the development of secondary sexual characteristics in the male, such as growth of beard, deepening of voice, and maturation of genital organs. Although the adrenal cortex produces both male and female sex hormones, the male hormones predominate. If for any reason, for example, tumor formation, the adrenal cortex of a female increases the secretion of its hormones, the individual will develop masculine characteristics, such as hair on the chest and face and a deeper voice, and will experience a suppression of menstruation.

Control of adrenal cortex hormone production is centered in the pituitary gland, which secretes an **adrenocorticotropic hormone** (ACTH). When adrenal cortex secretions fall below their threshold levels, the pituitary gland secretes ACTH, which stimulates the adrenal cortex to increase its synthesis of hormones. When adrenal hormones rise above their threshold levels, they cause the pituitary gland to reduce its ACTH production.

ISLANDS OF LANGERHANS

The islands of Langerhans are specialized groups of hormone-secreting cells within the pancreas (Fig. 12-22). There are two types of cells in each "island": **alpha cells,** which secrete the hormone glucagon, and **beta cells,** which secrete the hormone insulin. The hormone **glucagon** stimulates the liver cells to convert their stored glycogen to glucose and release it into the blood. Secretion of glucagon is elicited by a drop in the blood glucose level below 60 to 80 mg/100 ml of blood.

The hormone **insulin** acts as an antagonist of glucagon. Insulin promotes the passage of glucose from the blood into the cells of the liver, there to be converted to glycogen for storage. Secretion of insulin is stimulated by a rise in blood glucose level, especially after a meal.

A deficiency of insulin production results in **diabetes.** In diabetics, there is a decrease in the rate at which blood glucose is taken into the cells of the liver and other parts of the body. This results in a lack of energy in the cells and is followed by the use of lipids for energy. The increased breakdown of lipids leads to an accumulation of partially degraded fatty acids that are called **ketone bodies.** Ketone bodies are toxic to the body. They are acidic and decrease the normal alkaline level of body fluids. In severe cases, the accumulation of these acids in the blood causes the individual to lapse into a coma, and, if the condition is not corrected by the administration of insulin, death follows.

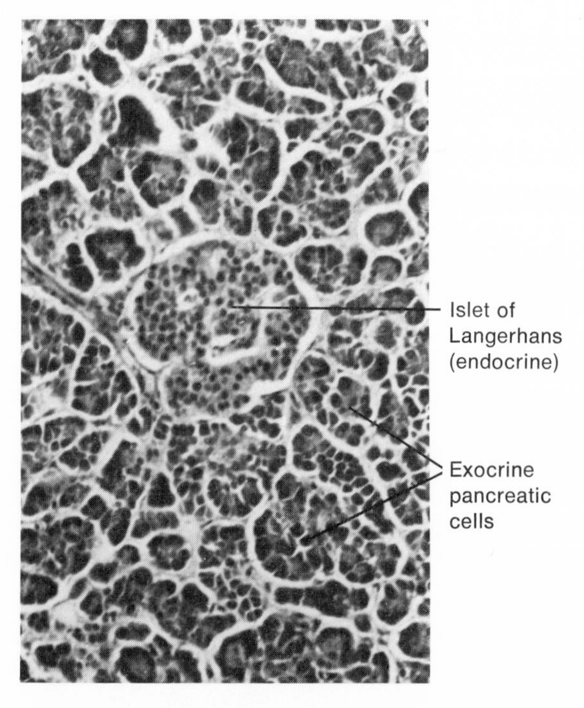

Fig. 12-22
One pancreatic island (islet) within exocrine pancreas. Islet cells produce insulin and glucagon; exocrine cells produce digestive juices. (From Phillips, J. B. 1975. Development of vertebrate anatomy. The C. V. Mosby Co., St. Louis.)

Diabetes is a rapidly increasing health problem. The number of diabetics in the United States is estimated to be 5 million, and the total number of new cases is growing at the rate of 1 million every 3 years. Treatment with insulin, strict diets, and oral antidiabetic agents has increased the life expectancy of diabetics. But despite treatment, diabetes is still the eighth leading cause of death in this country (about 45,000 people/year) and the second leading cause of blindness (mainly from cataracts and retinal degeneration).

■ THYMUS

The thymus (Fig. 12-19) is a two-lobed structure located in the upper part of the chest just behind the breastbone (sternum). The thymus is very large in early life, but becomes reduced in size after puberty. This change is correlated with its lessened function as a person grows older.

In Chapter 6, the role of antibodies in protecting individuals from foreign substances (bacteria, etc.) was discussed. Antibodies are produced by lymphocytes (see Fig. 6-1) that are formed in the lymph glands (spleen, tonsils, etc.) and bones of the body. There are two types of lymphocytes. One type of lymphocyte produces antibodies that react specifically against the antigens associated with bacteria; the lymphocytes producing this type of antibody are called **B cells.** The other type of lymphocyte reacts specifically to the antigens associated with viruses and transplanted cells; these lymphocytes are called **T cells.**

The developmental pathways of B cells and T cells are quite different. From their point of origin, B cells migrate directly to lymph nodes, which are distributed throughout the body. In the lymph nodes, the B cells produce antibodies until they die, usually within a few days after they are formed. T cells, on the other hand, first migrate to the thymus and thence to the lymph nodes of the body. While in the thymus, T cells become sensitized for their function of attacking viruses and foreign cells. How the sensitization process occurs is unknown.

The development of the individual's immunity system occurs shortly before birth. At that time, most of the various B cell and T cell lines of lymphocytes are formed, and it is then that the thymus gland is very large. Once the basic immunity system has become established, the production of new lines of lymphocytes is greatly reduced, and the thymus diminishes in size.

■ PINEAL GLAND

The pineal gland (Fig. 12-19) is a small lobe projecting upward from the rear portion of the forebrain, but completely covered by the cerebral hemispheres. The pineal gland produces the hormone **melatonin,** which appears to function in delaying sexual development. Children whose pineal glands develop tumors experience a delay in their sexual development as a result of increased hormone production by the gland.

It appears that the pineal gland is stimulated to produce melatonin by the light that falls on the retina of the eye. Girls that are blind from birth become sexually mature at an earlier age than their normal counterparts. The pineal gland apparently functions to convert nervous system information obtained through the eye into endocrine system activity.

PITUITARY GLAND

The pituitary gland (Fig. 12-19) has been called the "master gland of the body," because it regulates the activities of many of the other endocrine glands. It is located in a small depression on the inside surface of the floor of the skull, just below the hypothalamus of the brain, to which it is attached by a narrow stalk. The pituitary gland is a double gland consisting of an anterior and a posterior lobe. The anterior lobe develops embryonically, before the skull forms, as an upward outgrowth of the roof of the mouth, whereas the posterior lobe grows down from the floor of the brain. As the embryo develops, the anterior lobe loses its original connection with the mouth, and the skull bones form between this lobe and the mouth, but the posterior lobe retains its connection with the hypothalamus. The two parts of the gland have quite different functions.

The **anterior pituitary gland** secretes at least seven hormones. One of these is a **growth hormone,** called somatotropic hormone (STH), that plays a critical role in the growth process of the individual. If the supply of this hormone is seriously deficient during early development, growth will be stunted, and the child will be a **midget.** An oversupply of this hormone during development causes a child to grow into a **giant.** In both cases, the various parts of the body will retain their normal proportions to each other.

If an oversecretion of the growth hormone occurs during adult life, only the bones of the face, fingers, and toes respond with additional growth. This results in disproportionately enlarged hands, feet, jaw, cheekbones, eyebrow ridges, and nose—a condition called **acromegaly** (Fig. 12-23). The individual suffers from lethargy and severe headaches. If oversecretion is caused by a pituitary tumor, the progress of the disease can be slowed down by surgical removal of the tumor.

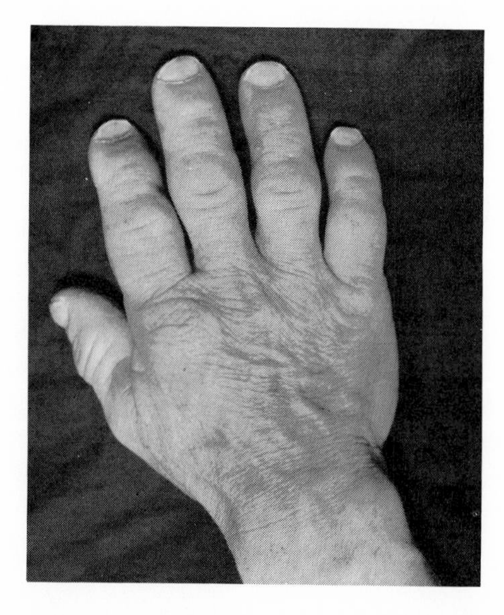

Fig. 12-23
Hand showing characteristics of acromegalic condition. (From Schottelius, B. A., and Schottelius, D. D. 1973. Textbook of physiology, ed. 17. The C. V. Mosby Co., St. Louis.)

A second anterior pituitary secretion is the **melanocyte-stimulating hormone.** It acts on pigment cells, causing the skin to darken (Chapter 10).

A third hormone, **prolactin,** is also called the lactogenic hormone. It stimulates milk production by the female mammary glands shortly after birth of a baby. In the absence of this hormone, milk secretion ceases soon after the birth of a child.

The remaining four hormones of the anterior pituitary regulate the activity of other endocrine glands. Two of these are called **gonadotropic hormones** (a follicle-stimulating hormone [FSH] and a luteinizing hormone [LH]) and act on the gonads. They will be discussed in Chapter 13. The last two secretions of the anterior pituitary are **adrenocorticotropic hormone** (ACTH), which stimulates the adrenal cortex to secrete its various hormones, and **thyrotropic hormone** (TTH), which controls the secretion of thyroxin. The regulation of hormone secretion of the gonads, adrenal cortex, and thyroid glands by the anterior pituitary is under **negative feedback control.**

The **posterior pituitary gland** is both structurally and functionally an outgrowth of the **hypothalamus** of the brain. Originally, it was thought that the posterior pituitary gland produced two hormones—oxytocin and vasopressin. However, it is now known that these two hormones are synthesized by the hypothalamus and are merely stored in the posterior lobe of the pituitary gland. The hormone-producing role of the hypothalamus parallels that of the adrenal medulla, where a portion of the nervous system has developed an endocrine function.

Oxytocin acts mainly on the muscles of the uterus, causing them to contract. This hormone functions most importantly in childbirth. **Vasopressin** causes constriction of the arterioles, resulting in a rise in blood pressure. This hormone also stimulates the kidney tubules to reabsorb water.

Summary

Coordination of body activities is a function of both nervous and endocrine systems. The nervous system is concerned with the rapid transmission of impulses (messages) from one part of the body to another, whereas the endocrine system is generally involved in the relatively slow processes of growth, development, and metabolism.

Neurons are the units of structure of the nervous system. They are arranged in complex circuits that unite receptors with their complementary effectors through various pathways involving the central, peripheral, and autonomic nervous systems. The simplest nervous system circuit, the reflex arc, can involve as little as one sensory neuron and one motor neuron, although most reflex arcs also contain association neurons. The passage of an impulse along a neuron involves a wave of depolarization of the cell membrane that is the result of the movement of sodium ions into and potassium ions out of the cell.

Information about the outside environment is gathered by sensory structures located in the skin, tongue, and nose and by such sense organs as the eyes and ears. This information is sent to the central nervous system by way of the peripheral nervous system. In the brain and spinal cord, the nerve impulses are sorted and relayed in a manner that permits an interpretation of the information received and the initiation of an appropriate response. A similar situation holds for information about our internal organs (heart, lungs, intestines, etc.), except that this is handled by the autonomic nervous system in conjunction with the central and peripheral nervous systems.

The nervous system is sensitive to many chemicals. Some of these, called drugs, are deliberately used by some people to obtain mind-altering experiences. A number of drugs act as depressants, slowing down or anesthetizing the central nervous system. Others are narcotic in their effects, deadening pain and inducing sleep. Still other drugs act as stimulants, increasing the level of activity of the nervous system. Finally, there is the hallucinogenic type of drug that produces strong and bizarre mental reactions. Individuals using drugs may become either psychologically or physically dependent on them. In addition, these peoople may become tolerant of the drug and require ever increasing doses of the drug to achieve the desired effect.

The other coordinating system of the body, the endocrine system, utilizes special chemical messengers, called hormones, to control body activities. Hormone production is regulated by a feedback system that responds either to the blood level of the substance controlled by the hormone or to the blood level of another hormone, for example, pituitary gland hormone.

Two of the thyroid gland hormones, thyroxin and triiodothyronine, control general body metabolism, whereas the third hormone, calcitonin, regulates blood calcium levels. The parathyroid gland hormone parathormone also regulates calcium metabolism, but its effect is opposite to that of calcitonin.

The adrenal medulla secretes two hormones, epinephrine and norepinephrine, that produce the emergency response of the body to a stressful situation. The adrenal cortex produces many hormones. They may be grouped into three categories on the basis of their functions: (1) glucocorticoids, which help regulate carbohydrate and protein metabolism; (2) mineralocorticoids, which help control salt and water balance; and (3) sex hormones, which affect primarily the development of secondary sexual characteristics.

The islands of Langerhans in the pancreas secrete two hormones, glucagon and insulin, that have opposite effects on the conversion of glucose to glycogen by the liver. The thymus gland is important in the maturation of the type of lymphocyte that functions in attacking viruses and foreign cells. The pineal gland produces the hormone melatonin that functions in delaying sexual development.

The anterior lobe of the pituitary gland secretes at least seven hormones: (1) the growth hormone that regulates body size, (2) the melanocyte-stimulating hormone that controls skin pigmentation, (3) prolactin that determines milk production by the female mammary glands, (4) and (5) two gonadotropic hormones (FSH and LH) that affect the development and functioning of the reproductive system, (6) the adrenocorticotropic hormone that regulates the secretion of the adrenal cortex hormones, and (7) the thyrotropic hormones that control the synthesis of thyroxin.

In contrast to the anterior pituitary gland, the posterior lobe does not synthesize the two hormones that it secretes, but acts solely as a storage organ. The two hormones are produced by the hypothalamus and are oxytocin, which causes smooth-muscle contraction, especially that of the uterus, and vasopressin, which causes constriction of arterioles.

The complexities of the nervous and endocrine systems reflect the intricacies of human structure, function, and development, and the necessity for careful coordination of all body activities. In the next chapter, we shall study the process by which we reproduce ourselves and how we develop from fertilized egg to adulthood.

SUGGESTED READINGS

Eccles, J. C. 1965. The synapse. Sci. Am. **212**:56-66. Review of the structure and function of the synapse, including a description of the experimental techniques used in its study.

Guillemin, R., and Burgus, R. 1972. The hormones of the hypothalamus. Sci. Am. **227**:24-33. Interesting discussion of the isolation and subsequent synthesis of two "releasing factors" that control anterior pituitary activity.

Heimer, L. 1971. Pathways in the brain. Sci. Am. **225**:48-60. Description of the technique used to study neural pathways.

Klemm, W. R. 1972. Science, the brain, and our future. The Bobbs-Merrill Co., Inc., Indianapolis. Discussion of nervous system function and the social significance of brain research.

Rasmussen, H., and Pechet, M. M. 1970. Calcitonin. Sci. Am. **223**:42-50. Very readable discussion of the discovery and action of this calcium-regulating hormone.

Ray, O. S. 1972. Drugs, society, and human behavior. The C. V. Mosby Co., St. Louis. Authoritative, rational discussion of drug use and abuse.

United State Environmental Protection Agency. 1971. Effects of noise on people, Publication NTID 300.7. U.S. Government Printing Office, Washington, D.C. Excellent discussion of the physiological, psychological, and sociological effects of the noises to which we are exposed in our daily lives.

Valenstein, E. S. 1973. Brain control. John Wiley & Sons, Inc., New York. Historical review of psychosurgery, including the social and ethical arguments for and against brain control.

Werblin, F. S. 1973. The control of sensitivity in the retina. Sci. Am. **228**:70-79. Lucid explanation of how the retina functions over widely ranging light conditions.

Wilkins, L. 1960. The thyroid gland. Sci. Am. **202**:119-129. Historical review of the discovery of the function of the thyroid gland.

Human fetus 11 weeks after conception; photographed within the intact amniotic sac. (Courtesy Dr. Landrum B. Shettles; from Whaley, L. F. 1974. Understanding inherited disorders. The C. V. Mosby Co., St. Louis.)

CHAPTER 13 Reproduction and development

LEARNING OBJECTIVES

- What are the parts of the male reproductive system, and what is the function of each structure in the reproductive process?
- Which hormones control male sexual development, and how does each function?
- How does the nervous system control male sexual response?
- What medical problems are associated with the reproductive system in older men?
- What are the parts of the female reproductive system, and what is the function of each structure in the reproductive process?
- Which hormones control female sexual development, and how does each function?
- How does the nervous system control female sexual response?
- What are the contraceptive practices in use today, and what are the characteristics of each one?
- What are the venereal diseases, how prevalent are they, and can they be cured?
- What factors affect the ability of a sperm to fertilize an egg?
- What are the stages of human development, and what are the characteristics of each one?
- How can we determine some of the genetic characteristics carried by a developing fetus?
- What are the three stages of the birth process, and what are the characteristics of each?
- What changes must occur in the circulatory system of the newborn in order to complete the transition from fetal to postnatal life?
- What developmental changes are involved in the onset and termination of the reproductive period of life?
- What is aging, and what has been the changing pattern of causes of death in our modern society?

We began our study of biology by pointing out that on both personal and group levels, human beings are involved in achieving **biological success.** For the species, success includes reproduction by members of each generation, and for the individual, success is the extension of life for the longest possible period of time. In this chapter, we shall examine the processes of reproduction and development, and see how man makes the transition from one generation to the next.

SEXUAL REPRODUCTION

Sexual reproduction involves the union of two cells (gametes): sperm and egg. In Chapter 9, we studied meiosis, the process by which genetic material is distributed to the gametes. Now, we want to examine the process by which sperm and egg are brought together to form a new individual.

In man, and most other land animals, the union of sperm and egg (fertilization) occurs within the body of the female. Internal fertilization requires both physiological and behavioral synchronization of the participants and involves extensive hormonal control of the development and functioning of the male and female reproductive systems.

☐ Male reproductive system

The male reproductive system (Fig. 13-1) consists of a pair of gonads (testes), sperm ducts, and accessory glands. The **testes** are formed in the abdomen of the embryo from the same tissue that goes to make up the ovaries in females. In fact, when it is first formed, the gonad is called an **ovotestis,** because at that stage, it can, under the influence of different hormones, become specialized into either an ovary or a testis. However, unlike the ovaries, which remain in the abdominal cavity, the testes descend shortly before or after birth into the **scrotal sac,** a loose pouch of skin that is an extension of the body wall. The cavity within the scrotal sac is part of the abdominal cavity and is connected to it by a passageway, the **inguinal canal.** After the testes have descended, the inguinal canal is usually filled by the growth of connective tissue, thereby separating the abdominal and scrotal cavities.

In some males, the inguinal canal fails to close completely, or even when it does, the canal can be broken open again by excessive strain, as may occur when an individual lifts a heavy weight. Regardless of how it may have occurred, an open canal is called an **inguinal hernia.** The danger associated with an inguinal hernia lies in

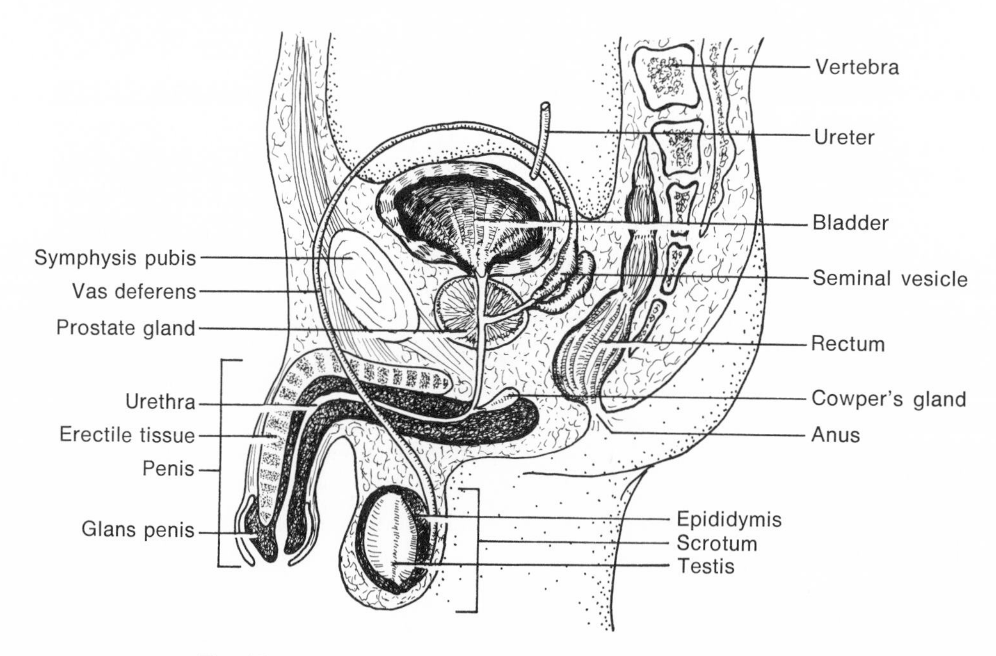

Fig. 13-1
Pelvic region of human male, showing reproductive tract.

the possible slipping of a loop of the intestine into the scrotal sac. Should this occur, the blood supply to that portion of the intestine may be cut off, necessitating an operation to correct the situation before gangrene (death and decay of tissue) sets in.

Each testis (Fig. 13-2) contains two functionally different tissues: **seminiferous tubules,** which produce the sperm cells, and **interstitial cells,** which secrete the male sex hormone testosterone. The temperature in the scrotal sac is about 3° F less than that of the abdominal cavity. For reasons that are not well understood, the cells of the seminiferous tubules will not undergo meiosis and become sperm at normal body temperature. Individuals whose testes have remained in the abdominal cavity are found to be sterile. Modern medicine has perfected a relatively simple surgical procedure by which undescended testes can be placed in the scrotal sac, and if done before the individual is age 5 years, the operation will result in normal adult fertility (Fig. 13-3).

In normally functioning seminiferous tubules, the sperm cells undergo meiosis and then pass from the tubules into a highly coiled tube, called the **epididymis,** that lies in the scrotal sac next to the testis. The sperm are stored in the epididymis until they are released during copulation.

From each epididymis, a duct, the **vas deferens,** passes from the scrotal sac through the inguinal canal into the abdominal cavity. There each duct loops over and behind the urinary bladder, joining the **urethra** immediately past the point where the urethra leaves the bladder. In the human male, therefore, the urethra

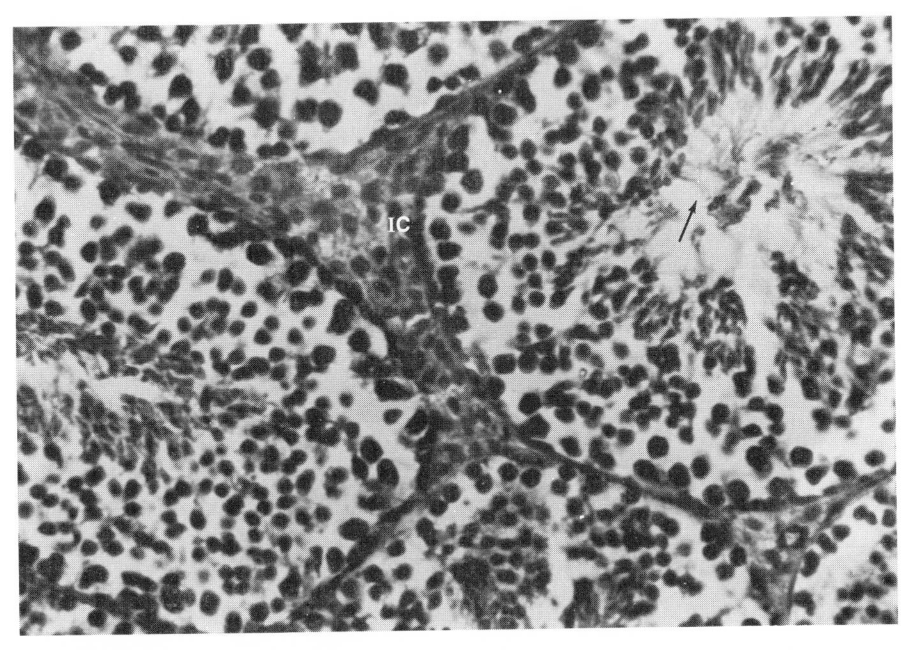

Fig. 13-2
Section through mammalian testis showing interstitial cell nests *(IC)* lying among seminiferous tubules. Sperm flagella are visible in lumen of seminiferous tubules (arrow). (×375.) (Courtesy Victor Eichler, Wichita State University, Wichita, Kan.; from Lane, T., ed. 1976. Life the individual the species. The C. V. Mosby Co., St. Louis.)

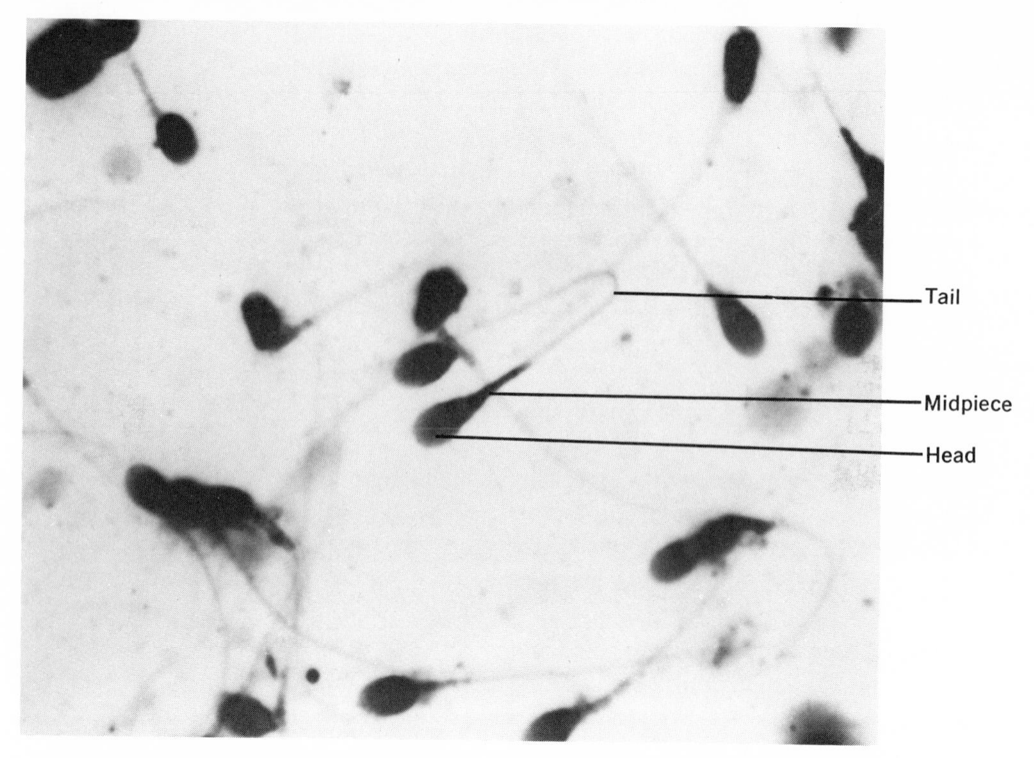

Fig. 13-3
Human sperm as seen in a smear of semen. (×2000.) (Courtesy Dr. Judith Ramaley, University of Nebraska Medical Center, Lincoln, Neb.; from Nagle, J. 1974. Heredity and human affairs. The C. V. Mosby Co., St. Louis.)

becomes a common passageway for both urine from the excretory system and sperm from the reproductive system. Nervous system control over both systems prevents the simultaneous discharge of sperm and urine into the urethra.

As sperm pass through the vasa deferentia and the urethra, various types of fluids, collectively called **seminal fluid,** are added to the sperm, forming a mixture called **semen.** The seminal fluid is secreted by three sets of accessory glands: **seminal vesicles,** which are joined to the vasa deferentia at a point just before these ducts join the urethra; **prostate gland,** which empties its contents into the urethra at the point where the urethra fuses with the vasa deferentia; and the **Cowper's glands,** which join the urethra at the base of the penis. The seminal fluid serves several functions: a transporting medium for the sperm, a lubricant of the passageways through which the sperm must travel, an alkaline solution to neutralize the acidity of the female genital tract, and a source of nutrients, especially sugar, for the highly active sperm.

The urethra carries the sperm from the vasa deferentia to the outside of the body. The last part of the urethra runs through the **penis,** the copulatory organ, which consists of a body, or **shaft,** that terminates in an expanded rim, the **glans.** A skinfold, the **foreskin,** normally passes forward over the glans unless removed by circumcision. The bulk of the penis consists of three cylinders of erectile tissue that run the length of the organ. During sexual excitement, a great deal of blood is pumped into the spaces of the erectile tissue, causing the penis to increase greatly in size and become hard. This prepares the penis for insertion into the female vagina during sexual intercourse.

Among the various parts of the male reproductive system, the organ that causes the most medical problems is the prostate gland. The prostate gland is relatively small during childhood, but begins to grow at puberty. It reaches its normal adult size by the age of 20 years and remains that way to the age of 40 or 50 years. At that time, in some men, it begins to decrease in size along with a diminished production of testosterone by the interstitial cells of the testes. However, in many older men, a benign growth of the prostate gland may occur. The enlarged prostate gland may cause urinary obstruction and require surgery to relieve the situation. This type of growth is not affected by the amount of testosterone in the body.

Unfortunately, the cells of the prostate gland can become cancerous and cause the death of the affected individual. This type of cancer is responsible for approximately 2% to 3% of all male deaths. Cancerous prostate cells are stimulated to more rapid growth by testosterone, but their growth can be slowed by the removal of the testes, thereby eliminating further formation of the hormone. In addition, the growth of prostatic cancer can be inhibited by the administration of the female hormone estrogen.

Hormonal control of male sexual development. In the discussion of the pituitary gland in Chapter 12, it was pointed out that the anterior pituitary gland secretes, among others, two hormones, collectively called **gonadotropic hormones,** that act on the gonads. These hormones affect both the male and female reproductive systems.

Before puberty, the testes produce no sperm and very little testosterone. At puberty, the anterior pituitary gland begins to produce **follicle-stimulating hormone** (FSH), which stimulates the seminiferous tubules to produce sperm, and **luteinizing hormone** (LH), which stimulates the interstitial cells to produce testosterone.

Once the testes begin producing testosterone in appreciable quantities, there occurs a complex series of events that we commonly associate with puberty. There is considerable growth of the penis, scrotal sacs, and testes. These constitute the effects of the hormone on primary sexual characteristics. The effects of testosterone on secondary sexual characteristics include a masculine distribution of body hair, an enlargement of the larynx with an accompanying deepening of the voice, a thickening of the skin, an increase in melanin production by the skin cells, an increased activity of the oil glands of the body, and a general increase in protein metabolism. As a result of this increase in protein metabolism, males usually have larger muscles and longer bones than females. In this situation, testosterone enhances the effect of the anterior pituitary growth hormone on the body's protein production.

The level of testosterone in the body is regulated by negative feedback control. If the testosterone level in the bloodstream drops, the anterior pituitary gland produces more FSH and LH. These hormones then cause the testes to produce more testosterone, which, in turn, leads to a reduction in the formation of gonadotropic hormones (FSH and LH). Although production of sperm and testosterone may oscillate somewhat, their overall levels in the bloodstream of the mature male remain relatively stable.

Male sexual response. The male sexual response begins with the erection of the penis. The flow of blood into the erectile tissue of the penis is caused by impulses from parasympathetic nerves of the autonomic nervous system. This can occur as a result of direct stimulation of the penis, a sexually stimulating sight, or erotic thoughts. Although thought processes usually play an important part in initiating the male sexual response, the cerebrum is probably not necessary for its completion. Appropriate genital stimulation can cause ejaculation in a human male whose spinal cord has been cut above the lumbar region and whose brain cannot therefore receive impulses from the genital region.

During sexual stimulation, **parasympathetic** impulses, in addition to promoting erection, cause the Cowper's glands to secrete mucus. This mucus passes through the urethra to the outside, thereby removing any urinary residues from the urethral canal and providing an alkaline medium for sperm passage.

Ejaculation is the culmination of male sexual response. When the sexual stimulus becomes extremely intense, impulses from the **sympathetic** division of the autonomic nervous system pass to the genital organs. Initially, there are peristaltic contractions of the ducts in the testis, epididymis, and vas deferens. These contractions push the sperm into the urethral canal. Simultaneously, rhythmic contractions of the seminal vesicles and prostate gland force their secretions into the vas deferens and the urethra, thereby mixing these seminal fluids with the sperm and the mucus of the Cowper's glands and forming the semen. Semen formation is followed immediately by ejaculation, which occurs as a result of contractions of skeletal muscles that surround the base of the erectile tissue of the penis. It is the contraction of these skeletal muscles with its accompanying stream of impulses to the spinal cord and brain that constitutes the sensation or orgasm that occurs during ejaculation.

☐ Female reproductive system

The female reproductive system (Fig. 13-4) consists of a pair of gonads (ovaries), a pair of oviducts (fallopian tubes), a uterus (womb), and a vagina. The **ovaries** are located in the lower part of the abdominal cavity and are held in place by ligaments. Like the testes, the ovaries have two functions: gamete production (egg cells) and hormone secretion (estrogen and progesterone).

The outer surface of the ovary consists of a specialized layer of cells, the **germinal epithelium,** from which the eggs develop. At birth, a female's ovary contains all the eggs (ova) that will be formed during her lifetime, estimated to be about 200,000 in each ovary. This is a far greater number than will be needed during the childbearing years. Normally, a woman produces one egg every 28 days, or approximately 13/year. This goes on for about 30 years, resulting in only about 400 eggs ever reaching maturity and leaving the ovary. The rest of the eggs degenerate and are absorbed by the other cells of the ovary. The ovaries alternate in releasing their eggs, but the alternation is irregular and unpredictable (Fig. 13-5).

In contrast to the male reproductive system, the ovary of the female is not connected to the series of tubes that form the reproductive tract. When an egg is discharged from an ovary (ovulated), it is released into the abdominal cavity. But lying close by is one of the **fallopian tubes (oviducts)** whose funnel-like opening partially surrounds the ovary. Normally, the egg enters the fallopian tube and is propelled to the uterus by the action of ciliated cells that line the inner surface of the oviduct.

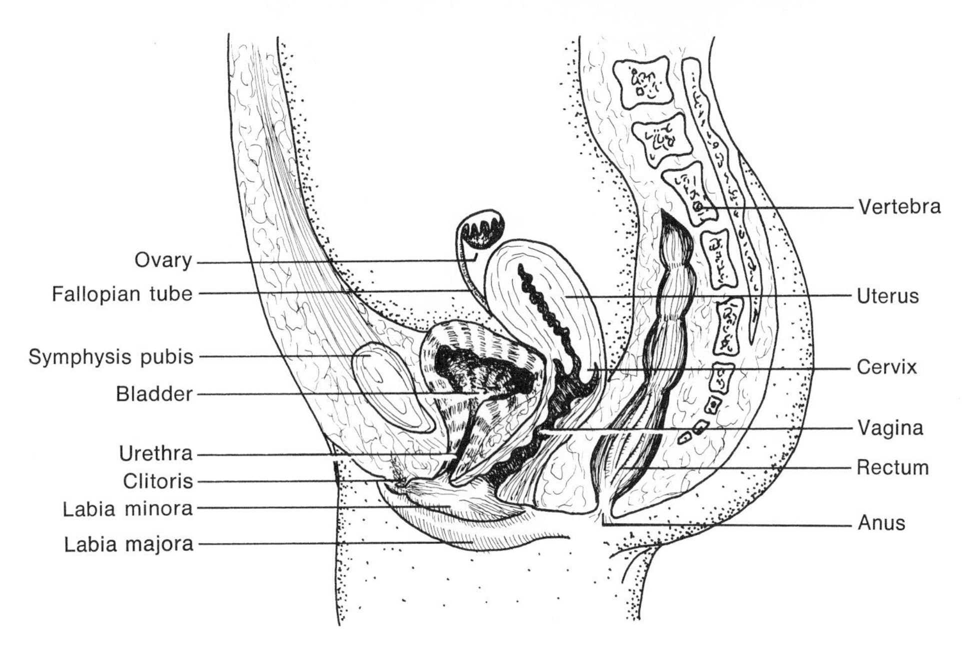

Fig. 13-4
Pelvic region of human female, showing reproductive tract.

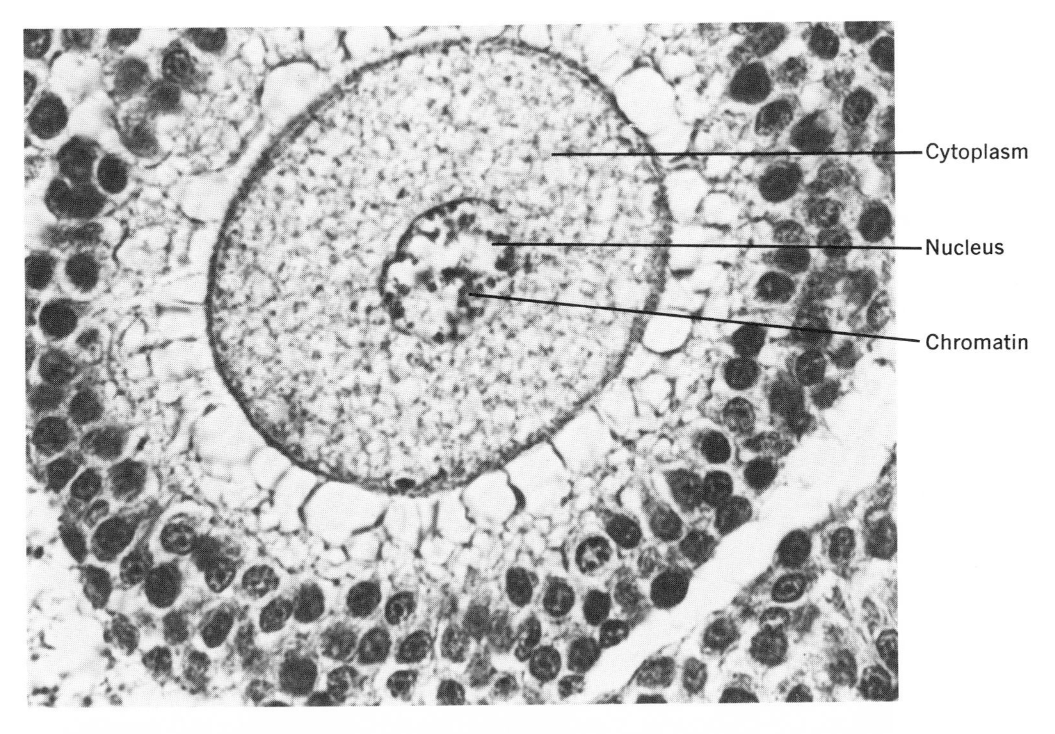

Fig. 13-5
Potential human egg (secondary oocyte) in the follicle of the ovary prior to ovulation. (×2000.) (Courtesy Dr. Judith Ramaley, University of Nebraska Medical Center, Lincoln, Neb.; from Nagle, J. 1974. Heredity and human affairs. The C. V. Mosby Co., St. Louis.)

Fertilization normally occurs during passage of the egg through the fallopian tube.

The two fallopian tubes connect to the upper corners of the uterus (womb). It is the uterus that houses the developing embryo until the time of birth. This organ is located in the middle of the lower part of the abdominal cavity, between the urinary bladder and the rectum. The womb has thick muscular walls, and its inner lining is richly supplied with blood vessels. At its lower end, the uterus terminates in a muscular ring, the **cervix.** Connected to the uterus is a muscular tube, the **vagina,** that leads to the outside of the body. The vagina acts as the receptacle for the penis during sexual intercourse and, together with the uterus, forms the birth canal.

In a young female, the opening of the vagina is partly closed by a thin membrane, the **hymen,** which has a central opening for the flow of the products of menstruation. Later in life, as sexual intercourse occurs, the hymen becomes reduced to shreds of tissue that are sloughed off from the inner margin of the vagina.

Unlike the structural arrangement found in the male, the genital tract of the female is completely separate from the urinary duct. The urethra of the female carries only excretory materials and has its own opening to the outside above the vagina. Urethral and vaginal openings are located close to one another in a region between the thighs called the **vestibule.**

Surrounding the vestibule are two sets of skin folds: an outer pair of skin folds, the **labia majora,** which are covered with hair and fused into a single layer of tissue below the vaginal opening, and an inner, smaller pair of skin folds, the **labia minora,** which are free of hair and are fused into a single layer of tissue above the urethral opening. At the point of fusion of the labia minora is a sensitive, erectile organ, about the size and shape of a small pea, the **clitoris.** This organ is **homologous** to the penis of the male, that is, it develops from the same embryonic tissue as does the penis. The clitoris is composed of spongy tissue and, like the penis, becomes engorged with blood during sexual excitement. However, unlike the penis, the clitoris does **not** contain any portion of the urethra.

The homology between male and female reproductive systems is seen in the similarities in development of testis and ovary as well as penis and clitoris. Additional examples of homology were found in the discovery that the embryonic tissue that forms the labia majora in females forms the scrotal sacs of males and the tissue that forms the labia minora in females forms that portion of the urethra that is included in the penis of males.

Hormonal control of female sexual development. The functioning of the female reproductive system, like that of the male, is controlled by the gonadotropic hormones of the anterior pituitary gland. At puberty, usually between the ages 9 and 17 years (average 13 years), there is a rise in the secretion of FSH and LH. These hormones cause the ovaries to increase in size and to begin secreting the female sex hormones estrogen and progesterone. The sex hormones, particularly estrogen, stimulate the development of the female secondary sexual characteristics. These include development of breasts, increase in size of the uterus and vagina, broadening of the hip bones, and beginning menstrual cycles. The first menstrual discharge of blood (**menarche**) from the vagina marks the beginning of a female's reproductive life. When menstruation ceases (**menopause**), at about age 45 years, a woman's reproductive life is at an end.

Unlike sperm production, which is a continuous, day-by-day process, the production of eggs is cyclical and intermittent. This cyclical pattern applies not only to egg production (oogenesis), but also to the functioning of virtually the entire female reproductive system and the hormones that control it.

The process of oogenesis begins with an increased secretion of FSH by the anterior pituitary gland. FSH stimulates a number of the potential egg cells in the **germinal epithelium** of each ovary to increase in size. As these cells become larger, a layer of other cells accumulates around each one. The aggregate of egg and surrounding cells is called a **graafian follicle** (Fig. 13-6).

Development of the graafian follicle is characterized by an increase in size of the egg cell and a proliferation of the surrounding cells. As the follicle grows, the cells surrounding the egg secrete a fluid that accumulates and forms a liquid-filled chamber around the cells.

At this time, the anterior pituitary gland dramatically increases its flow of LH. This hormone acts to enhance the effect of FSH and results in a more rapid increase of follicle cells and a greater secretion of follicular fluid. Under the stimulating influence of FSH and LH, the cells of the graafian follicle begin to secrete the ovarian sex hormone **estrogen.** This hormone stimulates the cells lining the inside of the uterus to grow and divide, thereby increasing its thickness for possible use by a developing embryo.

As the follicle grows in size, its diameter increases to about 12 millimeters (mm) (½ inch), and it protrudes above the surface of the ovary. The growth and development of the graafian follicle takes about 9 days. By this time, the egg cell has undergone its first meiotic division, producing a large cell and a small polar body (Chapter 9). During each menstrual cycle, about 20 to 30 graafian follicles are formed, but usually only one completes its development. On occasion, however, two or more complete their development, leading to the possibility of multiple births, although, as we shall see later, this is not the only way that multiple births can occur.

The increasing pressure of the growing graafian follicle finally causes the adjacent ovary wall to burst, and the follicular fluid escapes, carrying with it the egg

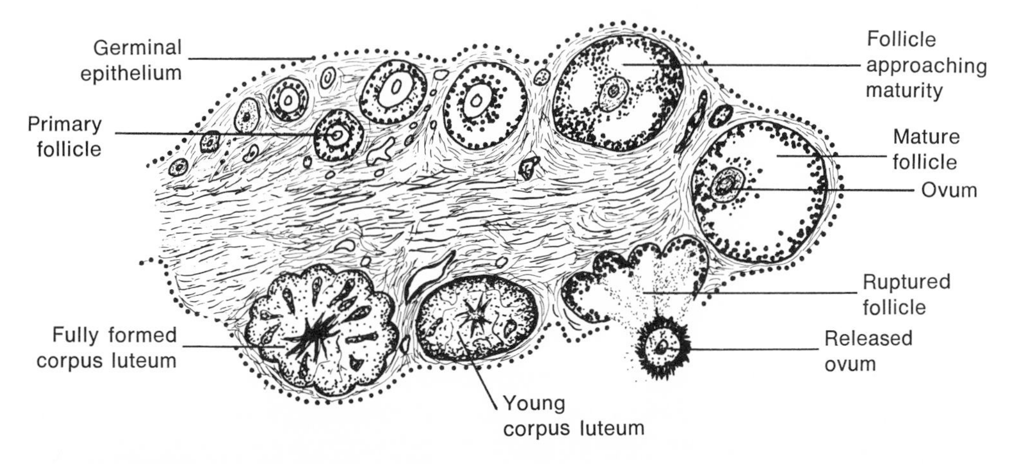

Fig. 13-6
Development of a graafian follicle and its replacing structure, the corpus luteum.

and surrounding follicle cells. The discharge of the egg from the graafian follicle is called **ovulation.** As discussed earlier, the egg is swept into the fallopian tube and is moved toward the uterus. If sperm are present in the fallopian tube, fertilization and pregnancy may follow.

Shortly after ovulation, the walls of the graafian follicle collapse. Its cells change in appearance, taking on a yellowish color, and the follicle is called a **corpus luteum.** It, too, acts as an endocrine gland, secreting both estrogen and another ovarian sex hormone, progesterone. Progesterone has a number of effects: (1) it acts on the anterior pituitary gland to suppress production of FSH, thereby preventing the development of new graafian follicles; (2) it acts on the cells lining the inside of the uterus, causing them to continue dividing and thus increase still further the thickness of the uterine lining (up to 5 mm) and also to increase the uterine lining blood supply; and (3) it acts on the muscular walls of the uterus, preventing their contraction. Both estrogen and progesterone prepare the uterus for the possibility that the egg will be fertilized and a pregnancy will follow.

If pregnancy does occur, the corpus luteum grows and persists until close to the birth of the child. However, if pregnancy does not occur, the corpus luteum degenerates rapidly, starting about 14 days after its formation. This decrease in size of the corpus luteum is accompanied by a sharp drop in its progesterone secretion. When this occurs, the thickened lining of the uterus can no longer be maintained, and it is sloughed off with an accompanying flow of blood (menstrual flow) for about 5 days. The decline in progesterone production also releases the anterior pituitary gland to produce FSH and start the next menstrual cycle.

Although the menstrual cycle actually begins with the development of new graafian follicles, it is customary to consider the first day of menstruation as the start of the cycle (Fig. 13-7). The first 5 days after the start of menstruation is the period of menstrual flow (**flow phase**), during which the thickened lining of the uterus is discarded. At the end of the flow phase, the uterine lining is at its thinnest. Next comes a **follicular phase,** lasting about 9 days, during which FSH and LH stimulate the development of the graafian follicle. The follicle, in turn, secretes estrogen that causes the uterine lining to thicken. At the end of the follicular phase, ovulation occurs, and the **luteal phase** begins. It lasts about 14 days and is characterized by high levels of estrogen and progesterone that stimulate the uterine lining to grow to its greatest thickness. If pregnancy does not occur, the flow phase of the next menstrual cycle takes place.

Female sexual response. The female sexual response begins with the secretion of a lubricating fluid by the cells of the inner lining of the vagina. In addition, the wall of the vagina becomes engorged with blood, resulting in a lengthening and broadening of the structure. These are the initial reactions to some form of sexual stimulation such as erotic thoughts or direct stimulation of the breasts or genitals. The clitoris is especially sensitive for initiating sexual sensations, having a richer nerve supply than any other organ of the body. Once the nerve impulses from the clitoris or other area of the body enter the spinal cord, they are transmitted to the cerebrum, where they contribute to the sense of sexual excitement.

As sexual arousal continues, heart rate and blood pressure increase. In addition, the uterus becomes elevated and pulls on the vagina, making it wider.

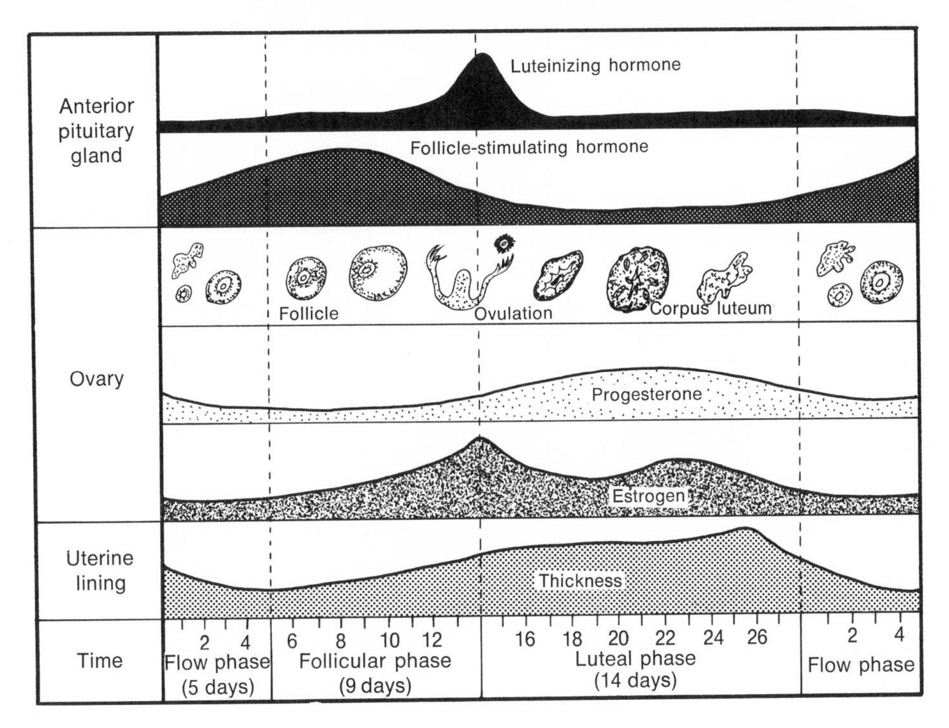

Fig. 13-7
Sequence of events in the human menstrual cycle.

Just inside the vagina, and surrounding its entrance, is erectile tissue that is almost identical to the erectile tissue of the penis. Vaginal erectile tissue, like that of the penis, is controlled by parasympathetic nerves that pass from the spinal cord to the genital region. As a result of sexual stimulation from the introduction of a penis into the vagina, impulses from these nerves produce a dilation of blood vessels in the erectile tissues, causing them to expand and tighten around the penis. This greatly aids both participants in attaining sufficient sexual stimulation for orgasm.

Orgasm in women is characterized by contractions of muscles in the outer third of the vagina. These occur rhythmically at about one second intervals, until three to fifteen contractions have occurred. The uterus also contracts in muscular waves. It is the contraction of both vaginal and uterine muscles with its accompanying stream of nerve impulses to the spinal cord and brain that supplies the peak of physical pleasure experienced at this time.

□ Contraceptive practices

For at least the past 3500 years, based on available evidence, people have attempted to prevent unwanted pregnancies. Some of the methods used over this period of time were quite effective, for example, penile sheaths and vaginal tampons, while others depended on superstition and magic, for example, charms and incantations, and resulted in failure. The reasons for preventing pregnancies were undoubtedly as varied in the past as they are today. However, in our modern society, a new factor has appeared. This is a concern with the quality of life as affected by the

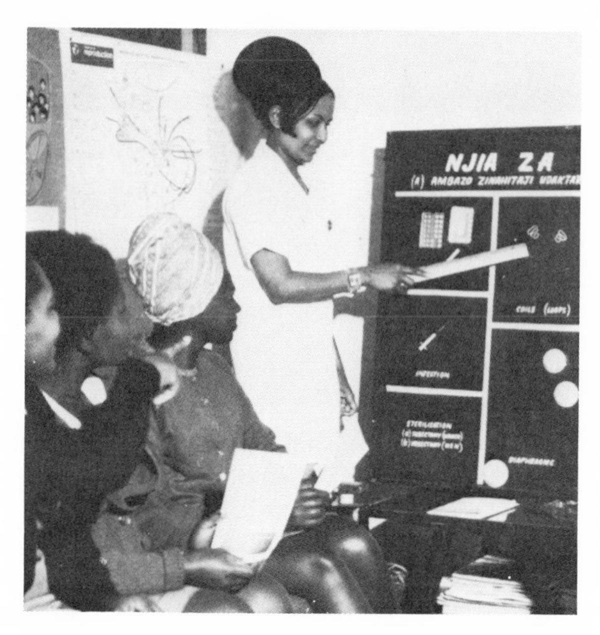

Fig. 13-8
Nurse describes various contraceptives to mothers at Kenyan health clinic. (From World Population Growth and Response, U.S. Agency for International Development.)

Fig. 13-9
Young Indian stands beside India's well-known sign promoting the small family. (From World Population Growth and Response, U.S. Agency for International Development.)

current world population explosion and the limits of the world's resources (Figs. 13-8 and 13-9).

Various nations have made family planning an integral part of their national policy. Very recently, the Mexican government, faced with a projected increase in population from 58 million in 1975 to over 100 million by 1990, a rate of increase of 3.43%/year, decided that its economy would not be able to support the anticipated additional burden on its health care programs, schools, and other public services. It therefore embarked on a family planning program involving the dispensing of contraceptive materials, without charge, and a publicity campaign for the program, with such slogans as, "Let's make ourselves fewer to live better," and "The small family lives better."

There are about half a dozen contraceptive methods that people use. One method, called the **rhythm method,** has as its biological basis a limitation of sexual intercourse to a period when the egg is not fertilizable. This method relies on two facts: (1) an egg is fertilizable for only about 24 hours after ovulation; and (2) sperm can survive inside the female reproductive tract for only about 48 hours. By avoiding sexual relations during the so-called "unsafe period" (3 days before and 3 days after ovulation), conception can be avoided.

The rhythm method makes certain assumptions: (1) that only one egg will be ovulated during each menstrual cycle and (2) that we can be certain, in advance, of the time of ovulation. Unfortunately, there is no assurance that only one egg will be ovulated per menstrual cycle or that, if two or more eggs are produced, they will be released simultaneously from the ovary. There is even evidence that some women ovulate following orgasm, which means that these women cannot hope to effectively use this method.

The success of the rhythm method also requires that the time of ovulation be known in advance, in order to be certain of the "unsafe period" before ovulation. Ovulation is usually accompanied by a rise in body temperature of about ½° F, which is maintained until the next menstruation. This is a reliable method of telling a woman when ovulation has occurred, but it cannot tell a woman when ovulation will occur in future menstrual cycles. Some women are quite regular in their ovulation dates, but others are not. How safe is the rhythm method? It is estimated to be about 74% effective. This means that of every 100 women who use this method for 1 year, 26 will become pregnant.

Other contraceptive methods can be grouped into those that are designed to prevent the sperm from reaching the egg and those that affect egg ovulation, movement, or implantation.

Sperm may be prevented from entering the male's urethra, the female's vagina, or the female's uterus. Each of these methods of avoiding pregnancy requires a different contraceptive device. The most effective approach is that which prevents the sperm from entering the male's urethra. This involves a cutting and tying off of the sperm duct (**vasectomy**) from each testis, so that the sperm cannot pass from the epididymis into the urethra. The sperm produced by the testes are resorbed by the cells of the epididymis. In this operation, the interstitial cells of the testes are not affected; thus testosterone production and the male sexual response remain the same. Vasectomy is a form of male sterilization and in most cases is permanent, since the rejoining of the cut sperm ducts is possible in only a small fraction of the cases. A vasectomy, properly performed, is a 100% effective method of contraception.

The prevention of sperm from entering the female's vagina involves the use of a rubber sheath (**condom**) around the penis to collect the ejaculated sperm. This method has been in widespread use for about 250 years and is considered to be about 85% effective. The lack of absolute protection from unwanted pregnancies usually involves carelessness in handling the condom rather than any defect in the device itself.

Preventing ejaculated sperm from entering the uterus can be accomplished by placing a cup (**diaphragm**) inside the vagina over the cervix. When used properly, the diaphragm should provide complete protection against conception. Its general effectiveness, however, is rated at 87%.

Contraception methods that are directed at the egg include cutting and tying off the fallopian tubes, the use of an intrauterine device, and the pill. Cutting and tying off the oviducts (**tubal ligation**) is comparable to vasectomy in the male. In the case of the female, the surgical operation prevents ovulated eggs from moving toward the uterus, and the eggs soon disintegrate in the fallopian tubes. This procedure does not affect hormone production by the ovary, and thus the menstrual cycles continue as before. As in the male, the cutting of the oviducts is 100% effective in preventing pregnancies and should be considered a permanent form of contraception.

The **intrauterine device** (IUD) is a flexible, thin plastic rod that is inserted into the uterus of a female. As the IUD enters the uterus, it curls into a double S shape. Although the exact mechanism by which an IUD prevents pregnancy is unknown, its presence in the uterus is believed to have two effects: (1) it hinders the movement of sperm through the uterus, thereby reducing the chance for conception, and (2) it prevents any embryo that may have been formed from implanting into the thickened lining of the uterus. The IUD is 98% effective in preventing pregnancy. This higher rate of effectiveness, compared to condom and diaphragm, is the result of the fact that once it is inserted, the IUD requires no attention or adjustment each time sexual intercourse occurs. The lack of 100% effectiveness is caused by the occasional spontaneous ejection of the IUD that, for whatever reason, goes unnoticed by the woman.

All the contraceptive practices discussed thus far have concentrated on preventing sperm and egg from meeting and fusing. The **pill,** on the other hand, is specifically designed to prevent ovulation, thus eliminating egg production. The suppression of ovulation is accomplished through the use of hormones. As was discussed earlier in this chapter, ovulation is induced by the anterior pituitary hormones FSH and LH. Secretion of these gonadotropic hormones is suppressed by high levels of estrogen and progesterone. The pill contains sufficient amounts of synthetically prepared estrogen and progesterone to inhibit ovulation, while permitting other aspects of the menstrual cycle to proceed normally.

A woman using this contraception method takes one pill a day, starting on day 5 after the start of her menstrual flow. The pill is taken for 20 days and is then discontinued. The next menstrual period starts about 3 days after the last pill is taken. This procedure establishes an artificial 28-day menstrual cycle. The pill is 98% effective in preventing unwanted pregnancies. The lack of complete effectiveness is thought to reflect situations in which, for whatever reason, a pill is not taken each day during the period required by the procedure.

Most of the contraceptive practices discussed do not involve any increased risk of illness or death to their users. However, in the case of the pill, recent evidence has shown that users of this contraceptive device are subject to a greater risk of both nonfatal and fatal heart attacks than women who use some other means of contraception. Among women aged 30 to 39 years who do not use the pill, the incidence of **nonfatal heart attacks** is 2.1/100,000. But for those taking the pill, the rate rises to 5.6/100,000. With respect to **fatal heart attacks,** the rate for the two groups are 1.9 and 5.4, respectively. Among women aged 40 to 44 years, the nonfatal heart attack rates for the two groups are 9.9 and 56.9/100,000, respectively, whereas the fatal heart attack rates are 11.7/100,000 for nonusers and 54.7/100,000 for users. In general, older women and women with circulatory system problems, for example, high blood pressure, must view this method of contraception with caution, especially as other methods of preventing pregnancy are available to them.

In closing this discussion of contraceptive practices,we should ask one more question. What is the rate of pregnancy in the absence of any contraceptive procedure? The pregnancy rate is about 40%. This means that out of every 100 women who do **not** use any contraceptive device for 1 year, 40 will become pregnant. It should also be pointed out that full-term pregnancy and delivery carries a risk of death to the mother from various causes. There are about 14 **deaths** of mothers/100,000 **live** deliveries of babies.

☐ Venereal diseases

Venereal diseases (VD) are those diseases caused by microorganisms and transmitted from person to person either by sexual intercourse or by close body contact involving the sex organs, mouth (kissing a person with a syphilitic sore in the mouth or throat), or rectum (in the case of male homosexuals). There are about half a dozen known venereal diseases. In the United States, two of these are widespread: gonorrhea and syphilis.

It is estimated that each year about 1.5 million Americans contract gonorrhea, and 120,000 contract syphilis. Venereal disease is the second most communicable disease in the country, outranked only by common measles.

The death rate directly attributable to syphilis is 3000/year. However, 1 in 15 victims of syphilis will develop heart disease, 1 in 50 will become insane, and 1 in 100 will become blind. Gonorrhea also causes serious physical damage, such as sterility, heart trouble, arthritis, and blindness.

Venereal diseases are most prevalent among young people. More than 50% of all venereal diseases occurs in those 13 to 25 years of age, although they comprise only 23% of the United States population. Even more disturbing are the data for the group aged 20 to 24 years, which constitutes 8% of the population, but contains more than 25% of all venereal disease cases.

Gonorrhea, also called "clap," "a dose," or "strain," is caused by a spherically shaped bacterium known scientifically as *Neisseria gonorrhoeae.* In the male, infection usually begins in the urethra; in the female, it may begin in the urethra, vagina, or cervix. If not treated, the infection can spread through the entire reproductive system to the gonads, causing sterility in both sexes. Invasion of other systems of the body can lead to arthritis, heart disease, and degenerative diseases of various organs.

If a woman has an active gonorrhea infection in her vagina when giving birth, the bacteria may infect the infant's eyes. If nothing is done to protect the child's eyes, blindness usually results. The laws of most states require that a few drops of silver nitrate be placed in the eyes of all newborn infants to kill any gonorrhea bacteria that may be present.

Treatment of adults infected with gonorrhea is relatively simple. There are a number of antibiotics and other drugs that are most effective against the bacterium. Unfortunately, there is **no** immunity developed against the disease. A person can be reinfected again and again and again.

Syphilis, also called "pox," "lues," "siff," or "bad blood," is caused by a spiral-shaped bacterium known scientifically as *Treponema pallidum*. The disease has three distinct phases. The first stage, **primary syphilis,** is characterized by the appearance of a sore, called a **chancre,** at the site of infection. Depending on the mode of transmission of the disease, the chancre can appear on the penis, vagina, lip, palm of the hand, or rectum. Although painless, the sore itself is highly infectious and contains a tremendous number of bacteria. The chancre usually appears 2 to 5 weeks after infection. If left untreated, the chancre will disappear, but invasion of the various body systems continues (Fig. 13-10).

Secondary syphilis appears 3 to 6 weeks after the chancre first erupts and is characterized by a rash of raised red spots on the skin anywhere over the body, often on the palms of the hands and the soles of the feet. Each red spot contains thousands of the bacteria and is highly infectious. The rash disappears spontaneously, and a latent period of variable length follows.

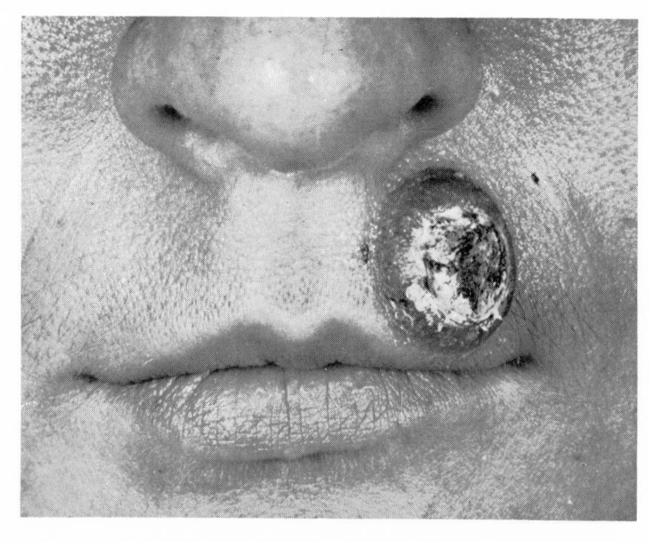

Fig. 13-10
Chancre, primary lesion of syphilis. Note extragenital location on lip. (From Top, F. H., Sr., and Wehrle, P. F. 1972. Communicable and infectious diseases, ed. 7. The C. V. Mosby Co., St. Louis.)

Tertiary syphilis develops months or even years after the initial infection. In this stage, degeneration of tissue occurs in the skin, blood vessels, liver, brain, and spinal cord, leading to heart disease, insanity, and death.

If a pregnant female has syphilis, the bacteria can be transmitted from mother to developing embryo any time after the third month of gestation. If the mother's infection is not treated in the first 3 months, the probability of the pregnancy resulting in a stillbirth (dead newborn) is quadrupled, and for those born alive, the chance of the infant's death during the first year of life is doubled. Of those born alive, 20% will suffer from **congenital** (existing at birth) **syphilis.** These children may subsequently become blind, deaf, paralyzed, or insane or may even die.

Syphilis, much like gonorrhea, can be treated simply and effectively. As with gonorrhea, there is no immunity built up to syphilis, and reinfection is always possible. Both gonorrhea and syphilis are to some extent preventable through the use of a condom by the male during sexual intercourse. The condom acts as a **prophylaxis** (protection from disease) against venereal diseases transmitted through this type of contact. However, other forms of close body contact with infected persons will continue to spread the disease.

■ DEVELOPMENT

The life of an individual begins with fertilization and ends with death. In the interval between these events, the fertilized egg and its descendant cells will undergo growth and division, resulting in a multicellular complex of organ systems whose integrated activities permit survival and reproduction. At each stage of this process that we call **development,** the individual is constantly changing: certain body structures increase in size and level of activity, while others decrease; rates of metabolism of some cell types rise, while others fall; resistance to some diseases is strengthened, while to others, it is weakened. In general, there is an overall buildup of body structure and function from fertilized egg to adulthood; after that, there is a consistent loss of both structure and function until death. We shall now examine the continuing process of development that has not only brought each of us from a fertilized egg to our present state, but will also determine our health and well-being until we die.

□ Fertilization

When a man ejaculates during sexual intercourse, about 300 million sperm are deposited at the inner end of the vagina. They pass through the cervix into the uterus and up into the fallopian tubes. The mortality rate of the sperm is very great; only a few thousand reach the point in the oviduct where an egg may be present and ready to be fertilized.

At ovulation, the egg with its surrounding follicle cells enters the fallopian tube and is moved slowly toward the uterus. Because both sperm and egg have very short periods of time during which they can unite, fertilization can only occur during a very brief part of the menstrual cycle. It is only because copulation is so frequent that fertilization occurs as often as it does.

The movement of sperm from the vagina to the site of fertilization is very rapid, and the first sperm arrive within 30 minutes of ejaculation. Although only one sperm is involved in fertilizing an egg, thousands of sperm must swarm around the egg before any one of them can enter it. The need for many sperm results from the presence of follicle cells around the egg. These follicle cells are held together by a complex organic substance known as **hyaluronic acid,** which must be broken down before a sperm can reach the egg. Each sperm cell has a very small amount of the enzyme **hyaluronidase,** which can act on hyaluronic acid. It requires the combined enzyme supply of several thousand sperm cells to separate the follicle cells sufficiently for one sperm to enter and achieve fertilization. It has been found that males whose ejaculates contain less than 100 million sperm are almost completely sterile.

Once a sperm makes contact with the egg, there is a fusion of their cell membranes, and the contents of the sperm cell pass slowly into the egg cytoplasm. Penetration of the egg by the sperm stimulates the egg to go through its second meiotic division. After oogenesis is completed, the nuclei of the sperm and egg unite, forming a fertilized egg that contains a full complement of 46 chromosomes. The entry of a sperm into the egg causes the egg to secrete a substance around itself that prevents other sperm from penetrating the egg.

□ Early cell division

The fertilized egg undergoes its first division about 36 hours after a sperm has entered. A second mitosis follows within 10 hours. The rate of cell division then increases, and, by 72 hours after conception, a solid clump of 16 to 32 cells, called a **morula,** is formed (Fig. 13-11). During these first 3 days, the developing individual is moved down the oviduct to the uterus by the ciliated cells of the oviduct's inner surface.

After entering the uterus, the morula floats about in the intrauterine fluid (from which it obtains nourishment) for about 3 days. During this time, cell division continues, and the cell mass hollows out, forming a cavity. This stage of development is called a **blastocyst** (**blastula**) and the cavity within, a **blastocoele** (Fig. 13-11).

Shortly after blastocyst formation, the cells on one side of the blastocoele divide very rapidly, forming a cluster that projects into the cavity. This cluster of cells is called the **inner cell mass,** while the cells forming the outer layer of the blastocyst are collectively called the **trophoblast** (Fig. 13-11). The inner cell mass and the trophoblast will play different roles in development, and their formation is the first specialization of tissues that occurs during development. With the formation of specialized tissues, the developing individual is referred to as an **embryo.** After the embryo has developed all the external features that make it easily recognizable as belonging to a particular species, it is called a **fetus** until the time of birth. For humans, the term "embryo" applies to the period from blastocyst formation (about day 5 after fertilization) only until the end of the first 2 months of gestation.

□ Implantation

At about 7 days after conception, the embryo becomes attached to some point on the inner surface of the uterus. Then the cells of the trophoblast secrete enzymes that destroy the cells and blood vessels of the thick inner lining of the uterine wall at the point of attachment. This produces a small cavity into which the embryo sinks, thereby becoming **implanted** in the lining of the uterus. The embryo sinks below the surface, and the lining closes over it.

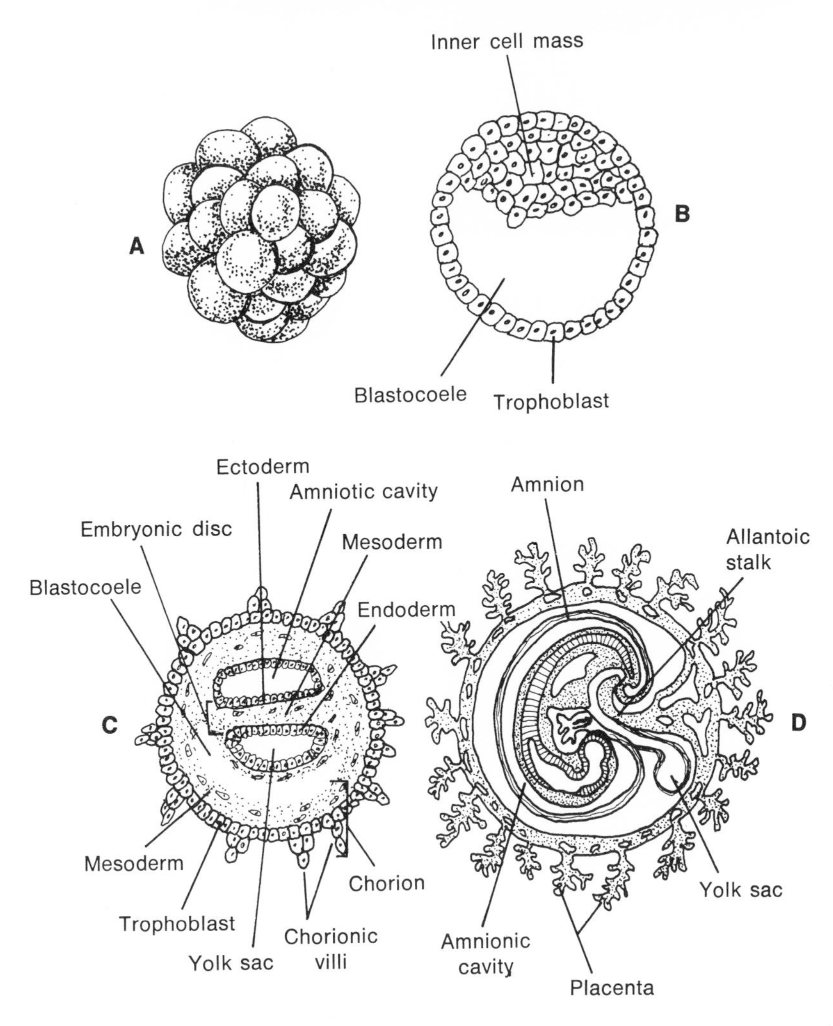

Fig. 13-11
Human embryonic development. **A,** Morula (external view). **B,** Blastocyst (cross section). **C,** Gastrula (cross section). **D,** Four-week-old embryo (cross section).

It is important to note that the time of implantation (7 days after conception) corresponds to the period of greatest thickness of the uterine wall (21 days after the beginning of menstruation). Thus while the egg is being fertilized and early development is taking place, the lining of the uterus is being stimulated by estrogen and progesterone to increase in size and blood supply for proper implantation of the embryo.

The destruction of uterine tissue that occurs during the implantation process serves a useful purpose. The destroyed cells are sources of food materials for the embryo. In addition, the destruction of blood vessels causes the embryo to lie in a pool of its mother's blood, from which additional nutrients are obtained and into which waste materials are deposited. Clotting of the mother's blood in the uterine wall is prevented by an anticoagulant that is secreted by the embryo.

Placenta formation and functions

After implantation, some of the cells from the inner cell mass divide and move over the surface of the blastocoele, forming an inner coat adjacent to the trophoblast. These cells, called **mesoderm** cells (to be discussed later), together with the **trophoblast** form a rather thick outer protective layer, the **chorion,** around the embryo (Fig. 13-11).

At this time (about 2 weeks after fertilization), the cells of the trophoblast grow and divide rapidly, forming projections that are called **chorionic villi.** These obtain nourishment from the surrounding maternal blood. The embryo also forms blood vessels within the villi, and a **placenta** is established. The placenta is made up of the villi plus the tissues of the uterine wall in which they are embedded.

The blood vessels of the placental villi are mainly capillaries that receive blood from the embryo by way of two **umbilical arteries** and return blood to the embryo by way of the **umbilical vein.** The umbilical arteries are secondary branches of the systemic artery, and the umbilical vein joins the inferior vena cava (Chapter 6). Umbilical arteries and the umbilical vein pass through a tube, called the **umbilical cord,** and connect the placental capillaries to the rest of the embryo's circulatory system.

Once the placenta is established, nutrients and waste products will be exchanged between embryonic and maternal circulatory systems across its membrane, but there will be **no** mixing of blood between mother and child. For the embryo, the placenta serves as its digestive tract, kidneys, and lungs, while those same organs of the mother are functioning to maintain both individuals.

The placenta functions only during the gestation period. At birth, the umbilical cord is cut, and the part attached to the newborn shrivels to form the **navel.** The remainder of the umbilical cord is attached to the placenta itself and is expelled from the uterus, along with the placenta, as the **afterbirth.** Cutting the umbilical cord involves a severing of the umbilical arteries and vein. The segments of these blood vessels that are attached to the embryo's circulatory system do not function after birth and become filled with connective tissue.

In addition to acting as an embryo's link to the outside world (through the mother's body), the placenta also functions as an endocrine gland. However, it is unique in that its hormones affect the mother's tissues rather than the child's. One of the placental hormones is **chorionic gonadotropin,** which is secreted into the maternal bloodstream. This hormone stimulates the corpus luteum of the mother's ovary to continue producing estrogen and progesterone; these hormones, as described earlier in this chapter, stimulate the buildup and maintenance of the lining of the uterine wall. In the absence of chorionic gonadotropin, the corpus luteum ceases its hormone production within 14 days after ovulation, and the uterine lining is sloughed off. Thus by preventing menstruation, the embryo affects the mother's body in a manner that promotes the maintenance of the developing child within the uterus.

Placental secretion of chorionic gonadotropin does not continue throughout the entire gestation period. It reaches a maximum approximately 7 weeks after conception and then decreases to a relatively low value some 9 weeks later. Accompanying the reduction in level of chorionic gonadotropin, the corpus luteum decreases in size and ceases to secrete estrogen and progesterone. However, at this time, the placenta itself begins to secrete both these hormones, and menstruation continues to be prevented. In this connection, it has been found that if the ovaries have to be removed anytime during the first 3 months of pregnancy, spontaneous abortion usually occurs. However, removal of the ovaries during the last 6 months of gestation has no effect on the pregnancy at all.

☐ Embryonic development

While the placenta is being formed, dramatic changes are taking place in the **inner cell mass.** About 10 days after fertilization, these cells divide rapidly and rearrange themselves so as to form two cavities within the inner mass (Fig. 13-11). The movement of the inner mass cells to establish these cavities is called **gastrulation,** and the embryo, at this stage, is referred to as a **gastrula.**

Primary germ layers. The upper cavity formed during gastrulation is called the **amniotic cavity.** Within it, a sac (**amnion**) will be formed that will enclose and protect the developing embryo. The cells of the lower margin of the amniotic cavity form the outer layer of cells of the developing embryo and are called **ectoderm.**

The lower cavity formed during gastrulation is called the **yolk sac** and functions in fish, amphibians, reptiles, and birds during development as a storehouse of food, but in man, it serves no such purpose. However, the yolk sac is important in human development because the cells of its upper margin become the inner layer of cells of the developing embryo and are called **endoderm.**

A third layer of cells, called **mesoderm,** is formed within the inner cell mass. Some of these cells grow over the inner surface of the blastocoele and, together with the trophoblast, form the chorion (as discussed earlier). Other mesodermal cells move between the ectoderm and endoderm and, together with them, form the **embryonic disc** (Fig. 13-11). It is from the embryonic disc that virtually all tissues of the body are formed. Because of this, ectoderm, endoderm, and mesoderm are called the **primary germ layers.**

Each of the three primary germ layers gives rise to certain specific tissues of the body. (See Chapter 7 for discussion of cells and tissues.) **Ectoderm** forms the epidermis of the skin and its derivatives (hair, nails, and various glands), the entire nervous system (including receptor cells), and the adrenal medulla (Chapter 12). **Endoderm** gives rise to the lining of the digestive tract, respiratory tract, urethra, and urinary bladder, and, in addition, the liver, pancreas, and thyroid, parathyroid, and thymus glands. **Mesoderm** produces the muscles, circulatory system (including blood, heart, arteries, veins, capillaries, and lymphatic vessels), the connective tissues (hence the dermis of the skin and the skeletal system), the kidneys and ureters, the reproductive system, and the adrenal cortex.

Embryonic membranes. In addition to the primary germ layers, early development results in the formation of four **embryonic membranes:** chorion, amnion, yolk sac, and allantois. As described earlier, the **chorion** develops from a fusion of the trophoblast layer of the blastocyst and mesoderm and produces the embryonic portion of the placenta. The chorion encloses completely the embryo and all other embryonic membranes.

The **amnion** develops from a combination of the cells surrounding the amniotic cavity (formed during gastrulation) and cells from the embryo. The amnion is a sac that completely surrounds the embryo (and fetus) and becomes filled with about a quart of a watery fluid that acts as a protective cushion to absorb any shocks to that part of the body while permitting the developing embryo (and fetus) freedom of movement. During the birth process, the amnion is broken by contractions of the uterus, releasing the "waters." The amnion is usually left behind in the birth process and is subsequently discharged from the mother's uterus as part of the afterbirth. If the child is born still surrounded by its amnion, the attending physician or midwife (still used in many parts of the world) removes it.

As described earlier, the **yolk sac** is formed during gastrulation. Shortly thereafter, cells from the endoderm of the embryo migrate to its periphery, and the yolk sac is enlarged. The yolk sac of man (and all other mammals) contains no food supply, but is important as the initial source of blood cells. However, it soon becomes nonfunctional and reduced in size.

The **allantois** develops as an outpocketing of the yolk sac. In reptiles and birds, the allantois grows to a considerable size and functions as a combined respiratory and excretory organ for the developing animal. In humans, however, the allantois is much reduced and functions solely to produce the blood vessels of the placenta.

Identical twin formation. One of the interesting events that can occur during early embryonic development is the formation of **identical twins.** These may be formed either through the separation and independent development of the daughter cells resulting from the first mitosis of the fertilized egg or, more usually, by a splitting of the developing embryo. If the embryo splits within the first 14 days after fertilization, the separation of twins is complete. However, if the embryo splits thereafter, the separation of the twins is usually incomplete, and the twins will be joined to one another at the back, head, chest, or side. Such twins are called **conjoined twins.** Approximately 1 in every 120,000 births involves conjoined twins. The best known example of conjoined twins were the so-called Siamese Twins (they were actually Chinese). They were born in Siam (Thailand) in 1811 and died, 2 hours apart, in North Carolina in 1874. They were joined at the chest by a rather simple band composed mainly of muscle and a thin sliver of liver tissue. In these days of modern surgery, it would be relatively simple to separate this type of conjoined twin. Unfortunately, most conjoined twins share very critical organs such as the heart or brain and cannot be successfully separated. In most cases, both twins die.

Identical twins have identical genotypes, they are of the same sex, and they are identical with respect to such genetic markers as blood groups. They are less similar in traits that are greatly influenced by the environment. As an example, identical twins may be quite different in birth weight, because of differences in prenatal nutrition.

Organ development. A 4-week-old embryo contains a heart, brain, tail, and most internal organs. At 5 weeks of development, the embryo is almost complete, although its size and weight are about the same as an aspirin tablet. The heart begins to pulsate, the arm and leg buds are formed, and the outline of the eyes appears. One week later, the jaw is well developed, the earliest reflexes appear, and the diaphragm is formed. By the seventh week, the embryo begins to assume a more human form. The rudimentary tail is absorbed by the body, the fingers and toes are formed, and the development of the face is evident. At this stage, the embryo, about to become a fetus, is 1 inch long (2.5 cm) and weighs $^1/_{10}$ ounce (3.3 gm).

☐ Fetal development

By the eighth week of development, all the human features are evident, and the developing child has become a fetus. Table 13-1 gives the length and weight of the average fetus at various ages of development. Between weeks 9 and 12, fingernails, toenails, and hair appear; tooth buds and bone form; and the sex of the fetus is recognizable externally by the presence of penis or vagina. At this stage, the kidneys begin to function (the urine being passed into the amniotic sac), and the motions of breathing and eating are evident (although neither function is actually performed by the fetus). Most of the rest of fetal development consists of growth in size, as all body structures have been formed.

One of the more exciting advances in medical science has been the ability to

Table 13-1
Fetal growth

Age (weeks)	Length (inches)	Weight
8	1	1/10 ounce
12	4	1 ounce
20	10	12 ounces
28	14	2 pounds
36	18	5 pounds
40	20	7 pounds

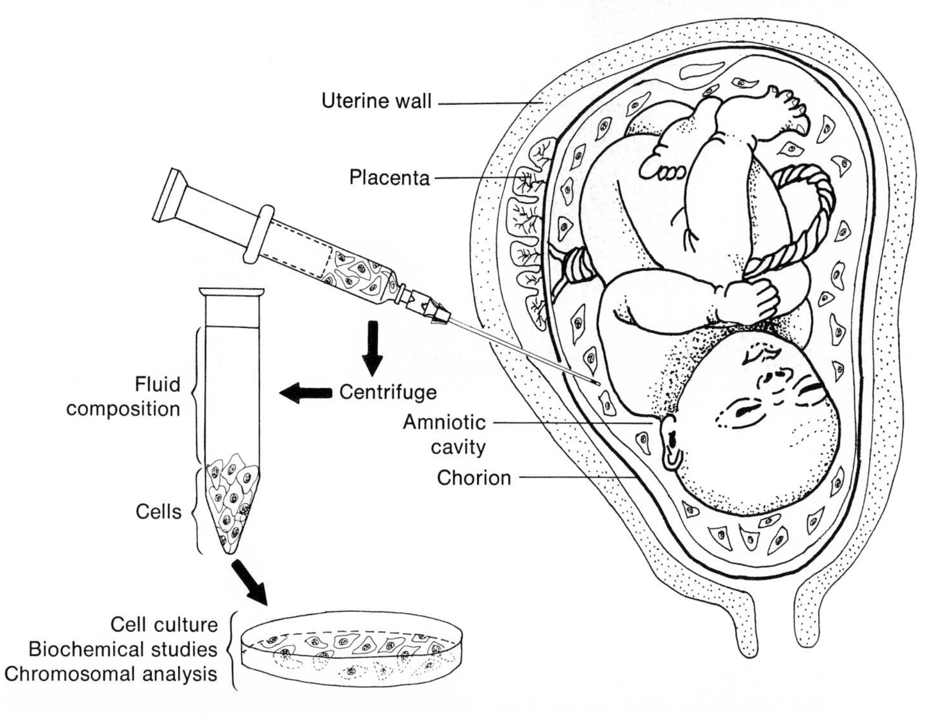

Fig. 13-12
Amniocentesis procedure for detection of genetic disorders.

determine many of the genetic characteristics of an unborn fetus, especially the presence of genetic disorders. The process through which this information is obtained is called **amniocentesis** (Fig. 13-12). Under sterile surgical conditions, a needle is passed through the mother's abdominal wall and into the amniotic sac of the fetus (usually sometime between 12 and 16 weeks of pregnancy), and some of the amniotic fluid is withdrawn. Suspended in the fluid are viable cells shed from the fetal skin and respiratory tract. These cells can be grown in tissue culture and analyzed for chromosome composition, giving evidence of Down's syndrome, Turner's syndrome, or Klinefelter's syndrome, for example, and for enzyme activity, showing Lesch-Nyhan syndrome or Tay-Sachs disease, among others. In many countries, abortion may be legally induced to prevent the birth of a malformed and/or mentally retarded child.

□ Abortion

Abortion is the termination of a pregnancy before the fetus is capable of surviving outside the uterus. Normally, a fetus must be 7 months old to have a minimum chance for life on its own. Abortions may be spontaneous (**miscarriages**) or induced. Spontaneous abortions are usually caused by abnormalities of fetus or placenta. About 10% of pregnancies end in spontaneous abortions, the majority occurring during the first 3 months of development. As was discussed in Chapter 9, 40% of all miscarriages during the first 90 days of pregnancy involve embryos with abnormal chromosome numbers.

Induced abortions are generally limited to the first 5 months of pregnancy and, depending on the age of the fetus, are performed in one of three ways. During the first 3 months of gestation, either suction or dilatation and curettage (D and C) are used. In the **suction** method, the cervix is dilated sufficiently to permit introduction of a suction tube, and the uterus is emptied by gentle suction. The **dilatation** and **curettage** method involves dilation of the cervix and insertion of a curette (a loop of steel attached to a special handle). The curette is used to scrape the inside of the uterus, thereby removing the implanted embryo and the surrounding uterine tissue.

Beyond 3 months of gestation, there is a risk of hemorrhage in using the suction or dilatation and curettage methods of abortion. During this later period of pregnancy, a **saline abortion** can be employed. In this method, a needle is inserted through the abdominal wall of the mother and into the amniotic sac of the fetus (much the same as in amniocentesis). A small amount of amniotic fluid is withdrawn and is replaced with a concentrated solution of salt water. The concentrated salt solution puts an end to fetal life, labor contractions are induced, and the fetus and the placenta are expelled from the uterus within 24 to 72 hours after the injection.

As with all surgical and semisurgical procedures, there is a risk of death to the pregnant women undergoing induced abortion. The rate is about 8 fatalities/100,000 abortions, most of them occurring in those abortions performed after the third month of pregnancy. As mentioned earlier in this chapter, the death rate of mothers in full-term childbirth is 14/100,000 deliveries.

□ Birth

The duration of pregnancy averages about 280 days from the beginning of the last menstrual period. About 90% of all births occur within a week before or after 280 days. Three main factors are responsible for the onset of birth: (1) there is a sharp drop in the production of progesterone by the placenta, thereby removing the previous inhibition of uterine wall muscle contraction; (2) the posterior lobe of the pituitary gland releases oxytocin, a hormone that stimulates the contraction of smooth muscle (especially of the uterine wall); and (3) the increasing weight of the fetus stretches the uterus and causes it to contract.

Labor. Childbirth (**parturition** or **labor**) involves a long series of involuntary contractions of the uterus, experienced as "labor pains." Labor, which usually lasts about 12 to 14 hours, may be divided into three periods: the first period lasts about 12 hours, during which time contractions of the uterus move the fetus toward the cervix and cause the rupturing of the amnion; the second period lasts from 20 minutes to 1 hour and results in the fetus passing through the cervix and vagina, thus being born or "delivered"; and the final period of labor lasts about 10 or 15 minutes after the birth of the child, during which time the placenta and other fetal membranes are loosened from the lining of the uterine wall and expelled as the **afterbirth.** The pulling away of the placenta from the uterine wall leaves a rather wide area of broken blood vessels that account for the considerable bleeding that accompanies parturition. The loss of a pint or so of blood is usual. Losses greater than that may require a transfusion to correct for the depletion of the mother's blood supply (Figs. 13-13 and 13-14).

Being born produces sudden and important changes in the functioning of the individual. Certain organ systems (digestive, respiratory, excretory) that have been nonfunctional up to that point must suddenly begin functioning, while other systems (circulatory, nervous, endocrine) are faced with making major adjustments to new conditions. As an example of the changes caused in organ system function by parturition, we shall examine the circulatory system before and after birth.

Fetal circulation. A diagram of human fetal circulation is shown in Fig. 13-15. Because the fetus does not use its lungs at all and has very little need for its liver, the flow of blood through both these organs is cut to a minimum by the use of three fetal shunts: the ductus venosus, the foramen ovale, and the ductus arteriosus. The **ductus venosus** carries blood from the umbilical vein into the hepatic vein, from which the blood enters the inferior vena cava. Because the ductus venosus also receives some blood directly from the hepatic artery and the hepatic-portal vein, very little blood is sent through the capillaries of the liver.

On reaching the heart, the inferior vena cava (and also the superior vena cava) empties into the right atrium. However, instead of all the blood going into the right ventricle and on to the nonfunctional lungs, a good deal of it passes from the right atrium to the left atrium through an opening, the **foramen ovale.** The remainder of the blood passes into the right ventricle and out through the pulmonary artery. However, here another shunt, the **ductus arteriosus,** directs most of the blood into the systemic artery, thus bypassing the capillaries of the lungs. Only about 10% of the fetus's blood is sent to the lungs.

At birth, the circulatory routing just discussed must be quickly changed. The foramen ovale in the heart closes off, and the ductus arteriosus and ductus venosus become filled with connective tissue, thus making pulmonary and systemic circulatory routes completely separate. If by chance the circulatory system's lung bypasses do not close, insufficient blood is sent to the lungs, and a "blue baby" results.

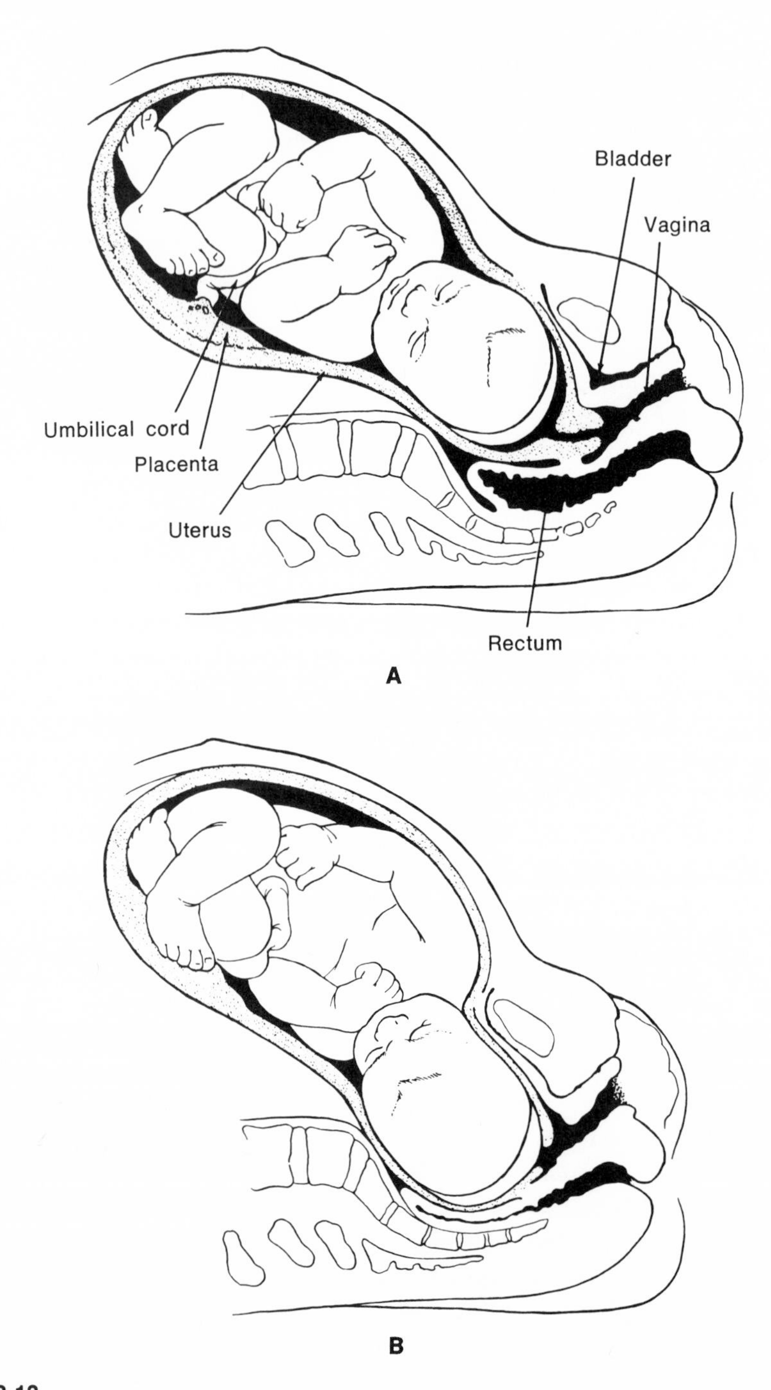

Fig. 13-13
Mechanisms of labor. **A**, Engagement. **B**, Descent with flexion. **C**, Internal rotation. **D**, Extension. **E**, External rotation. (From Iorio, J. 1975. Childbirth: family centered nursing, ed. 3. The C. V. Mosby Co., St. Louis.)

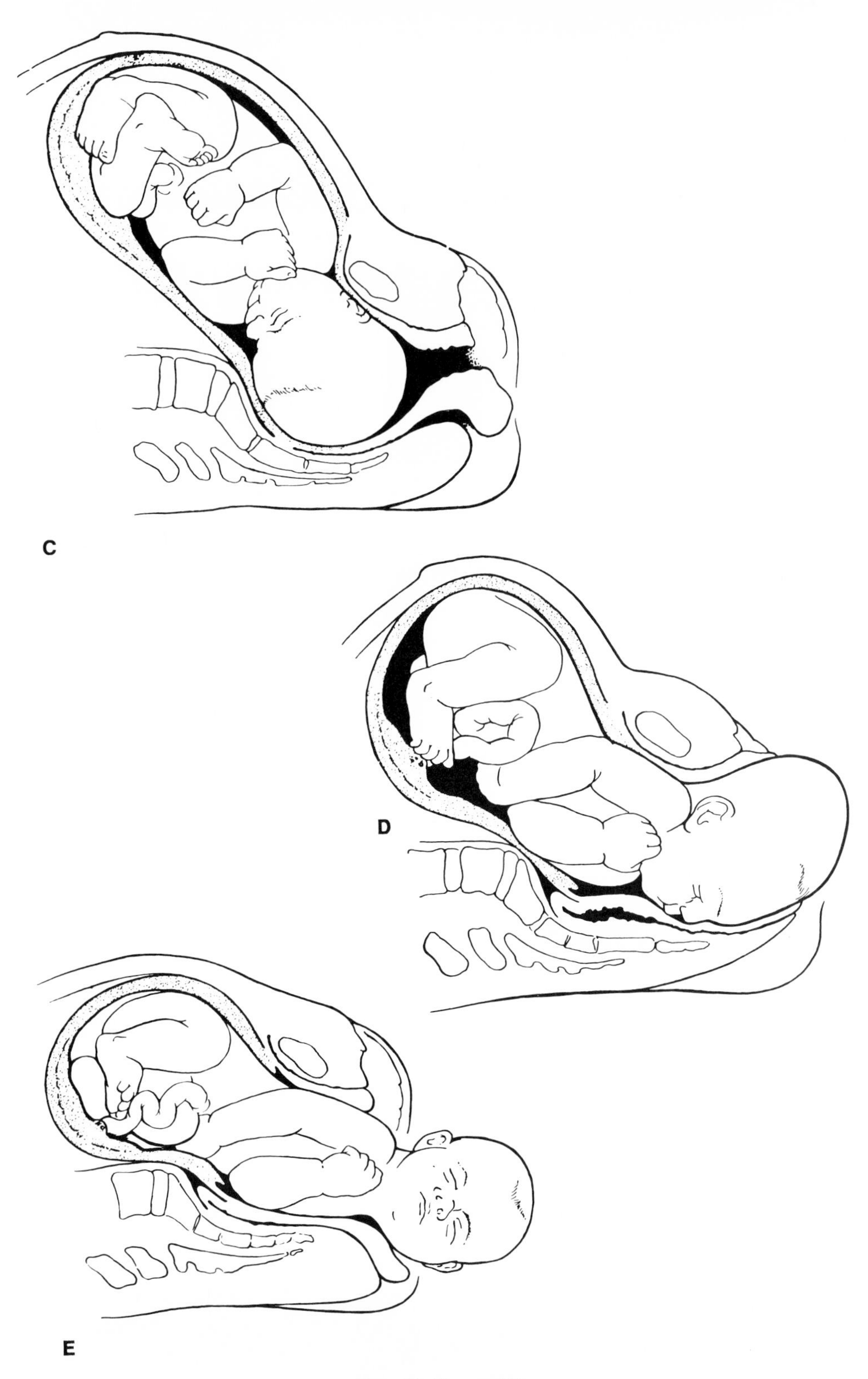

Fig. 13-13, cont'd

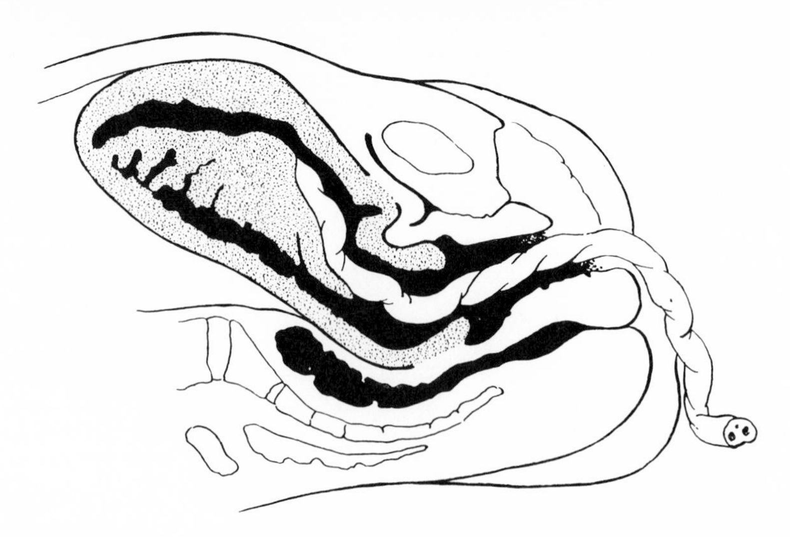

Fig. 13-14
Separation of the placenta. (From Iorio, J. 1975. Childbirth: family centered nursing, ed. 3. The C. V. Mosby Co., St. Louis.)

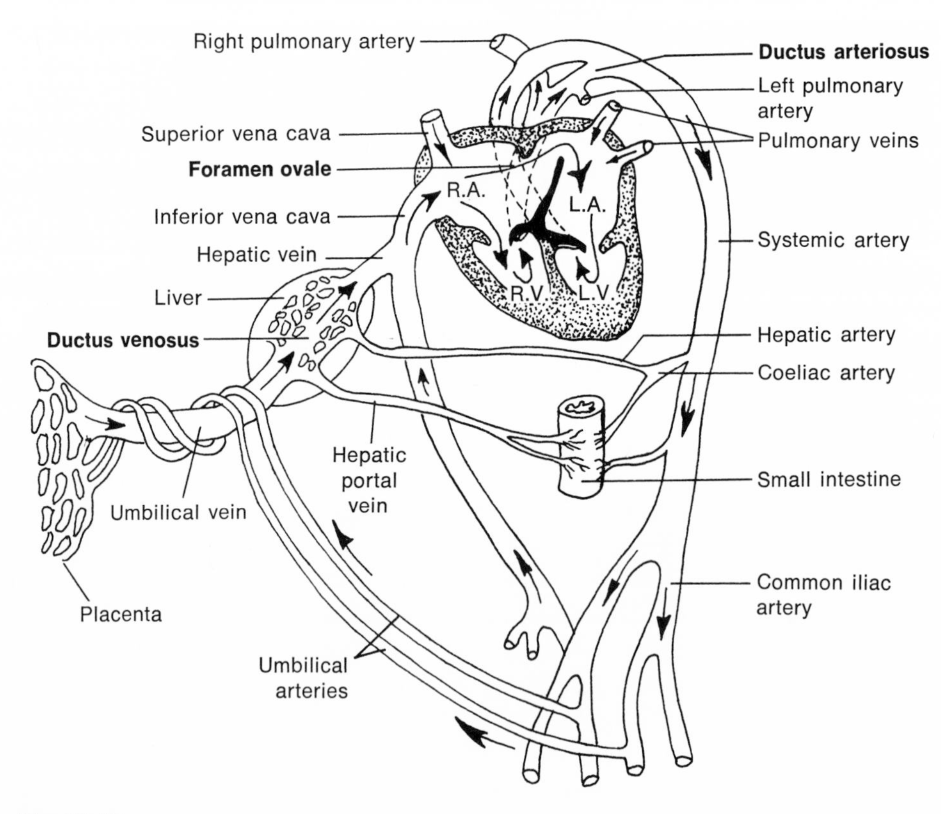

Fig. 13-15
Fetal circulation, showing three fetal shunts: ductus venosus, foramen ovale, and ductus arteriosus.

Modern surgery can correct this situation by sealing the foramen ovale and tying off the ductus arteriosus. With the cutting of the umbilical cord, as described earlier, the severed umbilical arteries and vein and the ductus venosus are no longer part of any continuous route of the circulatory system. Thus they become nonfunctional and the cells of their inner linings grow and divide, producing connective tissues that fill these blood vessels.

□ Lactation

After childbirth, the necessity of meeting the newborn's need for food becomes crucial. Fortunately, the human female body (as is true of all other mammals) has provision for meeting this need through the mammary glands. In Chapter 7, we discussed mammary glands as an example of a specialized skin gland. Mammary glands consist of two types of tissues: epithelial tissue that secretes the milk and connective tissue that surrounds and supports the glandular epithelium. Most women have about the same amount of milk-producing tissue. Differences in breast size are the result of differences in the amount of connective tissue.

All through pregnancy, as discussed earlier, the placenta produces large quantities of both estrogen and progesterone. In addition to their effect on the uterus, these hormones cause growth of the breast's glandular tissue, which forms a branching network of tubules. The epithelial cells lining these tubules are stimulated to synthesize milk by the action of **prolactin,** an anterior pituitary hormone (Chapter 12). By the end of pregnancy, the mother's breasts are fully developed for nursing, but only a few milliliters of fluid, called **colostrum,** are secreted each day until after the infant is born.

The passage of milk from the tubules of the breast through its nipple depends on an interaction between nervous and endocrine systems. When the infant suckles the breast, sensory impulses are transmitted to the spinal cord and from there to the hypothalamus. The hypothalamus is stimulated to secrete the hormone oxytocin, which flows by way of the bloodstream to the breasts. Oxytocin causes the cells surrounding the tubules to contract, thereby forcing the milk from the tubules into the ducts that lead to the nipple. Thus within 30 seconds to 1 minute after an infant begins to suckle the breast, milk begins to flow. This process is called milk "letdown."

□ Maturation

The infant at birth has by no means completed its development. Even its body parts are not proportionally the same as they will be as an adult. During the first 25 years of growth, body weight will increase 20 times, heart weight 12 times, and brain weight only 2 times. This explains why infants appear to have large heads. In fact, the head of an infant is roughly one fourth the length of the body, while in the adult, the fraction is closer to one eighth (Fig. 13-16). We can appreciate the complexities of postnatal development if we examine the behavioral progress of an infant during the first year of life. This is given in Fig. 13-17. The progressive nature of the various activities indicates the development of skeletal, muscle, and nervous systems.

Menarche. A very dramatic period of maturation occurs at **puberty** (about age 13 years), during which time the individual becomes sexually mature. The reproductive aspects of puberty were covered earlier in this chapter. However, there are also

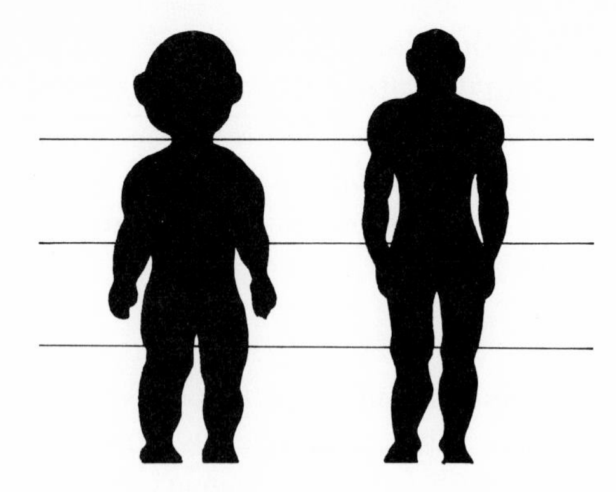

Fig. 13-16
Relative proportion of body parts of newborn and adult.

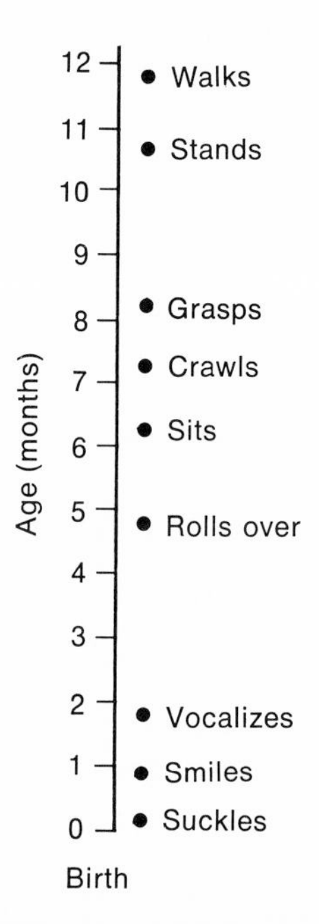

Fig. 13-17
Behavioral development of an infant.

developmental aspects that are important for our highly urban, modern society. Studies of European medical records for the last century have shown a consistent trend toward increased average adult height (1 inch) and weight (3 to 4 pounds) per generation (30 years). During this same time, the age at which girls begin to menstruate, **menarche**, has decreased by 1 year per generation. Presumably there has been a comparable reduction in age of sexual maturity for boys, but this is more difficult to measure. However, it is known that boys have experienced an increase in average height and weight each generation that corresponds to that of girls.

We cannot extrapolate this trend either backward or forward indefinitely. In fact, there is evidence that menarche occurred earlier in the 14th, 15th, and 16th centuries (Renaissance) than during the late 18th and early 19th centuries. This means that the general health and development of Europeans was poorer 200 years ago than it is today, undoubtedly because of the low standard of living of most urban groups at the beginning of the Industrial Revolution. Looking to the future, recent studies have indicated that the trend to larger body size and earlier menarche may be at an end. A survey of mothers and daughters, all of whom attended the same college and hence for whom medical records were available, found that the average age of menarche and the average height and weight for both groups was the same. A very large study of many diverse groups of girls indicates that it is body weight that is crucial for menarche. Regardless of age, menstruation will not begin unless a girl weighs 106 pounds (48 kilograms [kg]). At present, this weight is attained, in the United States, when the average girl is 13 years old.

Menopause. An individual's sexual fertility does not extend indefinitely, and at some time in life (for those living to an old age), a person becomes sterile. Although this situation includes both men and women, it is more recognizable and definite in women than in men. In a woman, the termination of menstruation (**menopause**) marks the end of her reproductive life. However, menstruation does not end suddenly. There is a period of time, usually lasting several years, during which the woman's sexual cycle becomes irregular and gradually comes to a complete halt.

The cause of menopause is a lack of the ovarian hormones estrogen and progesterone. These hormones are produced by the graafian follicle (estrogen) and by its replacing structure, the corpus luteum (estrogen and progesterone). The hormones stimulate uterine wall development and also affect the anterior pituitary gland in a manner that produces the normal female reproductive cycle (see earlier discussion in this chapter).

Over the fertile period of a woman's life, all of her egg cells are either ovulated or degenerate. Without these cells, there is no stimulus to produce graafian follicles, hence no ovarian hormone production. As a woman grows older, to about age 45 to 50 years, the number of eggs becomes extremely small, and their development becomes sporadic. This period of menstrual irregularity is called the **female climacteric.** The loss of ovarian hormones, especially estrogen, during the climacteric period and thereafter causes marked physiological changes in the body, including "hot flashes" (extreme flushing of the skin), irritability, fatigue, anxiety, and, in some cases, mental derangement. For women who are experiencing severe symptoms of menopause, daily injections of estrogen in small and gradually decreasing doses usually alleviates the situation. However, such treatment does prolong the duration of the climacteric.

Unfortunately, the use of estrogen to alleviate the symptoms of menopause increases a woman's chance of developing cancer of the lining of the uterus. The frequency of this type of cancer in all postmenopausal women is about 1/1000/year. Among the users of estrogen, the frequency of occurrence of this disease is, on the average, six times greater. Furthermore, the longer a woman takes estrogen, the greater is her chance of developing the disease. Thus the **relative risk ratio** (risk for users divided by that of nonusers) of this type of cancer increases from 5.6 in women taking estrogen for 1 to 5 years, to 7.2 for those taking the hormone for 5 to 7 years, to

13.9 for those taking estrogen for more than 7 years. Here, we have another example of a problem that is becoming increasingly important for our modern society, namely, that a therapeutic measure designed to correct one situation may cause a different and possibly more serious defect.

☐ Aging

Aging is the progressive deterioration, with time, of the structures and functions of an adult individual. The amount of loss of structure or function, over a given period of time, varies with the system of the body. Table 13-2 lists some typical examples of the decline that occurs between the ages of 30 and 75 years. It is quite clear that the heart of an older person is working at only 70% of what was its capacity earlier in life, while the kidney contains only 56% of the gomeruli with which to rid the body of metabolic wastes. For all other bodily structures and functions, a similar situation exists.

The underlying cause for much of the loss of body functions, with time, is a decline in the body's **basal metabolic rate.** This is the rate of energy utilization in the body during absolute rest, but while the person is awake. The basal metabolism produces the energy for normal body functions; thus its rate is a measure of the amount of energy available for those functions. A newborn child has a basal metabolic rate that is almost 1½ times that of a 25-year-old person, who, in turn, has a 10% higher rate than a person aged 50 years and a 20% higher rate than an 80-year-old person. Thus we have the biochemical basis for the slowness of actions of elderly people.

What causes aging? There is no substantiated answer to this question, but there are a number of theories. One theory stresses the observed accumulation of waste materials in nondividing cells, for example, muscles and nerves. As these cells grow older, their plasma membranes become less permeable, and there is a buildup of poisonous wastes. Cells that continuously grow and divide, for example, liver and bone marrow, do not exhibit the effects of aging.

Another theory of aging stresses the continuous occurrence of genetic mutations, especially as a result of exposure to radiation, for example, cosmic rays and x rays, and chemicals. Because mutations generally lead to altered proteins (Chapter 8), the affected body cells would be reduced in efficiency and might even die. Under this theory, aging is the result of the accumulation of deleterious mutations, with an accompanying loss of cells or a reduction in their function.

Although these theories, and others, do not explain the cause of aging, they do stimulate research workers to perform experiments to test particular hypotheses. The

Table 13-2
Average decline in various body structures and function from ages 30 to 75 years

Characteristic	% Decline
Weight of brain	44
Number of taste buds	64
Output of heart with body at rest	30
Number of glomeruli in kidney	44
Maximum oxygen uptake by blood during exercise	60

information gathered from these experiments gives us a greater understanding of the phenomenon of aging, and, hopefully, this knowledge will permit society to provide for the physical and mental health of its older citizens.

□ Death

The ultimate result of aging is death. **Death,** in the normal usage of the word, refers to the fact that the organism as a whole is no longer a functioning unit. It does **not** imply that every cell of the body has died. It is well established that after an individual has died, various cells survive for short periods of time: nerve cells are among the first to die because of their critical need for oxygen, whereas some of the skin cells are among the last.

When is an individual certified as having died? Traditionally, life and death have been linked to **heartbeat.** Although other organs may fail first, death in man has usually been linked to the time the heart stops beating. However, as discussed in Chapter 12, more recently, **brain waves** have become a criterion of whether a person is alive or dead. The absence of detectable brain waves is taken as proof that an individual has died, because without brain function, a person is little more than an organized group of cells.

What are the leading causes of death in our modern society? The 10 major causes of death in the United States today are given in Table 13-3. These account for 85% of the 2 million deaths that occur each year. Circulatory system diseases account for fully 50% of all deaths, while cancer causes another 19% of them. A comparison of the three leading causes of death today with those in 1900 shows that in the earlier period, infectious diseases predominated (influenza and pneumonia, tuberculosis, and gastroenteritis), whereas today, the chronic diseases (heart, cancer, and stroke) are the most prevalent. Modern medicine has made truly dramatic advances in the fight against infectious diseases. The problems for medicine in the future lie in chronic diseases (Fig. 13-18).

■ SUMMARY

Biological success for the species depends on the ability of its members to reproduce. In man, as in most multicellular organisms, the sexes are separate, and repro-

Table 13-3
Major causes of death in the United States

Cause of death	Death rate/ 100,000 persons*
Heart disease	307
Cancer	170
Stroke	91
Other circulatory diseases	57
Accidents	55
Influenza and pneumonia	31
Diseases of infancy	21
Diabetes mellitus	20
Cirrhosis of liver	15
Bronchitis, emphysema, and asthma	14

*Multiply each figure by 2200 to obtain the actual number of deaths from each cause per year.

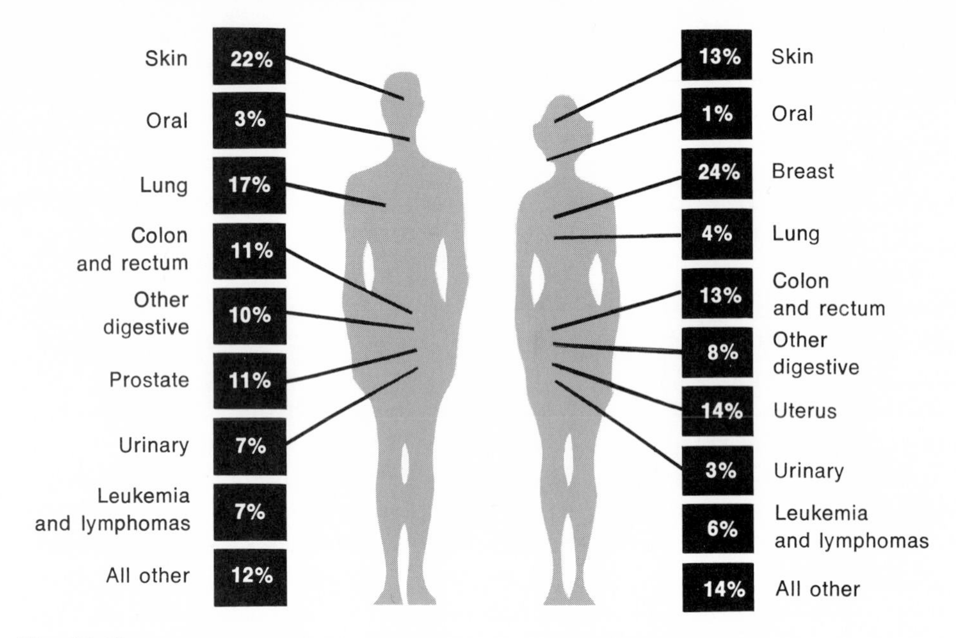

Fig. 13-18
Cancer incidence by site and sex. (Courtesy American Cancer Society; from Turner, C. E. 1971. Personal and community health, ed. 14. The C. V. Mosby Co., St. Louis.)

duction requires a high degree of both physiological and behavioral synchronization of the participants. This is achieved through extensive hormonal control of the development and functioning of both male and female reproductive systems.

The anterior pituitary gland plays a major role in sexual development. There are two hormones involved: (1) follicle-stimulating hormone, which stimulates the production of sperm in the male and the development of the graafian follicle in the female, and (2) luteinizing hormone, which stimulates the production of testosterone in the male and the development of the corpus luteum in the female.

Sperm production is a continuous, day-by-day process. The formation of fertilizable eggs, however, is cyclical and intermittent, as seen in the menstrual cycle. This cycle consists of a sequence of three phases, each usually lasting a rather specific number of days: (1) flow phase (5 days), (2) follicular phase (9 days), and (3) luteal phase (14 days). The sequence is repeated in regular fashion throughout the fertile period of a woman's life.

Contraceptive practices may involve one or more of the following methods: rhythm method, vasectomy, condom, diaphragm, tubal ligation, intrauterine device, and the pill. Use of the pill involves an increased risk of heart attack to women of all ages, but especially to those over 40 years of age.

A large and growing disease problem in modern societies involves the venereal diseases gonorrhea and syphilis. Both are highly contagious, and, in the case of a pregnant female, they can also affect the developing child. Although both diseases can be treated simply and effectively, there is no immunity built up, and reinfection is always possible.

Biological success for the individual rests on his or her ability to develop from fertilized egg through the birth process to adulthood and beyond to old age. The prenatal process involves early cell division (resulting in morula and blastocyst [formation of embryo] stages), gastrulation and subsequent embryonic development, and fetal development, terminating in the birth of the individual.

Included in prenatal development is the implantation of the embryo into the uterine wall and the formation of four embryonic membranes (chorion, amnion, yolk sac, and allantois), each with its own specific function.

The termination of a pregnancy before the fetus can survive outside the uterus is called an abortion. It is usually limited to the first 5 months of gestation and is performed in one of three ways: suction, dilatation and curettage, or saline abortion.

Childbirth generally occurs 280 days after the beginning of the last menstrual period. Labor is at least partially under hormonal control. The newborn's respiratory, excretory, and digestive systems are required to begin functioning at once, while the circulatory, nervous, and endocrine systems must undergo sharp adjustments to new conditions.

Growth and development continue throughout postnatal life. A dramatic stage is puberty, which, in females, is marked by menarche, indicating the beginning of the fertile period. This period of a woman's life will end with menopause. Although males also experience puberty and eventually become sterile, the changes are more gradual and not as clearly defined.

Aging and death are parts of the sequence of the life process. In our modern society, most people are dying because of chronic diseases (heart disease, cancer, etc.). Some years ago, infectious diseases (influenza, pneumonia, tuberculosis, etc.) were the prime killers. The problems for medicine in the future lie in the cure of chronic diseases and in the improvement of the physical and mental condition of older people.

In the next chapters, we shall consider some quite different aspects of man and his place in this world. First, we shall look at the historical development of organisms (evolution) and see where man fits into the picture. Following this, we shall consider man's present relationship to other organisms and to his physical environment (ecology).

SUGGESTED READINGS

Dahlberg, G. 1951. An explanation of twins. Sci. Am. **184**:48-51. Fine discussion on the occurrence of multiple births in man.

Friedmann, T. 1971. Prenatal diagnosis of genetic disease. Sci. Am. **225**:34-42. Well-presented description of amniocentesis and the genetic diseases that can be identified using this procedure.

Jaffe, F. S. 1973. Public policy on fertility control. Sci. Am. **229**:17-23. Review of the changes in public attitude and national policy toward contraception in recent years.

Odell, W. D., and Moyer, D. L. 1971. Physiology of reproduction. The C. V. Mosby Co., St. Louis. Authoritative and detailed book on human reproduction.

Smith, A. 1968. The body. Penguin Books, Inc., Baltimore. Informative account of human reproduction, including pregnancy, birth, and twinning.

Smith, C. A. 1963. The first breath. Sci. Am. **209**:27-35. Very readable discussion of the birth process and the physiological changes that accompany this event.

Tietze, C., and Lewit, S. 1969. Abortion. Sci. Am. **220**:21-27. In-depth discussion of highly controversial topic.

Wood, C. 1969. Human fertility: threat and promise. World of Science Library, Funk & Wagnalls, Inc., New York. Well-illustrated book on human reproduction, birth control, population problems, and the future of mankind.

World Health Organization. 1965. Physiology of lactation. WHO Tech. Rep. Ser. No. 305, Geneva. Detailed review of the factors that affect lactation.

Cro-Magnon man. The most recently extinct form of *Homo sapiens*. Cro-Magnon is best known for his cave art in France and Spain, which seems to have been closely tied to his spiritual life. (Courtesy the American Museum of Natural History; from Nagle, J. J. 1974. Heredity and human affairs. The C. V. Mosby Co., St. Louis.)

CHAPTER 14 Patterns of evolution

LEARNING OBJECTIVES

- What is evolution?
- What are the various types of evidence for the occurrence of evolution?
- How do we determine the age of rocks and their contained fossils?
- On what basis do we construct a "paleontological time table?"
- What is the mechanism of evolution?
- How are organisms classified scientifically and on what basis?
- What are the three "kingdoms" into which most organisms can be fitted, and what are the characteristics of each one?
- What group of organisms does not fit into any of the recognized kingdoms, and what are the characteristics of this special group?
- What is the overall picture of plant evolution, and how does each phylum fit into this picture?
- What is the overall picture of animal evolution, and how does each phylum fit into this picture?
- What is the overall picture of human evolution?
- What is the evolutionary future of man?

In previous chapters, we have looked at man as an individual and a species and have examined how he meets his needs for survival, development, and reproduction. However, a study of man is not complete until we ask, "From where did mankind come?" and also, "Where is mankind going?" It will prove relatively easy to discuss the evidence relating to man's origin and place among other species. It will be difficult to speculate on mankind's future, especially as man himself has become a controlling factor in his own further development.

Any change in the characteristics of organisms, in successive generations, is called **evolution.** These changes may be structural, physiological, developmental, or behavioral, and they reflect changes in the genetic structure, that is, mutation, chromosomal rearrangement, or changes in chromosome number, of the organisms involved. When it is postulated that members of the ancestral and modern populations would not be able to interbreed, even if they were contemporary, a new **species** is said to have developed. Speciation may also develop among contemporary populations of a species that are separated from one another for a long period of time **and** whose members become so different from one another that they are no longer able to interbreed.

■ EVIDENCE FOR EVOLUTION

There are a number of excellent sources from which scientists have obtained information for both the occurrence and the pattern of evolution. These include studies of comparative anatomy, comparative development, vestigial organs, blood proteins, and fossils (paleontology). Each of these lines of investigation utilizes a different approach to the study of evolution, yet they all corroborate Charles Darwin's important statement that all present-day species are descended from previously existing species.

☐ Comparative anatomy

The underlying assumption in this approach to the study of evolution is that organisms whose body systems are similarly organized are related to one another through descent from a common ancestor. This is to say, the body plan of an ancestral type has been inherited and variously modified by all descendant groups. One of the best examples of this type of investigation is to be seen in a comparison of the bones of the forelimb (arm) of man, frog, lizard, bird, and whale in contrast with the parts of a leg of the honeybee (Fig. 14-1).

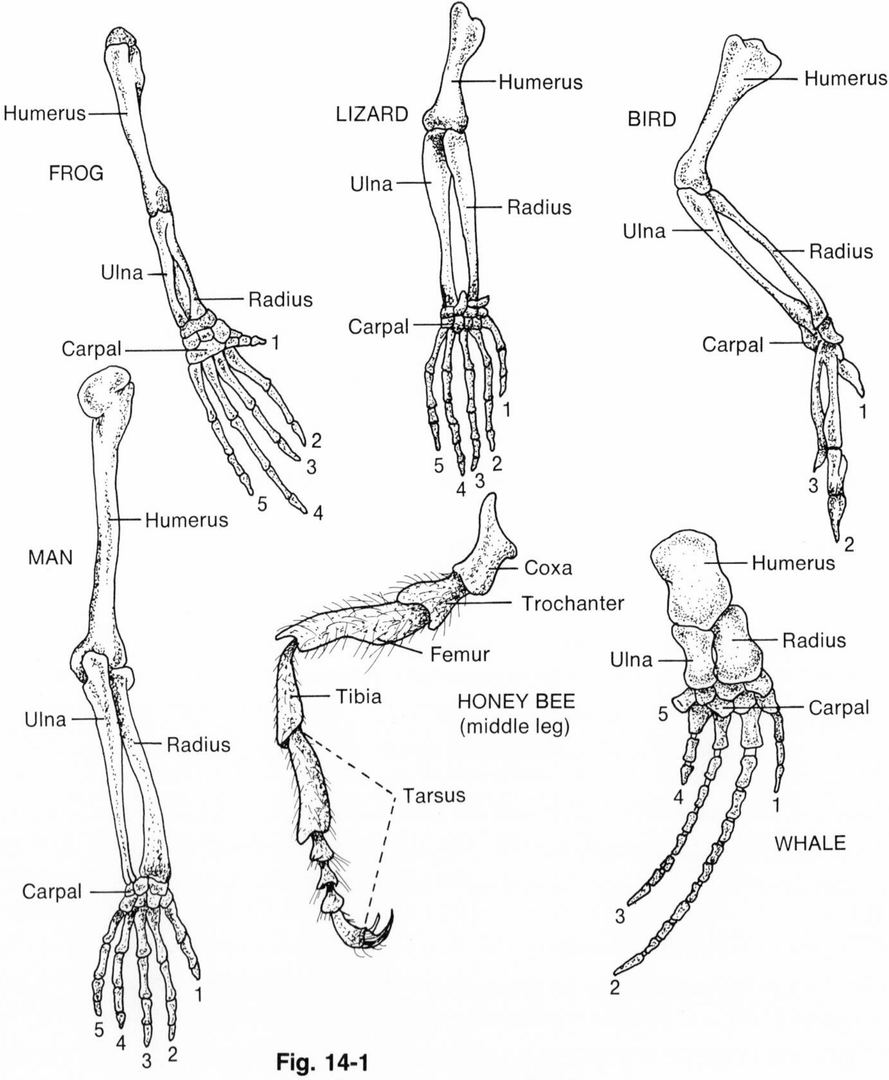

Fig. 14-1
Homology and analogy of limbs.

All the forelimbs shown in Fig. 14-1 have a roughly triangular arrangement of their bones: the upper arm segment contains a single bone, the lower arm segment has two bones, and the wrist, hand, and finger bones radiate out from that point. In contrast to this triangular arrangement, the leg of the honeybee consists of five segments arranged as a linear series.

The similar skeletal plans of man, frog, lizard, bird, and whale are best understood if we assume that they are descended from a common ancestor. This explanation is reinforced by the realization that the same basic bone arrangement has been modified in the different organisms to serve for grasping (man), support (frog), walking (lizard), flying (bird), and swimming (whale). Similarities of body plan in different organisms, regardless of the function served by the structures, is called **homology** and is considered strong evidence for the occurrence of evolution within that group.

A comparison of a leg of the honeybee with the forelimbs of the other organisms shows no similarity of body plan. We therefore assume that the honeybee and man (or any of the others) are **not** directly descended from a common ancestor, despite the fact that the leg of the honeybee and that of the lizard are both used for walking. Similarity of function of body parts is called **analogy.** Analogy by itself does not prove evolutionary relationship, although when analogous structures are also homologous, for example, the hind legs of man, frog, lizard, and bird, descent from a common ancestor is indicated.

□ Comparative development

In comparative development, as was the case in comparative anatomy, similarity of pattern and organization indicates a common evolutionary origin. Fig. 14-2 shows various stages in the development of fish, chick, pig, man, and fruit fly. There are tremendous similarities, especially in the early stages, in fish, chick, pig, and man, indicating descent from a common ancestor. The fruit fly, however, shows no developmental resemblance to any of the others, implying that it has a different evolutionary origin.

The more closely related two organisms are, the longer do they resemble one another during development. Thus in Fig. 14-2, the top row of figures of fish, chick, pig, and man look very similar. In the middle row, the fish is clearly distinguishable from the others, whereas in the bottom row, each type has achieved its characteristic appearance. If we had examined intermediate stages between the second and third rows, we would have found a developmental period during which we could have distinguished the chick embryo from pig and man, but not the latter two from each other.

In a comparison of this sort, the organism that is distinguishable first is the one that is closest to the ancestral type; in this case, it is the fish. The fact that the embryos of forms that evolved later resemble the embryos of those closer to the ancestral type led to an interesting theory that has been summed up in the phrase "*ontogeny recapitulates phylogeny.*" In its original form, it expressed the idea that the individual in its development (ontogeny) goes through the evolutionary stages of the species (phylogeny). However, this is not correct. Although it is not easy to differentiate an early human embryo from a fish embryo, at no stage in its devel-

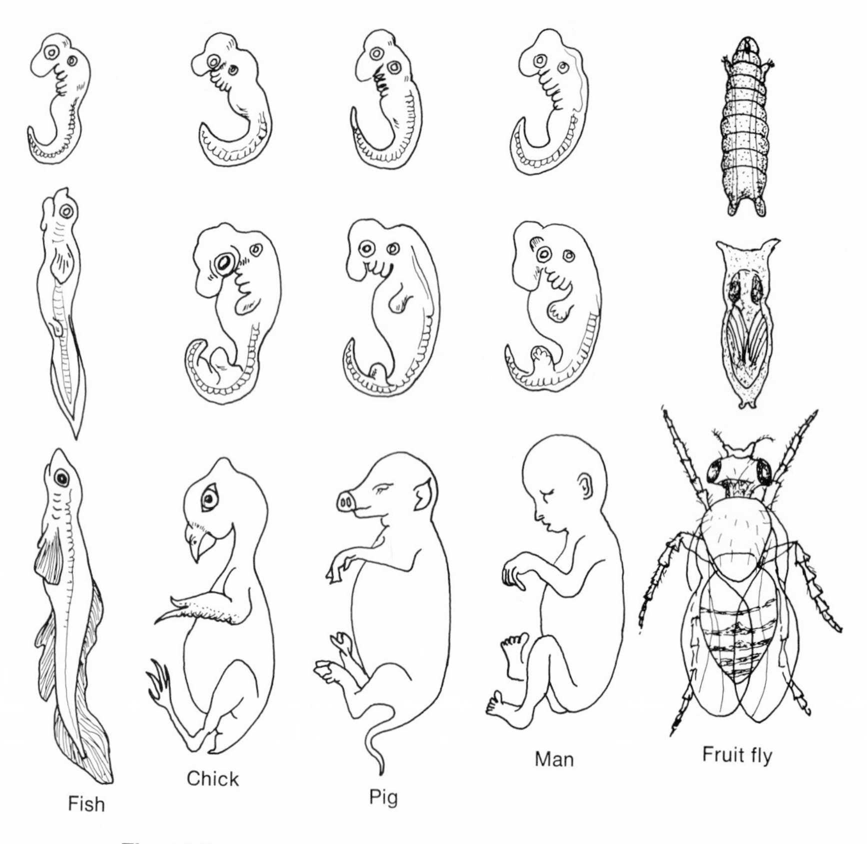

Fig. 14-2
Comparative development of related and unrelated organisms.

opment does a human embryo look like an adult fish. As more properly understood today, human embryos and fish embryos (and the others as well) go through similar developmental stages, indicating a common ancestor.

□ Vestigial organs

Another kind of evidence for evolution is the presence of organs that are useless and degenerate. In the human skeleton, we find that the vertebral column contains a **coccyx** (fused tail vertebrae). The coccyx is vestigial in man and represents an evolutionary remnant of our fishlike ancestor.

A second example of a vestigial organ in man is the **appendix** (Chapter 5). The appendix is an attachment of the cecum, a large and important digestive tract structure in herbivorous animals. It would appear that in our evolutionary history, there was a switch from a totally vegetarian diet to one that contained a large amount of meat.

Man is certainly not alone in having vestigial organs. Whales, who do not have hind legs, have vestigial posterior limbs embedded in the flesh of the body wall. Another example of vestigial organs is found in the python (one of the larger snakes). Here, too, vestigial hind legs are present in the flesh of the body wall.

☐ Blood proteins

In addition to finding evidence for the occurrence of evolution in comparative anatomy and comparative development, we can also look for similarities and differences in the proteins of the body. The most used source of body proteins for this type of study has been the blood. In Chapter 6, we discussed antibodies as a defense mechanism of the body. Antibodies against the blood proteins of one species can be used to check for the presence of similar proteins in other species.

If a small amount of human blood is injected into a rabbit, the plasma of the rabbit's blood will soon contain antibodies against human proteins. When some of the rabbit plasma is then mixed with a sample of human blood, a heavy precipitate is formed as these antibodies react with the human blood proteins. If, however, the rabbit plasma is mixed with gorilla blood, the precipitate is not as heavy. An even weaker reaction occurs with orangutan blood, and a very weak reaction occurs with baboon blood. When the rabbit plasma is mixed with ox blood, there is virtually no precipitation. The amount of precipitation can be used as a measure of protein similarity of different organisms, indicating their evolutionary relationship to one another. On this basis, human blood proteins appear to be very similar to gorilla blood proteins, less similar to orangutan and baboon proteins, and almost completely different from the blood proteins of an ox.

☐ Paleontology

Paleontology—the study of the fossil record—provides not only evidence of the occurrence of evolution, but also indicates the pathways of many lines of descent. The fossil record tells us which type of organism was the ancestral stock of a given line of evolution and which groups were derived from it.

A **fossil** is any evidence of the previous existence of an organism. Fossils are of various kinds: (1) actual remains of extinct forms, for example, frozen bodies of mammoths (extinct elephants) found in Siberia; (2) traces of extinct forms, for example, molds and casts of the shells of organisms that have been embedded in mud or sand that later became rock; and (3) petrification, a process in which the original body material has been replaced by minerals from the surrounding rock (actual rock in the form of a bone, tree trunk, etc.).

Fossil formation. Fossils cannot be found in all rocks. Of the three types of rock that can be formed (igneous, sedimentary, metamorphic), only sedimentary rock contains fossils. **Igneous rock** is formed by volcanic activity and in its molten state is called **lava.** It does **not** contain fossils, because any organism that falls into volcanic lava is completely incinerated.

Sedimentary rock is formed at the bottom of lakes and oceans from particles of soil, sand, and stone that are carried into these bodies of water by streams and rivers. The rate of deposition of these particles at the bottom of the lakes and oceans varies from season to season and from year to year, thus forming distinct layers (**strata**) of sediment. Any animals and plants that die in the lakes or oceans or are carried into them by streams and rivers will drop to the bottom of the water and be covered subsequently by sediment. Under these conditions, petrification can occur, and a fossil can be produced.

The height to which sedimentary rock can be formed is limited only by the depth of the lake or ocean at the point of sedimentation. The Grand Canyon is 1 mile deep, and the mountains that form its sides are gigantic deposits of sedimentary rock that were formed millions of years ago under the Pacific Ocean. The deepest layers were deposited first and contain the oldest fossils. The upper layers were deposited at later times and contain fossils of organisms that are descended from the earlier ones.

As the layers of sedimentary rock increase, those on the bottom are subjected to tremendous pressure. This has the effect of generating a great deal of heat, which can cause the bottom rock to melt and flow. A rock in which this has occurred is called **metamorphic rock.** The melting and flowing of the rock obliterates any fossils that may have been present.

Paleontological time table. A study of sedimentary rock over the world has found that there are five major rock strata, each of which is composed of lesser strata. On this basis, a paleontological time table of 5 **eras** and 11 **periods** was constructed (Table 14-1). The various eras and periods can be dated by taking advantage of the fact that certain radioactive elements are transformed into other elements at a constant rate, usually specified as a **half-life.** As an example, half of a given amount of uranium will be converted into lead in 4.5 billion years. By measuring the amounts of uranium and lead in a given rock, the age of the rock can be calculated. Not all rocks contain uranium and lead. Fortunately, however, enough rocks containing both elements have been found to permit an accurate estimate of the age of the various eras and periods.

Geologists estimate that the earth is about 4½ billion years old. Most of the rock at that time was probably igneous in origin, the earliest sedimentary rock being formed subsequently. No fossils are found in the oldest strata of sedimentary rock,

Table 14-1
Paleontological time table

Era	Period	Age of rock (millions of years)	Organisms showing greatest biological success
Cenozoic	Quaternary	2	Man, mammals, flowering plants
	Tertiary	62	
Mesozoic	Cretaceous	130	Birds, reptiles, evergreen plants
	Jurassic	180	
	Triassic	230	
Paleozoic	Permian	280	Amphibians
	Carboniferous	340	
	Devonian	400	Fishes, fern plants
	Silurian	450	
	Ordovician	500	Invertebrates, algae
	Cambrian	570	
Proterozoic		3000	First multicellular organisms, first unicellular organisms
Azoic		4000	No fossils found
Formation of the earth		4500	

indicating a period of about 1½ billion years during which there were no living organisms on the earth. Starting about 3 billion years ago, fossils of unicellular organisms are found, followed shortly thereafter by fossils of multicellular forms. Starting about 570 million years ago, there appear most of the major groups of plants and animals. Geologically speaking, the appearance of the major groups of organisms in the fossil record is relatively recent, having occurred only in the most recent 12% of the time since the formation of the earth.

Of great interest to us is the fossil record of man and his close evolutionary groups. As will be discussed later in this chapter, the organisms to which man is related include the reptiles, birds, amphibians, and fish, of which the fish are the ancestral type. Fish appear in the fossil record about 400 million years ago, followed by amphibians and reptiles. The reptiles lead to both birds and mammals through separate evolutionary pathways. Man is a most recent arrival in the paleontological time table, appearing only about 2 million years ago. His biological success in so short a period of time is probably without equal in all of evolution.

MECHANISM OF EVOLUTION

How does evolution occur? Evolution takes place only if two conditions are present in a population: (1) there is genetic variation, which provides the raw materials for evolution, and (2) there is differential reproduction, which has the effect of selecting the genetic variations that will be transmitted to succeeding generations.

Every population is characterized by a certain amount of **variation** in the traits of its members. The traits involved may be structural, physiological, or behavioral. Some of the variation among individuals is caused by differences in environment during development, for example, food supply, infectious diseases, etc. Environmentally produced variation in one generation will not be transmitted to the next generation through the reproductive process. This type of variation will only appear in succeeding generations if the environment remains constant. However, a good deal of the variation in a population is the result of differences in the genetic makeup of its members. These genetic differences can be transmitted to succeeding generations through the reproductive process, thus providing the opportunity for evolutionary change. Without genetic variability, all members of a population would be identical (except, of course, for environmentally produced variation), and every generation would be just like the preceding one. Under such conditions, there would be no evolution, because in order to have evolution, there must be genetic variability in a population.

The kinds of genetic changes that can occur in an individual, for example, mutation, chromosomal rearrangement, and changes in chromosome number, were discussed in Chapters 8 and 9. However, the occurrence of genetic variations in various members of one generation does not mean that they will be automatically transmitted to the next generation. The factor that determines to what extent a particular mutation (or other genetic change) will be found in a succeeding generation is **differential reproduction.** If an individual who carries a mutation does not have any offspring, then that allele will not be present in the next generation, and no evolutionary change will have occurred. However, if such an individual has a larger than usual number of offspring and many or all of them carry the allele, the next generation will be characterized by some slight change in the number of individuals with a particular structural, physiological, or behavioral trait, and an evolutionary change will have occurred.

What determines differential reproduction, or, put another way, what determines whether or not a mutation will be transmitted to the next generation? The answer to this question involves two factors: the reproductive capability of organisms and the resources available in the environment to support the offspring produced. In general each mating pair of any species has a large **reproductive capability.** As examples, we can consider the following: a single female cockroach will produce about 500 offspring during her lifetime, a rainbow trout, about 30,000, and an oyster, about 80 million. If this rate of reproduction went unchecked, the world's land area would soon be completely covered by cockroaches, the lakes and streams would be clogged with rainbow trout, and the oceans would be completely filled with oysters. However, this does not occur, and, in fact, the number of individuals of any species remains remarkably constant over considerable periods of time (man, of course, is presently an exception to this statement). What keeps the number of individuals of a species constant? This is determined by the **available environmental resources,** which include food, water, and space. These resources are quite limited for the huge number of offspring produced each generation.

The situation produced by the presence of a large number of offspring in an area with insufficient food, water, etc. leads to competition for the limited amount of available resources. Insofar as genetically different organisms will vary in their efficiency in obtaining food, etc., there will be an unequal survival rate among the carriers of different alleles (**survival of the fittest**). Those who do survive to sexual maturity must then face the problems associated with finding a mate and reproducing. Here, too, carriers of different alleles will vary in their rate of success. The various obstacles to survival and reproduction will result in a differential rate of reproduction among the carriers of different genetic variations. The phenomenon of differential reproduction is often called **natural selection** and is the force that acts on the transmission of genetic variation to succeeding generations, thereby causing evolutionary change to occur.

■ DIVERSITY OF ORGANISMS

When we think about the various kinds of organisms on the earth, we are faced with a vast array of different types. Yet, we can identify certain large groups, all of whose members have a number of characteristics in common, for example, plants, animals, bacteria, protozoa, and viruses. Within each large group, there are identifiable subgroups, each subgroup having certain unique traits, but all the subgroups having the common characteristics of the larger unit. Working with various structural, physiological, developmental, and biochemical traits, the biologist has set up a classification system for extinct as well as present-day organisms.

□ Taxonomy of man

The science of classification, **taxonomy,** uses the following categories in specifying the evolutionary relationship of organisms to one another: kingdom, phylum, class, order, family, genus, and species. A **kingdom** is the largest category into which organisms are placed. The kingdom is divided into phyla, each phylum into classes, and so on until we reach species. A **species** is an assemblage of organisms whose members can mate among themselves and produce fertile offspring. Thus all human beings belong to the same species, *Homo sapiens*, and no other organism on the earth fits into this group. The complete classification of man is given in Table 14-2, which also demonstrates how taxonomy can be used to indicate the extent of evolutionary relationship.

Table 14-2 lists a wide variety of organisms within the kingdom **Animalia** (animals). One of the phyla within this kingdom, namely, the phylum **Chordata,** includes man, apes and monkeys, dogs, cats, etc., but **not** grasshopper, clam, earthworm, or starfish. Each of these latter four types belongs to a different phylum and hence does not bear as close an evolutionary relationship to man as do the others in the table. Within the phylum Chordata, there is a group that has as one of its characteristics mammary glands. These organisms belong to the class **Mammalia,** and, as can be seen in the table, this group does **not** include reptiles (lizard), amphibians (frog), and fish. Thus we see that the bat, whale, dog, cat, etc. bear a closer evolutionary relationship to man than do the fish, amphibians, and reptiles. Within the mammal group, there are a number of organisms that have five digits (fingers or toes) on both hands and feet, with the innermost finger, or toe, or both, opposable to the second finger for grasping and climbing. These organisms belong to the order **Primates** and include the monkeys, apes, and man, but not the other mammals. Within the primate group, there is **one** line of evolution that resulted in organisms with a tremendously increased brain size, a very flat face, legs about 30% longer than arms, and the big toe **not** opposable to the second toe. This was the line of evolution that led to man, and the organisms of this type belong to the family **Hominidae** (men), the genus ***Homo*** (man), and the species ***Homo sapiens*** (thinking man). Because there was only one primate group that evolved in this fashion, mankind alone occupies these last three taxonomic categories.

Table 14-2
Classification of man

Kingdom (Animalia)	Phylum (Chordata)	Class (Mammalia)	Order (Primates)	Family (Hominidae)	Genus *(Homo)*	Species *(Homo sapiens)*
Man	Man	Man	Man	Man	Man	Man
Ape	Ape	Ape	Ape			
Monkey	Monkey	Monkey	Monkey			
Cat	Cat	Cat				
Dog	Dog	Dog				
Whale	Whale	Whale				
Bat	Bat	Bat				
Lizard	Lizard					
Frog	Frog					
Fish	Fish					
Grasshopper						
Clam						
Earthworm						
Starfish						

□ Classification of organisms

Taxonomists generally agree that the organisms on earth can be divided into three kingdoms: Monera, Plantae (plants), and Animalia (animals). The criteria for establishing these three kingdoms are a number of very significant differences in cell structure, which reflect distinct lines of evolutionary development. However, before considering the various kinds of organisms in these kingdoms, we shall discuss a group that does not fit into any of the categories. These are the viruses.

Viruses. As we shall see later in this chapter, organisms are classified as belonging to a particular kingdom, depending on the structural characteristics of their cells. Viruses are not given a place in this type of classification, because their component parts are not organized into cells. A virus consists of protein and nucleic acid. The protein acts as a "coat" that surrounds the nucleic acid "core." The nucleic acid is a single molecule of either DNA or RNA, depending on the particular virus.

Viruses have a very restricted type of life cycle that does **not** include growth or division, but **does** include reproduction. They have no metabolic machinery of their own, and, as a result, they cannot carry on an independent existence. Viruses are parasites that must rely completely on the metabolic machinery of their host cells for all their activities. Among the many human diseases caused by viruses are measles, chickenpox, mumps, smallpox, influenza, viral pneumonia, poliomyelitis, the common cold, and infectious hepatitis.

When a virus invades a cell, the virus follows one of two patterns: (1) it takes control over the cell's metabolism and directs the cell to synthesize viral nucleic acid and protein and to assemble these components into complete viruses, or (2) the nucleic acid of the invading virus is inserted into the nucleic acid of the host cell and becomes part of the host's genetic composition, functioning like any other set of genes.

Where a viral invasion results in the production of new viruses, we have a characteristic infection pattern. The new viruses are released from the host cell and, in turn, invade other cells. The host cells are usually killed when the virus particles are released, because this often involves a splitting of the host cell's plasma membrane. Poliomyelitis (infantile paralysis) is an example of this type of viral infection.

Where a viral invasion is followed by insertion of the viral nucleic acid into the host's genetic system, there follows a replication of both viral and host nucleic acid during each cell cycle and a transfer of the viral genes to each cell derived by mitosis from the original host cell. The viral genes function in the host cell in the production of proteins with important consequences for some human diseases.

One of the types of food poisoning has been attributed to a bacterium, called *Clostridium botulinum.* However, there are some strains of this bacterium that do not cause the disease. A study of toxic (disease-causing) and nontoxic strains has shown that the toxic strains contain a virus that the nontoxic strains do not. There is little doubt that this type of food poisoning is caused by a toxin that is specified by the genes of a virus and not by its bacterial host.

Another disease that can develop following viral nucleic acid incorporation into the host cell's genetic system is **cancer.** There is good evidence that some cancers are caused by viruses and a growing suspicion that all types of cancer are virally induced. The viral genes direct the production of proteins that in some manner, which is as yet unknown, transform a normal cell into a fast-growing, fast-dividing, tremendous energy–consuming type of cell. These cells also have a tendency to break away from the group that formed them, travel through the body by way of the circulatory system, and establish themselves at new sites. Complicating the whole study of virally induced cancer is the fact that some of these viruses can be acquired early in life without causing a cancer until, in some cases 20 to 25 years later, induced by an irritant, such as certain chemicals, tobacco smoke, frequent abrasion, or radiation.

Monera. The kingdom **Monera** includes all those organisms whose cells do **not** possess a nuclear membrane, mitochondria, endoplasmic reticulum, golgi apparatus, and lysosomes (Chapter 7). The absence of a nuclear membrane results in the cell's genetic material (in these organisms, a single chromosome in the form of a ring) lying directly in the cytoplasm. This is considered a very primitive type of cellular organization, and the Monera are referred to as **procaryotes,** that is, before the nucleus. This kingdom consists solely of single-celled organisms that fall into one of two phyla: Schizophyta (bacteria) and Cyanophyta (blue-green algae).

Bacteria are most important to man in a number of ways. They obtain their food through the chemical breakdown of the dead bodies of other organisms. In doing so, bacteria return the elements contained in dead bodies to the world's **food cycle,** thus acting as one of the **decomposers** in food chains (see Chapter 2 and Fig. 2-1). Bacteria are also useful in many technological processes. They are used in the production of vinegar, several vitamins, leather goods, cheeses, and many antibiotics. Unfortunately, some bacteria and their poisons, called **toxins,** also cause a number of diseases: cholera, diphtheria, syphilis, gonorrhea, tetanus, tuberculosis, typhoid fever, bacterial pneumonia, meningitis, and many others. Modern medicine uses the antibiotics produced by some bacteria to combat those that cause disease.

Blue-green algae are chlorophyll-containing Monera that are important as one of the photosynthetic food **producers** on earth. These algae are not eaten directly by man, but they do contribute to the various **food pyramids** from which mankind benefits (Chapter 2). The fact that bacteria and blue-green algae are placed in the same kingdom indicates that the presence or absence of chlorophyll is not the most important factor in classifying these organisms. The Monera are considered the most primitive single-celled organisms on the earth today and are thought to be close to the ancestral stock from which all other groups evolved.

Plantae (plants). The kingdom **Plantae** includes all those organisms whose cells **do** possess a nuclear membrane, mitochondria, endoplasmic reticulum, golgi apparatus, and lysosomes, **and,** in addition, have a nonliving **cell wall** outside the plasma membrane. The presence of a nuclear membrane separates the cell's genetic material (in these organisms, more than a single chromosome, usually in the form of rods) from the cytoplasm, thus establishing a nucleus. Organisms with a distinct nucleus, which also includes the Animalia, are called **eucaryotes,** meaning having a true nucleus. This kingdom consists of three phyla: Thallophyta (algae and fungi), Bryophyta (liverworts and mosses), and Tracheophyta (ferns, evergreens, and flowering plants).

Thallophyta. The **thallophytes** are the least complex of all plants. Organisms belonging to this group include both single-celled and multicellular forms. However, the multicellular thallophytes have very little tissue formation, all their cells tending to be similar in structure and function. This phylum is divided into a chlorophyll-containing group, called Algae, and a nonphotosynthetic group, called Fungi (yeasts, molds, and mushrooms).

Algae belonging to the kingdom Plantae are of three types: green, brown, and red. Although, unfortunately, the names are similar, these algae are not closely related to the blue-green algae discussed earlier. Green, brown, and red algae are important as photosynthetic food **producers. Fungi** are thallophytes that lack chlorophyll. They, like the bacteria, are **decomposers** and, together with the bacteria, often cause spoilage of food, leather goods, fabrics, paper, and lumber. Some fungi are parasitic on man and cause such skin diseases as ringworm and athlete's foot. On the other hand, fungi are beneficial in a number of ways. Yeasts are important in the manufacture of alcohol products and in the baking of bread. The antibiotic penicillin is obtained from a mold, while mushrooms are used as food (Fig. 14-3).

Although some of the fungi live on land, for example, mushrooms and bread mold, most thallophytes live either in fresh or salt water. In their sexual reproduction, the water-dwelling (aquatic) species use the surrounding water as the medium through which their gametes travel to fuse with one another. A great deal of plant evolution has involved the development of land-dwelling (terrestrial) forms. The

Fig. 14-3
Green mold shown here is *Penicillium chrysogenum*, a mutant form from which almost all the world's supply of penicillin is obtained. (Courtesy Chas. Pfizer & Co., Inc., New York, N.Y.; from Arnett, R. H., and Braungart, D. C. 1970. An introduction to plant biology, ed. 3. The C. V. Mosby Co., St. Louis.)

great problem for land plants is the development of a mechanism for achieving sexual reproduction. The two phyla we shall now discuss represent two stages in the evolution of plants from water to land.

Bryophyta. The **bryophytes** are the simplest land plants. They are multicellular with a moderate amount of tissue differentiation. Together with all other types of photosynthetic organisms, the liverworts and mosses function as **producers** in the food cycle. In the bryophytes, as will also be true of the tracheophytes, the life cycle includes two distinct phases: haploid and diploid. Both parts of the life cycle have characteristic plant structures, and together they constitute an **alternation of generations** (Fig. 14-4).

The **haploid phase** of the bryophyte life cycle is relatively simple, consisting in some species of a thick sheet of cells and in others of a short upright structure. Gametes, **sperm** and **egg,** are produced in special compartments on the plant, the sperm being motile whereas the eggs are sessile. Sexual reproduction is accomplished by the sperm swimming to the egg and fusing with it, forming a **zygote.** The zygote is the beginning of the **diploid** phase of the life cycle. The zygote divides by **mitosis** and forms a relatively simple diploid structure that remains attached to the haploid parental plant. Specialized diploid cells, called **spore mother cells,** are produced and undergo **meiosis,** forming haploid **spores.** The spore is the beginning of the haploid phase of the life cycle. The spore divides by mitosis, producing the next haploid plant of the alternation of generations.

In the bryophyte life cycle, sperm must swim to the egg to effect sexual reproduction. As a result, liverworts and mosses are generally restricted to moist localities, especially the sides of rivers, lakes, and ponds. The bryophytes represent an early stage in the evolution of plants from water to land.

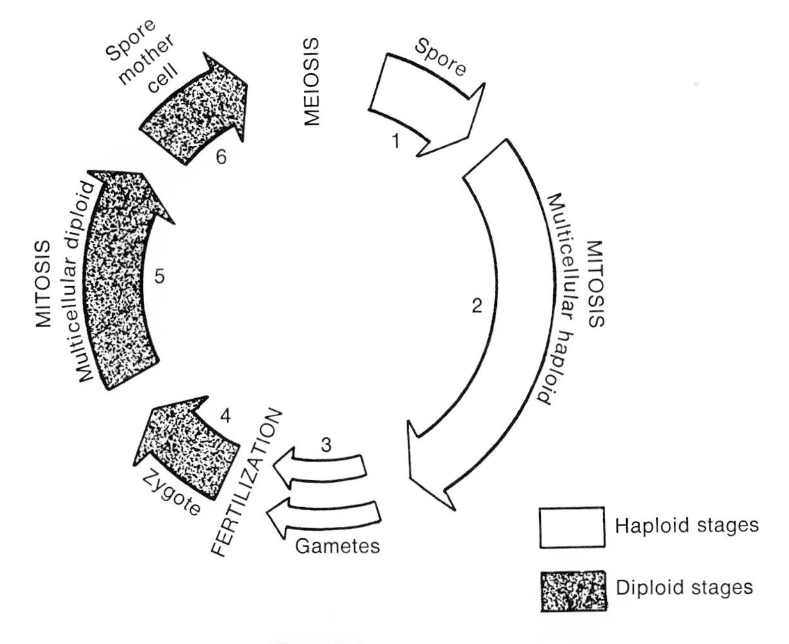

Fig. 14-4
Life cycle of a bryophyte.

Tracheophyta. The **tracheophytes** are the most advanced land plants. They are multicellular and have a tremendous amount of tissue specialization in the **diploid** phase of their life cycle, forming roots, stems, and leaves. The most characteristic type of tissue that they have evolved is **vascular tissue,** which permits the development of very large plants that can conduct both raw materials (water, nitrates, etc.) and manufactured compounds (carbohydrates, amino acids, etc.) to all their cells.

With regard to their life cycle, we find two situations: (1) **ferns** follow the bryophyte pattern just discussed and, as a result, are restricted to relatively moist localities in order to effect sexual reproduction and (2) **evergreens** and **flowering plants** have evolved an independence from water for sexual reproduction.

In the alternation of generations of **advanced tracheophytes** (evergreens and flowering plants), shown in Fig. 14-5, most of the life cycle is diploid, the haploid stage being severely reduced. As stated earlier, the diploid phase of the life cycle consists of what we normally consider a typical plant, namely, root, stem, and leaves. Most of the leaves are photosynthetic, but some are modified for reproduction and may be found in clusters, called **cones** in the case of evergreens and **flowers** in the case of flowering plants. In both cases, specialized structures produce diploid **spore mother cells** that undergo **meiosis,** forming haploid **spores.** Unlike the bryophytes and the ferns, advanced tracheophytes retain the spore that will form the **female haploid plant** within the cones or flowers, as the case may be. The female haploid plant develops within the diploid parent and produces eggs. The spores that will produce the **male haploid plant** are shed from the cones or flowers in the form of **pollen**

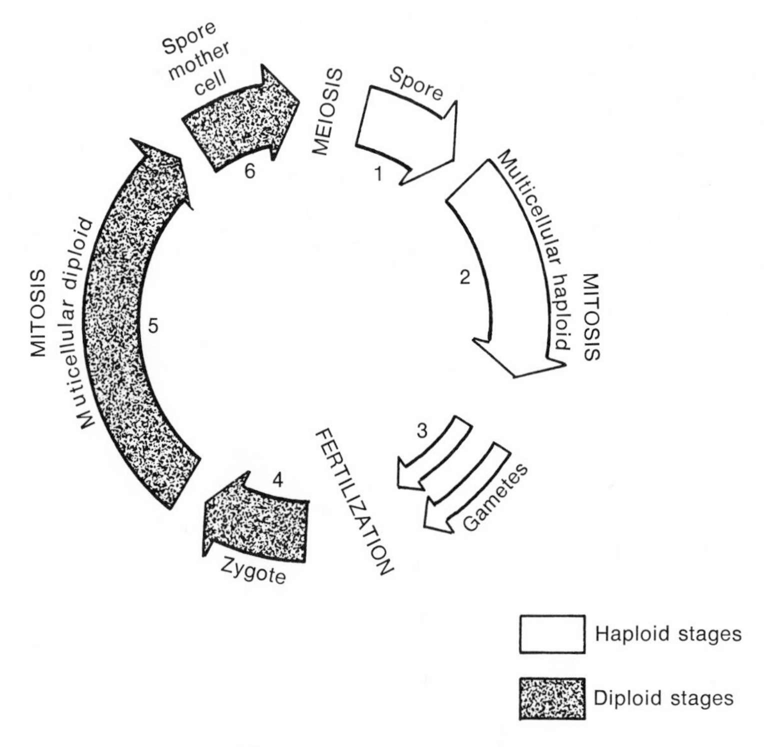

Fig. 14-5
Life cycle of a tracheophyte.

grains and are carried about by wind or insects. Those pollen grains that land on a plant of the same species will develop a **pollen tube** that enzymatically dissolves its way to the female haploid plant. There a fusion of **sperm** and **egg** cells that results in the formation of a **zygote.** The zygote divides by mitosis and forms an embryo plant that is called a **seed.** In the case of flowering plants, the base of the flower may develop into a large and tasteful **fruit** that will attract animals that, in eating the fruit, will help disperse the seeds.

The importance of tracheophytes to our well-being cannot be overstated. In the case of **ferns,** their contribution to our present welfare began more than **300 million years** ago in the Carboniferous period of the Paleozoic era (Table 14-1). At that time, most of the central United States was moist and warm, and fern forests flourished. The growth rate of ferns was so rapid that complete decay and recycling of elements did not occur. A good deal of the dead fern plants became incorporated into sediments in swamps and were eventually converted into **coal.** Thus today's coal reserves represent the stored chemical energy of photosynthesis and is a nonrenewable resource.

Evergreens, especially the pines, firs, and spruces, are of great commercial value as a source of **wood** for lumber, paper, and plastics. Wood comes from the stems of trees and is actually millions of dead empty cell walls of the vascular tissue that is so characteristic of tracheophytes.

Flowering plants literally feed, clothe, and house mankind. Most of our plant food comes from this group. Such basic parts of our diet as corn, wheat, rye, oats, barley,

Fig. 14-6
Fibers of the cotton boll, a form of fruit, supply much of the clothing and other fabrics used throughout the world. (USDA photograph; from Arnett, R. H., and Braungart, D. C. 1970. An introduction to plant biology, ed. 3. The C. V. Mosby Co., St. Louis.)

and rice are the seeds of their respective plants. Apples, oranges, grapes, peaches, and pears are fruits. In addition, we eat roots (carrots, potatoes, radishes), stems (celery, rhubarb), and leaves (lettuce, cabbage, spinach).

Much of our clothing is made from **cotton,** probably the most important plant fiber. Cotton has been spun, woven, and dyed since prehistoric times. The seeds of the cotton plant are formed in capsules, or bolls; each seed is surrounded with downy fiber, white or creamy in color and easily spun. It is this fiber that is processed and made into clothing (Fig. 14-6). Another plant fiber that has been used since prehistoric times to make cloth is **flax.** In this case, it is the stem of the plant that contains the desired fibers.

Although most of our commercial wood comes from evergreens, the flowering plants also provide us with materials for our houses. Furniture is often made from the stems of oak, walnut, ebony, and teak trees.

Plant evolution. The main branches of plant evolution are shown in Fig. 14-7. It should be noted that the origins of the various groups are only generally indicated, without specifying which species was the ancestral stock for any of the evolutionary lines of development. The uncertainty in this case is the result of the lack of a fossil record of every species that ever existed. Most of our fossils come from organisms with bones, shells, or hard woody parts. Any soft-bodied, especially microscopic, organism has only a remote chance of being fossilized. This fact is all the more significant in drawing up an evolutionary tree, because many of the dramatic steps in evolution appear to have involved soft-bodied forms.

Animalia (animals). The kingdom Animalia includes all those organisms whose cells possess a nuclear membrane, mitochondria, endoplasmic reticulum, golgi apparatus, and lysosomes, **but** are **without** a cell wall. Members of this kingdom have a distinct nucleus and, like the Plantae, are **eucaryotes.**

There are a great many more animal than plant phyla. This indicates that animal evolution has been more diverse than plant evolution. The difference in number of evolutionary pathways is probably linked to the basic problem that animals have of acquiring their food either from plants or other animals. The need to find (or hunt) and obtain (sometimes capture and kill) diverse kinds of foods has undoubtedly stimulated the evolution of different ways of life, each carrying with it a certain characteristic arrangement of body parts, embryonic development, and even biochemistry.

As was discussed earlier, most of plant evolution can be seen as a migration from water to land, with the emergence of the diploid phase of the life cycle as the dominant part of the alternation of generations. The production of root, stem, and leaf by diploid tissue is essentially an adaptation to life on land and is quite similar in most tracheophytes (vascular plants). In animals, by contrast, there is **no** alternation of generations. The only part of the life cycle that is haploid is the gamete (sperm or egg), and it does not have an independent existence in any animal species. Animal evolution has involved the diploid part of the life cycle almost exclusively. Based on the various types of evidence of evolution discussed earlier, it is generally agreed that animal evolution had an early stem line, consisting of relatively simple organisms (Protozoa, Coelenterates, Platyhelminthes). Further evolutionary change resulted in two distinct lines: invertebrate (Annelida, Mollusca, Arthropoda) and vertebrate

(Echinodermata, Chordata). An overall picture of animal evolution is shown in Fig. 14-8. There are many more animal phyla than those eight shown in Fig. 14-8, but these will indicate the general evolutionary trends of the Animalia.

Early stem line of animal evolution. There are three phyla—Protozoa, Coelenterata, and Platyhelminths—that represent the initial stages of evolutionary development within the animal kingdom. The phylum **Protozoa** is made up of single-celled animals. Because they have no chlorophyll, most protozoan nutrition is accomplished by ingesting solid foods. In order to do this, protozoa must be capable of movement, and three types of locomotion are common: (1) fingerlike pseudopodia (false feet), (2) whiplike flagella, and (3) hairlike cilia. Each type of locomotion is characteristic of a separate group of these organisms. There is a fourth group that has

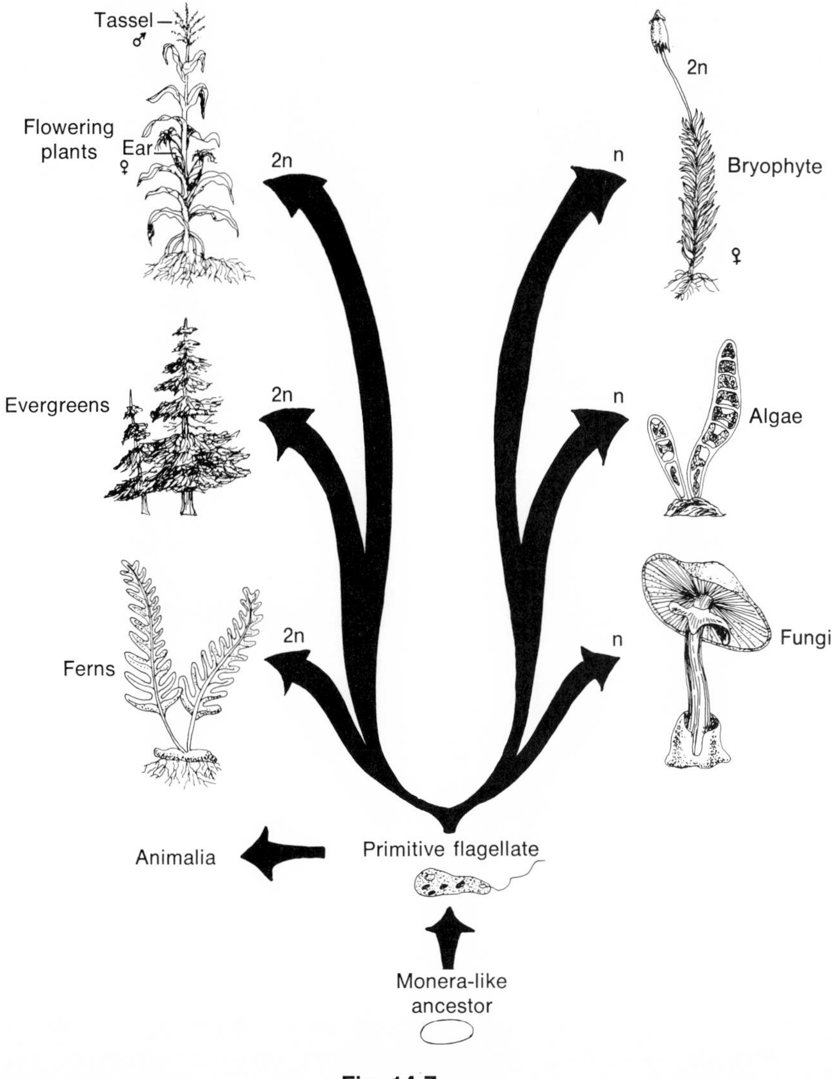

Fig. 14-7
Plant evolution.

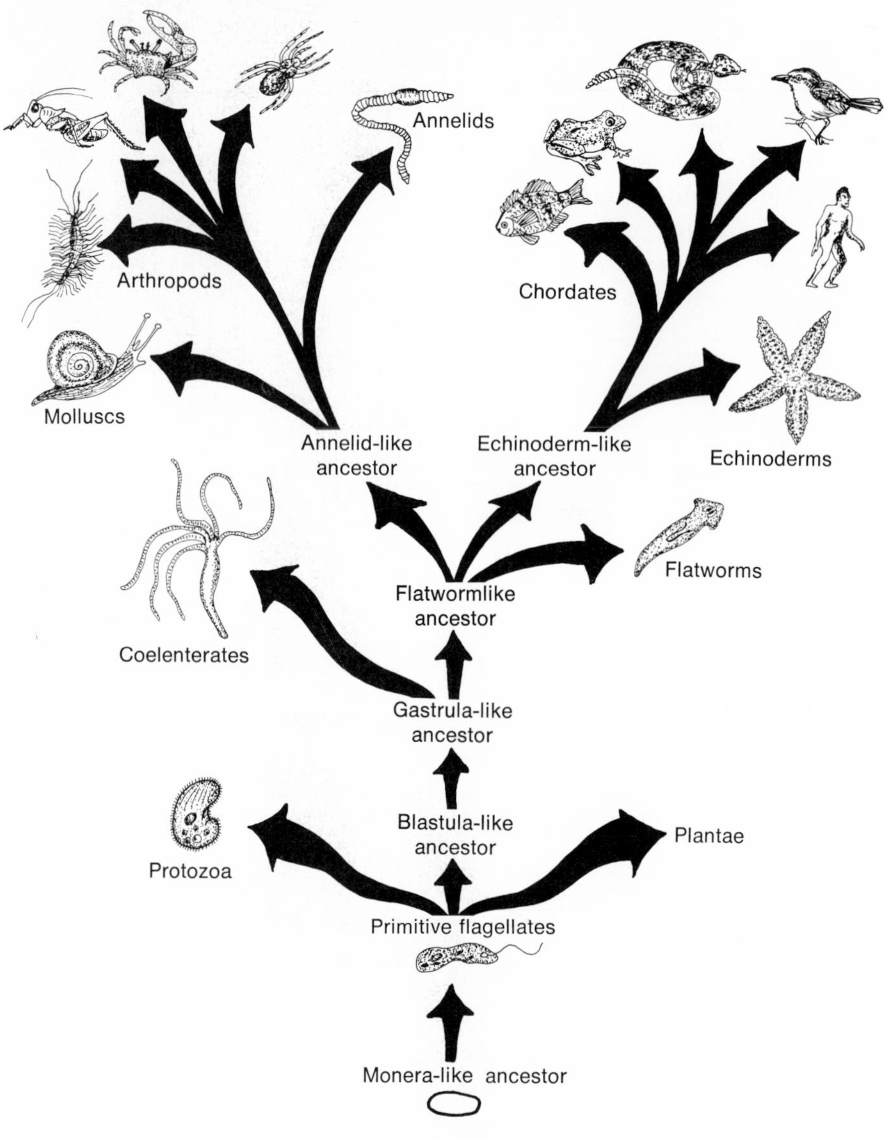

Fig. 14-8
Animal evolution.

no means of locomotion. Its members are parasitic on other animals; perhaps the best known are the causative agents of various types of **malaria:** *Plasmodium vivax, P. malariae,* and *P. falciparum.* Other types of protozoans also cause human diseases. One of the pseudopod-forming protozoans, *Entamoeba histolytica,* invades the intestinal tract and causes **amebic dysentery,** whereas one of the flagella-type protozoans, *Trypanosoma gambiense,* invades the nervous system, causing **African sleeping sickness.**

The phylum **Coelenterata** is the simplest of the multicellular animals. Its body plan, shown in Fig. 14-9, consists of two layers of cells: **ectoderm** on the outside and

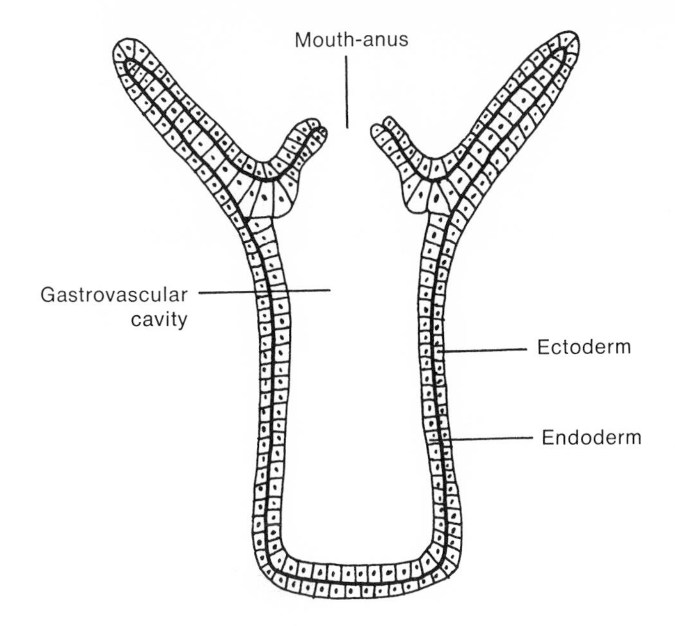

Fig. 14-9
Body plan of a coelenterate.

endoderm on the inside. There is a single opening into the central **gastrovascular cavity,** which acts as both a digestive and a circulatory tract. The coelenterate body plan has a strong resemblance to the early gastrula stage of some organisms and has been part of the basis for the theory of "ontogeny recapitulates phylogeny." Included in this phylum are the jellyfishes, sea anemones, and corals.

The phylum **Platyhelminthes,** commonly called the flatworms, has an overall body plan similar to that shown in Fig. 14-9. However, a third layer of cells, called **mesoderm,** is located between the ectodermal and endodermal cells. This is the first phylum of animals that has the three primary germ layers that are so important in the development of the various organ systems of the body. The resemblance of the flatworm body plan to the later gastrula stage of many organisms also supports the idea of evolutionary recapitulation during development. Some of the flatworms are parasitic in man. A group known as **flukes** have various species, one of which (blood fluke, *Schistosoma hematobium*) invades the bloodstream, whereas another (liver fluke, *Clonorchis sinensis*) invades the liver, and yet another (lung fluke, *Paragonimus westermani*) invades the lungs. Another group of flatworms is the **tapeworms.** They are human intestinal parasites that gain entrance to the digestive system through man's eating of raw or "rare" beef, pork, or fish, most of which carry these parasites.

Invertebrate line of animal evolution. The invertebrate line of evolution can be studied through three of its phyla: Annelida, Mollusca, and Arthropoda. The phylum **Annelida** consists of organisms with ectoderm, endoderm, and mesoderm **and** with two openings for its digestive tract (separate mouth and anus). The annelid body is segmented into nearly identical units going from mouth to anus. These animals have a circulatory system whose main pumping organ is located dorsally in the body and a

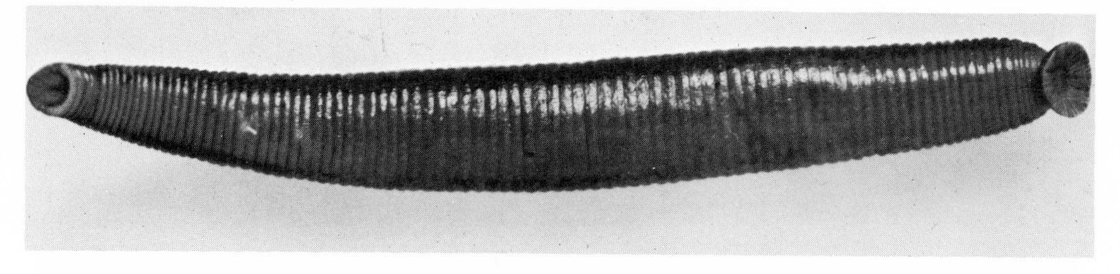

Fig. 14-10
Hirudo medicinalis, medicinal leech. This form was once widely used in bloodletting and is about 4 inches long, but capable of great contraction and elongation. (From Hickman, C. P., Sr., Hickman, F., and Hickman, C. P., Jr. 1974. Integrated principles of zoology, ed. 5. The C. V. Mosby Co., St. Louis.)

nervous system whose nerve cord is solid, that is, without a neural canal, and located ventrally. The annelids include, among others, the earthworms and the leeches. Earthworms live and burrow in the soil, making it porous and thereby improving the soil for gardening. Leeches have been of historical importance in medicine, because at one time, it was believed that certain illnesses were caused by the presence of excess blood in the body. Leeches, whose normal diet is blood, were used for extracting blood from those thought to be suffering from excess blood. During those times, pharmacies kept a supply of leeches for this purpose (Fig. 14-10).

As seen in Fig. 14-8, two phyla evolved from an annelid-like ancestor. One of these is the phylum **Mollusca.** Among the best-known molluscs are snails, clams, oysters, squids, and octopuses. These organisms have most of the annelid body characteristics, with the exception that they have an **unsegmented body** that in most cases is covered by a **hard, inflexible shell.** The unsegmented molluscan body illustrates an interesting point about evolution, namely, there is no one line of development that all groups follow. The segmented annelid body arose from an unsegmented flatworm-type body and, in turn, gave rise to an unsegmented molluscan body.

The second phylum to have evolved from an annelid-like ancestor is the **Arthropoda.** This phylum is by far the largest on earth and contains more than 80% of the approximately 1.2 million described animal species. As was discussed in Chapter 1, every species is an evolutionary adaptation to a particular environment, and the members of different species cannot normally mate with one another. The presence of a large number of species indicates that the phylum has evolved many different groups, each one adapted to a somewhat different way of life. The multiplicity of species is called **adaptive radiation** and is considered a measure of the biological success of phyla.

What accounts for the high degree of biological success of the arthropods? The answer appears to lie in the arthropod body plan, which was derived from the annelids by a simple but important modification. Whereas the segments of annelids closely resemble each other, the segments of most arthropods vary greatly in design and function, depending on the part of the body involved. At the anterior end of the body, the segments are specialized for sensory functions and ingestion of food. The middle portion of the body is usually specialized for locomotion, whereas the posterior part of the animal is involved with excretion and reproduction. Each section of the body may become modified to meet a particular environmental need independently of the others. This versatility in specialization of body parts has permitted the arthropods to evolve a vast array of species, each one adapted to a particular way of life.

Another important evolutionary development of the arthropods is the formation of a flexible nonliving **exoskeleton** derived from ectoderm. The exoskeleton consists of a series of thick, hard, overlapping plates, joined together by thin, soft, connecting strips. The hard plates serve both for protection and as points of attachment for muscles, whereas the soft connecting strips provide for the flexibility of movement that is so important in obtaining food and escaping enemies. The presence of a nonliving exoskeleton presents a problem for growth. In order to grow, an arthropod must periodically shed its exoskeleton (a process called **molting** or **ecdysis**) and form a new one. During the molting process, the organism is quite defenseless and unable to use its muscles effectively. However, as seen in the tremendous amount of adaptive radiation of the group, the shortcomings of a nonliving exoskeleton have not been a great handicap in arthropod success.

The phylum Arthropoda consists of a number of classes. These include the **Crustacea** (crabs, lobsters, crayfish), **Myriapoda** (centipedes and millipedes), **Arachnida** (spiders, scorpions, ticks, and mites), and **Insecta** (cockroaches, flies, termites, lice, grasshoppers, beetles, butterflies, moths, silverfish, and bugs). Of these classes of arthropods, the insects have the greatest number of species (more than 70% of all described animal species). There are those insect species whose activities are beneficial to man, but these are outweighed by those that are harmful. The beneficial insects include, among others, the bees that produce honey, the lac insects that secrete a wax from which shellac is made, and those insects that serve as food for fish and birds, thereby contributing to a food pyramid. Unfortunately, insects are harmful to man in many ways. Many insects destroy or damage plants, lumber and fruit trees, stored grains, fruits, and vegetables. Other insects are destructive to houses, while still others attack clothing, carpets, and fabrics. Finally, various diseases are transmitted by insects: malaria (mosquito), dysentery (flies), bubonic plague (fleas), and typhus fever (body lice).

Vertebrate line of animal evolution. The pattern of animal evolution shown in Fig. 14-8 indicates that two lines of evolution developed from a flatworm-like ancestor. In what ways do the invertebrate and vertebrate body plans, and hence ways of life,

Table 14-3
Differences in body plan of invertebrates and vertebrates

	Features	
Body structure	*Invertebrates*	*Vertebrates*
Skeleton	Nonliving, external, derived from ectoderm	Living, internal, derived from mesoderm
Nerve cord	Solid, located ventrally in body	Hollow, located dorsally in body
Heart	Located dorsally in body	Located ventrally in body

differ? There are a considerable number of differences between the two groups. A few of these are listed in Table 14-3. Variations in parallel evolutionary lines derived from a common ancestral stock illustrate an important phenomenon in evolution, called **opportunism in evolution,** which refers to the fact that when two or more possibilities for meeting a need exist, for example, skeletons can be internal or external, different evolutionary lines will often follow different pathways in meeting the need.

There are two phyla that comprise the vertebrate line of evolution: Echinodermata and Chordata. The **echinoderms** include the starfish, brittle stars, sea urchins, sea cucumbers, and sea lilies. Their importance to man lies mainly in their destruction of large numbers of molluscs, particularly oysters and clams, which they use as food, but which diminishes man's food supply.

The phylum **Chordata,** to which man himself belongs, contains five classes: fish, amphibians, reptiles, birds, and mammals. The evolutionary relationship of these groups to one another are shown in Fig. 14-11. It should be noted that the term "fish" includes four different groups:

1. **Jawless fish,** which are the first chordates to be found in the fossil record. Their mouths are circular because of the lack of a lower jaw. These forms probably lived on the bottom of streams and ponds, feeding on small, soft-bodied invertebrates. There are two living examples of jawless fish. They are the **lampreys** and the **hagfishes,** whose importance for man lies in their mode of feeding. They attach themselves to other fish, bore into their skin to obtain

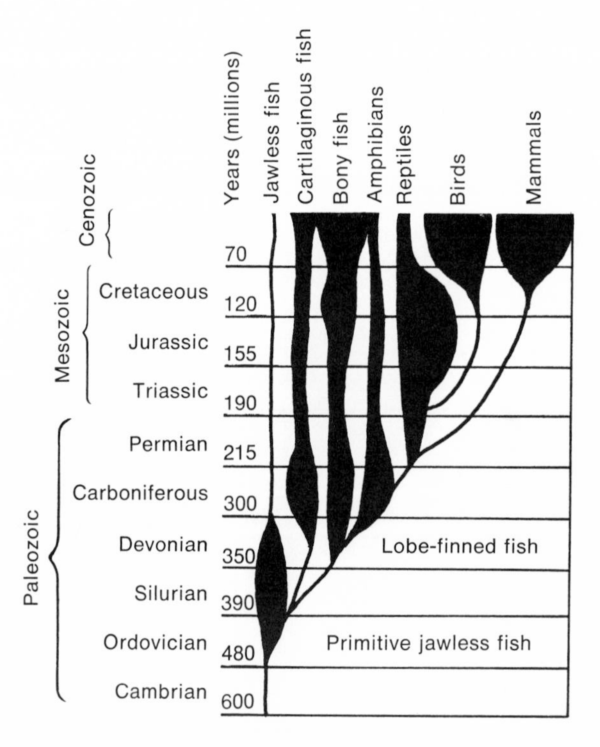

Fig. 14-11
Evolution of the phylum Chordata. Width of each group represents relative number of species.

blood and soft tissues, and thereby cause the destruction of the fish and, as a result, the loss of these fish from man's food supply (Fig. 14-12).

2. **Cartilaginous fish,** whose living members include the sharks, rays, and skates. They are a very old group that evolved from the jawless fish, but, for reasons unknown, have their skeletons as cartilage rather than bone.
3. **Bony fish,** which includes most of our edible fish and which have been very successful in their adaptive radiation of the earth's hydrosphere (oceans, lakes, rivers).
4. **Lobe-finned fish,** of which there is only one species alive today, which was the ancestral group of the first land vertebrates—the amphibians.

The amphibians, in turn, were the ancestral group of the reptiles. As shown in Fig. 14-11, one group of reptiles evolved into birds, while another group evolved into mammals. The mammals have been very successful as a land-living group. Their adaptive radiation has produced species that now occupy most regions of the earth. Mammals have displaced reptiles as the dominant land group, and man, more recently, through his dramatic population expansion and technology has caused the near extinction of many species of reptiles, birds, and even other mammals, including the primate group to which he himself belongs.

The original mammal type was very close in structure, etc. to the contemporary group we call the **shrews** (somewhat resembling a mouse in size, living in trees, and feeding on insects). From this group, there evolved such diverse types as whales, bats, elephants, camels, horses, rabbits, rats, mice, cats, dogs, cows, monkeys, apes,

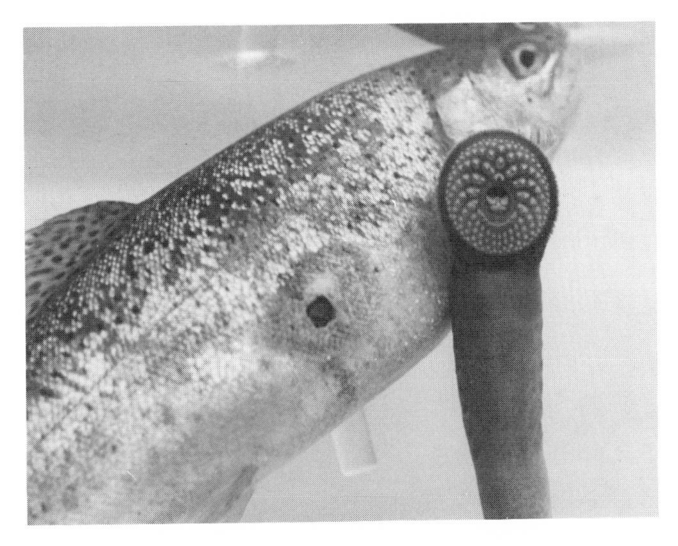

Fig. 14-12
Lamprey detached from rainbow trout to show feeding wound that had penetrated body cavity and perforated gut. Trout died from wound. Note chitinous teeth on underside of lamprey head. (Courtesy United States Bureau of Sport Fisheries and Wildlife, Fish Control Laboratory, La Crosse, Wis.; photograph by L. L. Marking; from Hickman, C. P., Sr., Hickman, F., and Hickman, C. P., Jr. 1974. Integrated principles of zoology, ed. 5. The C. V. Mosby Co., St. Louis.)

and man. The last three types (monkeys, apes, and man) are members of the order Primates.

Primate evolution is not as well documented as we would like to have it. This is because primates evolved in the moist, tropical forests of Africa where conditions for fossil formation are very poor. Recently, a great deal of intensive research in Africa has uncovered a good deal of information on man's origin.

EVOLUTION OF MAN

The human species is one branch in the evolution of the primate group. Primates were derived from shrewlike, insect-eating mammals, called **insectivores,** about 65 million years ago. The primitive primates, called **prosimians,** meaning before the apes, became abundant in number and worldwide in distribution about 55 million years ago. Most of the prosimians became extinct, however, and their present-day

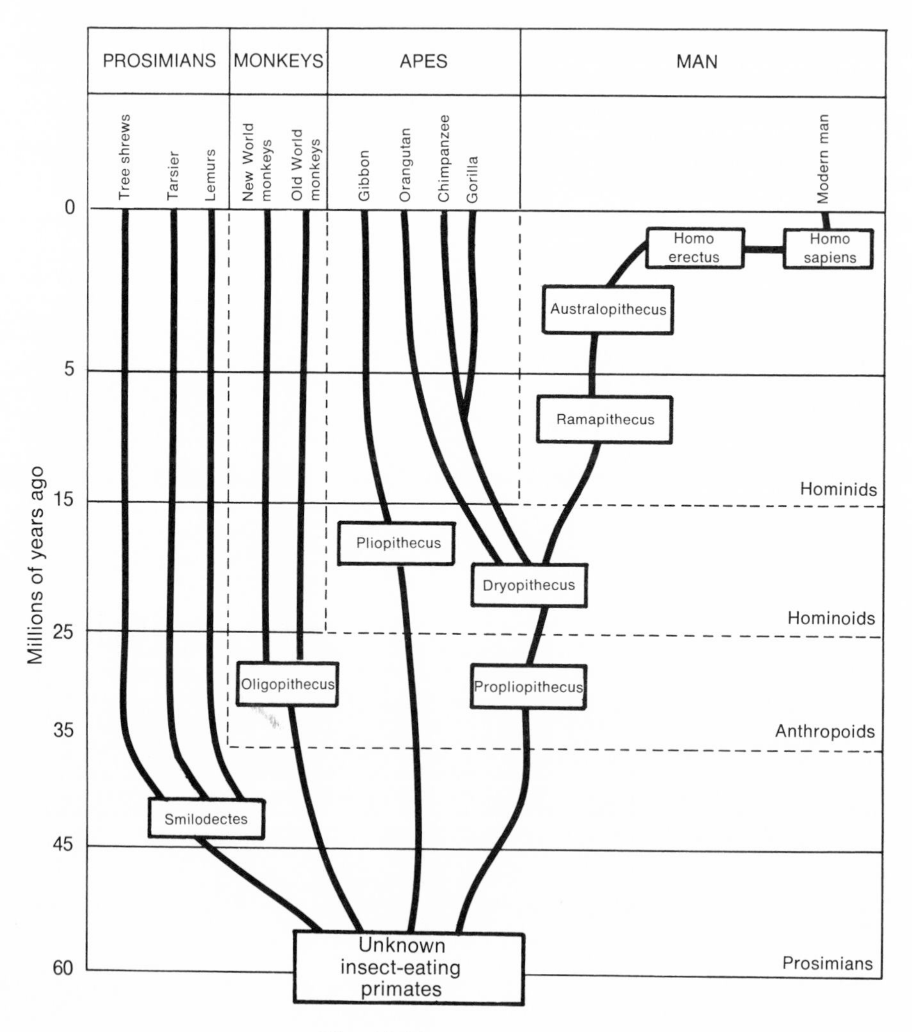

Fig. 14-13
Evolution of the primates.

descendants (tree shrews, lemurs, tarsiers) are found only in tropical Africa, Asia, and the island of Madagascar. Prosimians have small brains, but grasping-type (first digit opposable to second) hands and sometimes feet. Monkeys, apes, and man, collectively called **anthropoids,** arose from prosimian ancestors (Figs. 14-13 to 14-15).

Anthropoids evolved from prosimians about 35 million years ago. The early anthropoids were monkeylike in body structure and appearance. Their modern descendants consist of the **new-world monkeys** of Central and South America (marmoset, spider monkey, howler monkey, capuchin monkey) and the **old-world monkeys** of Africa, Asia, and Europe (rhesus monkey, baboon, proboscis monkey, mandrill). An interesting difference between the two types of monkeys lies in their ability to use their tails in grasping objects (prehensile use). The tails of new-world monkeys are generally prehensile, while those of old-world monkeys are not. There is no generally accepted explanation for this evolutionary difference. It is another example of divergence in related lines (Figs. 14-16 and 14-17).

Some of the monkey types evolved into **hominoid** primates (apes and man) about 25 million years ago. Hominoids are characterized by the complete absence of a tail at birth. The early hominoids were apelike in structure and appearance. Their modern descendants are the gibbon (Asia), orangutan (Asia), gorilla (Africa), and chimpanzee (Africa). As shown in Fig. 14-13, the gibbon is a very primitive hominoid, being more directly descended from the original prosimian stock than the other apes. This closeness to the ancestral group can be seen in the fact that the gibbon spends most of its

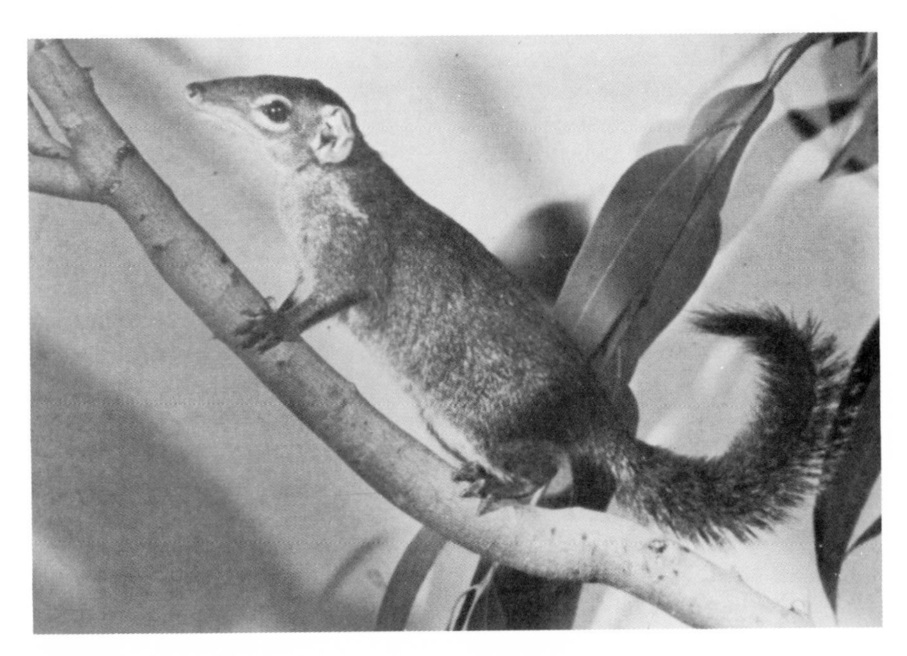

Fig. 14-14
Tree shrew. A native of Borneo, this creature still resembles the small insectivore from which the primates evolved. Although its thumb and big toe are opposable, note that claws are retained on its digits. (San Diego Zoo photograph; from Nagle, J. J. 1974. Heredity and human affairs. The C. V. Mosby Co., St. Louis.)

Fig. 14-15
Tarsier. This prosimian derives its name from the greatly elongated tarsal bones in its feet. Although only about the size of a chipmunk, the powerful leverage provided by its long tarsals enables it to make leaps of 4 to 6 feet through the trees. Its hands possess broad disks on the fingertips that aid in grasping, and fingernails replace claws. Being nocturnal, its night-adapted eyes are the largest of all primates. If human eyes were proportionately as large as those of the tarsier, they would be the size of a softball. (San Diego Zoo photograph by Ron Garrison; from Nagle, J. J. 1974. Heredity and human affairs. The C. V. Mosby Co., St. Louis.)

Fig. 14-16
Guatemalan red howler, a new-world monkey characterized by its prehensile tail, which serves as a "fifth hand." (San Diego Zoo photograph; from Nagle, J. J. 1974. Heredity and human affairs. The C. V. Mosby Co., St. Louis.)

Fig. 14-17
Macaque family. These representatives of old-world monkeys, commonly called rhesus monkeys, are used extensively in biological research. Note the comma-shaped nostrils. (San Diego Zoo photograph; from Nagle, J. J. 1974. Heredity and human affairs. The C. V. Mosby Co., St. Louis.)

time in trees and is far more agile than the other apes in swinging from branch to branch (Figs. 14-18 to 14-21).

☐ Hominid evolution

About 15 million years ago, the **hominid** line of evolution (apelike man and man) developed from the hominoid group (manlike ape). The differences between man and the apes are many. Perhaps the most noticeable one is that man is **bipedal** and walks upright on the soles of his feet, whereas the apes are four footed (quadrupedal) and

Fig. 14-18
Gray gibbon, a member of the most primitive hominoid family surviving today. Note its extremely long arms. Although awkward on the ground, gibbons travel so gracefully through the trees that they appear to be flying as they swing from branch to branch. (San Diego Zoo photograph; from Nagle, J. J. 1974. Heredity and human affairs. The C. V. Mosby Co., St. Louis.)

Fig. 14-19
Female and male orangutans. These are members of the great ape family that also includes the chimpanzees and gorillas. Orangutans display marked sexual dimorphism. The male (right) develops peculiar cheek flanges at sexual maturity that, along with a high cranial crest, make the face appear concave. Females lack these features and look quite similar to chimpanzees. (San Diego Zoo photograph by Ron Garrison; from Nagle, J. J. 1974. Heredity and human affairs. The C. V. Mosby Co., St. Louis.)

walk in stooped fashion. A second and more important difference is **brain size.** The size of a gorilla brain is about 500 ml and that of man is 1500 ml. Increased brain size carries with it a larger number of neurons and pathways, hence greater mental ability. Man's modern society with its complicated technology is excellent proof of the importance of brain size in evolution.

The evolutionary path from apelike ancestor *(Dryopithecus)* to man *(Homo sapiens)* has become well documented by recent fossil discoveries. Perhaps the most interesting of these was the apelike man *(Australopithecus)*. This individual was short (about 4 feet tall), weighed about 75 pounds, and was **bipedal.** The fact that he walked on his hind legs made him manlike. However, his **brain size,** about 600 ml, was close to that of the modern gorilla, meaning that his mental ability was apelike or, probably, slightly better. *Australopithecus* illustrates the very important point that the different aspects of human evolution developed at different times. The bipedal

Fig. 14-20
Chimpanzee, which has many likenesses to humans, including same length of gestation period, same menstrual cycle phases, a large repertoire of vocal and facial expressions, and a great ability to use tools. (San Diego Zoo photograph by Ron Garrison; from Dillon, L. 1973. Evolution: concepts and consequences. The C. V. Mosby Co., St. Louis.)

Fig. 14-21
The gorilla is the largest of the apes, mature males attaining a height of 6 feet and a weight of 600 pounds, whereas the females are much smaller and are usually not heavier than 280 pounds. Gorillas do vocalize a great deal, but do not show any ability in using or making tools. (San Diego Zoo photograph.)

nature of man became established about 3 million years (120,000 generations) before his brain size reached its present dimension (Fig. 14-22).

More recent hominid evolution has centered on increased brain size. This occurred gradually and is imperceptible when viewed only from one generation to the next. An important stage of brain development was reached about 2 million years ago with the emergence of *Homo erectus* (Java man, Peking man), who is considered the first "true man." His brain capacity was only about 850 ml, but, based on the artifacts found in his caves, we know he was an expert hunter, used fire, and constructed primitive tools of stone (Fig. 14-23).

The most recent development in man's evolution is the emergence of *Homo sapiens* about 1 million years ago. By that time, his brain capacity (1500 ml) and technology (toolmaking) had developed to the point that we can consider him an early form of modern man (Fig. 14-24).

One of the interesting aspects of hominid evolution is its geographical location. Although the ancestral apelike hominoids (*Dryopithecus* and *Pliopithecus*) were distributed over Africa, Asia, and Europe, the apelike man (*Australopithecus*) has been found only in Africa. In the absence of any contradictory evidence, this places man's early evolutionary development in Africa. The first true man, *Homo erectus*, however, had a wide distribution that included Africa, Asia, and Europe, as did early *Homo sapiens*. The earliest migration of man to North and South America occurred

Fig. 14-22
Australopithecus. Known as the South African ape man, he was the earliest user of tools and fire in man's ancestry. (Courtesy The American Museum of Natural History; from Nagle, J. J. 1974. Heredity and human affairs. The C. V. Mosby Co., St. Louis.)

only about 20,000 years ago and was accomplished by groups of *Homo sapiens* living in Asia.

What has happened to the early forms of *Homo sapiens*? Why have they not persisted to the present time? The disappearance of groups of organisms is one of the most interesting aspects of the study of evolution. To say that a group has become **extinct** merely describes a phenomenon, but does not explain how it occurred. In evolution, extinction can take two forms: (1) **absolute extinction,** the death of every single member of a group, or (2) **transformation,** the continuation of the group, but with its members having new or altered characteristics. Both types of extinction have occurred in human evolution. Undoubtedly many of the early forms of *Homo sapiens* were hunted for food, and subsequently wiped out, by newly emerging superior human types. (There is ample evidence of cannibalism occurring at this and later times.) However, the presence of modern man, looking quite different from his remote ancestors, indicates that extinction through transformation also occurred.

Fig. 14-23
Homo erectus. The form depicted, known as Java man, represents a species so similar to ourselves that it is classified in the same genus, *Homo.* Sites at which the fossil remains of this "man" have been collected range throughout the European and African continents. These sites indicate that *Homo erectus* led a communal life in bands of 30 or more who cooperated in the hunt for food and their mutual survival. (Courtesy The American Museum of Natural History; from Nagle, J. J. 1974. Heredity and human affairs. The C. V. Mosby Co., St. Louis.)

Fig. 14-24
Neanderthal man. A member of our own species, *Homo sapiens*, Neanderthal man was widely distributed throughout the Old World. He is no longer thought to have appeared as "primitive" as this restoration indicates. Rather, he was probably quite human looking and not as hunched over as we typically imagine. It is now thought that if he were bathed and shaved and dressed in contemporary garb, no one would pay him any particular attention as a passerby on a city street. (Courtesy The American Museum of Natural History; from Nagle, J. J. 1974. Heredity and human affairs. The C. V. Mosby Co., St. Louis.)

□ Races of man

All human beings belong to the same species, *Homo sapiens*. Our species has a wide distribution over the earth, but is broken up into a number of geographically separated populations that vary in the amount of contact they have with one another. The members of some of these populations are easily distinguishable from one another based on skin color, hair color and texture, physique, and other genetically determined characteristics. Populations that differ from one another in the frequency of their various alleles for particular traits, for example, melanin-producing genes and blood group genes, are called **races.**

Members of one population can migrate and become incorporated into another population. If a great deal of migration occurs, the populations will become very similar in the frequency of their various alleles. The members of the two populations will then be indistinguishable from one another and will be considered members of the same race.

Although the subdivision of the human species into distinct races may in time disappear, at present several genetically distinguishable populations are recognizable: (1) Negroid, (2) Caucasoid, (3) Mongoloid, (4) Bushmen, (5) Australoid, and (6) Polynesian. The original criteria used in designating these racial groups included skin, hair, and eye colors; stature; hair form; conformation of nose and lips, and shape of head. The resulting racial classification is not perfect, because the differences between the races are relative rather than absolute, and each race contains types that are considered typical of some other group.

□ Biological future of man

Some aspects of man's biological future appear to be relatively simple to predict; others are most difficult. First, let us consider the possibility that mankind will diverge into a number of species sometime in the future.

Races of a species are genetically distinguishable populations living in different territories. The genetic differences between populations will result in structural, physiological, and behavioral differences in the members of one race as compared to another. When these differences become so great that they prevent the members of two races from breeding with one another, we say that we have two species.

The separation of modern man into geographical races probably took place about 500,000 years ago. Despite the evolutionary divergence that took place between the races since that time, the members of all human races can breed with one another. In more recent times, modern transportation has permitted very widespread travel of people, with the inevitable consequence of marriage between persons from different populations. This will result in the races of man being fused into larger groups, whose members will vary tremendously in those traits that previously distinguished the different races. The net effect of this amalgamation of the races will be to eliminate completely the possible divergence of mankind into a number of species sometime in the future. Man will undoubtedly remain as a single species.

Much of man's more recent evolution has stressed increased brain size. Will there continue to be an evolutionary development toward larger brains? This is a possibility. Whether it will occur or not will depend on whether those with larger brains have more offspring, on the average, than those with smaller brains. If this does occur, succeeding generations should, very gradually and initially imperceptibly, have larger brains. Are there any problems that might interfere with such a projected evolutionary trend? Yes, we can anticipate that increased brain size might interfere with the birth process and result in the death of the child, possibly together with the mother. However, medical science has already provided a solution for this problem. Babies that are too large for natural birth can be delivered through a **cesarean operation,** in which a cut is made through the abdominal and uterine walls, and the infant is removed directly from the womb. Thus modern medical technology has become a factor in man's future evolutionary development. It is also important to realize that only those societies that have the advanced medical facilities necessary to perform cesarean operations will be able to evolve larger brains.

We know that advances in medical science have permitted individuals with genetic defects to survive and reproduce. Will this result in the human species having a greater frequency of genetic disorders in the future? The answer is emphatically,

"Yes." By eliminating the deleterious effects of genetic defects, our modern society has increased the incidence of these diseases and has also increased the requirement for certain types of medical facilities and services in future generations. This is a known and inevitable consequence of our progress in understanding and alleviating this form of human suffering.

Is it possible that mankind has reached the end point of its evolutionary development? This does not seem likely. Although on the one hand man, through his technology, for example, air conditioning and central heating, has eliminated many environmental factors, for example, heat and cold, that would have promoted evolutionary changes, his own technology has introduced new environmental factors, for example, pollutants of all sorts, including chemical exhausts from factories, and automobiles, noise from industrial processes, and radioactive materials from nuclear reactors, that will promote future evolutionary development and determine its direction.

■ SUMMARY

A study of evolution provides us with an understanding of the origin and development of the vast array of organisms we find on the earth today. Present-day animals and plants are descended from organisms that lived millions and billions of years ago. The evidence for the occurrence of evolution comes from a number of different sources: (1) comparative anatomy with its emphasis on homology, (2) comparative embryology with its guiding principle of "ontogeny recapitulates phylogeny," (3) vestigial organs whose uselessness indicates a change in way of life, (4) blood proteins that indicate biochemical relationships, and (5) paleontology that gives us the historical sequence of changes as seen in the fossil record.

The various organisms on the earth can be grouped into categories that reflect their evolutionary relationship to one another. This classification includes three kingdoms: (1) Monera, organisms whose cells lack a nuclear membrane and other structures; (2) Plantae, organisms whose cells have a nuclear membrane, etc., and also a cell wall outside the plasma membrane; and (3) Animalia, organisms whose cells are like those of the Plantae, except that they lack a cell wall.

Plant evolution has, to a large extent, reflected the movement of plants from water to land. This has brought with it a changing emphasis on both the haploid and diploid phases of the plant life cycle. For the Thallophytes and Bryophytes, it is the haploid phase of the alternation of generations that is predominant. In the case of the Tracheophytes, the diploid part of the life cycle is emphasized, with its production of root, stem, and leaf.

Animal evolution has more variability than plant evolution, an apparent effect of the need of animals to find and capture their food, which in some cases involves plants, in other cases animals, and for some species both. Although a number of animal groups moved from water to land, the main emphasis in animal evolution is on different body plans.

There are three main lines in animal evolution: (1) early stem line, involving the transition from unicellular forms (Protozoa) to increasingly complex multicellular types (Coelenterata and Platyhelminthes); (2) invertebrate line, characterized by a nonliving exoskeleton, a dorsally located heart, and a ventrally located nerve cord

(Annelida, Mollusca, and Arthropoda); and (3) vertebrate line, identified by a living endoskeleton, a ventrally located heart, and a dorsally located nerve cord (Echinodermata and Chordata).

Mankind represents the current end point of one branch of chordate evolution. The earliest known members of this phylum are the fish, followed by the amphibians, reptiles, birds, and mammals. The fish-amphibian-reptile sequence represents a movement from water to land. It is one of many such sequences within the animal kingdom, most of the others having occurred within different invertebrate phyla.

Within the phylum Chordata, the class Mammalia is the most successful land-living group. One branch of this adaptive radiation (Primates) started out as tree-living forms able to use both hands and feet for grasping (prosimians and monkeys). From a monkeylike ancestor, there evolved one group that spent a good deal of their time walking on the ground, but in a quadrupedal, stooped manner (apes). From an apelike ancestor, there developed a line of evolution in which the individuals became bipedal, and their brain size increased tremendously (man).

Although the human species is broken up into a number of distinguishable races, it is quite certain that man will remain as a single species in the future. Man's biological future will most probably include evolutionary changes. The direction of any future evolutionary development in man's structure, physiology, or behavior, will undoubtedly be determined by new factors that man, through his technology, has introduced into the environment.

SUGGESTED READINGS

Clark, J. D. 1958. Early man in Africa. Sci. Am. **199**:76-83. Discussion of the types of early men that evolved in Africa.

Cold Spring Habor Symposium on Quantitative Biology. 1959. Genetics and twentieth century Darwinism, vol. 24. Cold Spring Harbor, N.Y. Analysis of the mechanisms of evolution as formulated from the study of genetics, ecology, and paleontology.

Dobzhansky, T. 1962. Mankind evolving. Yale University Press, New Haven, Conn. View of human evolution as an interaction of biological and cultural forces.

Eckhardt, R. B. 1972. Population genetics and human origin. Sci. Am. **226**:94-103. Illuminating article on the mechanism of evolution.

Goldsby, R. A. 1971. Race and races. Macmillan Publishing Co., Inc., New York. Well-written introductory survey to the study of human races, with emphasis on the social and psychological aspects of this area of biological investigation.

Pilbeam, D. 1972. The ascent of man: an introduction to human evolution. Macmillan Publishing Co., Inc., New York. In-depth review of the development of our species and the factors that appear to have influenced it.

Roslansky, J. D., ed. 1966. Genetics and the future of man. Appleton-Century-Crofts, New York. Series of discussions on various aspects of genetics as they affect the future of man.

Sahlins, M. D. 1960. The origin of society. Sci. Am. **203**:76-86. Provocative discussion of the factors that led to the establishment of large groups of humans.

Temin. H. 1972. RNA-directed DNA synthesis. Sci. Am. **226**:25-33. Explanation of how viruses that are composed of RNA can enter the cells of our body and transform them into cancerous cells at some later time.

Washburn, S. L. 1960. Tools and human evolution. Sci. Am. **203**:63-75. Thoughtful discussion on the role of early technology in determining man's subsequent evolution.

Representation of European city of Middle Ages. Physical as well as biological conditions in these cities contributed greatly to spread of bubonic plague in middle 1300's. (From Lane, T., ed. 1976. Life the individual the species. The C. V. Mosby Co., St. Louis.)

CHAPTER 15 Our place in nature

LEARNING OBJECTIVES

- What is ecology?
- What is an ecosystem, and which unique ecosystem has man formed?
- What are the two parts of a species' biotic environment, and what are the characteristics of each?
- What is the relationship between the prevalence of malaria in Africa and sickle cell anemia?
- What is the relationship between the prevalence of blood fluke infections in Egypt and the Aswan High Dam?
- Why do epidemics of bubonic plague occur in densely populated cities during periods of food shortage?
- What are the factors that make up a species' physical environment, and how does each factor affect the ability of a species to inhabit a given area?
- What is a "circadian rhythm," and what role does it play in the adjustment of a species to its environment?
- Why must the molecules contained in the bodies of all organisms be recycled?
- How is carbon recycled in nature?
- How is nitrogen recycled in nature?
- How is water recycled in nature?
- What is a poison?
- How has man through his technology poisoned his own environment?

Ecology is the study of the interrelationships of organisms both with each other (biotic environment) and with the nonliving world (physical environment). The basic unit of study in ecology is the **ecosystem,** which is a stable association of organisms in a given area involving a cyclic interchange of materials between the ecosystem's living and nonliving components. Examples of ecosystems are large lakes, forests, islands, mountaintops, and deserts. Within each type of ecosystem, as shown in Fig. 2-1, there is a complex interaction of green plants (producers); herbivores, carnivores, and omnivores (consumers); and bacteria, molds, and yeasts (decomposers). **Energy** is **not** recycled in an ecosystem, but is continuously lost in the form of heat and replaced in the form of sunlight. **Matter,** on the other hand, **must** be recycled and reused if the ecosystem is to survive.

■ ECOSYSTEM WITHOUT BOUNDARIES

Each type of ecosystem is distributed widely over the earth. However, in any one place, each occupies a limited area. Although man can be, and frequently is, part of the geographically limited ecosystems, he has also formed an ecosystem that does **not** have any describable geographic boundary. This man-made ecosystem is the **modern city.**

A city, such as New York, has its own local complex ecosystem (wooded areas, rivers, lakes, parks, etc.) with its food pyramids and recycling of matter. Man, however, lives independent of the local ecosystem, using it only as part of his living and recreational space and not for its productive capacity. The food consumed in New York represents the productivity of the soils of tropical Africa, the fisheries of the North Sea, the plantations of Central America, the orchards of Australia, the tea-growing hillsides of Ceylon and India, the groves of Florida and California, the fields of Nebraska, Ohio, North Dakota, and South Dakota, etc. Even the water that is consumed in New York is not, to any appreciable extent, that which falls on the city as rain or snow. Rather, the water has been brought to the city from outlying areas through a complicated pipe system that extends as far as 200 miles away. The lack of involvement with the local ecosystem also applies to the wastes produced in New York. They are not distributed to the soil in and around New York or returned to the soils that produced the food, but are sent through sewage disposal plants into the Hudson River to enter new food pyramids in the ocean. Thus our modern city does **not** participate in the natural recycling of matter so important in maintaining an ecosystem.

In our modern technological world, most people are urban dwellers, and their percentage is constantly increasing (see Tables 1-6 and 1-7). This means that the "ecosystem without boundaries" will utilize an ever greater proportion of the world's resources and will create more and more imbalances in the other types of ecosystems. The increase in number of modern multimillion cities in different countries can easily lead to conflict between nations as their populations compete for food and other resources in far distant places on the earth. If we are to be able to make intelligent decisions about our role as a nation on this earth, we must understand the characteristics and global implications of this man-made ecosystem—the modern city.

■ BIOTIC ENVIRONMENT

All species of plants, animals, bacteria, viruses, etc. of an ecosystem have either direct or indirect relationships with one another. There are two important kinds of direct relationship that can exist between species: (1) members of two species can live together permanently or temporarily with one or both deriving advantage from the relationship (**symbiosis**) or (2) one species may act as the source of nutrients and energy for another species in a food pyramid (**feeding relationship**).

□ Symbiosis

Symbiosis, living together in close association, may take one of three forms: mutualism, commensalism, or parasitism.

Mutualism. In the type of symbiotic relationship termed mutualism, both species are benefited by the close association. An example of mutualism is seen in the relationship between a human being and the bacteria that live in the large intestine. The bacteria obtain nourishment from the unused food residues and synthesize vitamin B_{12} (cobalamin) as a byproduct of their metabolism. As a result of this association, the human being obtains a vitamin necessary for normal cell division (see discussion of pernicious anemia in Chapter 6), while the bacteria receive protection, a stable environment, and a continuing supply of food.

Commensalism. When members of different species live together so that one species benefits from the relationship and the other is neither harmed nor benefited, the association is called **commensalism.** In our modern society, the relationship of man to his "pets" (dogs, cats, fish, birds) represents this type of symbiosis. In this association, the pet clearly benefits, and the person is not affected to any significant extent. It might be argued that the relationship between man and his pets represents mutualism because the pets do give their owners "pleasure" and, in the case of some dogs, "protection." However, we might just as convincingly argue that man is being harmed by sharing his food with his pets and is inconvenienced by his need to care for them.

The most conspicuous example of urban commensalism is the association of man and dog. In New York City, there are about 8 million people and 1 million dogs. This relationship has produced a number of problems for the community as a whole. One of these is dog bites. Bites are the major cause of dog-related human casualties. More than 47,000 dog bites are reported in New York City annually, and it is estimated that at least as many bites go unreported. About 40% of the reported bites involve children under 11 years of age. In addition, more than 40 diseases and infestations can be transmitted from dog to man, including rabies, tapeworms, ticks, and fleas.

A growing problem associated with the urban dog is fecal littering on the city streets. In New York City, the 1,000,000 dogs deposit an estimated 300,000 pounds of feces and 180,000 gallons of urine each day. This litter is left to decompose on the streets and serves to nourish the cockroach and rat populations of the city. However, despite the problems that are involved in this commensal relationship, an increasing number of people living in cities are forming this type of symbiotic association with dogs.

Parasitism. In the type of symbiotic relationship termed parasitism, individuals of one species benefit, while those of the other species are harmed. Some parasites live on the outside of the host's body and are called **ectoparasites,** whereas others live within the host's body and are called **endoparasites.** Human ectoparasites include such organisms as fleas and lice, both of which are insects. Human endoparasites include some of the protozoa, bacteria, viruses, flatworms, and roundworms. Many of the chronic infectious diseases, for example, syphilis, gonorrhea, and tuberculosis, represent a parasitic type of association.

Parasitism might be viewed as an ideal way of life. The parasite, especially the endoparasite, is protected from natural enemies, has relatively little exposure to changes of temperature, light, moisture, etc., and has access to all the nourishment it needs. There is, however, one drawback to this way of life. *When the host dies, so does the parasite.* This limitation on the parasitic way of life requires that some mechanism be evolved for the transfer of at least some of the parasites from one host to another before the death of the first host. As we shall see, parasites usually have complicated life cycles, involving a number of hosts and both asexual and sexual forms of reproduction.

Malaria. Malaria is an infectious disease of the red blood cells caused by a protozoan parasite that is transmitted from human being to human being by a mosquito (Fig. 15-1). The disease is characterized by intermittent chills and fever, and, in at least one of its forms, it is usually fatal.

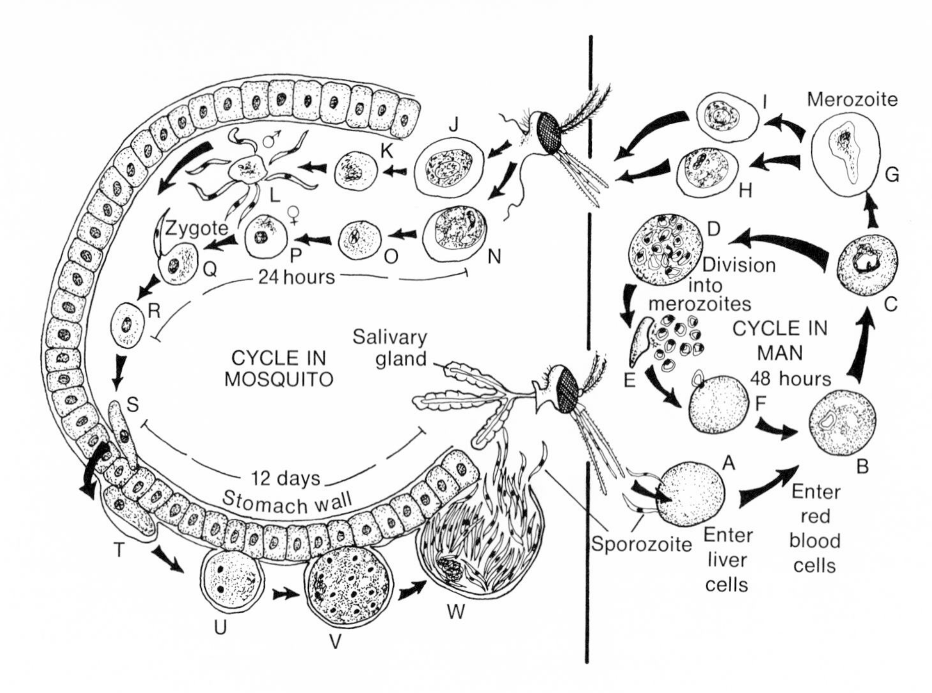

Fig. 15-1
Life cycle of the malarial protozoan.

When an infected female *Anopheles* mosquito (the males are perfectly harmless) pierces a human, she injects the protozoan parasite along with some of her saliva into the bloodstream. The parasite, at this time, is in the **sporozoite** stage of its life cycle. While in the bloodstream, the sporozoites travel throughout the body and enter the cells of the liver. There they undergo multiple divisions, eventually leaving the liver cells and entering the red blood cells. Inside the red blood cells, the sporozoites grow and divide to form the next stage of their life cycle, **merozoites.** These break out of their red blood cells into the blood plasma and invade other erythrocytes, where they grow and divide to form more merozoites, repeating the cycle many times. After a number of these merozoite-increasing cycles, the parasites are so numerous that the shock of their nearly simultaneous release from the red blood cells produces a chill, followed by a high fever as part of the body's response to the toxins released by the parasites. The chill-and-fever cycle varies with the species of parasite: every 48 hours in the disease caused by *Plasmodium vivax,* every 72 hours when *P. malariae* is the parasite, and, in the often fatal form of the disease caused by *P. falciparum,* the cycle is lacking or irregular.

After a period of time, some merozoites become either male or female sexual forms, called **gametocytes.** These do not undergo any further changes in the human host. If, at this time, the human being is again pierced by a female *Anopheles* mosquito, some of the gametocytes are taken into the digestive tract of the mosquito along with the person's blood. When this occurs, the female gametocyte forms a **macrogamete,** and the male gametocyte divides into six spermlike **microgametes.** Two gametes of opposite sex fuse to form a zygote, which invades the stomach wall of

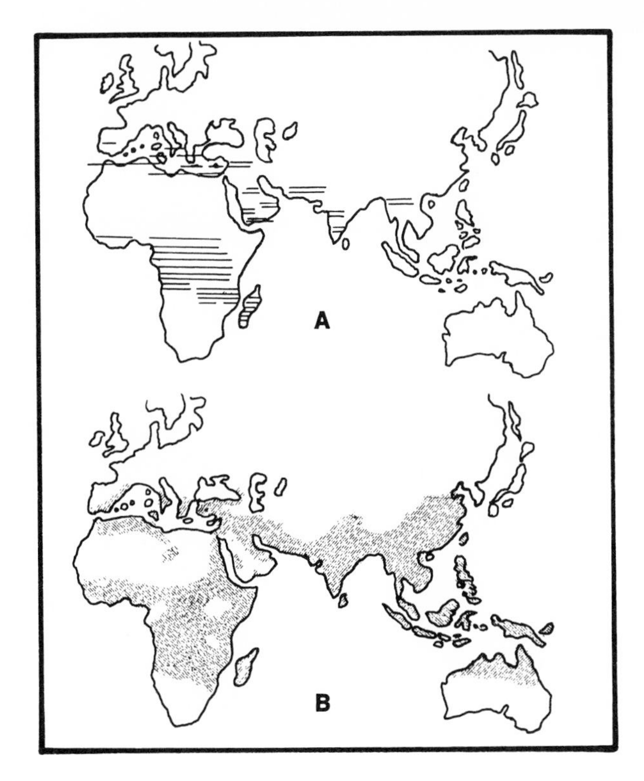

Fig. 15-2
Distribution of sickle cell hemoglobin (bars in map **A**) and falciparum malaria (shaded areas in map **B**). (From Levine, L. 1973. Biology of the gene, ed. 2. The C. V. Mosby Co., St. Louis.)

the mosquito and forms a cyst. Within the cyst, thousands of **sporozoites** are formed. These break out of the cyst, migrate through the mosquito's body, and enter its salivary glands to await transfer to a human host and a repetition of the life cycle.

If a person is infected only once, the disease will remain in an active state for about 1 to 2 years. Unfortunately, most victims are reinfected periodically, and the disease becomes chronic. However, there are drugs available (atabrine and quinine) that kill the malarial parasites while they are in the blood.

An important interaction occurs between the ecology and genetics of those human populations that live in the areas of Africa, Asia, and Europe where *P. falciparum* exists. (This is the protozoan that causes the fatal type of malaria.) In these areas, we find that the gene for sickle cell anemia is present in high frequency in the human populations. Sickle cell anemia is a genetic disease in which the red blood cells have a reduced oxygen-carrying capacity (Chapter 6). Even people heterozygous for sickle cell anemia are characterized by erythrocytes with a below-normal oxygen-carrying capacity. The malarial parasite requires a great deal more oxygen for its growth and division than is available in the red blood cells of sickle cell homozygotes or heterozygotes. As a result, human beings carrying the sickle-cell gene and living in areas where falciparum malaria is present have a greater chance of surviving and reproducing than those without this gene. Thus the sickle cell gene protects its carriers from the effects of falciparum malaria. In some *P. falciparum* areas, as much as 40% of the population are sickle cell heterozygotes. The role of falciparum malaria (ecology) in determining the presence or absence of the sickle cell gene in a population (genetics) is seen in Fig. 15-2. The relationship between falciparum malaria and the sickle cell gene is also seen in the case of the American Negro whose forced transfer to the United States resulted in his removal from a falciparum malaria region. In the absence of this disease for the past 300 years (10 generations), the frequency of sickle cell heterozygotes has dropped to 8% and will continue to diminish in the future.

Until recently, malaria was the most important parasitic disease of man, killing over 3 million people each year. Effective control of this disease, involving the widespread use of DDT against the mosquitoes that spread the disease, did not begin until 1945. By 1958, the number of deaths was less than 1 million, and it has continued to decrease. The number of people in the world with active cases of malaria was estimated to be 250 million in 1955, 140 million in 1962, less than 100 million in 1967, and about 20 million in 1975. Most of the current cases of malaria are to be found in South Asia (India, Pakistan, Bangladesh, Indonesia), where an initially successful antimalarial campaign was phased out, and a resurgence of the disease has occurred. As we shall discuss later in this chapter, the use of DDT for the eradication of disease-carrying and crop-destroying insects has produced problems in our worldwide ecosystem.

Blood fluke infections. With the generally successful worldwide campaign against malaria, blood fluke infection has become the most important human disease caused by animal parasites. Well over 100 million people in the world have it, and the recent growth of irrigation projects, for improved agricultural production, has increased the incidence of the disease.

The causative agent of this infection is the flatworm (Platyhelminthes) *Schistosoma hematobium.* It is most prevalent in Africa and Asia. In some areas of Egypt, up to 90% of the population suffer from these flukes. The life cycle of these parasites includes both asexual and sexual stages and the use of a snail as an intermediate host in going from one person to another (Fig. 15-3).

In the human being, the blood fluke lives in the blood vessels of the intestinal tract and urinary bladder. The male fluke is the larger organism and has a groove in his body in which the female is held. Eggs of the blood fluke are passed into the intestine or bladder and pass out of the person's body with the feces or urine. If the feces or urine are deposited in a lake or other body of fresh water, the eggs will hatch into a ciliated stage, called **miracidia.** These can survive in the water for only 24 hours and, within that time, must find a snail of an appropriate species and bore into it. Within the snail, each miracidium forms a saclike structure, inside of which it reproduces asexually to form the **sporocyst** stage. These, in turn, form yet another stage, called **cercariae.** Cercariae leave the snail and swim through the surrounding water. They can live for about 3 days in the water. If, within that time, they come in contact with a person standing in the water, they burrow through the skin and into the bloodstream. They travel through the body, finally settling in the small blood vessels of the intestine and urinary bladder. There they mature into adults, mate, and begin laying eggs, thus beginning the infective cycle again.

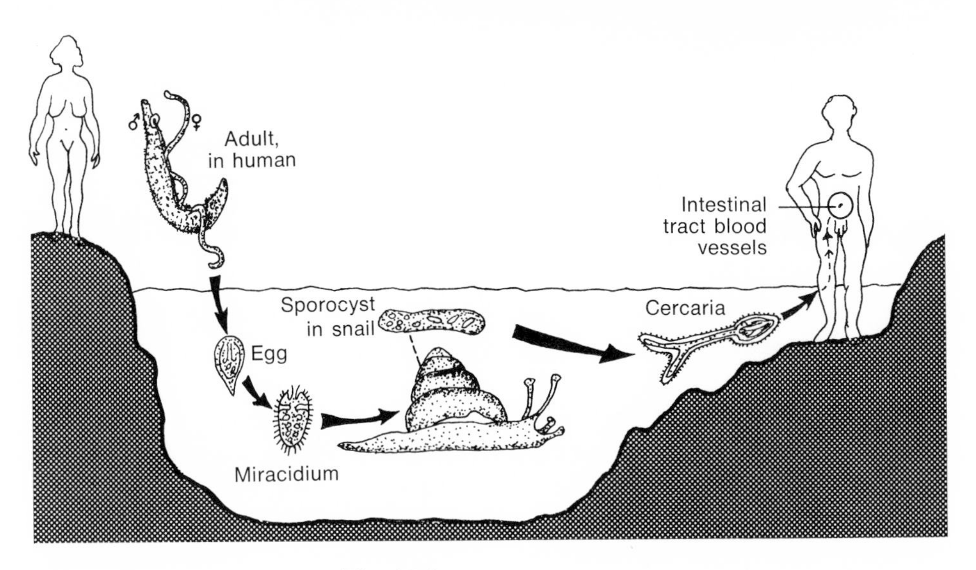

Fig. 15-3
Life cycle of the blood fluke.

In the intestine, the production and release of eggs cause abdominal pain and bloody diarrhea, whereas in the urinary bladder, there is pain and the passage of bloody urine. Repeated infections of the intestines by blood flukes can lead to severe dysentery and death. The continued presence of flukes in the urinary bladder may cause an inability to urinate and the formation of tumors and may also prove fatal.

As mentioned earlier in this discussion of blood flukes, various projects designed to improve agricultural production have served to increase the incidence of this disease. As a case in point, in 1971 the Aswan High Dam in Egypt was completed. It was designed to provide hydroelectric power and to increase crop yields. By damming the Nile, excess water from the Abyssinian mountains is stored, and the controlled release of water increases agricultural production. However, the Aswan High Dam has created a massive health problem. The year-round storage of water has provided a breeding ground for the snails that transmit the human blood fluke. These snails have now migrated into the irrigation canals served by the dam and act as a continuing source of the blood flukes that now infect 90% of the human population living along the canals. Such situations make it clear that a modern society must balance its need for higher crop yields against the health hazards produced by its agricultural advances.

Bubonic plague. Bubonic plague has been historically important as one of the most devastating diseases of mankind. It has been estimated that during the 14th century, the population of Europe was reduced from 85 to 60 million by this disease, and a smaller reduction from 110 to 100 million occurred during the 16th century. Bubonic plague has always been associated with densely populated cities, poor sanitation, and famine (usually associated with war). The disease is still prevalent in India, Africa, and parts of South America.

The disease is caused by a bacterium, *Pasteurella pestis*, that is transmitted to a person by a flea, an ectoparasite that bites its hosts to obtain blood. On entering the bloodstream, the bacteria multiply and spread to all the organs, especially the lungs and lymph nodes. Death occurs in more than half the cases unless antibiotics are available.

Man is not the usual host for the bacterium causing bubonic plague. The organism is normally found in the bloodstream of wild rodents (rats, squirrels, chipmunks, etc.) that act as reservoirs of the bacteria. Wild rodents usually have fleas as one of their ectoparasites. The fleas are present in large numbers and bite the rodents frequently to obtain blood, which is their normal food. In drawing blood from an infected animal, the flea also ingests large numbers of the disease-causing bacteria, which may then be transmitted to the same or other animals that the flea bites. Fleas move from animal to animal and can thus infect an entire population of rodents if they have fed on a single diseased individual.

During periods of food shortage, the rats of a city are forced to seek their food in people's homes, usually at night when the people are sleeping. This brings the rats in close contact with human beings. The fleas on the rats' bodies are thus able to transfer, by jumping, from one food source (rat) to another (human). The fleas proceed to bite the person in order to obtain blood, and in the process, bacteria that cause bubonic plague are passed into the individual's bloodstream.

Fig. 15-4
Brown rat (*Rattus norvegicus*). Originally from the tropical forests of Asia, this species and the less pugnacious tree-living black rat have spread throughout the world. Living all too successfully beside man in his habitations, the brown rat not only causes great damage to man's food stores, but also spreads disease, including bubonic plague (carried through infected fleas), which greatly influenced human history in medieval Europe. (Photograph by L. L. Rue, III; from Hickman, C. P., Sr., Hickman, F., and Hickman, C. P., Jr. 1974. Integrated principles of zoology, ed. 5. The C. V. Mosby Co., St. Louis.)

In the case of bubonic plague, we have a bacterium that acts as an endoparasite in man, flea, and rat. The transmission of the bacterium from rat to man is effected by the flea, which is an ectoparasite on both rat and man. Here, too, we see an example of parasitism requiring a complicated life cycle that involves a number of hosts (Fig. 15-4).

☐ Feeding relationship

The biotic environment of all species in an ecosystem involves both the various types of symbiotic associations we have just discussed and the relationship that species have to one another as sources of nutrients and energy. In a land-based ecosystem, the producers are the grasses and trees; the consumers consist of herbivores (cows, horses, rodents, birds, worms, insects, etc.), carnivores (lions, hawks, foxes, wolves, etc.), and omnivores (primates, including man); and the decomposers are the bacteria and fungi. In a water-based ecosystem, the producers are mainly the blue-green and other algae; the consumers consist of herbivores (worms, molluscs, and certain fish), carnivores (seals, birds, and most fish), and omnivores (whales and man); and the decomposers, here, too, are the bacteria and fungi.

In many parts of the world, man is merely another organism in the particular ecosystem. This situation exists in those areas characterized by "subsistence agriculture," that is, the people subsist only on the food they produce, not buying or selling any agricultural products. Subsistence agriculture still exists in many parts of the world, for example, jungles of the Amazon in Brazil, isolated mountain villages in

Fig. 15-5
An Ifugao tribesman of Banawe, in the Philippines, looks out across the terraces of his people. The rice terraces of the Ifugao mountain tribe have existed for 3000 years and are carefully tended by each succeeding generation. (FAO photograph; from Williams, S. R. 1973. Nutrition and diet therapy, ed. 2. The C. V. Mosby Co., St. Louis.)

Afghanistan, and small island populations in the various oceans. However, the number of people involved in this way of life is continuously decreasing (Fig. 15-5).

For most of the world's populations, that is, those living in cities, food is produced far from the consumer, and the waste materials are not recycled. In addition, there are many intermediate energy-consuming steps between food production and consumption. These steps involve the handling, packaging, transportation, and selling of the food. In all of these processes, there is a tremendous use of energy, for example, the manufacturing of trucks and trains, the use of gasoline or other power sources to run these vehicles, and the manpower involved in the food industry. By the time food has reached our tables, more energy has been expended on its transportation and distribution than is contained in the food (Fig. 15-6). The source of this additional expended energy is the fossil fuels (coal and oil). Without the energy from fossil fuels, man's predominant ecosystem—the modern city—could not survive. These fossil fuels are nonrenewable, and the amount of oil available throughout the earth at the present rate of consumption will be expended within 50 to 100 years. In making plans for its own future, our modern society must take into account the fact that the amount of sunlight falling on the earth is not replacing the energy lost by the earth's ecosystems and that the amount of fossil fuel available to meet the earth's present energy needs is being used at a very rapid rate.

■ PHYSICAL ENVIRONMENT

In every ecosystem, there are physical factors that limit the distribution of organisms within it. These factors include temperature, light, and available chemical molecules. Most species have a rather limited tolerance to changes in temperature.

Fig. 15-6
The transportation and distribution of food in our modern society expends more energy than is contained in the food itself. (Copyright © 1976 by Theodore R. Lane.)

Especially critical points are 0° C, at which water freezes, and 100° C, when water evaporates. Because protoplasm contains a great deal of water, these temperatures can cause the death of most organisms. Two groups of chordates, however, have a relatively wide **range of tolerance** of temperature changes. They are the birds and the mammals, which maintain a constant body temperature and hence are referred to as **warm-blooded** organisms. All other groups of organisms have body temperatures that fluctuate with the temperature of the surrounding air or water and are called **cold-blooded** organisms. The maintenance of a constant body temperature permits the birds and the mammals to live in such places as the South Pole, for example, penguins, and the North Pole, for example, polar bears.

Light can be a critical factor in determining which species are successful in an ecosystem. This is especially true of plants, which require sunlight for the manufacture of food (photosynthesis). Light is also important for animals, because it determines when they will be awake and active. Nowhere is the adaptation of organisms to their physical environment better demonstrated than in the cyclical pattern of those activities that match the daily changes of light and darkness. A daily pattern that follows a 24-hour cycle is called a **circadian rhythm.** Examples of circadian rhythms that are linked to light and darkness can be seen in rats and mice, which are active during the night and sleep by day. Man, on the other hand, has the opposite sleep-awake pattern. In addition, a person's body temperature, adrenal gland activity, and red blood cell population all vary with the time of day.

A third limiting factor in every ecosystem is the availability of necessary chemical molecules. If a particular chemical is vital to a species, for example, iodine to make the hormone thyroxin, its total absence would exclude a species from the area. In the case of man, however, his modern technology, including planes and trucks, has made it possible for him to live in all regions of the earth regardless of the local availability of adequate amounts of the materials required for life.

All organisms are composed of various combinations of about 35 chemical elements (Chapter 2). However, there is little correlation between the elements needed by organisms and the availability of these elements in the nonliving world. Thus a recycling of these materials is required for life to continue. This absolutely necessary process is accomplished by the bacteria and fungi, which serve as **decomposers** (organisms of decay). Although bacteria and fungi do **not** contain chlorophyll, their method of obtaining food material is different from that of animals. Animals engulf their food into some type of digestive tract. In contrast, bacteria and fungi secrete digestive enzymes into the dead bodies of plants, animals, other bacteria and fungi and allow the digested material to diffuse into their cells. However, in this process, a great deal of excess digested material is formed, which becomes available to producers as a source of needed chemical elements. The central role of decomposers in the recycling of matter can be seen in the carbon, the nitrogen, and the water cycles.

□ Carbon cycle

Carbon is the central element in living material. The ability of carbon to form molecules of very diverse size and composition has resulted in the whole array of compounds (carbohydrdates, lipids, proteins, and nucleic acids) that form the structural and functional characteristics of protoplasm.

The recycling of carbon occurs in both land-based (terrestrial) and water-based (aquatic) ecosystems and involves three basic processes: photosynthesis, respiration, and metabolism (Fig. 15-7). In photosynthesis, chlorophyll-containing plants take carbon, in the form of carbon dioxide, and "fix" it, that is, attach other atoms to it, to form glucose as per the following reaction:

$$\underset{\text{Carbon dioxide}}{6CO_2} + \underset{\text{Water}}{6H_2O} \xrightarrow{\text{Energy from sun}} \underset{\text{Glucose}}{C_6H_{12}O_6} + \underset{\text{Oxygen}}{6O_2}$$

Plants also carry on respiration, during which process, they return some of the carbon of glucose to the air as carbon dioxide (see Fig. 8-13). However, photosynthesis extracts more carbon from the air than plant respiration returns, the excess carbon being present in the organic molecules making up the body of the plant.

When a herbivore or omnivore eats part or all of a plant, the animal obtains carbon atoms in its food. When the herbivore is eaten by a carnivore or omnivore, there is another transfer of carbon atoms. All consumers return some carbon to the air, as carbon dioxide, when they respire. However, a great deal of carbon remains in the bodies of uneaten plants and animals. When these organisms die, the decomposers enter the carbon cycle, for they degrade the dead organic matter into carbon dioxide. Thus respiration and decomposition return carbon for reuse time and time again.

Most of the carbon in organisms is recycled in continuous fashion, but some may escape recycling for varying periods of time. An example of delayed carbon recycling that is most important for our modern society occurred 300 million years ago and led to the formation of **coal** (Chapter 14). Another example of delayed carbon recycling occurred in the formation of **petroleum,** which is the remains of marine organisms. As these fossil fuels (coal and oil) are burned by man, carbon dioxide is released into the atmosphere, and the recycling of the previously bound carbon begins once again.

Carbon may also be removed from recycling by marine organisms that form skeletons or shells out of calcium carbonate ($CaCO_3$), which they produce by chemically combining calcium and carbon dioxide. The hard remains of these organisms sink to the ocean floor and are eventually compacted into rock layers, called **limestone,** by the pressure of the material that continuously settles on top of them. The chalk cliffs of Dover in England exemplify such a rock, formed on what was long ago the ocean floor by millions on millions of skeletons of protozoa. Eventually, the carbon locked up in limestone may be returned for recycling by the gradual destruction of the rock by rain and wind (**erosion**).

□ Nitrogen cycle

Nitrogen is an essential element of both amino acids and nucleic acids and hence is required by all organisms. Unfortunately, most organisms cannot utilize nitrogen in its most available form, namely, as a gas in the atmosphere. Therefore the nitrogen cycle is quite complex, involving a number of different organisms and biochemical processes (Fig. 15-8).

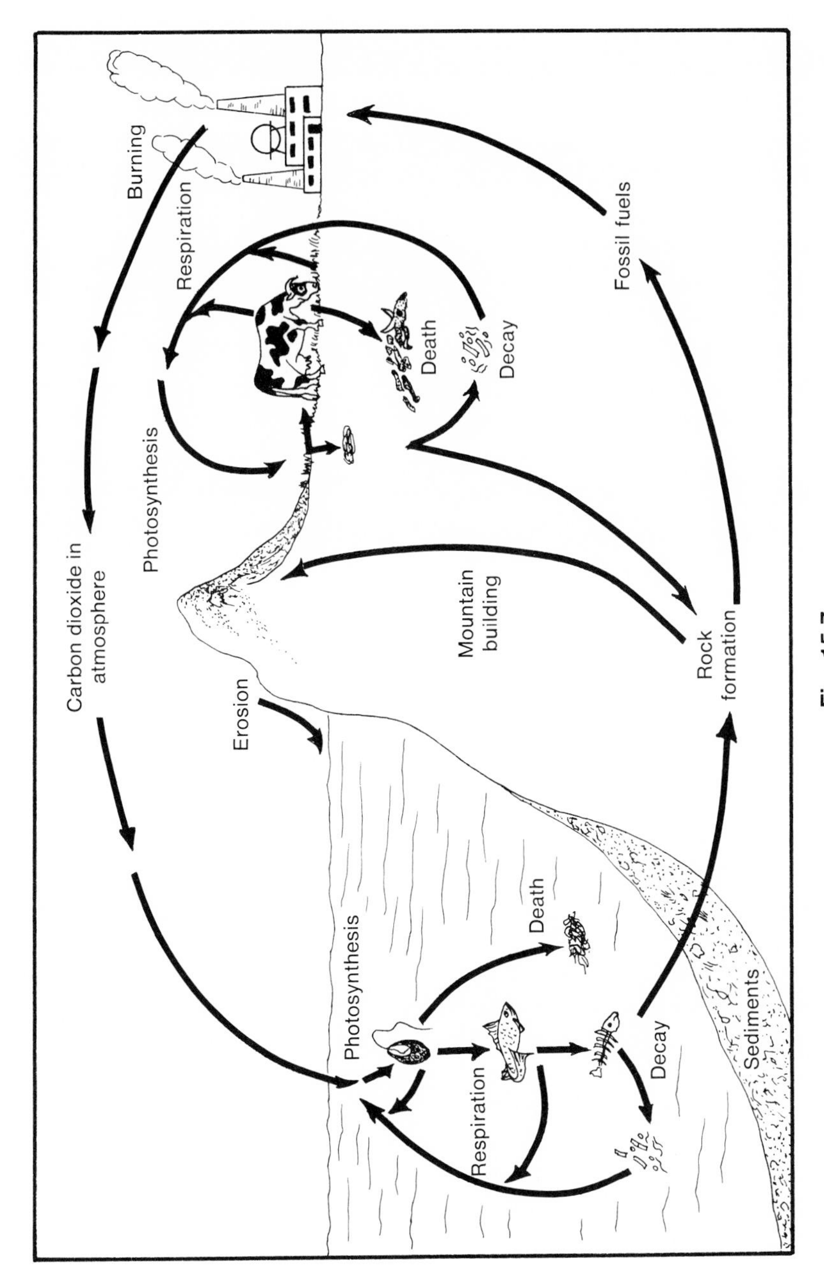

Fig. 15-7
Carbon cycle.

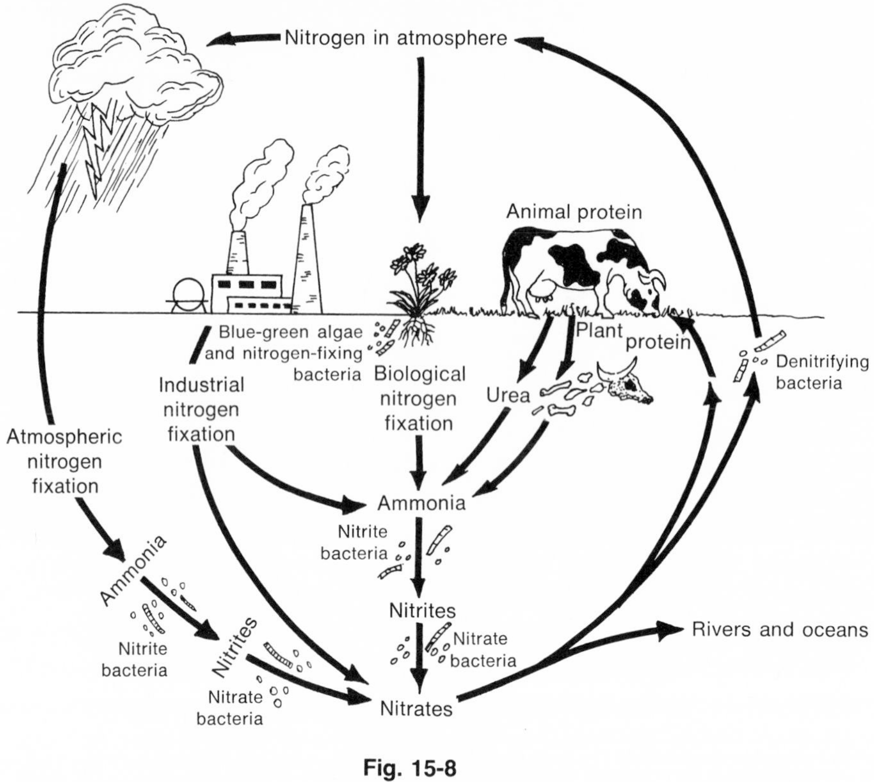

Fig. 15-8
Nitrogen cycle.

Most of the atmosphere consists of nitrogen (78%), but only certain bacteria and blue-green algae can utilize nitrogen gas in their metabolic pathways. These organisms can convert nitrogen gas into ammonia ($N_2 \rightarrow NH_3$), a process called **nitrogen fixation.** The ammonia is then used as a source of nitrogen for the synthesis of amino acids and nucleic acids. When these bacteria and blue-green algae die, the nitrogen-containing compounds of their cells are released into the surrounding environment by decomposers and can be utilized by other organisms in the ecosystem.

Many nitrogen-fixing microorganisms live in soil or water and are a continuing source of usable nitrogen for the algae and green plants that form the bases of various food pyramids. A number of nitrogen-fixing bacteria have evolved a **mutualism** type of symbiotic relationship with leguminous plants (beans, peas, alfalfa, clover). The bacteria live on and in the roots of the **legumes** and obtain carbohydrates from the host plant, which is used as a source of energy for the fixing of atmospheric nitrogen. The bacteria, in turn, fix nitrogen into amino acids that are released into the host plant. As a result of this association, legumes can grow well in soils that are poor in nitrogen.

In addition to providing the host plant with all the fixed nitrogen it needs, the nitrogen-fixing bacteria release excess ammonia into the surrounding soil, thereby enriching it for plant growth. Much before it was realized that bacteria even existed, farmers were periodically planting legumes in their fields in order to obtain higher yields from subsequently planted crops.

Once nitrogen has been incorporated into the amino and nucleic acids of plants, it can be transferred to herbivores, carnivores, and omnivores along with the other elements. When plants and animals die, the decomposition of the nitrogenous compounds of their bodies by bacteria and fungi results in the release of ammonia (NH_3). A special group of **nitrite bacteria** converts the ammonia to nitrite ($NH_3 \rightarrow NO_2$). Subsequently, another group of bacteria, **nitrate bacteria,** acts on the nitrite to produce nitrate ($NO_2 \rightarrow NO_3$). The nitrate is usable by algae and green plants as a source of nitrogen.

The completion of the nitrogen cycle involves the return of nitrogen to the atmosphere. This is accomplished by still another group of bacteria, the **denitrifying bacteria,** that degrades nitrates, nitrites, and ammonia and, in the process, liberates gaseous nitrogen into the atmosphere.

There are two nonbiological processes by which atmospheric nitrogen can be fixed: atmospheric nitrogen fixation and industrial nitrogen fixation. In **atmospheric nitrogen fixation,** nitrogen and hydrogen gases are combined to form ammonia during storms, the energy for this chemical reaction coming from the electrical discharge that constitutes lightning. The amount of nitrogen fixed by this process is very small and of little agricultural significance.

In **industrial nitrogen fixation,** nitrogen and hydrogen gases are combined to form ammonia under conditions of high temperature and pressure. A portion of the ammonia is caused to react with oxygen, producing nitric acid. Nitric acid and ammonia can then be combined to form ammonium nitrate, a widely used fertilizer. The energy used in this process is obtained from fossil fuels and is greater than the energy gained from the increased plant growth that the fertilizer promotes. Because of the need to produce more food for the ever increasing world human population, it has not been possible to develop any long-range plans for the conservation of nonrenewable fossil fuels.

☐ Water cycle

All plants and animals consist mostly of water (about 70%), and without it, living organisms cannot survive. The principal bodies of water on our earth are the oceans and the seas, making up about 97% of the hydrosphere. Of the remaining 3%, three fourths are locked up in the polar ice caps and in glaciers, the rest being found in streams, rivers, lakes, and ponds. A very small fraction of the earth's total water is found in the atmosphere (water vapor). Although the relative amount of atmospheric water is very small, it is extremely important in the water cycle (Fig. 15-9), and without it, there would be no weather.

Whenever rain falls on the earth, some of it quickly evaporates into the atmosphere. Of the water that does not evaporate, some runs off the surface of the ground into streams and lakes and some sinks into the soil and is absorbed by plants; the rest of the soil water continues to percolate through the soil until it reaches solid rock and becomes part of the **water table** (accumulated water below ground). The water in the streams, lakes, and water table eventually flows into the ocean. From the oceans, streams, lakes, and the bodies of plants and animals, there is constant evaporation of water, resulting in a continuous cycling of water between the earth and the atmosphere. Although the recycling of water does not involve any specific action of decomposers, all organisms contribute to the process.

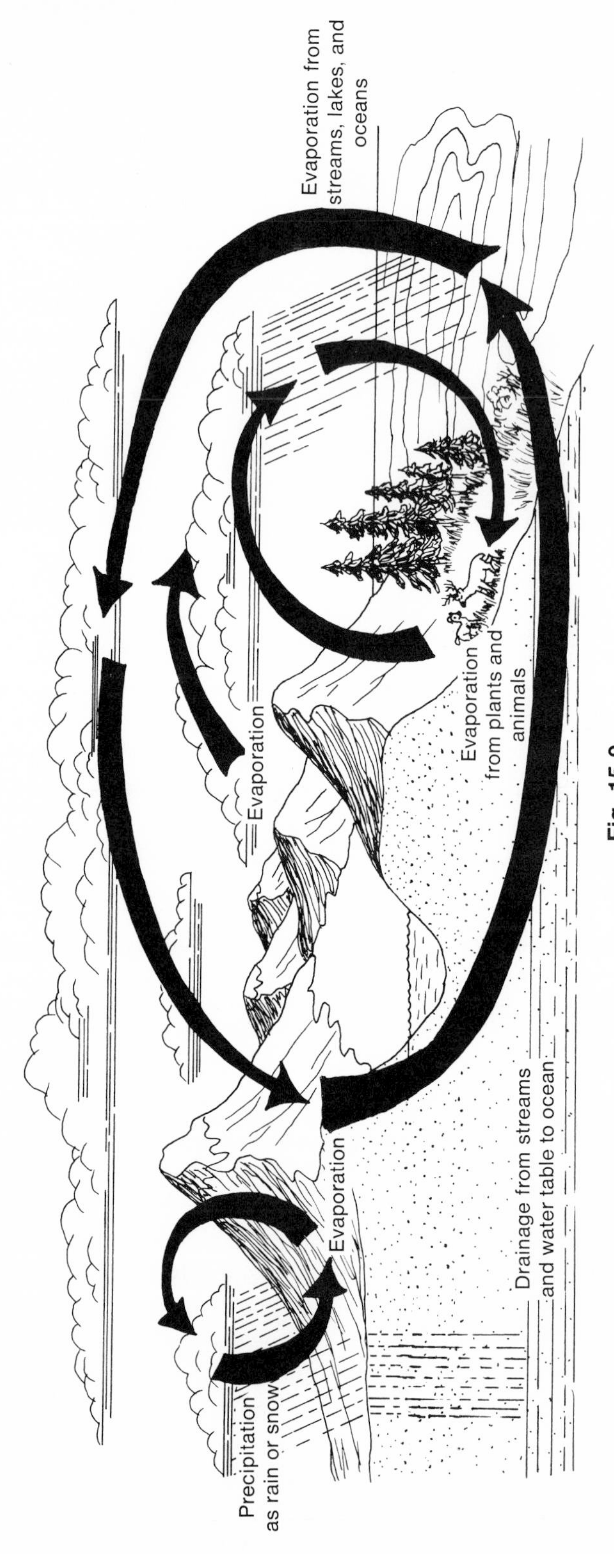

Fig. 15-9
Water cycle.

The amount of water vapor in the air is not the same everywhere, either geographically or seasonally. It is greatest at and near the equator, least at the polar regions, and in between at the middle latitudes. Even within geographic areas, there is great variation in atmosphere water, based on wind currents and the surface features of the land. As a result, some regions of the world will contain dense forests, while other areas are barren deserts.

It takes tremendous amounts of water to produce food, for example, 60 gallons of water is needed to produce 1 pound of wheat and 200 to 250 gallons of water is needed to produce 1 pound of rice. Our industrial society is also a great consumer of water. When we average water consumption over the entire United States population, we find that in 1960, each person used about 1500 gallons/day, and it is estimated that by 1980 the daily consumption will be 2500 gallons/day. Although water is a renewable resource that can be recycled again and again, long-range plans must be developed to ensure its availability and proper use in light of an ever increasing demand for it.

■ OUR POISONED ENVIRONMENT

Incredible as it may sound, many of the technological advances that have done so much to ensure the biological success of man, both as an individual and a species, have also added substances to man's environment that threaten his future well-being. As will be discussed subsequently, some of these substances are compounds that man has released into the environment for some specific purpose, for example, DDT, some are waste products of industrial processes, for example, mercury, and others are inhaled or ingested by workers in a particular industry, for example, asbestos. In certain instances, the environmental pollution is limited to the relatively small area occupied by a factory or mine, while in other cases, virtually the entire world has been affected.

The problem of environmental pollution becomes especially severe when a substance enters a food pyramid and becomes progressively more concentrated as it is transferred through a number of organisms. Many pollutants are neither used nor eliminated by the body, but are simply stored in various tissues. Because only about 20% of the food ingested by an animal is used in the construction of the organisms' tissues, the remainder of the food being used as a source of energy, the lack of elimination of unusable compounds results in their becoming progressively more concentrated at each succeeding level of the food pyramid. When the concentration of a substance exceeds the range of tolerance of the organism, the compound exerts a damaging effect on the organism and is called a **poison.**

□ Asbestos

An example of a poisoned environment that is associated with a particular industry can be found in mines, factories, and construction projects that involve **asbestos.** Asbestos is a naturally occurring mineral that is widely used in insulation materials and is therefore present in most buildings in the United States. A study was made of two groups of men, matched for age, height, weight, etc., but differing in the fact that one group had worked with asbestos for 10 years. Of a total of 458 deaths that occurred in both groups, 56% were in the asbestos-exposed group, while 44% were

Table 15-1
Carcinogenic effects of asbestos

Type of cancer	% Occurrence	
	Asbestos-exposed group	*Control group*
Lung	18	3
Stomach	11	5
Chest lining	5	—

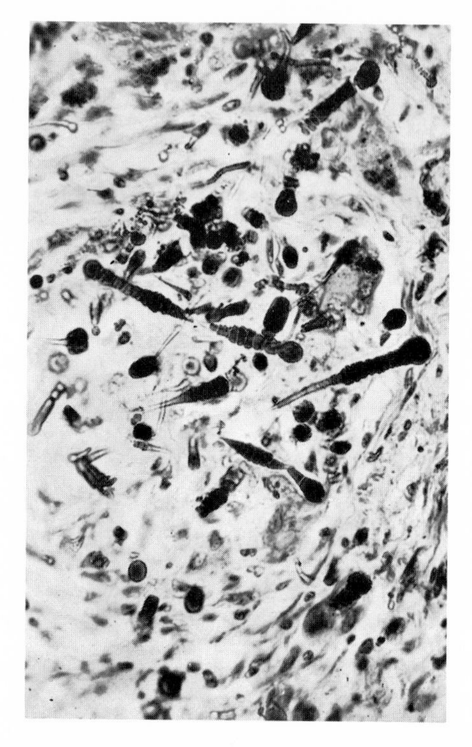

Fig. 15-10
Asbestos particles in alveoli of lung of asbestos miner. The asbestos fibers have become encrusted with calcium deposits, giving them a beaded shape with drumstick ends. (From Anderson, W. A. D., and Kissane, J. M., eds. 1971. Pathology, ed. 7. The C. V. Mosby Co., St. Louis.)

in the nonasbestos group. An analysis of the deaths caused by various types of cancer in both groups is shown in Table 15-1. There is a significantly higher incidence of all types of cancer in the asbestos-exposed group (34%) than in the control group (8%). Most significant is the occurrence of cancer of the cells of the chest lining (mesothelioma) **only** in the asbestos group, indicating a direct cancer-producing effect. The importance of asbestos as a health hazard applies to a considerable number of people, since about 5% of the total work force in the United States works with asbestos at some time during their careers.

The health problem presented by the asbestos industry is not limited to those working in the industry itself. An x-ray study of populations living in areas surrounding asbestos mines showed that about 8% of the people had calcium deposits in their lungs, a condition associated with asbestos workers (Fig. 15-10). Populations living in adjacent areas not containing asbestos mines were free of this condition. The mining of asbestos results in the release of asbestos particles into the surrounding air. Winds distribute these particles over the nearby area, where they are inhaled by people. These particles lodge in the lung tissues, and calcium deposits form around them. The physical and chemical irritation of the surrounding cells often induces cancer. The solution to this problem is **not** the elimination of the asbestos industry, which contributes greatly to our well-being, but rather a **modification of the technology** used in the industry, so that both its workers and the surrounding population will be protected from the health hazards presently suffered because of the industry.

□ Mercury

Over 80 different industries use mercury in producing such things as plastics, electronics, and fungicides. In the case of plastics and electronics, inorganic mercury is used as a catalyst in the production process. Some of the mercury is lost in the procedure and is discharged from factories along with other waste materials, usually into some adjacent river, lake, or bay. Microorganisms in the water convert the inorganic mercury to **methyl mercury** as part of one of their metabolic pathways. Although inorganic mercury is poorly absorbed through the intestinal tract and thus can be ingested with no adverse effects, methyl mercury is easily absorbed and hence extremely dangerous. Methyl mercury can cause brain damage and chromosome breakage and can even cross the placenta and attack the nervous system of a developing embryo.

Because methyl mercury passes easily across cell membranes, it can enter the bodies of fish either through their gills in breathing or through their intestines from food containing the poison. Once inside the fish, little of the methyl mercury is excreted, and it accumulates as the fish absorbs more of the compound. The poison is further concentrated in the bodies of fish eaters, which may include man. Fish may concentrate methyl mercury as much as 5000 times over the surrounding water.

There have been at least two tragedies involving the eating of fish with high levels of methyl mecury. Both these instances occurred in Japan and involved the wastes discharged from plastics factories. One took place in 1953, and of the 52 people whose cases were studied, 17 died and 23 were permanently disabled. A second occurrence took place in 1964, and of 26 people who became ill, 5 died.

The concentration of mercury in unpolluted waters is about 1 part per billion (ppb), but in waters around plastic plants, the concentration can be as high as 1800 parts per million (ppm) (1,800,000 ppb). Symptoms of methyl mercury poisoning in man are apparent when the blood concentration reaches 0.2 ppm, but in order to maintain this level, a 70 kg (154-pound) person would have to ingest 9 pounds of fish containing 0.5 ppm each week. The U.S. Food and Drug Administration has established a concentration of 0.5 ppm as the allowable limit in fish used for human consumption.

A more direct addition of methyl mercury to our environment results from its use as a fungicide. It is used directly on grain to prevent its spoilage and is also used in the production of paper to prevent fungi from developing and clogging the machinery

Fig. 15-11
Japanese family at dinner. The evening meal consists of rice, vegetables, pickles, and seafood. Sometimes chicken is prepared. The favorite dish is sashimi, pieces of raw fish dipped in soy sauce. (WHO photograph by T. Takahara; from Williams, S. R. 1973. Nutrition and diet therapy, ed. 2. The C. V. Mosby Co., St. Louis.)

in the paper mills. In the United States, the Great Lakes have levels of mercury that are comparable to those found in Japan. Americans have not suffered a mercury-poisoning tragedy, because they do not eat as much fish as the Japanese (Fig. 15-11). However, the presence of increasing amounts of mercury pollution in our environment poses a problem that requires our attention.

□ DDT

Although DDT (dichloro-diphenyl-trichloroethane) was first synthesized in the laboratory more than 100 years ago, its insect-killing properties were not discovered until 1930, and its use did not become worldwide until 1945. Since that time, in many countries, DDT spraying has eliminated such insect-transmitted diseases as typhus, malaria, yellow fever, and plague. It has also increased crop yields in many areas by killing the insects that feed on the plants (Fig. 15-12).

DDT acts on the central nervous system of **insects,** causing uncontrolled discharges of nerve impulses. This results in convulsions and paralysis, followed by death. Unfortunately, DDT can also kill **fish** and other animals. In fish, DDT interferes with gas exchange across the gills, and, instead of developing a nervous system disorder, the fish suffocate to death. Yet a different effect of DDT is found in **birds.**

Fig. 15-12
Spraying lettuce fields by plane. The use of insecticides in modern agriculture has enabled farmers to produce increasing crop yields to supply population needs. (Shell photograph, USDA; from Williams, S. R. 1973. Nutrition and diet therapy, ed. 2. The C. V. Mosby Co., St. Louis.)

The compound induces the synthesis of certain enzymes in the liver that break down sex hormones. The bird's sex hormones control the deposition of calcium in the eggshell, and in the absence of a sufficient amount of these hormones, thin-shelled eggs are produced. These eggs are fragile and are usually broken when sat on in the nest during incubation. In the case of birds, DDT causes the death of the developing chick rather than the adult.

More than 6 billion pounds of DDT have been released into the world since 1945. Unfortunately, the compound has been distributed over the entire world by ocean currents and winds. Even the birds and seals that live their entire lives in the Antarctic are found to contain DDT, although the area has never been sprayed with the compound.

As was the case with methyl mercury, there is an increased concentration of DDT in the organisms of a food pyramid as we go from the smaller, more numerous food organisms to the larger ones. Thus in a particular aquatic ecosystem, the plankton (microscopic plants and animals) were found to contain 0.04 ppm; the fish (minnows), 1 ppm; and the birds that feed on the minnows, 75 ppm.

When human tissue is examined, we find that the fat cells in some persons contain as much as 12 ppm, and human breast milk has been found to contain as much as 5 ppm. It is important to realize that the U.S. Food and Drug Administration has set the permissible DDT level of food sold for human consumption at 0.05 ppm. Under this criterion, the human breast milk just discussed should be declared unfit for human consumption.

DDT toxicity in man has yet to be unequivocally demonstrated. However, there is evidence that it is involved in such human disorders as hypertension, cerebral hemorrhage, cirrhosis of the liver (destruction of cells and their replacement by fibers), and cancer.

At present, the use of DDT in the United States has been banned by the Environmental Protection Agency (EPA), except for certain special situations. However, there is a continuing campaign to influence the EPA to once more permit the widespread use of DDT because of its effectiveness as an insecticide.

☐ Sulfur dioxide

Soft coal and some types of fuel oil contain sulfur as well as carbon (Fig. 15-13). When they are burned, **sulfur dioxide** as well as carbon dioxide is produced. By itself, sulfur dioxide is harmless. However, in moist air, it combines with water to form sulfur trioxide and sulfuric acid. When inhaled, these compounds cause throat irritation, hoarseness, coughing, shortness of breath, nausea, and a feeling of constriction of the chest. In extreme cases, continued exposure to these chemicals results in death from heart failure or suffocation.

Most areas of the earth are swept by a rather continuous stream of air currents. In industrial regions, these winds move the smoke from factories and homes to other

Fig. 15-13
Next to the automobile, the burning of coal having a high sulfur content is the greatest source of air pollution in many cities. (Courtesy Harold M. Lambert; from Turner, C. E. 1971. Personal and community health, ed. 14. The C. V. Mosby Co., St. Louis.)

areas, thereby preventing the buildup of noxious or poisonous gases. However, during periods of time without winds, stagnant air masses form and become loaded with suffocating smoke and gases. Over the years, a number of tragedies caused by these conditions have occurred in different places. The greatest loss of life occurred in London in 1952. The air over the city became still and loaded with moisture from a fog. Smoke from the homes and factories combined with the fog to form **smog,** which accumulated large amounts of sulfur dioxide. The fog turned yellow, then amber, and then black. This condition lasted for 4 days, during which time the "black fog" caused over 4000 deaths.

□ Nitrogen dioxide

Nitrogen makes up about four fifths of the air around us. When gasoline is burned in a car engine, the energy released not only causes the car to move, but also causes oxygen to combine with nitrogen to form **nitrogen dioxide.** This gas is emitted from the car engine and rises into the air, where it may absorb ultraviolet light from the sun and, through a series of chemical reactions, form a number of compounds, including ozone, formaldehyde, arolein, and peroxyacetylnitrate. The presence of all these compounds in the air causes the formation of a yellowish brown haze called **photochemical smog.** Chronic exposure to this type of polluted air may result in bronchitis, asthma, emphysema, and cancer (Fig. 15-14).

Photochemical smog predominates in the western United States. Sulfur dioxide smog is common in the eastern part of the country. The difference results from the fact that very little coal or fuel oil are burned in the West, whereas they are the primary sources of energy in the East. In the western United States, homes are heated and factories are powered by the burning of natural gas, which produces very few pollutants. Thus we find that the air around us is filled with the waste products of our technology. The predominant type of pollution may vary in different regions, but one thing is clear: more attention must be given to cleaning up our land, waters, and air.

■ SUMMARY

The biological success of man depends on his ability to interact with both the living and nonliving segments of his ecosystem. His biotic environment includes both those organisms with which he has some form of symbiotic relationship and those with which he has a feeding relationship. It is to man's advantage to maximize the number of associations involving mutualism with other species and minimize those involving parasitism. The major death-causing parasitic diseases in the world today are malaria, blood fluke infections, and bubonic plague. They are found mainly in those areas of Africa and Asia that have poor sanitary and medical facilities. In the case of blood fluke infections, man's attempt to improve agricultural production in Egypt has led to a tremendous increase in the incidence of the disease in that country.

The feeding relationship that man has with other species is complicated by the fact that we draw our food from many different ecosystems, distributed throughout the world. A further problem is that we do not return our waste materials to the ecosystem from which our food was obtained. As a result, man must replace the

Fig. 15-14
Smog in Los Angeles. **A,** Photograph taken on a clear day. **B,** Photograph taken when there was a temperature inversion layer 200 feet above the surface of the ground. **C,** In this photograph, the temperature inversion layer was about 1500 feet above ground level. (Courtesy Los Angeles County Air Pollution Control District, Los Angeles, Calif.; from Turner, C. E. 1971. Personal and community health, ed. 14. The C. V. Mosby Co., St. Louis.)

removed chemicals with fertilizers. Both the production of fertilizers and the transportation of food from their various sources to man have required an increasing use of nonrenewable fossil fuels.

Man's physical environment includes such factors as temperature, light, and available chemical molecules. Our technology has reduced appreciably, but not eliminated, the roles of temperature and light as important factors in determining man's distribution over the earth. The need for a continuing supply of necessary chemical molecules can be met, in special situations (Antarctica), through our advanced means of transportation. However, on a worldwide level, there is a need for the recycling of all the basic elements. This is especially true for agricultural areas and involves the important role of decomposers in the recycling of matter. In general, man's unique ecosystem without boundaries, the modern city, has upset many of the ecological patterns that have sustained organisms up to now. Unfortunately, we have done very little to correct the imbalances, which we ourselves have produced in the world's ecosystem.

One of the ironies of our civilization is that many of our technological advances have directly or indirectly polluted our environment and threatened our well-being. The polluting factors have included some of our industrial products, some of the waste products of our industrial processes, and some of the compounds we have used to combat disease-carrying and crop-destroying insects. Man must come to realize that the movement of matter in the worldwide ecosystem includes the poisons he himself has produced. In the worldwide ecosystem, man is but another organism, and he must learn to live intelligently within it if he is to survive and continue being successful.

SUGGESTED READINGS

Carson, R. 1962. Silent spring. Houghton Mifflin Co., Boston. Book that has alerted the world to the dangers that will befall human beings as a result of the widespread use of pesticides.

Clapham, W. B. 1973. Natural ecosystems. Macmillan Publishing Co., Inc., New York. Book that stresses both the biotic and physical components of the world's ecosystems and how man is altering them.

Cloud, P., and Gibor, A. 1970. The oxygen cycle. Sci. Am. **223**:110-123. Detailed description of the role of oxygen in the world's ecosystem.

Deevey, E. S., Jr. 1970. Mineral cycles. Sci. Am. **223**:148-158. Detailed review of the role of phosphorus and sulfur in the world's ecosystem.

Farvar, M. T., and Milton, J. D., eds. 1972. The careless technology. Natural History Press, Garden City, N.Y. Series of articles, including two on the Aswan High Dam, showing that the nutrition and health of much of Africa is declining rapidly as a result of technological improvements in agriculture.

Hawking, F. 1970. The clock of the malaria parasite. Sci. Am. **222**:123-131. Discussion of the life cycle of the malaria parasite and how it is carefully timed to the feeding habits of the mosquito.

Maddox, J. 1972. The doomsday syndrome. McGraw-Hill Book Co., New York. In-depth consideration of the problems that will arise as a result of the combination of overpopulation and pollution.

Newell, R. E. 1971. The global circulation of atmospheric pollutants. Sci. Am. **224**:32-47. Very readable discussion of the dangers posed to the entire world by pollutants produced in any one place on earth.

Odum, E. P. 1971. Fundamentals of ecology, ed. 3. W. B. Saunders Co., Philadelphia. Authoritative book on the basic principles of ecology.

Weiner, J. S. 1971. The natural history of man. Universal Books, New York. Book that deals with the interaction of man with his physical environment.

Glossary

In this glossary, you will find a list of the technical terms used in this book, together with an explanation of each term and the condition it describes. In most cases, the need to look up a term will not occur when it is first used, because there is an explanation of the term at that point. However, when the same term is repeated in later chapters, it may require a review of its meaning and relevance. In order to maximize the usefulness of this glossary, at the end of each explanation, I have included the chapter or chapters in which the term is discussed at some length.

abortion Termination of a pregnancy before the fetus is capable of surviving outside the uterus. (Chapter 13)

acetyl-CoA First compound in the Krebs' citric acid cycle. The compound results from the joining of acetic acid and coenzyme A. (Chapter 8)

activation energy Extra energy that must be gained by a molecule before it will undergo a chemical reaction. (Chapter 8)

active transport Transportation of substances across a cell membrane with the expenditure of energy. (Chapters 5 and 7)

adenosine diphosphate (ADP) Compound consisting of adenine, ribose, and two phosphate groups, of which only the terminal group is attached to the rest of the molecule by an energy-rich phosphate bond. (Chapter 8)

adenosine triphosphate (ATP) Compound consisting of adenine, ribose, and three phosphate groups, of which only the two terminal groups are attached to the rest of the molecule by energy-rich phosphate bonds. ATP has been called the energy currency of the body. (Chapter 8)

ADH See *Antidiuretic hormone.*

ADP See *Adenosine diphosphate.*

aerobic Living and functioning in the presence of free oxygen (O_2). (Chapter 8)

afterbirth Placenta and fetal membranes expelled from the uterus after childbirth. (Chapter 13)

albumin Protein composed of carbon, hydrogen, nitrogen, oxygen, and sulfur. It is found in eggs, milk, and many vegetables and is one of the three types of proteins found in blood plasma. (Chapter 6)

alleles Various forms of a given gene. Each allele transcribes a different messenger RNA (mRNA). (Chapter 9)

allergy Hypersensitivity to certain substances as a result of producing an excessive number of antibodies against the substances. (Chapter 6)

all-or-none law Rule that applies to certain body structures to the effect that when stimulated, the structure responds to its fullest extent or not at all. This rule applies to single skeletal muscle fibers, to the heart as a whole, and to single nerve fibers. (Chapter 10)

alternation of generations Type of plant life cycle in which a sexually reproducing (haploid) stage alternates with an asexually reproducing (diploid) stage. (Chapter 14)

amino acid Organic compound containing an amino (NH_2) and a carboxyl (COOH) group. Amino acids are the building blocks for the synthesis of proteins and the end products of protein digestion. (Chapter 2)

amniocentesis Withdrawal of fluid and cells contained within the amniotic sac of fetus to search for any indication of certain genetic diseases. (Chapter 13)

anaerobic Capable of living and functioning in the absence of free oxygen (O_2). (Chapter 8)

analogy Similarity in function, but not in structural plan, as, for example, the wing of an insect and that of a bird. (Chapter 14)

anemia Pathological condition characterized by a reduction in the number of red blood cells, or the amount of hemoglobin in each red blood cell, or both. (Chapter 4)

antibodies Specific types of protein that are produced by lymphocytes and function to neutralize foreign substances in the body. (Chapter 6)

anticodon Group of three nucleotides of a transfer RNA (tRNA) molecule that allows it to recognize a specific messenger RNA (mRNA) codon during its translation on the ribosome. (Chapter 8)

antidiuretic hormone (ADH) Also called **vasopressin.** Hormone secreted by the posterior lobe of the pituitary gland. ADH stimulates the reabsorption of water from kidney tubules into the bloodstream. (Chapters 11 and 12)

antigen Substance that will stimulate the production of antibodies and will, in turn, react with them. (Chapter 6)

appendix Small, blind tube at the distal end of the cecum in man. (Chapters 5 and 14)

arthritis Inflammation of a joint or joints. (Chapter 10)

atherosclerosis Form of heart disease involving a narrowing of arterial passageways, which is caused by the accumulation of large amounts of cholesterol just inside the innermost layer of the arterial wall. (Chapter 3)

atmosphere Envelope of air surrounding the earth. (Chapter 2)

atom Smallest particle of an element that can exist alone or in combination with other atoms. (Chapter 2 and Appendix A)

ATP See *Adenosine triphosphate.*

autoimmunity Condition in which a person develops antibodies that react against the individual's own proteins. (Chapter 6)

autolysis Chemical breakdown of a cell that occurs after a cell has died. The destruction of the cell is caused by intracellular enzymes that are present in the cell's lysosomes. (Chapter 7)

autosomes All chromosomes except the sex (X or Y) chromosomes. (Chapter 9)

basal metabolic rate Rate of energy utilization in the body during absolute rest, but while the person is awake. (Chapter 13)

biosphere Portion of the earth inhabited by living organisms. The biosphere involves land masses (lithosphere), bodies of water (hydrosphere), and air (atmosphere). (Chapter 2)

blastula Stage in animal development in which the cells are arranged in the form of a hollow ball with a single layer. The blastula stage precedes the gastrula. (Chapter 13)

blood pressure Force exerted on the blood by pumping action of the heart. (Chapter 6)

bolus Chewed and moistened mass of food that is formed in the mouth and then swallowed. (Chapter 5)

calorie Small calorie (cal) is the amount of heat required to raise the temperature of 1 gram of water 1° C. Kilocalorie (kcal), or large calorie (Cal), is the unit of heat used in the study of body metabolism and the energy content of foods. (Chapter 2)

carbohydrate Compound that contains carbon, hydrogen, and oxygen, with the latter two elements in the ratio of 2:1, as in water. (Chapter 2)

carcinogen Any substance that induces the transformation of normal cells into cancerous ones. Known carcinogens include various types of radiation and numerous chemicals. (Chapter 11)

caries Cavities formed in teeth as a result of the chemical destruction of their enamel and possibly also dentine. (Chapter 5)

cecum Blind pouch that forms the first section of the large intestine. (Chapter 5)

cerumen Earwax, a waxlike substance formed by the skin glands that line the canal of the outer ear. (Chapter 10)

Chediak-Higashi syndrome Genetic disease, inherited as an autosomal recessive trait, in which the white blood cells contain defective lysosomes. (Chapter 7)

chromosomes Bodies found in the cell's nucleus that contain DNA, the hereditary material, and its associated proteins. (Chapters 8 and 9)

circadian rhythm Pattern of biological activity that repeats itself every 24 hours. (Chapter 15)

codon Any group of three nucleotides (triplet) in messenger RNA (mRNA) that specifies the amino acid to be inserted in a specific position in a polypeptide formed during translation. (Chapter 8)

coenzyme Nonprotein organic molecule required for the activity of some enzymes. Coenzymes act as donors or acceptors of groups of atoms that have been added to or removed from the substrate. Some coenzymes are derivatives of vitamins. (Chapter 8)

complementarity Association of nitrogenous bases that occurs between two chains of DNA or between DNA and RNA: adenine-thymine, guanine-cytosine, and adenine-uracil. (Chapter 8)

compound Substance composed of two or more elements combined in definite proportions. (Chapter 2 and Appendix A)

condensation Chemical combination of molecules with an accompanying production of water. Condensation is the reverse of the process of hydrolysis. (Chapter 5)

connective tissue Associations of cells in which the cells are separated from one another, and the intercellular spaces are filled with materials secreted by the cells: fibrous connective tissue, tendons, ligaments, fat, cartilage, and bone. (Chapter 7)

consumer Organism that does not produce its own food, but obtains its energy and building materials by engulfing and digesting all or parts of other organisms. (Chapter 2)

covalent bond Bond formed between the atoms of a compound through the sharing of electrons. (Chapter 2 and Appendix A)

crossing-over Process that can occur during meiosis, in which breaks occur in homologous chromosomes, and there is a mutual exchange of genetic material. (Chapter 9)

cystic fibrosis Hereditary autosomal recessive disease in which the mucus-secreting epithelial cells of the body produce a thick, slow-moving type of mucus that blocks small passageways and ducts of the body. (Chapter 7)

cytochrome system Mitochondrial complex of enzymes that transfers hydrogen through a sequence of decreasing energy levels, thereby producing 34 energy-rich phosphate bonds. The hydrogen is ultimately combined with oxygen to form water. (Chapter 8)

decomposer Organism that does not produce its own food, but obtains its energy and building materials by chemically breaking down the dead bodies of other organisms and absorbing the predigested material. (Chapter 2)

deoxyribonucleic acid (DNA) Molecule that is the hereditary material of the individual. A DNA molecule consists of two sugar-phosphate chains from which purines and pyrimidines project and bind the chains together. (Chapter 8)

diabetes Disorder of carbohydrate metabolism resulting from an insufficient production of insulin by the islands of Langerhans. (Chapters 3 and 12)

diffusion Movement of atoms, ions, or molecules from a region of higher concentration to one of lower concentration. (Chapters 5 and 7)

digestion Chemical process by which large, complex food molecules are broken down into simpler molecules that can be absorbed by an organism. (Chapter 5)

diploid (2n) Having double the number of chromosomes present in a gamete of the species, or having double the haploid number of chromosomes. (Chapters 9 and 14)

disaccharide Also called **double sugar.** Type of carbohydrate that is composed of two monosaccharide molecules. The disaccharides include lactose, maltose, and sucrose. (Chapter 3)

disjunction Moving apart of chromosomes to opposite ends of the cell during cell division. Disjunction occurs both in mitosis and meiosis. (Chapter 9)

DNA See *Deoxyribonucleic acid.*

Down's syndrome Also called **mongolism.** Genetic disease caused by the presence of an extra chromosome 21. (Chapter 9)

drug Any chemical agent, other than food, that significantly affects body structure or function. (Chapter 12)

ecology Study of the relationships between organisms and those between organisms and the environment. (Chapter 15)

ecosystem Stable association of organisms in a given area, involving a cyclic interchange of materials between the living and nonliving components of the area. (Chapter 15)

edema Excessive accumulation of fluid in tissue spaces. (Chapter 4)

effector Structure that responds to stimulation. An effector may be a muscle or a gland. (Chapter 12)

element Pure substance consisting of identical atoms. (Chapter 2)

embryo Organism in the early stages of development, that is, from the first specialization of tissues until it is easily recognizable as belonging to a particular species. (Chapter 13)

emphysema Disease of the lungs in which the alveoli become enlarged, and their walls become torn. (Chapter 11)

endocrine system System of ductless glands

that control metabolism through the secretion of hormones into the bloodstream. It is one of the two coordinating systems of the body. (Chapter 12)

energy Capacity to perform work. Energy is found in various forms: chemical, mechanical, electrical, thermal, or radiant. (Chapters 2 and 8)

enzyme Protein molecule that can speed up the rate of a chemical reaction without becoming incorporated in the end product of the reaction. (Chapters 3 and 8)

epithelium Type of tissue in which the cells are tightly fitted together. Epithelium covers the surface of our bodies and lines all our internal cavities and ducts. (Chapter 7)

erythroblastosis fetalis Red blood cell disease of certain infants caused by the development of antibodies by the mother, who in this situation must be Rh negative, against the red blood cells of the fetus, who in this situation must be Rh positive. (Chapter 6)

eucaryotes All organisms whose cells have their hereditary material located in a true nucleus, that is, separated from the cytoplasm by a nuclear membrane. (Chapter 14)

evolution Process by which different kinds of organisms have developed from ancestral forms. Evolution is sometimes defined as "descent with modification." (Chapter 14)

excretion Elimination or discharge of waste products of cell metabolism from the body. (Chapter 11)

extinction Disappearance of a species from the earth. Extinction will occur if every single member of the species dies or if the species has evolved into a new form (transformation). (Chapter 14)

fatty acid Chain of carbon atoms with a carboxyl (COOH) group at one end. When one, two, or three fatty acid molecules are joined to a molecule of glycerol, a lipid molecule is formed. (Chapter 3)

fetus Organism in the later stages of development, that is, from the time it is easily recognizable as belonging to a particular species until birth. (Chapter 13)

fibrinogen Protein present in blood plasma that, in forming a blood clot, is converted into fibrin. (Chapter 6)

follicle-stimulating hormone (FSH) Anterior pituitary gonadotropic hormone that stimulates sperm production in the male and graafian follicle development in the female. (Chapter 13)

fossil Any remains, impression, cast, or trace of an organism of a past geological period. (Chapter 14)

FSH See *Follicle-stimulating hormone.*

galactosemia Hereditary autosomal recessive disease caused by the inability of the individual to convert galactose to glucose. (Chapter 3)

gametogenesis Processes by which gametes (sperm and eggs) are formed. On the chromosome level, gametogenesis involves meiosis. (Chapter 9)

ganglion Group of nerve cell bodies lying outside the brain and spinal cord. (Chapter 12)

gastrula Stage in embryonic development that follows the blastula. In humans, the gastrula stage initially consists of two cavities, amnion and yolk sac, surrounded by ectodermal and endodermal cells, respectively. (Chapter 13)

gene Section of DNA that, in its sequence of nucleotide bases, contains the code that specifies the amino acid sequence of a protein. (Chapter 8)

genetic code Sequence of nucleotide base triplets of DNA and messenger RNA (mRNA) that specify the amino acid sequence of a protein. (Chapter 8)

genotype Genetic constitution of an organism, as distinguished from its physical appearance (its phenotype). (Chapter 9)

globulin Type of protein found in blood plasma. There are three kinds of globulin proteins: alpha, beta, and gamma. Gamma globulin contains most of the antibodies formed by the lymphocytes of the body. (Chapter 6)

glycerol Three-carbon chain which a hydroxyl group (OH) is attached to each carbon atom. When one, two, or three fatty acid molecules join to a molecule of glycerol, a lipid molecule is formed. (Chapter 3)

glycolysis Chemical breakdown of glucose to pyruvic acid, with the accompanying production of two energy-rich phosphate bonds that are used in the synthesis of two molecules of adenosine triphosphate (ATP). (Chapter 8)

goiter Enlargement of the thyroid gland. Goiter occurs commonly in areas where the food and water supplies are deficient in iodine. (Chapter 4)

gonadotropic hormones Two anterior pituitary hormones that act on the gonads: (1) follicle-stimulating hormone (FSH) and (2) luteinizing hormone (LH). (Chapter 13)

haploid (n) Having half the number of chromosomes present in the general body cells; having one of each of the pairs of homologous chromosomes characteristic for the species; having the number of chromosomes in a gamete of the species. (Chapters 9 and 14)

hemoglobin Iron-containing, oxygen-carrying protein present in red blood cells. (Chapter 6)

hemophilia Genetic disease, inherited as an X-linked recessive trait, in which the clotting time of the person's blood is excessively long, and, as a result, prolonged bleeding results from even minor cuts. (Chapter 9)

homeostasis Maintenance of a constant and optimal internal environment in an organism. (Chapter 11)

homologous chromosomes Pair of chromosomes that have either identical genes or their alleles located at corresponding points on their chromosomes. (Chapter 9)

homology Similarity in structural plan and developmental origin, but not necessarily in function, as, for example, the human hand and the bird wing. (Chapter 14)

hormone Substance secreted by an endocrine gland and transported by the circulatory system to other parts of the body, where it evokes a response in some specific "target cells." (Chapters 3 and 12)

Huntington's chorea Genetic disease, inherited as an autosomal dominant trait, in which there is involuntary muscle contraction and progressive mental deterioration. (Chapter 7)

Hurler's syndrome Genetic disease, inherited as an autosomal recessive trait, that is characterized by an excessive production of mucopolysaccharides by the fibrous connective tissue cells of the body. (Chapter 7)

hydrogen bond Weak bond formed between a hydrogen atom that is already covalently bonded to an oxygen or nitrogen atom and a second atom. (Chapter 2 and Appendix A)

hydrolysis Chemical splitting of a molecule into two smaller molecules with the addition of the hydrogen of water to one of the smaller molecules and the addition of the hydroxyl group of water to the other. Hydrolysis is the reverse of the process of condensation. (Chapter 5)

hydrosphere Aggregate of all the bodies of water on the earth: oceans, lakes, streams, subterranean water, and water vapor in the atmosphere. (Chapter 2)

immunity Ability to resist infection or to overcome the effects of infection. (Chapter 6)

implantation Developmental process by which an embryo attaches to and becomes embedded within the wall of the uterus. (Chapter 13)

interphase Period between successive mitoses. Interphase consists of three phases: G_1, in which proteins are synthesized and cell growth occurs; S, in which both proteins and DNA are formed; and G_2, in which more protein synthesis and growth occur. (Chapter 9)

ionic bond Bond formed between the atoms of a compound through the transfer of one or more electrons from one atom to another. (Chapter 2 and Appendix A)

ketone Chemical compound formed as a result of the partial chemical breakdown of lipids. (Chapters 4 and 12)

kidney threshold level Concentration of a substance in the blood, below which the substance will be reabsorbed from the kidney tubules into the surrounding capillaries. (Chapter 11)

Klinefelter's syndrome Genetic disease caused by the presence of an extra X chromosome in the cells of a male, resulting in the individual being sterile. (Chapter 9)

Krebs' citric acid cycle Series of energy-producing reactions that occur in the mitochondria of cells. Pyruvic acid (product of glycolysis) is broken down to carbon dioxide (CO_2) and hydrogen, with the accompanying formation of two energy-rich phosphate bonds. (Chapter 8)

kwashiorkor Severe nutritional disorder in children caused by a deficiency of protein in the diet. (Chapter 2)

lactation Secretion of milk by the mammary gland. (Chapter 13)

leukemia Cancerous disorder of the leukocyte-forming tissues of the body. (Chapter 6)

LH See *Luteinizing hormone.*

ligament Type of connective tissue in which the intercellular fibers are all oriented in the same direction. Ligaments bind bone to bone. (Chapter 7)

lipid Compound that contains carbon, hydrogen, and oxygen, with the latter two elements in a ratio that is **greater than** 2:1. A molecule of lipd consists of one molecule of glycerol attached to one, two, or three molecules of fatty acid. (Chapters 2 and 3)

lithosphere Solid component of the earth. (Chapter 2)

luteinizing hormone (LH) Anterior pituitary gonadotropic hormone that stimulates testosterone production in the male and corpus luteum development in the female. (Chapter 13)

lymph Fluid found in lymphatic vessels. Lymph originates from the tissue fluid that surrounds the cells of the body. (Chapter 6)

marasmus Severe nutritional disease in children caused by a deficiency of calories in the diet. (Chapter 2)

meiosis Type of cell division in which the chromosome number is reduced from the diploid (2n) to the haploid (n) number. Meiosis occurs in animals during gametogenesis and in plants during sporogenesis. (Chapter 9)

melanin Brown pigment responsible for the coloration of skin, hair, and iris of the eye. (Chapter 10)

menarche Occurrence of a female's first menstrual cycle. Menarche marks the beginning of the reproductive period of a female's life. (Chapter 13)

menopause Termination of the occurrence of a female's menstrual cycles. Menopause marks the end of the reproductive period of a female's life. (Chapter 13)

messenger RNA (mRNA) Type of RNA that functions as the intermediary between the genes located in the nucleus and the ribosomes located in the cytoplasm. The mRNA carries the genetic code for polypeptide chains that is contained in the sequence of the DNA's bases. (Chapter 8)

metabolism Sum total of all the chemical reactions by which protoplasm is produced, by which energy is made available to the organism, and by which waste products are removed. (Chapter 3)

mineral Inorganic substance that has distinctive physical properties and a definite chemical composition. (Chapter 4)

mitosis Type of cell division in which there is an exact division and distribution of chromosome material, so that the daughter cells will have the same genetic constitution as the cell from which they arose. (Chapter 9)

molecule Smallest particle of an element or a compound that can exist separately and still retain the characteristic properties of the element or compound. (Chapter 2 and Appendix A)

monosaccharide Also called **simple sugar.** Type of carbohydrate containing three or more carbon atoms that **cannot** be broken down by hydrolysis. (Chapter 3)

morula Stage in animal development in which the cells are arranged as a solid clump of 16 to 32 cells. The morula stage precedes the blastula. (Chapter 13)

mosaicism Condition in which an individual is composed of cells of different genetic composition. Mosaicism may occur as a result of a mutation in a cell of the body or because of nondisjunction of chromosomes during one or more cell division(s). (Chapter 9)

mRNA See *Messenger RNA.*

muscle Type of tissue that functions primarily in contraction. In humans, there are three types of muscle: smooth, striated, and cardiac. (Chapters 7 and 10)

muscular dystrophy General name given to a number of closely related diseases that result in muscle degeneration. One type, Duchenne, is inherited as an X-linked recessive trait. (Chapter 7)

mutation Any alteration of the nucleotide base sequence of a DNA molecule. A mutation can occur through the addition, deletion, or substitution of a nucleotide base. (Chapter 8)

nephritis Disease of the kidney, often the result of bacterial infection of the glomeruli. (Chapter 11)

nerve tissue Type of tissue that possesses the properties of irritability and conductivity. Nerve tissue is composed of neurons and neuroglia, which comprise the nervous system. (Chapters 7 and 12)

nervous system One of the two coordinating systems of the body, consisting of: (1) the central nervous system, (2) the peripheral nervous system, and (3) the autonomic nervous system. (Chapter 12)

neuroglia One of the two types of nervous system cells. Neuroglia nourish and support the neurons of the nervous system. (Chapter 12)

neuron One of the two types of nervous system cells. Neurons are the impulse-conducting cells of the nervous system. (Chapters 7 and 12)

neurotransmitter Chemical, released by an axon, that travels across a synapse and stimulates either a second neuron or a muscle. Neurotransmitters include acetylcholine, epinephrine, and norepinephrine. (Chapter 12)

nitrogen fixation Process by which free, atmospheric nitrogen (N_2) is converted into ammonia (NH_3). Nitrogen fixation is accomplished by nitrogen-fixing bacteria. (Chapter 15)

nondisjunction Failure of chromosomes to separate and migrate to opposite ends of the cell during cell division. Nondisjunction may occur either in mitosis or meiosis. (Chapter 9)

nucleic acid Type of organic compound present in the nuclei and cytoplasm of cells either as DNA or RNA. (Chapters 2 and 8)

nucleotide Basic unit or building block of both DNA and RNA. Every nucleotide consists of three subunits: (1) base, (2) sugar, and (3) phosphoric acid. (Chapter 8)

nutrient Any substance that promotes growth or provides energy for physiological processes. (Chapter 3)

nutrition Total of the processes involved in the taking in, absorption, and utilization of foods. (Chapters 2 and 4)

ontogeny Development of the individual from fertilization to maturity. (Chapter 14)

osmosis Passage of water through a semipermeable membrane as a result of the differences in the concentrations of the solutions on both sides of the membrane. (Chapter 7)

ovotestis Early developmental stage of the human gonad before it becomes differentiated into either an ovary or a testis. (Chapter 13)

ovulation Discharge of an egg from a graafian follicle. In humans, ovulation occurs about 14 days after the beginning of the menstrual flow. (Chapter 13)

oxygen debt Amount of oxygen required to break down the lactic acid that has been accumulated in the blood and liver during sustained physical activity. (Chapter 10)

parturition Act of giving birth to young. (Chapter 13)

pedigree chart Diagram that indicates the ancestry of a family over a number of generations. (Chapter 9)

peptide bond Covalent bond that holds two amino acids together. A peptide bond is formed when the amino group of one amino acid is bonded to the carboxyl group of another amino acid with the accompanying production of water. (Chapter 3)

periodontal disease Destruction of the cells of the gums by bacterial enzymes and toxins. (Chapter 5)

peristalsis Progressive wavelike movements of the intestines and other tubular organs that move the contents of the organs along. (Chapter 5)

pernicious anemia Reduced production of red blood cells by the body as a result of a deficiency of cobalamin, vitamin B_{12}. (Chapter 6)

phagocytosis Engulfment of solid particles by cells, as, for example, the ingestion of microorganisms by leukocytes. (Chapter 6)

pharynx That area of the throat that serves as a common passageway for material from the mouth cavity to the esophagus and from the nasal cavities to the larynx. (Chapter 5)

phenotype Physical makeup or appearance of an individual in contrast to its genetic constitution (its genotype). (Chapter 9)

phenylketonuria Genetic disease, inherited as an autosomal recessive trait, in which the accumulation of phenylalanine and phenylpyruvic acid in the blood of children results in retarded mental development. (Chapter 3)

phosphorylation Combination of a compound with phosphoric acid, as occurs in glycolysis when a high-energy phosphate group is attached to a monosaccharide molecule. (Chapter 8)

photosynthesis Process in which green plants combine carbon dioxide and water to produce simple sugars and oxygen. In photosynthesis, light energy from the sun is converted into the chemical energy of the sugar molecule. (Chapter 2)

phylogeny Evolutionary development of a race or species. (Chapter 14)

placenta Organ that consists of embryonic and maternal tissues attached to the inner surface of the uterus and through which the embryo or fetus obtains its nourishment and discharges its wastes. (Chapter 13)

plasma Clear fluid portion of blood or lymph. (Chapter 6)

poison Any substance that exerts a damaging effect on the normal functioning of organs or tissues. (Chapter 15)

poliomyelitis Also called **infantile paralysis.** Sometimes fatal disease, caused by a virus, in which the motor neurons of the spinal cord are damaged or killed. (Chapter 7)

polymer Large molecule (macromolecule) that results from the combination of many smaller molecules of the same type, as, for example, proteins that are composed of amino acids or nucleic acids that are composed of nucleotides. (Chapter 8)

polynucleotide chain Polymer made up of nucleotides linked together by the bonding of the sugar of one nucleotide to the phosphate group of the next nucleotide. (Chapter 8)

polypeptide chain Polymer made up of amino acids linked together by peptide bonds. (Chapter 8)

polysaccharide Polymer made up of many monosaccharides. Polysaccharides include cellulose, starch, and glycogen. (Chapter 3)

procaryotes All organisms whose cells do **not** have their hereditary material located in a true nucleus, that is, separated from the cytoplasm by a nuclear membrane. (Chapter 14)

producer Organism that can produce its own food, as, for example, the green plants. (Chapter 2)

protein Polymer made up of amino acids that are organized into one or more polypeptide chains. (Chapters 2 and 3)

pyruvic acid Final compound formed in glycolysis. On its formation, pyruvic acid may either enter Krebs' citric acid cycle for further degradation or be converted to lactic acid. (Chapter 8)

receptor Structure that receives a stimulus, as, for example, a sense organ (eye, ear) or sensory cell (temperature sensitive cells of the skin). (Chapter 12)

reflex arc Nervous system pathway from point of stimulation to responding organ. A reflex arc usually includes a receptor, a sensory neuron, an association neuron, a motor neuron, and an effector. (Chapter 12)

renal corpuscles Filtering units of the kidneys. Each renal corpuscle consists of a glomerulus and its surrounding Bowman's capsule. (Chapter 11)

ribonucleic acid (RNA) Polymer of nucleotides that is of great importance in protein synthesis. There are three types of RNA molecules, each having a different function: ribosomal RNA, transfer RNA, and messenger RNA. (Chapter 8)

ribosomal RNA (rRNA) Type of RNA that, in association with proteins, forms the ribosomes, which are the sites of synthesis of new proteins. (Chapter 8)

ribosomes Cytoplasmic structures on whose surfaces protein synthesis takes place. Ribosomes are usually associated with the cell's endoplasmic reticulum. (Chapters 7 and 8)

RNA See *Ribonucleic acid.*

rRNA See *Ribosomal RNA.*

segmentation contractions Simultaneous contractions of the intestine at a number of different, equally spaced, points. Segmentation contractions serve to mix the food with the digestive juices. (Chapter 5)

sex chromosome One of a pair of chromosomes (XX in the female, XY in the male) that play a primary role in the determination of sex in the individual. X chromosomes also contain genes for traits that are not related to the sex of the person. (Chapter 9)

sickle cell anemia Genetic disease, inherited as an autosomal recessive trait, in which the hemoglobin molecules have a reduced oxygen-carrying capacity. (Chapter 6)

smog Mixture of fog and smoke. (Chapter 15)

species All the populations of individuals that can breed with one another, but that cannot breed with members of other species. (Chapters 1 and 14)

sporogenesis Process by which plants produce spores. On the chromosome level, sporogenesis involves meiosis. (Chapter 9)

substrate Compound acted on by an enzyme. (Chapter 8)

symbiosis Way of life in which two organisms of different species live in intimate association with each other. Depending on the nature of the association, the relationship is called mutualism, commensalism, or parasitism. (Chapter 15)

synapse Point of functional contact between the axon of one neuron and the dendrites of the next neuron in the particular circuit. (Chapters 7 and 12)

taxonomy Classification of animals and plants into groups based on their evolutionary relationships. (Chapter 14)

tendon Type of connective tissue in which the intercellular fibers are all oriented in the same direction. Tendons bind muscle to bone. (Chapter 7)

thalassemia Genetic disease, inherited as an autosomal recessive trait, in which there is a severe reduction in the amount of hemoglobin in each red blood cell. (Chapter 6)

thyroxin Iodine-containing hormone, produced by the thyroid gland, that regulates general body metabolism. (Chapters 4 and 12)

tonus State of partial contraction that characterizes all so-called resting muscles. (Chapter 10)

transcription Formation of messenger RNA (mRNA) from DNA. Transcription is controlled by the enzyme RNA polymerase. (Chapter 8)

transfer RNA (tRNA) Type of RNA that functions to carry amino acids to the ribosomes for incorporation into a growing polypeptide chain. (Chapter 8)

translation Process by which amino acids are combined into a polypeptide as specified by a messenger RNA (mRNA). (Chapter 8)

tRNA See *Transfer RNA.*

Turner's syndrome Genetic disease caused by the absence of one of the X chromosomes in the cells of a female, resulting in the individual being sterile. (Chapter 9)

ulcer Destruction of the cells of the inner lining of the stomach or intestines, accompanied by the degeneration and death of the underlying tissues. (Chapter 5)

venereal diseases Infections associated with sexual intercourse or other close body contacts. The most prevalent venereal diseases are gonorrhea and syphilis. (Chapter 13)

vitamin Organic compound, often functioning as a coenzyme, that is required in relatively minute amounts in the diet for normal growth, development, and maintenance of an organism's physiological activities. (Chapter 4)

APPENDIX A Atoms and how they combine with one another

Biology is the study of the characteristics of living things. What organisms can and cannot do is determined by their physical and chemical organization. If we are to understand how our bodies function, we must learn how they are constructed from certain basic molecules. In order to appreciate how we are able to take molecules from plants and other animals and reorganize them into human material, we must learn how atoms themselves are organized and how they can combine with one another.

■ ORGANIZATION OF ATOMS

Atoms are composed of three types of particles. These are protons, electrons, and neutrons. A **proton** carries a positive electric charge and is about 1836 times heavier than an **electron,** which carries a negative charge. A **neutron** has about the same mass as a proton, but is electrically neutral. The protons and neutrons are located in the center of an atom in what is called the **nucleus.** The electrons travel around the nucleus within definite regions called **orbits.** An atom contains the same number of protons and electrons and is, therefore, electrically neutral. We can compare an atom to our solar system. In such a comparison, the nucleus of the atom with its orbiting electrons resembles the sun with its orbiting planets. Fig. A-1 shows the structure of the four most common atoms in living material.

An analysis of the structure of atoms will indicate the manner in which they function in chemical reactions. One of the four most common elements in organisms is hydrogen. As shown in Fig. A-1, the nucleus of the hydrogen atom contains a single proton (p), while around the nucleus, there is a single electron (e) in orbit. The number of protons that an atom contains is called its **atomic number.** Hydrogen, therefore, has an atomic number of 1, which is written as a subscript before the symbol for hydrogen, that is, $_1H$. An atom contains the same number of electrons as it

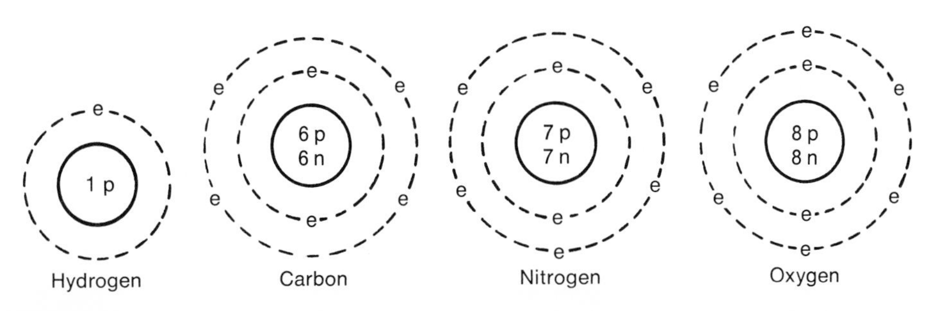

Fig. A-1
Diagrams of the structure of the four most common atoms in living material. *p*, Proton; *n*, neutron; *e*, electron.

does protons. Therefore we can also use the atomic number to tell us how many electrons an atom contains.

In addition to its atomic number, an atom can be characterized as to the sum of the number of protons (p) and neutrons (n) in its nucleus. This is called its **mass number,** or **atomic weight.** Since the hydrogen nucleus contains no neutrons, its mass number is also 1, which is written as a superscript before the symbol for hydrogen, that is, ^{1}H. Should we want to characterize hydrogen, both as to its atomic number and its mass number, we would indicate it as follows: $^{1}_{1}H$. In similar fashion, the other elements in Fig. A-1 would be indicated as follows: carbon, $^{12}_{6}C$; nitrogen, $^{14}_{7}N$; and oxygen, $^{16}_{8}O$.

Isotopes

While all atoms of any element contain the same number of protons, they can vary in their neutron number. Atoms that have the same atomic number (protons) but have different mass numbers (protons plus neutrons) are known as **isotopes.** Returning to our example of hydrogen, we find that the most common isotope of hydrogen in nature is $^{1}_{1}H$. However, a heavier isotope of hydrogen is known to occur, in rare amounts, and contains one proton and one neutron. This isotope of hydrogen is called **deuterium,** or heavy hydrogen, and is indicated as $^{2}_{1}H$. Another and still heavier isotope of hydrogen contains one proton and two neutrons. It is called **tritium** and has the designation $^{3}_{1}H$. In actual usage, when indicating an isotope of an atom, we specify only the mass number; hence deuterium is usually designated as ^{2}H and tritium as ^{3}H. Both deuterium and tritium can substitute for the common isotope of hydrogen, usually indicated simply as H, in chemical reactions. However, tritium has one characteristic that is not exhibited by deuterium. Tritium is "unstable," which means that it tends to emit high-energy particles from its nucleus until it reaches a stable form. Such unstable forms of an atom are known as **radioactive isotopes.** Since all isotopes have the same chemical properties as the commonly occurring atoms but have different weights, they may act as "tags" and can be identified. Tagged atoms make it possible to trace various chemicals through an organism. This permits us to determine what happens to certain substances in normal life processes. When the isotope is radioactive, its presence can be detected by a Geiger counter or other similar instrument. The detection of radioactive isotopes is easier than the detection of nonradioactive isotopes, which requires a more complicated laboratory procedure.

Although atoms of the same element may differ in their neutron number, they cannot differ in their proton number. When an atom gains or loses a proton, its chemical identity changes. Again, let us consider the hydrogen atom, which normally contains a single proton and a single electron. If hydrogen gains a proton, it is no longer hydrogen, but helium. Since an atom must contain the same number of electrons as it has protons, the helium nucleus would attract an electron, from somewhere in its surroundings, to itself. The helium atom formed as just described would be an isotope of helium, because the commonly occurring helium atom contains two neutrons, in addition to two protons and two electrons. It is of great consequence for our modern society that hydrogen atoms can be fused, at extremely high temperatures, to form helium atoms, with a concomitant release of a tremendous amount of energy. This occurs most importantly in two situations. First, it occurs in the sun, which is made up of a mixture of hydrogen and helium gases. Within the core of the sun, which is estimated to have a temperature of about 16 million degrees Celsius, hydrogen atoms are constantly fusing with one another to form helium. Some of the huge amount of energy released in this process reaches us in the form of sunlight, whose importance in food production is discussed in Chapter 2. The second situation in which the formation of helium from hydrogen atoms has great consequence for our modern society involves the tremendous destructive force of hydrogen bombs. The energy released in the explosion of hydrogen bombs results from the controlled fusion of the heavier isotopes of hydrogen (deuterium and tritium) to form helium.

□ Arrangement of electrons

Each atom behaves the way it does because of the number of protons and neutrons in its nucleus and the number of electrons in its orbits. We find that the arrangement of electrons, in orbits, around an atomic nucleus follows a definite pattern. For reasons not well understood, the various orbits around an atomic nucleus can contain only a certain maximum number of electrons. The orbit closest to the nucleus (orbit 1) can contain at most 2 electrons; the next orbit (2) 8 electrons; orbit 3, 18 electrons; orbit 4, 18 electrons; and orbit 5, 8 electrons. The location of any particular electron depends on its energy level. The less energy the electron has, the closer it will be found to the nucleus of the atom. In Table A-1, we find listed the chemical elements found in the human body (silicon has been added for later discussion), together with the distribution of their electrons in the various orbits. A study of Table A-1 will raise a question about potassium, calcium, and iron. These three elements contain electrons in orbit 4 despite the fact that in each case, orbit 3 contains less than its maximum of 18 electrons. This seemingly contradictory situa-

Table A-1
Structural characteristics of elements present in living matter

Element	Symbol	Atomic number	Mass number	Number of electrons in orbit				
				1	*2*	*3*	*4*	*5*
Hydrogen	H	1	1	1	—	—	—	—
Carbon	C	6	12	2	4	—	—	—
Nitrogen	N	7	14	2	5	—	—	—
Oxygen	O	8	16	2	6	—	—	—
Sodium	Na	11	23	2	8	1	—	—
Magnesium	Mg	12	24	2	8	2	—	—
Silicon	Si	14	28	2	8	4	—	—
Phosphorus	P	15	31	2	8	5	—	—
Sulfur	S	16	32	2	8	6	—	—
Chlorine	Cl	17	35	2	8	7	—	—
Potassium	K	19	39	2	8	8	1	—
Calcium	Ca	20	40	2	8	8	2	—
Iron	Fe	26	56	2	8	14	2	—
Iodine	I	53	127	2	8	18	18	7

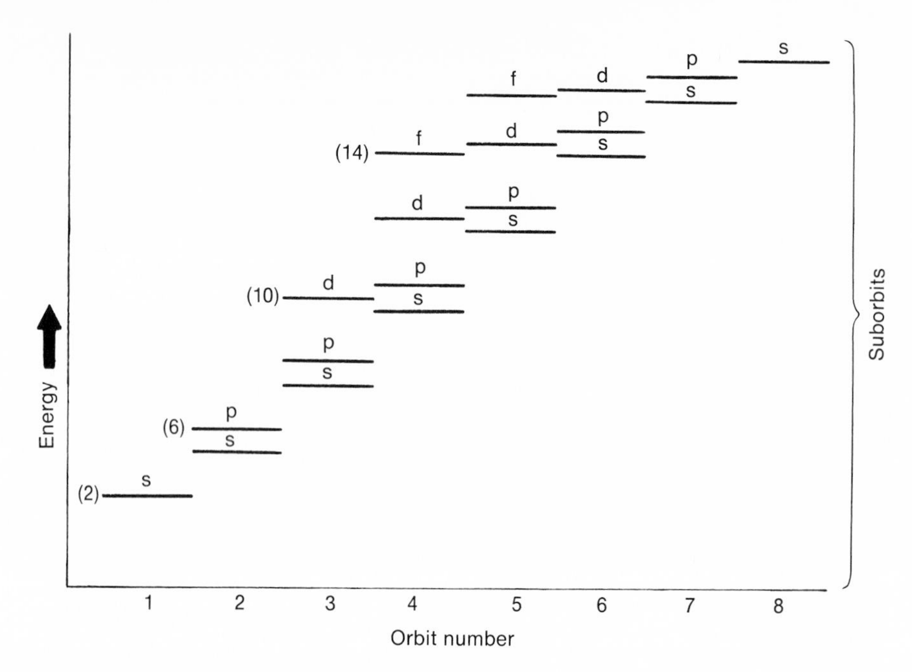

Fig. A-2
Relative energy relationships of electron orbits and suborbits. The number to the left of each representative suborbit is the number of electrons required to fill that type of suborbit.

tion results from the fact that, starting with orbit 2, the orbits are literally composed of suborbits. For reasons unknown, the innermost suborbit of orbit 4 requires electrons of less energy than the outermost suborbit of orbit 3. This results in electrons filling the innermost suborbit of orbit 4 before beginning to fill the outermost suborbit of orbit 3. The relative energy relationships of the various orbits and their suborbits are shown in Fig. A-2, together with the maximum number of electrons that can fill any suborbit.

BONDING OF ATOMS

Although, as stated earlier, all atoms are electrically neutral, they are by no means all chemically stable. Chemical stability of an atom appears to be mainly a function of the number of electrons in its outermost orbit, which is also called its **valence orbit.** If that orbit is **full,** the element is inert, that is, it tends not to react chemically with other atoms. An example of an inert element is helium, which has an atomic number of 2 and contains two electrons in its valence orbit (orbit 1). Another inert element is neon, which has an atomic number of 10 and contains eight electrons in its valence orbit (orbit 2). We find that all valence orbits, other than 1, behave as though they were complete when they contain eight electrons.

An examination of most atoms shows that they do not have the complete number of electrons in their valence orbit. These atoms can achieve chemical stability by losing, gaining, or sharing electrons. In order for this to occur, atoms must become joined to one another through one of three types of chemical bonding: ionic, covalent, or hydrogen. We shall now consider each of these types of chemical bonding and the kinds of molecules they form.

□ Ionic bonding

If we examine Table A-1 again and look at sodium, we find that an atom of this element has only one electron in its outermost orbit. Chemical stability can be achieved if the sodium atom either gains seven or loses one electron. In either case, the atom will have eight electrons in its outermost orbit, although, to be sure, the total number of orbits will be different in the two situations. As can be imagined, it is much easier for the sodium atom to lose one electron than gain seven. Therefore we find that in chemical reactions involving a loss or gain of electrons, sodium always loses an electron. However, in becoming chemically stable, the sodium atom is no longer electrically neutral. It now has 11 protons, but only 10 electrons. The atom now has an excess of one positive charge. This is indicated by a "+" sign written as a superscript to the right of the symbol for sodium, that is, Na^+. If we now consider chlorine, we find the reverse situation. In this case, the atom achieves chemical stability by gaining an electron. It will now have an excess of one negative charge, which is indicated by a "−" sign written as a superscript to the right of the chlorine symbol, that is, Cl^-. All electrically charged atoms are called **ions.** An atom that loses an electron is said to have been **oxidized,** while an atom that gains an electron is said to have been **reduced.**

We have discussed how atoms can achieve chemical stability by either the gain or loss of electrons. In order for this to occur, both an electron donor and a recipient are required. The process is most easily accomplished if an atom that needs to gain an electron unites with an atom that needs to lose an electron. In this procedure, both atoms will become chemically more stable than they were before. When atoms unite, they form a **molecule.** If the atoms of a molecule are of different elements, the molecule is called a **compound.** Molecules may have quite different properties from their constituent atoms, just as atoms are quite different from protons, neutrons, or electrons. It may not be possible to predict the characteristics of a molecule based on the properties of its atoms, for example, in the case of water versus hydrogen and oxygen, as indicated in Table A-2.

If we now return to our consideration of sodium and chlorine, we can visualize atoms of these two elements uniting to form the compound sodium chloride, which is ordinary table salt. The atoms that constitute the compound are in their **charged,** or **ionic,** states. The two ions are held together by the electrostatic attraction of their opposite charges, and the force of attraction between the ions is called an **ionic bond.** The formation of sodium chloride (Na^+Cl^-) proceeds according to the following formula:

$$Na + Cl \rightarrow Na^+Cl^-$$

Table A-2
Physical properties of water and its constituent atoms

Substance	Formula	Freezing point (°C)	Boiling point (°C)
Water	H_2O	0	100
Hydrogen	H_2	−259	−253
Oxygen	O_2	−218	−183

A further examination of Table A-1 will indicate that calcium can achieve a more stable chemical state by losing two electrons. This can be accomplished by uniting with two chlorine atoms. In general, an atom that contains one, two, or three electrons in its valence orbit will tend to be an electron donor, while an atom that contains five, six, or seven electrons in its valence orbit will tend to be an electron recipient. Elements whose atoms act as electron donors are called **metals,** while those whose atoms act as electron recipients are called **nonmetals.** The number of electrons lost or gained by an atom in forming an ionic bond is called its **valence.** Since the number of electrons lost or gained is a function of the number of electrons in the atom's outermost orbit, we can see why the outermost orbit of an atom is called its valence orbit.

An element of tremendous importance in biochemical reactions is **hydrogen.** As seen in Table A-1, the hydrogen atom contains a single electron in its valence orbit. There are two ways in which hydrogen can attain chemical stability. Since orbit 1 is complete when it contains two electrons, hydrogen can achieve chemical stability by obtaining one electron. It can do this by combining with an atom like calcium. However, it will be necessary for two hydrogen atoms to combine with one atom of calcium, because there are two electrons in the valence orbit of calcium. The formation of calcium hydride (CaH_2) proceeds as follows:

$$Ca + 2H \rightarrow Ca^{2+} + 2H^-$$

Hydrogen can also achieve chemical stability by giving up its one electron and remaining as a positively charged proton, that is, H^+. However, hydrogen cannot do this in its pure atomic form, because as a free proton, it is much too reactive. But it can give up its one electron and become an ion when it is in combination with some other molecule, such as water.

In becoming ions, atoms lose their electric neutrality and become charged. However, the molecule they form is generally electrically neutral, since the atoms are usually bonded together strongly enough to be considered a single entity. Important exceptions to this situation exist. Compounds like sodium chloride have a strong tendency to dissociate into separate ions when placed in a liquid medium. When in solution, NaCl forms two separate entities, a sodium ion (Na^+) and a chlorine ion (Cl^-). The charge on a particle greatly affects its ability to enter and leave an organism's body and also affects its functioning in biochemical reactions.

Our discussion of the bonding of atoms has centered on those atoms whose valence orbits contain one, two, three, five, six, or seven electrons. We have not considered those atoms whose valence orbits contain four electrons, for example, carbon and silicon. These atoms bind to other atoms by sharing electrons rather than by their transfer. Atoms that are capable of forming ionic compounds are also capable of forming molecules as a result of sharing electrons. We shall now examine the process of molecule formation through the sharing of electrons.

□ Covalent bonding

As was discussed previously, atoms can achieve chemical stability by transferring electrons from one to another. For this to occur, a **strong** electron acceptor must be available to pull the electron(s) away from the electrical attraction field of the donor atom. There are, however, many instances where the interacting atoms have either equal or near equal attraction for one another's electrons. Under these conditions, we find that the atoms complete their outermost orbits by sharing electrons, which acts to bind the atoms together; the bond thus formed is called a **covalent bond.** For an example of covalent bonding of atoms, we can turn to hydrogen. Hydrogen, as we noted earlier, can form ionic bonds either by gaining or losing electrons. If, as occurs in many instances, no strong electron acceptor or donor is present, two hydrogen atoms can bond to one another to form a molecule of hydrogen (H_2). This reaction is pictured as follows, in which the electrons are represented by dots:

$$H\cdot + H\cdot \rightarrow H:H$$

In the covalent bonding of two hydrogen atoms, the electrons are shared, so that each hydrogen atom has, in a sense, two electrons in its outermost orbit. Since there is no actual transfer of electrons in covalent bonding, there is no valence change in the atoms. In covalent bonding, the atoms have both electrical neutrality and chemical stability.

Another, and very important, compound formed as a result of covalent bonding is water, whose formation is pictured in the following diagram (only the outermost orbit of oxygen is shown):

$$2H\cdot + \cdot\ddot{\underset{\cdot\cdot}{O}}\cdot \rightarrow H:\ddot{\underset{\cdot\cdot}{O}}:H$$

Water also ionizes to some extent, depending on the circumstances surrounding it. Its unusual combination of characteristics helps explain water's importance in living material.

Many larger compounds are found to be formed as a result of both ionic and covalent bonding among their constituent parts. An example of this is as follows:

$$H_2SO_4 = H_2^{2+}SO_4^{2-} = H^+ + H^+ + \left[\begin{array}{c} :\ddot{O}: \\ :\ddot{\underset{\cdot\cdot}{O}}:\ddot{\underset{\cdot\cdot}{S}}:\ddot{\underset{\cdot\cdot}{O}}: \\ :\underset{\cdot\cdot}{O}: \end{array}\right]^{2-}$$

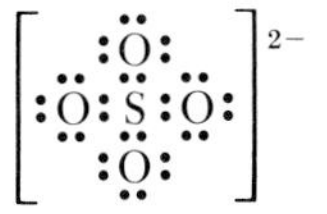

Sulfuric acid **Hydrogen ions** **Sulfate ion**

A most important group of compounds is formed through the covalent bonding of their constituent atoms. These are the organic compounds, which always contain both carbon and hydrogen and may, in addition, contain the atoms of other elements. The element that is of central importance in organic compounds is carbon. It has four electrons in its outermost orbit and, therefore, has very little tendency either to lose or to gain electrons. In fact, carbon forms compounds only by sharing electrons. An example of a very simple type of carbon compound is methane (CH_4), a gas used both as a fuel and for illumination. In this molecule, carbon shares each one of its outer electrons with a separate hydrogen atom. The compound is pictured in the following diagram (each **pair** of shared electrons is represented by a bond sign between the involved atoms):

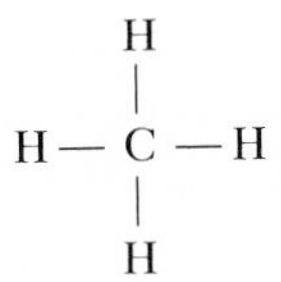

$$\begin{array}{ccc} & H & \\ & | & \\ H- & C & -H \\ & | & \\ & H & \end{array}$$

Methane

Carbon can also share electrons with atoms of other elements. An example of this is carbon tetrachloride (CCl_4), which is used in fire extinguishers and in the dry cleaning of clothes. It is formed as follows:

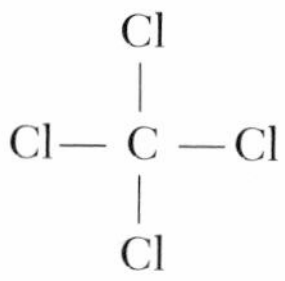

$$\begin{array}{ccc} & Cl & \\ & | & \\ Cl- & C & -Cl \\ & | & \\ & Cl & \end{array}$$

Carbon tetrachloride

Carbon can also share electrons with combinations of other atoms, as seen in chloroform ($CHCl_3$), which is used as an anesthetic and formed as follows:

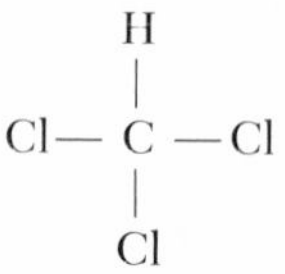

$$\begin{array}{ccc} & H & \\ & | & \\ Cl- & C & -Cl \\ & | & \\ & Cl & \end{array}$$

Chloroform

It is also possible for two or more carbon atoms to link up with one another in forming compounds. This is seen in ethane (C_2H_6), which is also a gas that can be used both as a fuel and for illumination and is formed as follows:

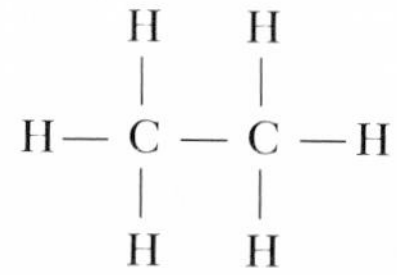

$$\begin{array}{cccc} & H & H & \\ & | & | & \\ H- & C & -\ C & -H \\ & | & | & \\ & H & H & \end{array}$$

Ethane

The ability of carbon atoms to bond covalently to one another permits the formation of tremendous-sized molecules with many different combinations of other atoms. Another source of variability in organic compounds results from the fact that the same atoms can form different bonding patterns. As a relatively simple example, we can consider the compound C_2H_6O. The atoms present in this molecule can be bonded together in two different arrangements, as follows:

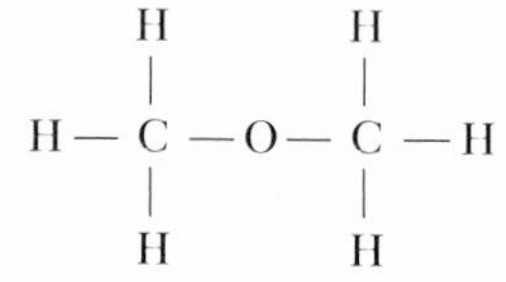

$$\begin{array}{ccccc} & H & & H & \\ & | & & | & \\ H- & C & -O- & C & -H \\ & | & & | & \\ & H & & H & \end{array} \quad \text{or} \quad \begin{array}{ccccc} & H & H & & \\ & | & | & & \\ H- & C & -\ C & -O- & H \\ & | & | & & \\ & H & H & & \end{array}$$

Dimethyl ether **Ethyl alcohol**

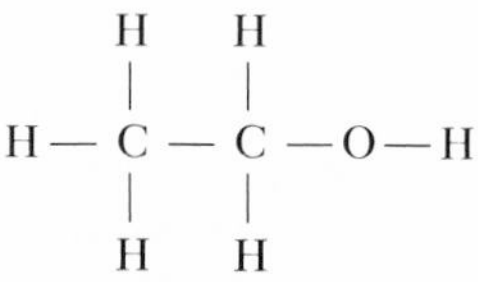

These two compounds have the same **molecular formula** (C_2H_6O), but different **structural formulas.** They also have quite different properties and uses. Two compounds that have the same molecular formula but different structural formulas are called **isomers** of one another. The occurrence of isomers is further indication that, in addition to constituent parts, the organization of materials determines their characteristics. It will be noted that, in the preceding diagram, there are four electrons of each oxygen atom that are not shared with other elements. These electrons have been omitted from the structural formulas to simplify the diagram; this procedure will be followed in all later diagrams as well.

All the organic compounds just discussed have been formed as a result of the sharing of a single **pair** of electrons between a carbon atom and each of four other atoms. It is also possible for a carbon atom to share two or even three pairs of electrons with another atom. The structural formulas that follow illustrate both situations:

```
H— C =O          H—C≡N
    |
    H
```

Formaldehyde **Hydrogen cyanide**

Table A-1 contains another element whose outermost orbit has four electrons. That element is silicon. It, too, forms molecules only as a result of covalent bonding. A question arises as to whether silicon rather than carbon might form the base of some other system of living organisms, possibly on some other planet. This possibility has intrigued writers of science fiction for some time. There are a number of limitations that the properties of silicon impose on such a system. Because of its size and weight, silicon does not readily form chains of atoms. As a result, most compounds having silicon as its central atom tend to be relatively small and simple. In addition, many, but not all, organic compounds are readily soluble in water. The same is not true of silicon compounds. In fact, most of the silicon of the world is bound up in relatively insoluble minerals in the earth's crust. Another aspect of silicon chemistry has to be mentioned. One compound of carbon, carbon dioxide, is critical for living material as we know it. Carbon dioxide exists as a gas at temperatures that permit organisms to function, and, as a gas, it is in a form that can be used by living systems. On the other hand, the comparable compound of silicon, silicon dioxide (quartz), is a solid at typical terrestrial temperatures, that is, its melting point is 1470° C. As a solid, it cannot be used by living systems as we know them. It may still be possible that some other system of living material has evolved elsewhere in the universe using silicon as its central atom. However, any such system would be quite limited in its diversity as compared to the organisms we find on earth.

□ Hydrogen bonding

A third type of bond formation between atoms occurs when a hydrogen atom that is covalently bound to one atom forms a weak linkage with an electron from a second atom. This weak linkage of hydrogen with a second atom is called a **hydrogen bond.** Hydrogen bonds usually form between hydrogen and either oxygen or nitrogen. Both these elements have a greater tendency to attract electrons than hydrogen and, therefore, have a strong tendency to share electrons with it. The formation of a hydrogen bond results in the hydrogen atom being shared simultaneously between two atoms and forming a bridge between them. Hydrogen bonding is particularly important in maintaining the structure of the organism's genetic material, as discussed in Chapter 8.

A somewhat different pattern of hydrogen bonding can be found in water, where hydrogen bonds are the cohesive forces that keep the molecules of water together. This is as follows:

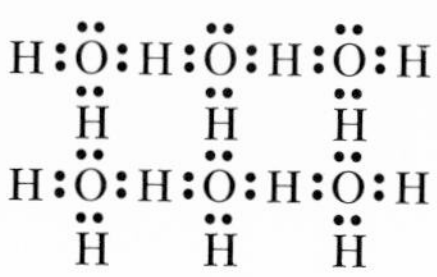

```
   ..    ..    ..
 H:O:H:O:H:O:H
   ..    ..    ..
   H     H     H
   ..    ..    ..
 H:O:H:O:H:O:H
   ..    ..    ..
   H     H     H
```

As seen in the diagram, the hydrogen atoms tend to be shared between two oxygen atoms. Under these conditions, it is difficult to say where one water molecule ends and another begins.

APPENDIX B Acids and bases

Two classes of **ionic compounds** are of great consequence in biochemical reactions. These are **acids** and **bases.** An acid is defined as any compound that can act as a hydrogen ion (proton) donor, while a base is defined as any compound that can act as a hydrogen ion acceptor. As an example of an acid, we can consider hydrogen chloride (HCl). In its pure form, hydrogen chloride is a gas at normal room temperatures, that is, its boiling point is $-84°$ C, and does not ionize. However, when it is dissolved in water, the hydrogen chloride molecules ionize by transferring their protons to water molecules, which become hydronium ions (H_3O^+). It is from the hydronium ions that the hydrogen ions become available for other chemical reactions. The chemical reaction of hydrogen chloride with water results in the formation of hydrochloric acid.

As an example of a base, we can consider sodium hydroxide (NaOH). In its pure form, it is a solid at room temperatures, that is, its melting point is 318° C, and does not ionize. However, when dissolved in water, sodium hydroxide ionizes to form sodium ions (Na^+) and hydroxyl ions (OH^-).

Our definition of acids and bases involved only the ability of a compound to act as a proton donor or acceptor. The compound involved can be either a neutral molecule or a charged ion. Examples of both types of acids and bases are given in the following list, together with their hydrogen ion donating or accepting reactions (although not indicated, in each case, the reaction occurs in a liquid medium):

Acids

$$HCl \rightleftharpoons H^+ + Cl^-$$

$$H_2CO_3 \rightleftharpoons H^+ + HCO_3^-$$

$$NH_4^+ \rightleftharpoons H^+ + NH_3$$

$$H_2O \rightleftharpoons H^+ + OH^-$$

Bases

$$NaOH + H^+ \rightleftharpoons Na^+ + H_2O$$

$$HPO_4^{2-} + H^+ \rightleftharpoons H_2PO_4^-$$

$$HS^- + H^+ \rightleftharpoons H_2S$$

$$H_2O + H^+ \rightleftharpoons H_3O^+$$

Acids and bases do not actually react independently of one another. In order for a compound to function as a proton donor, there must be some other molecule that will act as a proton acceptor. This is most clearly seen when we cause HCl and NaOH to interact with one another, resulting in the formation of water (H_2O) and sodium chloride (NaCl), as follows:

$$HCl + NaOH \rightarrow H_2O + NaCl$$

This chemical equation represents a combination of the reactions involving HCl, as indicated in the acids column, and NaOH, as shown in the bases column. The sodium chloride formed in the chemical equation falls into the category of compounds called **salts,** which are always formed when acids react with bases. Salts are molecules in which the hydrogen ion of an acid is replaced by some other positive ion. Salts play a very important role in many chemical reactions of living material.

An examination of the reactions listed under "acids" and "bases" indicates the double role played by water. Water ionizes to some slight extent and can act as a proton donor. However, if a large number of hydrogen ions are present in water, that is, derived from some compound such as HCl, water will act as a proton acceptor. Any substance that can function in such a dual capacity is said to be **amphoteric.**

Every ionic compound can be classified as an acid, base, or salt. Acids and bases vary significantly in the extent to which they dissociate into free ions in aqueous solutions. Those that dissociate completely are called **strong** acids or bases, while those that dissociate only to a limited degree are called **weak.** As an example, we can consider hydrochloric and acetic acids. Hydrochloric is a strong acid and dissociates completely into H^+ and Cl^-. Virtually no HCl remains in the form of the intact compound. However, when acetic acid (CH_3COOH), ordinary vinegar, dissolves in water, only a small fraction of the molecules dissociates into H^+ and CH_3COO^-. The rest of the molecules remain in the form of intact acetic acid. Most of the acids and bases of biological interest are weak. All salts, of both strong and weak acids, dissociate completely in water.

In many biological studies, it is important to know the **strength** of an acid or base in a solution of ionic compounds. This can be determined by an electrical apparatus that measures the number of free protons present in the solution. The information is expressed as a number, called the **pH** of the solution. Mathematically, pH has been defined by the following equation:

$$\mathrm{pH} = \log \frac{1}{[\mathrm{H}^+]}$$

In this equation, $[H^+]$ indicates how many grams of hydrogen ions are present in a liter of solution, that is, the concentration in grams per liter (gm/liter), and the log indicates that the value will be expressed in terms of a logarithmic scale. In the case of water, it is found that there are 0.0000001 gm of protons present in 1 liter. The pH of water is calculated as follows:

$$\mathrm{pH} = \log \frac{1}{0.0000001} = 7$$

The pH of water is taken as **neutrality.** A solution with a pH below 7 is **acidic,** while one with a pH above 7 is **basic,** or **alkaline.** Since the scale is logarithmic, a change of 1 pH unit indicates a 10-fold change in absolute concentration of hydrogen ions. This means that a solution whose pH is 4 is 10 times more acidic than a solution whose pH is 5 and 100 times more acidic than a solution whose pH is 6. In corresponding manner, a solution whose pH is 12 is 10 times more alkaline than a solution with a pH of 11 and 100 times more alkaline than a solution with a pH of 10. The pH value varies from 0 (1 gm of protons/liter of solution) to 14 (0.00000000000001 gm of protons/liter of solution). The pH of our blood and tissue fluids must be maintained within very narrow limits if our bodies are to function properly (discussed in Chapter 6).

APPENDIX C Measurements

LENGTH

English to metric

1 mile (mi) = 1.609 kilometers (km)
1 yard (yd) = 0.914 meter (m)
1 foot (ft) = 30.48 centimeters (cm)
1 inch (in.) = 2.54 centimeters (cm)
1 inch (in.) = 25.4 millimeters (mm)

Metric to English

1 kilometer (km) = 0.62 mile (mi)
1 meter (m) = 3.28 feet (ft)
1 centimeter (cm) = 0.394 inch (in.)
1 millimeter (mm) = 0.039 inch (in.)
1 micrometer (μm) = $^1/_{25,400}$ inch (in.)

Metric system

1 kilometer (km) = 1000 m
1 centimeter (cm) = 0.01 m
1 millimeter (mm) = 0.001 m
1 micrometer (μm) = 0.000001 m
1 nanometer (nm) = 0.000000001 m
1 angstrom (Å) = 0.0000000001 m

WEIGHT

English to metric

1 pound (lb) = 0.4536 kilogram (kg)
1 ounce (oz) = 28.35 grams (gm)

Metric to English

1 kilogram (kg) = 2.2 pounds (lb)
1 gram (gm) = 0.035 ounce (oz)

Metric system

1 kilogram (kg) = 1000 gm
1 milligram (mg) = 0.001 gm
1 microgram (μg) = 0.000001 gm

VOLUME

English to metric

1 gallon (gal) = 3.785 liters
1 quart (qt) = 0.946 liter
1 pint (pt) = 0.473 liter
1 ounce (oz) = 0.029 liter

Metric to English

1 liter = 0.264 gallon (gal)
1 liter = 1.056 quarts (qt)
1 milliliter (ml) = 0.03 ounce (oz)

Metric system

1 liter = 1000 milliliters (ml)
1 milliliter (ml) = 0.001 liter
1 microliter (μl) = 0.000001 liter

TEMPERATURE

English to metric

Degrees Celsius (°C) = (°F − 32) × $^5/_9$

Metric to English

Degrees Fahrenheit (°F) = (°C × $^9/_5$) + 32

Index

I